APPENDIX 4 AREAS FOR t DISTRIBUTIONS

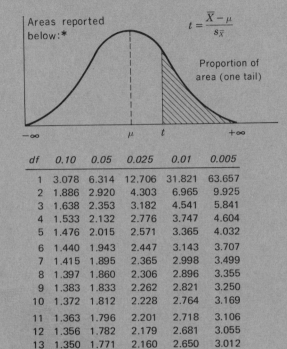

Areas reported below:*

$$t = \frac{\overline{X} - \mu}{s_{\overline{X}}}$$

Proportion of area (one tail)

df	0.10	0.05	0.025	0.01	0.005
1	3.078	6.314	12.706	31.821	63.657
2	1.886	2.920	4.303	6.965	9.925
3	1.638	2.353	3.182	4.541	5.841
4	1.533	2.132	2.776	3.747	4.604
5	1.476	2.015	2.571	3.365	4.032
6	1.440	1.943	2.447	3.143	3.707
7	1.415	1.895	2.365	2.998	3.499
8	1.397	1.860	2.306	2.896	3.355
9	1.383	1.833	2.262	2.821	3.250
10	1.372	1.812	2.228	2.764	3.169
11	1.363	1.796	2.201	2.718	3.106
12	1.356	1.782	2.179	2.681	3.055
13	1.350	1.771	2.160	2.650	3.012
14	1.345	1.761	2.145	2.624	2.977
15	1.341	1.753	2.131	2.602	2.947
16	1.337	1.746	2.120	2.583	2.921
17	1.333	1.740	2.110	2.567	2.898
18	1.330	1.734	2.101	2.552	2.878
19	1.328	1.729	2.093	2.539	2.861
20	1.325	1.725	2.086	2.528	2.845
21	1.323	1.721	2.080	2.518	2.831
22	1.321	1.717	2.074	2.508	2.819
23	1.319	1.714	2.069	2.500	2.807
24	1.318	1.711	2.064	2.492	2.797
25	1.316	1.708	2.060	2.485	2.787
26	1.315	1.706	2.056	2.479	2.779
27	1.314	1.703	2.052	2.473	2.771
28	1.313	1.701	2.048	2.467	2.763
29	1.311	1.699	2.045	2.462	2.756
30	1.310	1.697	2.042	2.457	2.750
40	1.303	1.684	2.021	2.423	2.704
60	1.296	1.671	2.000	2.390	2.660
120	1.289	1.658	1.980	2.358	2.617
∞	1.282	1.645	1.960	2.326	2.576

*Example: For the shaded area to represent 0.05 of the total area of 1.0, the value of t with 10 degrees of freedom is 1.812.
Source: Reprinted by Hafner Press, a division of Macmillan Publishing Company, from *Statistical Methods for Research Workers*, 14th ed., abridged Table IV, by R.A. Fisher. Copyright © 1970 by University of Adelaide.

STATISTICS

A Fresh Approach

STATISTICS

FOURTH EDITION

A Fresh Approach

Donald H. Sanders
Educational Consultant
Fort Worth, Texas

McGRAW-HILL PUBLISHING COMPANY

New York   St. Louis   San Francisco   Auckland   Bogotá
Caracas   Hamburg   Lisbon   London   Madrid   Mexico
Milan   Montreal   New Delhi   Oklahoma City   Paris
San Juan   São Paulo   Singapore   Sydney   Tokyo   Toronto

Page 99: From THE HOUSE AT POOH CORNER by A. A. Milne, 1928, p. 90. Used by permission of the Canadian Publishers, McClelland and Stewart, Toronto. Reprinted by permission of Methuen Children's Books. Copyright 1928 by E. P. Dutton, renewed 1956 by A. A. Milne. Reprinted by permission of the publisher, E. P. Dutton, a division of Penguin Books USA Inc.

Pages 165 and 214: From WINNIE-THE-POOH by A. A. Milne, 1926, pp. 45 and 100. Used by permission of the Canadian Publishers, McClelland and Stewart, Toronto. Reprinted by permission of Methuen Children's Books. Copyright 1926 by E. P. Dutton, renewed 1954 by A. A. Milne. Reprinted by permission of the publisher, E. P. Dutton, a division of Penguin Books USA Inc.

STATISTICS: A Fresh Approach

2 3 4 5 6 7 8 9 0 D O C D O C   9 4 3 2 1 0

ISBN 0-07-054881-1

Library of Congress Cataloging-in-Publication Data

Sanders, Donald H.
    Statistics: a fresh approach / Donald H. Sanders.—4th ed.
        p.   cm.
    Includes index.
    ISBN 0–07–054881–1   ISBN 0–07–835204–5 (software).
    ISBN 0–07–054882–X (instructor's manual).—ISBN 0–07–054883–8 (study guide)
    1. Statistics.   I. Title.
QA276.12.S26   1990   89–8077
519.5—dc20   CIP

This book was set in Caledonia by the College Composition Unit in cooperation with Ruttle Shaw & Wetherill, Inc.
The editors were Bonnie Binkert and Larry Goldberg;
the designer was Joan E. O'Connor;
the production supervisor was Salvador Gonzales.
Photo research was done by Elyse Rieder.
Permissions research was done by Lynn Mooney.
New drawings were done by Fine Line Illustrations, Inc.
R. R. Donnelley & Sons Company was printer and binder.

ABOUT THE AUTHOR

Donald H. Sanders is the author of eight books about computers and statistics. Twenty editions of these texts have been published, and over a million copies of these books have been used in college courses and in industry and government training programs.

Dr. Sanders has 20 years of teaching experience. After receiving degrees from Texas A & M University and the University of Arkansas, he was a professor at the University of Texas at Arlington, at Memphis State University, and at Texas Christian University. In addition to his books, Dr. Sanders has contributed articles to journals such as *Data Management, Automation, Banking, Journal of Small Business Management, Journal of Retailing,* and *Advanced Management Journal.* He has also encouraged his graduate students to contribute computer-related articles to national periodicals, and over 70 of these articles have been published. Dr. Sanders chairs the "Computers and Data Processing" Subject Examination Committee, CLEP Program, College Entrance Examination Board, Princeton, N.J.

TO THOSE WHO OPEN
THIS BOOK WITH DISMAY

CONTENTS

PART TWO SAMPLING IN THEORY AND PRACTICE

PART THREE   COPING WITH CHANGE

PREFACE

It's that time again—time to attempt once more to present the subject of
statistics in an interesting (and sometimes humorous) way so that a period
spent on the subject doesn't seem to students to represent the eternity
suggested by the above quote.

Actually, most readers of this book accept the fact that an educated citizen
needs an understanding of basic statistical tools to function in a world that's
becoming increasingly dependent on quantitative information. But most who
read this text have never placed the solving of mathematical problems at the
top of their list of favorite things to do. In fact, many probably don't care
much for math (may even be terrified of the subject and consider it a foreign
language) and have probably heard numerous disturbing rumors about statistics
courses.

A motivating force behind the preparation of this text is the distinct possibility that the misgivings and apathy implicit in the introductory quote are
related in some way to the unfortunate fact that many existing statistics books
are rigorously written, mathematically profound, precisely detailed—and excruciatingly dull!

THE PURPOSE OF THIS EDITION

The *main difference between this text and many others* is that an attempt is
made here to (1) communicate with students rather than lecture to them,

(2) present material in a rather relaxed and informal way without omitting the more important concepts, (3) show with integrated examples presented throughout the text how computer statistical software packages are used to eliminate computational drudgery and support analysis and decision-making efforts, (4) recapture student attention with occasional quotes, ridiculous names, and unlikely situations of a humorous nature, and (5) utilize an intuitive and/or commonsense approach to develop concepts whenever possible. In short, this book is written for students rather than statisticians, and its intent is to convince readers that the study of statistics can be a lively, interesting, and rewarding experience.

More specifically, the *purpose of this book* is to introduce students at an early stage in a college program to many of the important concepts and procedures they'll need to (1) evaluate such daily inputs as organizational reports, newspaper and magazine articles, and radio and television commentaries, (2) improve their ability to make better decisions over a wide range of topics, and (3) improve their ability to measure and cope with changing conditions both at home and on the job. And since users of this text may frequently be consumers rather than producers of statistical information, the emphasis here is on explaining statistical procedures and interpreting the resulting conclusions. However, the *mathematical demands are modest*—no college-level math background is required or assumed. (The treatment of probability and probability distributions, for example, is limited to the essentials.)

ORGANIZATION AND REVISION FEATURES IN GENERAL

This edition is organized into *four parts*. Each part is introduced by a brief essay that explains the purpose of the part and identifies the chapters included in the part. Each chapter, in turn, is introduced with *opening pages* containing the following features:

- An *opening vignette* that highlights some aspect of the contents of the chapter. These vignettes—many of which are new in this edition—provide statistical applications, cases, and items of interest.
- A *Looking Ahead section* that previews the chapter contents and lists the *Learning Objectives* for each chapter.
- A *Chapter Outline*.
  In the *body* of the chapters, you'll find:
- *Boxed inserts*—many newly selected for this edition—that are included to supply additional "real-world" applications and cases. These inserts help maintain student interest and stimulate discussions.
- Important new terms and concepts that are highlighted in **boldface type** and then defined when they appear for the first time in a chapter and in the text.
- Outputs of *statistical software packages*, which are often used to support computing, analysis, and decision-making efforts. Common input data are often supplied to two statistical programs—*Mystat* and *Minitab*. Students can then see that although output formats may differ the results produced are similar. The presentation and integration of this statistical software material draws on the experience I've gained from two decades of writing computer texts. These

books have been translated into German, French, Spanish, and Portuguese versions, and over a million of them have been used in college courses and industry and government training programs.

At the *end of each chapter* you'll find:

- A *Looking Back* section that addresses the chapter learning objectives by summarizing the main points found in the chapter.
- A listing of *Key Terms and Concepts* that includes the page numbers where the boldfaced new terms/concepts and formulas are first mentioned.
- Sections that present *Problems* and *Review and Discussion Topics*. Scores of new problems are added for this edition, and the *solutions to selected problems* are now included in Appendix 12 at the back of the book. Additional problems and questions are presented in the *Self-Testing Review sections* that are included in most chapters.
- A new *Projects/Issues section*, added for this edition. This section suggests topics for student research that are based on chapter material.
- *Answers to Self-Testing Review Questions*, supplied for student feedback. These problems and questions support the learning objectives of the chapter.
- A new *Closer Look Reading* that gives additional optional information to stimulate discussion and provide more in-depth coverage of selected topics.

A brief summary of the four parts of the text, along with some more specific comments about the revisions made in this edition, is presented below.

REVISING THE PARTS

Part One: Descriptive Statistics

The subject of the first four chapters is descriptive statistics. After introductory materials are presented in Chapter 1, the focus in that chapter turns to a new section on the role of computers in statistics. Computer software concepts and categories are discussed, and the methods people use to communicate with prewritten software packages are outlined. Two types of software packages used to analyze statistical data—the electronic spreadsheet and the statistical analysis program—are introduced. Several examples and illustrations give students a feel for the functions performed by spreadsheet and statistical packages. For example, procedures used to enter a data set into the *Mystat* and *Minitab* programs are shown, and some of the results obtained when these programs analyze the data set are presented. *Mystat* and/or *Minitab* operations are then integrated throughout most of the remaining chapters of the book.

Chapter 2 still introduces examples of how statistical methods have been used improperly. However, the proper use of statistical tables, line, bar, and pie charts, and statistical maps is shown with the help of new tables and charts. A new section on computer graphics programs is found in Chapter 2, and the full-color gallery of photographs showing the types of output produced by such graphics packages is also new.

Chapters 3 and 4 deal with measures of *central tendency* and *dispersion*. A new discussion of computer-generated histograms is introduced in Chapter 3, and new sections on exploratory data analysis are added in Chapters 3 and 4. For example, computer-generated stem-and-leaf displays are presented in Chapter 3, and box-and-whiskers displays are introduced and discussed in Chapter 4. Beginning in Chapter 3, important formulas are numbered and highlighted in boxes for emphasis, and this approach continues throughout the book. The ease with which statistical software packages produce central tendency values (in Chapter 3) and dispersion values (in Chapter 4) is demonstrated and explained.

Part Two: Sampling in Theory and Practice

Statistical inference concepts are considered in the seven chapters of Part 2. The foundation for the material on sampling applications is presented in Chapters 5 and 6. The first of many computer simulation examples (ones that imitate 200 tosses of a coin and 180 rolls of a die by a random process) are presented in Chapter 5. The material on probability computations is reworked and expanded in Chapter 5, and calculations of binomial, Poisson, and normal probabilities by a statistical software package are added. Central Limit Theorem concepts are demonstrated in Chapter 6 with new computer simulations.

Chapter 7 shows how sample data are used to estimate population parameters (new computer simulations help validate important concepts). Chapters 8 through 11 then focus on hypothesis-testing procedures. In Chapter 8, new statistical quality control concepts are introduced, and new examples with computer solutions are presented. The analysis of variance (ANOVA) testing procedure in Chapter 10 is simplified, and a new section on the one-way ANOVA table and the use of computers to ease ANOVA calculation efforts is added. New computer techniques to reduce chi-square testing efforts are also introduced in Chapter 11.

Part Three: Coping with Change

The three chapters in Part 3 focus on the *measurement and prediction of change*. Chapter 12 looks at how index number procedures are used to measure changes in economic conditions. Index number examples are updated to reflect the changes made in the base periods of popular series. Chapters 13 and 14 show how time-series analysis and regression and correlation techniques are used in forecasting. New time-series charts are found in Chapter 13, and new approaches are used to compute regression equation and standard error of estimate values in Chapter 14. A major new section on relationship tests and prediction intervals is now included in Chapter 14. The relationship tests include a t test for slope and an ANOVA test, and prediction intervals are prepared for both large and small samples. The use of *Mystat* and *Minitab* packages to carry out tedious regression and correlation calculations is demonstrated. A lengthy Closer Look reading on multiple linear regression and correlation—complete with example problem and relevant calculations—is now available at the end of Chapter 14 for those wishing to study this technique.

Part Four: Concluding Topics

Several additional quantitative tools available to the decision maker that haven't been considered in the preceding 14 chapters are included in Part 4. For example, Chapter 15 deals with some *nonparametric statistical procedures,* and this chapter now shows how statistical programs can help carry out selected procedures. And Chapter 16 is a brief essay that describes procedures that cannot be considered in any detail. New readings provide a general overview of several of these important tools.

SUPPLEMENTS FOR THIS EDITION

Several supplements have been prepared to make this *Statistics* package a complete teaching and learning tool. They include:

- *Computer Supplement.* A computer supplement is available for those interested in learning how to use a statistical computer package. The components of the computer supplement are: (1) *Mystat diskette:* An educational version of the *Systat* statistical package. The diskette contains the *Mystat* program and disk space for storing data. (2) *Using Mystat:* A complete user manual that teaches the reader how to use the *Mystat* statistical package.
- *The Student Study Guide.* The Study Guide presents problems and exercises to help students develop a better understanding of statistical reasoning and analysis. Each chapter includes (1) an overview of key terms and important formulas presented exactly as found in the textbook, (2) several illustrative problems, with detailed solutions, (3) set of exercises, some of which reinforce basic problem-solving skills while others explore variations on underlying themes, (4) solutions to problems and exercises.
- *The Instructor's Manual.* For each chapter of the text, the Instructor's Manual contains (1) solutions to the end-of-chapter problems that aren't presented at the end of the book, (2) a key to discussion questions, (3) possible true-false test questions, (4) possible multiple-choice test questions, and (5) transparency masters.

USE OF THIS BOOK

This book is written for use in an early one- or two-term course in statistics. As noted earlier, no college-level math background is required or assumed. The organization of this book into four parts permits a certain amount of *modular flexibility.* Although the order of presentation is logical and is used successfully by many instructors, there's no necessary reason why chapters and even parts must be covered in the sequence in which they appear in this volume. Depending on the objectives of the course, some (but certainly not all) of the ways this book might be used are as follows:

PART	PART	PART	PART	PART
1 ⎫	1 ⎫	1 ⎫ first	1 ⎫ one	1 ⎫
2 ⎬ one	2 ⎬ one	3 ⎭ term	2 ⎬ term	Chapter 12 ⎪
3 ⎪ term	3 ⎪ term		Chapter 14 ⎭	2 (selected ⎪ one
4 ⎭	4 ⎭	2 ⎫ second		chapters) ⎬ or
		4 ⎭ term		Chapter 15 ⎪ two
				Chapter 13 ⎪ terms
				Chapter 14 ⎭

The chapters in Part 3 are essentially free-standing and may be covered in sequence or independently. Chapter 15 may be included with the material in Part 2, or it may be used in place of some of the material in Part 2.

One inevitable limitation of a book such as this involves the priorities given to the various statistical topics. Some may feel that there's an appalling lack of coverage of topics that should be included. Certainly, many of the subjects briefly discussed in the last chapter—"Where Do We Go from Here?"—could each have been expanded to lengthy chapters with the possible result that the finished product would be twice its present size. And others may feel that too much space is devoted to some area—for example, sampling applications in Part 2. Since this book is written for beginners, though, the preference here is to err in the direction of keeping the size (and cost) of the book down.

ACKNOWLEDGMENTS

It's customary to conclude a preface by acknowledging the help and suggestions received from numerous sources. And certainly the many who have contributed to this edition are deserving of recognition. Useful comments and suggestions were provided by the following colleagues who responded to questions about earlier editions, and who reviewed this text during its development: Ernest A. Beasley, Okaloosa-Walton Junior College; Ronald L. Coccari, Cleveland State University; Joel Feiner, Suny—Old Westbury; Keith Moore, Park College—Missouri; Joseph B. Murray, Philadelphia Community College; Mark J. Ratkus, LaSalle University; Theodore W. Roesler, University of Nebraska—Lincoln; and Larry J. Schuetz, Linn Benton Community College.

I am grateful to the Literary Executor of the late Sir Ronald A. Fisher, F.R.S., to Dr. Frank Yates, F.R.S., and to Longman Group Ltd. London, for permission to reprint an abridgment of Table IV from their book *Statistical Tables for Biological, Agricultural and Medical Research* (6th edition, 1974). Another word of thanks goes to those who furnished the readings and photos reproduced in the text. Their individual contributions are acknowledged in the body of the book. Thanks also to Bob Eng and Frank Murph for their input in earlier editions.

The final tribute and greatest appreciation, though, is reserved for these few: to Bonnie Binkert, Larry Goldberg, Joan O'Connor, Shelly Langman, developer of *User's Manual to Mystat*, and Sal Gonzales, production supervisor; to Dr. Gary D. Sanders, Arizona State University, for his insight and problem contributions; and to Joyce Sanders for her continuing suggestions and encouragement.

Donald H. Sanders

STATISTICS

A Fresh Approach

PART ONE

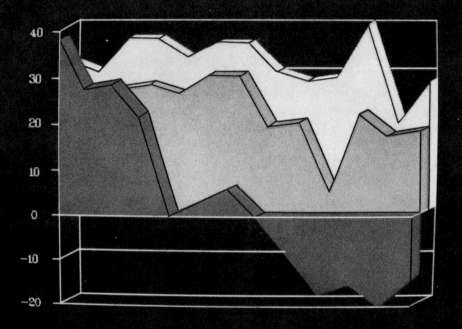

DESCRIPTIVE STATISTICS

The procedures for collecting, classifying, summarizing, and presenting quantitative facts are an important part of the subject of statistics. Earlier introductory statistics books, in fact, dealt almost exclusively with these descriptive procedures.

In Part 1 we'll see how various statistical measures are used (and misused) to describe and summarize relationships that exist between variables. And although the focus in Part 3 is on measuring and predicting change, the three chapters in that section of the book also deal largely with descriptive statistical procedures.

In contrast to earlier statistics books, however, current texts also emphasize the statistical inference procedures used to make decisions on the basis of computed measures taken from samples. (Statistical inference is the subject of the chapters in Part 2 of this book.) But even in statistical inference applications, computed descriptive measures are used as the basis for decision making. Thus, a knowledge of descriptive statistics is needed by all consumers of quantitative information—that is, by all educated citizens.

The chapters included in Part 1 are:

1   Let's Get Started
2   Liars, #$%*& Liars, and a Few Statisticians
3   Statistical Description: Frequency Distributions and Measures of Central Tendency
4   Statistical Description: Measures of Dispersion and Skewness

LET'S GET STARTED

The Romantic Statistician

In O. Henry's *The Handbook of Hymen,* Mr. Pratt is wooing the wealthy Mrs. Sampson. Unfortunately for Pratt, he has a poet for a rival. To compensate for his romantic disadvantage, Pratt selects a book of quantitative facts to dazzle Mrs. Sampson.

"Let us sit on this log at the roadside," says I, "and forget the inhumanity and ribaldry of the poets. It is in the glorious columns of ascertained facts and legalized measures that beauty is to be found. In this very log we sit upon, Mrs. Sampson," says I, "is statistics more wonderful than any poem. The rings show it was sixty years old. At the depth of two thousand feet it would become coal in three thousand years. The deepest coal mine in the world is at Killingworth, near Newcastle. A box four feet long, three feet wide, and two feet eight inches deep will hold one ton of coal. If an artery is cut, compress it above the wound. A man's leg contains thirty bones. The Tower of London was burned in 1841."

"Go on, Mr. Pratt," says Mrs. Sampson. "Them ideas is so original and soothing. I think statistics are just as lovely as they can be."

LOOKING AHEAD

Mr. Pratt is making effective use of "statistics" in the preceding vignette. But as you read Chapter 1, you'll see that the word statistics is used in several ways. You'll also learn the meaning of other important terms that we'll use throughout this book, have a better understanding of the need for statistics, and be introduced to the basic steps used to solve statistical problems. Finally, the role of the computer in statistics is explored.

Thus, after studying this chapter, you should be able to:

- Explain the meaning of the terms in boldface type such as **statistics, descriptive statistics, statistical inference, population, parameter, census, sample,** and **statistic.**
- Understand and explain why a knowledge of statistics is needed.
- Discuss the basic steps in the statistical problem-solving methodology.
- Outline the role of computers and data-analysis software packages in statistical work.

CHAPTER OUTLINE

WHAT TO EXPECT

It's possible (even likely) that you don't yet share the view of statistics expressed by Mrs. Sampson in the opening vignette. Oh, you may agree that an understanding of statistical tools is necessary in a modern world. But you've never placed the solving of mathematical problems at the top of your list of favorite things to do, you've possibly heard disturbing rumors about statistics courses, and you've not been eagerly awaiting this day when you must crack open a statistics book. If the comments made thus far in this paragraph apply to you, you needn't be apologetic about your possible misgivings. After all, many statistics books are rigorously written, mathematically profound, precisely detailed—and excruciatingly dull!

It isn't possible in this book to avoid the use of formulas to solve statistical problems and demonstrate important statistical theories. But a knowledge of advanced mathematics certainly *isn't* required to grasp the material presented here. In fact, you'll be relieved to know that a beginning-level high school algebra course prepares you for all the math required (to perform such tough operations as addition, subtraction, multiplication, division, and finding square roots).

You'll find in the pages and chapters that follow that the intent is to (1) communicate with you rather than lecture to you, (2) present material in a rather relaxed and informal way without omitting the more important concepts, (3) recapture your attention with occasional quotes, ridiculous names, and unlikely situations of a humorous nature, and (4) utilize an intuitive and/or commonsense approach to develop concepts whenever possible. In short, this book is written for beginning students rather than statisticians, and its intent is to convince you that the study of statistics is a lively, interesting, and rewarding experience. (If Mr. Pratt could convince Mrs. Sampson, then maybe you too can be converted.)

PURPOSE AND ORGANIZATION OF THE TEXT

Purpose of This Book

To do is to be—J.-P. Sartre
To be is to do—I. Kant
Do be do be do—F. Sinatra[1]

The purpose of this book is to introduce you to many of the important statistical concepts and procedures you'll need to (1) evaluate such daily in-

---

[1] Have you ever noticed that chapters and sections of chapters in learned books and academic treatises are often preceded by quotations such as these that are selected by the author for some reason? In some cases a quotation is intended to emphasize a point to be presented; in other cases (often in the more erudite sources) there appears to be no discernible reason for the message, and it forever remains a mystery to the reader. In this particular case, the quotations from the above philosophers unfortunately fall into the *latter category!* However, we will from time to time throughout the book attempt to use quotations (from such authorities as Aldous Huxley, Mark Twain, and Winnie-the-Pooh) for the more valid purpose of emphasizing a point.

puts as organizational reports, newspaper and magazine articles and polls, and radio and television commentaries, (2) improve your ability to make better decisions on such wide-ranging topics as the quality of a particular product, the public servant likely to be thrown out of office, or the salesperson to believe, and (3) improve your ability to measure and cope with changing conditions both at home and on the job.

But *the purpose of this book isn't* necessarily to make a professional statistician out of you, because it's recognized that you're unlikely to be seeking such a career. Therefore, since you'll more likely be primarily a *consumer* rather than a producer of statistical information, the emphasis of this book is placed on explaining statistical procedures and interpreting the resulting conclusions. In short, the following dialogue from K. A. C. Manderville's *The Undoing of Lamia Gurdleneck* concludes with an important message that's kept in mind throughout this text.[2]

> *"You haven't told me yet," said Lady Nuttal, "what it is your fiancé does for a living."*
>
> *"He's a statistician," replied Lamia, with an annoying sense of being on the defensive.*
>
> *Lady Nuttal was obviously taken aback. It had not occurred to her that statisticians entered into normal social relationships. The species, she would have surmised, was perpetuated in some collateral manner, like mules.*
>
> *"But Aunt Sara, it's a very interesting profession," said Lamia warmly.*
>
> *"I don't doubt it," said her aunt, who obviously doubted it very much. "To express anything important in mere figures is so plainly impossible that there must be endless scope for well-paid advice on how to do it. But don't you think that life with a statistician would be rather, shall we say, humdrum?"*
>
> *Lamia was silent. She felt reluctant to discuss the surprising depth of emotional possibility which she had discovered below Edward's numerical veneer.*
>
> *"It's not the figures themselves," she said finally, "it's what you do with them that matters."*

Definitions and Organization

Let's pause here just long enough to define a few terms. The word "statistics" commonly has two different meanings. Lady Nuttal or Mrs. Sampson would likely define the word in its *first context* as being essen-

---

[2] Frontispiece from Maurice G. Kendall and Alan Stuart, *The Advanced Theory of Statistics,* vol. 2: *Inference and Relationships,* Hafner Publishing Company, Inc., New York, 1967.

tially the same as "figures." That is, they might consider **statistics** to be a **plural** term that means numerical facts or data. In addition to meaning quantitative facts, however, "statistics" is also used in a *second way* and in a singular sense, as you'll see in the next paragraphs.

Early statistics books dealt primarily with the procedures for *describing* data by means of classifying, summarizing, and communicating methods. The emphasis was generally on *collecting* and *classifying* data, and then on the use of *summary* measures such as averages that would effectively describe the basic structure of the subject being studied. The preparation of charts and tables to show relationships and to interpret and *communicate* the measured values also received much attention.

Obviously, *statistical description is still an important part of the study of statistics.* Sales or revenue data, for example, may be classified or grouped by (1) volume, size, or quantity, (2) geographic location, or (3) type of product or service. To be of value, masses of data are often condensed or sifted—that is, summarized—so that resulting information is concise and effective. A general sales manager, for example, may be interested only in the average monthly total sales of particular stores. Although she could be given a report that breaks sales down by department, product, and sales clerk, such a report is more likely to be of interest to a department manager. Once the facts have been classified and summarized, it's then often necessary that they be presented or communicated in a usable form—perhaps through the use of tables and charts—to the final user.

Although we've seen that the term may refer in a plural form to numerical data, **statistics** is more generally used in a **singular** sense in this book to refer to the body of principles and procedures developed for the collection, classification, summarization, interpretation, and communication of numerical data, and for the use of such data. Thus, statistics refers to a subject of study in the same way that mathematics refers to such a subject.

This book is organized into four parts and 16 chapters to achieve its purpose. The subject of Part 1 (the first four chapters) is **descriptive statistics**—the procedures of data collection, classification, summarization, and presentation. With a knowledge of descriptive statistical procedures, you can evaluate information presented in reports, articles, and broadcasts, and you can improve your ability to measure and thus cope with changing economic conditions. In Chapter 2, for example, you'll see how statistical methods have been *improperly used* by individuals and groups to confuse or deliberately mislead people. Many of the invalid uses presented involve descriptive procedures. And in Chapter 3 we'll consider data collection and classification and the *measures of central tendency* (or averages) frequently used by decision makers. Other important measures used to describe the amount of *dispersion or spread* in the data are discussed in Chapter 4.

In addition to descriptive statistics, however, another topic which must receive extensive treatment in any modern statistics text is inferential statistics. **Statistical inference** is the process of arriving at conclusions about the total group under study on the basis of data obtained from only a por-

DESCRIPTIVE STATISTICS

The multipurpose statistics collected by government statistics bureaus are primarily descriptive statistics. The immediate occasion for their collection is usually a practical one, not the search for scientific truths. Nevertheless, these descriptive statistics are frequently incorporated into scientific theories or assist in the development of new knowledge. The population censuses, originally prescribed for the apportionment of congressional seats, have, together with the birth and death statistics developed to promote public health, formed the basis of the science of demography. The Census of Manufactures and our statistics of international trade, designed to support the infant economy of a new country, were the forerunners of an enormous array of economic data that are used not only for current monitoring of the economy but also to test economic theories and are incorporated into econometric models to analyze relationships. Weather statistics provide some of the raw data for the atmospheric sciences. Thus opportunities do arise for the development of new knowledge in connection with what may appear to be mundane data collection activities.

The processes of collecting, summarizing, analyzing, and of disseminating data, furthermore, afford ample opportunity, given sufficient resources, for the application of statistical methodology and the development of new statistical techniques.

—From Margaret E. Martin, "Statistical Practice in Bureaucracies," *Journal of the American Statistical Association*, March 1981, p. 3. Reprinted with permission.

---

tion of the total group. The total collection of measurements or observations being studied by a decision maker is called the **population** (or **universe**). Note, though, that "population" (as used in statistics) is a term that isn't limited to a group of people but instead refers to the total of any kind of unit under consideration. Thus, a population could be the parts assembled in a production run, the number of frozen chickens in a shipment, the number of credit accounts of a firm, or the number of people in an organization.

The portion or subset of the total group or population that supplies data to a decision maker is called a **sample.** In contrast to a **census,** which is a study of an entire population, a sample is some selected segment of the population that's examined to (1) *make estimates* about some unknown population characteristic such as the percentage of consumers who like a new product or (2) *make tests* to see if assumptions about an unknown population characteristic are likely to be acceptable—such as the claim of a salesperson that the average life expectancy of a product is greater than that of the product you're now using.

In these two examples, each of the *population* characteristics or measures being sought—the population percentage and the population average (arithmetic mean)—is called a **parameter.** If taken from a *sample*, each of these measures would be called a **statistic.**

To summarize, then, statistical inference typically involves using a sample statistic to make a judgment or decision about an unknown population parameter (see Figure 1-1). Thus, statistical inference goes beyond the

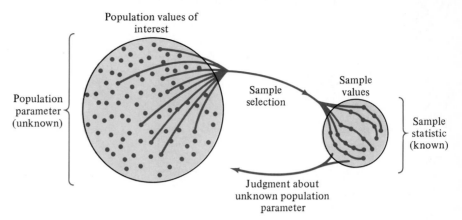

FIGURE 1-1
The statistical inference process involves the use of a known sample statistic to arrive at a judgment about an unknown population parameter.

mere description of sample data and becomes an analysis tool to aid the decision maker in reducing the level of uncertainty that would have existed without any sample data. Figure 1-2 summarizes several of the points made in the preceding paragraphs.

Part 2 of this book is entitled "Sampling in Theory and Practice" and includes Chapters 5 through 11. Chapters 5 and 6 present the conceptual foundations for the material on sampling applications. Then, in Chapters 7 through 11, the use of sample data to make estimates about, and test assumptions about, some unknown population characteristic is presented.

Part 3 ("Coping with Change") contains elements of statistical inference, but the material in Chapters 12, 13, and 14 is essentially descriptive. However, the focus of Part 3 is on the *measurement and prediction of change*. In Chapter 12 you'll learn about *index number procedures* that *measure* relative changes in economic conditions over time. For example, you'll be introduced to the Consumer Price Index, which tracks the average changes in prices of many consumer goods. You see references to this measure all the time on television and in the newspapers. Then, in Chapter 13, you'll examine a *time-series forecasting approach* that consists of (1) analyzing past data to detect reasonably dependable patterns and then (2) projecting these patterns into the future to arrive at future expectations. Finally, in Chapter 14, you'll study a forecasting approach that consists of identifying and analyzing reasonably predictable factors that may have an influence on the variable you're interested in projecting.

Part 4 ("Concluding Topics") includes Chapters 15 and 16. These final chapters discuss some of the additional tools available to the decision maker that haven't been considered elsewhere.

Most of the chapters in this book contain *self-testing review sections* following the presentation of important material. You're encouraged to test your grasp of the reviewed subjects by answering the questions and/or problems in these sections before moving to the next topic. Answers to self-testing reviews are given at the end of the chapters. You'll also find at the end of most chapters (1) a *Looking Back chapter summary*, (2) a list of

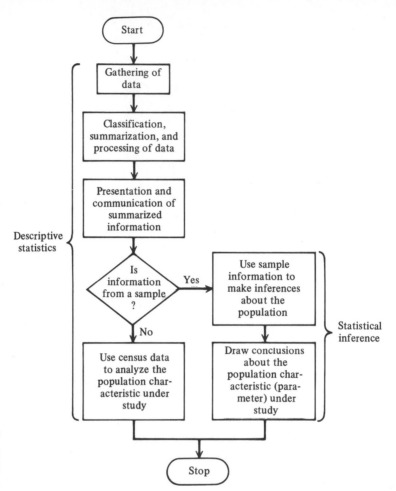

FIGURE 1-2
An overview of
descriptive statistics
and statistical
inference.

the *key terms and concepts* presented in the chapter, (3) *additional problems* (with answers to selected problems found at the back of the book), (4) *topics for review and discussion*, (5) *projects/issues* to consider, and (6) a *Closer Look reading* that provides additional information on some topic discussed in the chapter.

NEED FOR STATISTICS

At the beginning of this chapter the following sentence appeared: "Oh, you may agree that an understanding of statistical tools is necessary in a modern world." Perhaps that statement was premature; perhaps you don't agree at all. It *is* a fact, though, that you need a knowledge of statistics to help you (1) describe and understand relationships, (2) make better decisions, and (3) cope with change.

Describing Relationships between Variables

The amount of quantitative data that's gathered, processed, and presented to citizens and decision makers for one purpose or another has increased rapidly. Thus, it's necessary that you be able to sift through this almost overwhelming mass to identify and describe those sometimes obscure relationships between variables that are often so important in decision making. The following examples serve to illustrate the need for statistical analysis in understanding relationships:

- A businessperson can, by summarizing masses of revenue and cost data, compare the average return on investment during one time period with related figures from previous periods. (Several decisions might hinge on the outcome of this comparison of descriptive measures.)
- A public health official may come to conclusions about the relationship between smoking and/or obesity and a variety of diseases by applying statistical techniques to huge amounts of input data. These conclusions, in turn, may lead to decisions that affect millions of people.
- A marketing researcher may use statistical procedures to describe the relationships between the demand for a product and such characteristics as the income, family size and composition, age, and ethnic background of the consumers of the product. On the basis of these relationships, advertising and distribution efforts may then be directed toward those groups that represent the most profitable market.
- An educator may use statistical techniques to see if there's a significant relationship between scholastic aptitude test scores and the grade point averages of students at her school. If there's such a relationship, she can make predictions about the probable academic success of an applicant on the basis of the test score.

Aiding in Decision Making

An administrator can use statistics to make better decisions in the face of uncertainty. Consider the following examples:

- Suppose that you're the manager in charge of purchasing for a large food-processing plant that packages frozen fried-chicken dinners. You're responsible for purchasing dressed broilers in shipments of 10,000 birds. Standards have been established specifying that the average weight of the birds in a shipment should be 32 ounces. (Birds over that weight tend to be too tough; lighter birds are too scrawny.) The truck of a new supplier rolls up to the unloading dock with a shipment you've agreed to take provided that the weight and quality standards are met. The supplier's salesperson assures you that the shipment will meet your standards. Should you accept the shipment on the basis of this claim? Probably not. Rather, you could use statistical inference techniques to select an appropriate *sample* of, say, 100 birds from the *population* of 10,000 broilers. You could then weigh each bird in the sample, compute the average weight of the 100 birds, perform some other calculations, and then reach a conclusion about the average weight of the population of 10,000 chickens. Given this information, you could then make a more informed decision on whether or not to accept the shipment.

▶ Suppose again that you're the production manager of a plant that produces shotgun shells. It's known that certain variations will exist in the shells produced—there will likely be some variation in shot patterns produced and in shot velocity—but these variations can be tolerated if they don't exceed specified limits more than 1 percent of the time. By using a statistically designed sampling plan, you can arrive at reliable conclusions or inferences about the quality of a production run. Your conclusions or inferences are based on tests conducted by firing a relatively small number of shells randomly selected from those produced during the run.

▶ Finally, suppose that the manager of Big-Wig Executive Hair Stylists, Hugo Bald, has advertised that 90 percent of the firm's customers are satisfied with the company's services. If Polly Tician, a consumer activist, feels that this is an exaggerated statement that might require legal action, she can use statistical inference techniques to decide whether or not to sue Hugo. (The decision in this case is found in Chapter 8.)

In the first example above, you could have weighed all 10,000 birds to determine the average weight. That approach would have been expensive and time-consuming. In the second example, you could have determined the quality of the product by testing all the shells produced, but since such testing is destructive, there would have been nothing left for you to sell. (In both examples, of course, you received statistical information that permitted you to make better decisions.) And in the third example, Polly can test Hugo's claim before deciding about taking legal action (with the real possibility of a countersuit if it turns out that she's wrong).

Coping with Change

To plan is to decide in advance on a future course of action. And plans and decisions are based on expectations about future events and/or relationships. Thus, we are all required to employ some forecasting process or technique to arrive at a future expectation. Although statistical procedures will obviously not enable us to predict the future with unerring accuracy, there are, as the following examples indicate, useful statistical aids that may help to measure current change and improve the forecasting process.

▶ Government statisticians periodically gather price data on hundreds of different items from over 50 urban areas to compute a single monthly summary figure that measures overall changes in price levels between the current period and some period in the past. Thus, a personnel manager or union leader can use information on changes in price levels to see what has happened to the purchasing power of the dollar before entering into future wage negotiations.

▶ Suppose that a sales manager has sales figures on a product line extending over a 10-year period. If, after studying these time-series figures, she believes that an identifiable past pattern will persist, she can build a forecast of future sales by using statistical procedures to project the past pattern into the future. She can also adjust her future sales forecast to take into account seasonal variations such as the sales peaks that typically occur in December.

▶ Let's assume that a personnel manager has noted that job applicants who score high on a manual dexterity test later tend to perform well in the assembling of a product, while those with lower test scores tend to be less productive. By applying a statistical technique known as regression analysis (the subject of Chapter 14), the manager can predict how productive a new applicant will be on the job on the basis of how well he performs on the test.

A comparison of the above sections that show why you need a knowledge of statistics with the purpose and organization of this book stated earlier reveals similarities that can be summed up as follows: The purpose of the parts of this book is to help you acquire the statistical understanding you need to better describe and understand relationships, make better decisions, and more effectively cope with changing conditions.

STATISTICAL PROBLEM-SOLVING METHODOLOGY

One *could* approach a problem-solving situation in a way somewhat analogous to the following illustration:

In the comic dialect of the last century, a satirical almanac offers a solution to astrological signs:

Tew kno exackly whare the sighn iz, multiply the day ov the month bi the sighn, then find a dividend that will go into a divider four times without enny remains, subtrakt this from the sighn, add the furst quoshunt tew the last divider, then multiply the whole ov the man's body bi all the sighns, and the result will be jist what yu are looking after.[3]

Or, one could elect to follow Mark Twain's example in *Sketches Old and New:*

---

[3] Josh Billings, *Old Probability: Perhaps Rain—Perhaps Not*, G. W. Carleton and Company, Publishers, New York, 1879.

If it would take a cannon ball 3⅓ seconds to travel four miles, and 3⅜ seconds to travel the next four, and 3⅝ to travel the next four, and if its rate of progress continued to diminish in the same ratio, how long would it take to go fifteen hundred million miles?

—*Arithmeticus*
Virginia, Nevada

I don't know.

—*Mark Twain*

In most statistical problem-solving situations, though, it's appropriate to follow a more scientific approach. Several steps are followed to get rational answers to statistical problems, and when one of these steps is ignored, the final results may be invalid, inaccurate, or needlessly expensive. The basic steps in statistical problem solving are outlined below.

Identifying the Problem or Opportunity

The researcher must first understand and correctly define the problem or opportunity that exists. Quantitative information useful at this time includes data outlining the nature and scope of the problem such as production shortage figures and sales back orders. Facts about the population to be studied, and the impact of the situation on personnel, materials, and money may also be needed.

Gathering Available Facts

Data must be gathered that are accurate, timely, as complete as possible, and relevant to the problem being considered. Sources of data may be classified into internal and external categories.

Internal data are found in the departments of an organization. In a business, internal data are produced in accounting, production, and marketing departments; in a college or hospital, the registrar's office or admitting section generates internal facts. **External data** are facts produced by outside sources such as professional associations, government agencies, customers, and suppliers. External data are supplied by nongovernment publications such as the *American Economic Review, The American Statistician, BioScience, Business Week, Chemical Engineering, Forbes, New England Journal of Medicine, Science, Teaching of Psychology*—this list could go on and on. A wealth of external data are also supplied by government publications such as the *Census of the United States*, the *Census of Business*, the *Survey of Current Business*, the *Statistical Abstract of the United States*, the *Monthly Labor Review*, and the *Federal Reserve Bulletin*.

It's generally preferable to gather data from **primary sources**—those that initially gather the data and first publish them—rather than from **secondary sources**—those that republish the data. This is true because secondary sources may introduce reproduction errors and may not explain how the facts were gathered or what limitations exist to their use. Secondary sources may also fail to show how the primary source defined the vari-

PLAYING THE NUMBERS GAME

It's springtime. George Will celebrates baseball, and I celebrate the "Statistical Abstract of the United States." This seems a satisfactory division of labor. George can write for the 7 percent of Americans who eagerly await opening day, and I can write for the rest who breathlessly anticipate the newest edition of the "Abstract."* It's just out from the Census Bureau with 1,478 tables (29 fewer than last year) telling us who we are, what we do and the kind of world we live in.

The "Abstract" has endless uses. For example, it's ideal for humiliating your wife. You take an unknowable statistic and turn it into an unanswerable question. My wife plays this game because she tolerates my quirks and ignores the odds against her. Recently I asked her what proportion of Americans over 45 wear glasses. About 90 percent, she said. The correct answer is 89.9 percent. OK, smarty, what's the net worth of households under 35? About $5,000, she said. The answer is $5,764 (in 1984). After that I quit; I'd had enough humiliation for one night.

For a columnist, writing about the "Abstract" provides temporary relief from telling the rest of the world how to run its business. We are supposed to know all the right questions and most of the right answers. After a while, this is hard work even if you're opinionated and unburdened by a sense of your own ignorance. The "Abstract" reminds us of how little we know. The sheer variety of statistics on so many subjects is chastening.

No matter how often it's used, the "Abstract" always has new nuggets of information and insights. Consider these:

Item: Maybe an explosion of new engineers signals a resurgence of U.S. competitiveness. Between 1975 and 1984, the number of annual college and university engineering graduates nearly doubled—from 65,000 to 118,000. Higher pay may be one cause of the increase. Starting monthly salaries for electrical engi-

neers with bachelor degrees ($2,283 in 1985) are about a third higher than those for starting accountants ($1,697). One in eight new engineers is a woman, compared with fewer than one in 100 in 1970.

Item: It's true that Americans are losing their taste for beef. Since 1975, annual per capita beef consumption has fallen nearly 10 pounds, from 88 to 79 pounds. Chicken consumption has risen sharply (from 40 pounds annually in 1975 to 58 in 1985). Greater health consciousness? Well, maybe. But if so, it's selective. Since 1975 the average American has increased use of sugar and sweeteners 29 percent to 170 pounds annually. . . .

If statistics remind me of the complexity and confusion of everyday life, they're also one of the ways that we attempt to simplify these confusions. Statistics create an aura of certainty and neutrality. They're supposed to convey truth, settle debates and make it easier to control our environment. Up to a point, they do. But they can also be manipulated or misunderstood and, thereby, distort reality. Even the absence of statistics is sometimes a deliberate attempt to shape perceptions, as Princeton sociologist Paul Starr and Harvard demographer William Alonso argue in a new book.†

"For years after World War II Lebanon did not hold an official census, out of fear that a count of the torn country's Christians and Muslims might upset their fragile, negotiated sharing of power (which broke down anyway)," they write. "Saudi Arabia's census has never been officially released, probably because of the Saudis' worry that publishing an exact count (showing their own population to be smaller than supposed) might encourage enemies to invade the country or promote subversion."

LIVING STANDARDS

Many of our social statistics are inherently limited. Sociologist Christopher Jencks of Northwestern University points out (in another essay

*Statistical Abstract of the United States: 1987, 960 pages. Available from the Superintendent of Documents, U.S. Government Printing Office, Washington, D.C. 20402, $22 paperback, prepaid. GPO stock number 003-024-06572-0.

†The Politics of Numbers, edited by William Alonso and Paul Starr, Russell Sage Foundation, 480 pages. $18.95 paperback and $37.50 hard-cover. Forthcoming, July 1987.

from the same book) that precise measurement of changes in living standards may be impossible. When rising family incomes in the 1970s are adjusted for inflation, as reflected in the consumer price index, there's almost no increase. Using a less well-known—but equally good—inflation index called the "personal consumption deflator" results in a 1.0 percent average annual gain. That's about 11 percent during the 1970s: lower than the 1950s or 1960s but not (as commonly thought) stagnation.

Moreover, statistics on incomes and prices miss many of the most important changes. In 1978, Jencks writes, "my prematurely born son's life was saved by a drug that did not exist in 1970 . . . [M]y standard of living was 'infinitely' higher in 1978 than in 1970 had my son been born then." Of course, the process also works in reverse. As conventionally measured, living standards have risen recently. But the gains seem lessened by greater economic uncertainty and insecurity. Also, there are new threats—AIDS, for instance.

Even the most avid fan of statistics inevitably collides with a number that seems totally unbelievable. The newest "Abstract" has this one: in 1985, 48 percent of Americans over the age of 14 reported they did volunteer work. That caused my calculator to self-destruct. Granting all the Little League coaches, Scout leaders, PTA officials and church-choir singers, does every other person do volunteer work? It's not me and not (I think) nearly half my friends. Are all the nonvolunteers on the East Coast offset by the fact that everyone in Illinois, Iowa and Idaho volunteers?

These are good questions. Maybe some day I'll answer them. But right now there's a more pressing matter. I've got to figure out why my wife beat me at the numbers game. She must have cheated, but I just don't know how.

ables. During the depression month of November 1935, for example, the National Industrial Conference Board estimated the number unemployed at about 9 million; the National Research League estimated 14 million; and the Labor Research Association topped them all with a figure of 17 million. These estimates varied primarily because of differences in the way unemployment was defined. Sherlock Holmes summarized the importance of data gathering in *The Adventure of the Copper Beeches* when he said: "Data! Data! Data! I can't make bricks without clay."

Gathering New Original Data

In many cases the data needed by decision makers simply aren't available from other sources, and so there's no alternative but to gather them. There are advantages to gathering new data, for the decision maker who's aware of the problem can define the variables and determine how they will be measured so that the resulting facts will possess the properties needed to solve the problem.

There's a variety of methods for obtaining desired data. Common data-gathering practices make use of personal interviews and mail questionnaires. In a **personal interview**, an interviewer asks a respondent the prepared questions that appear on a *schedule* form and then records the answers in the spaces provided on the form. This data-gathering approach allows the interviewer to clarify any terms that aren't understood by the respondent, and it results in a high percentage of usable returns. But it's an expensive approach and is subject to possible errors introduced by the interviewer's manner in asking questions. Interviews are often conducted over the telephone. This is less expensive, but, of course, some households don't have telephones or have unlisted numbers, and this may bias the survey results.

When **mail questionnaires** are used, the questions are printed on forms, and these queries are designed so that they can be answered by the respondent with check marks or with a few words. The use of questionnaires is often less expensive than personal interviews. But the percentage of usable returns is generally lower. And those who do answer may not always be the ones to whom the questionnaire was addressed, and/or they may respond because of a nonrepresentative interest in the survey subject.

Classifying and Summarizing the Data

After the data have been collected, the next step is to organize or group the facts for study. Identifying items with like characteristics and arranging them into groups or classes, we've seen, is called classifying. Production data can be classified, for example, by product make, location of production plant, and production process used. Classifying is sometimes accomplished by a shortened, predetermined method of abbreviation known as *coding*. Code numbers are used to designate persons (social security number, payroll number), places (zip code, sales district number), and things (part number, catalog number).

Once the data are arranged into ordered classes, it's then possible to reduce the amount of detail to a more usable form by summarization. Tables, charts, and numerical descriptive values such as measures of central tendency (or averages) and measures of dispersion (the extent of the spread or scatter of the data about the central value) are summarizing tools.

Presenting and Analyzing the Data and Making the Decision

Summarized information in tables, charts, and key quantitative measures facilitates problem understanding. Such information also helps identify relationships, and allows an analyst to present important points to other interested parties.

The analyst must next interpret the results of the preceding steps, use the descriptive measures computed as the basis for making any relevant statistical inferences, and employ any statistical aids that may help identify desirable courses of action. The appropriateness of the options selected is, of course, determined by an analyst's skill and the quality of his or her information.

Finally, the analyst must weigh the options in light of established goals to arrive at the plan or decision that represents the "best" solution to the problem. Again, the correctness of this choice depends on analytical skill[4] and information quality.

Figure 1-3 expands Figure 1-2 and summarizes the steps in the statistical problem-solving methodology.

ROLE OF THE COMPUTER IN STATISTICS

Computers are efficiently used when data input isn't trivial, similar tasks are performed repeatedly, jobs must be completed accurately and quickly, and/or processing complexities present no practical alternative to computer use. Since many statistical problems have a relatively large volume of input data, are repetitive, and their solution produces information that needs to be accurate and timely, the computer is a vital tool to those who solve these problems. Procedures that take hours, days, or weeks with a desk calculator are accurately completed in seconds or minutes with a computer.

The problems you'll encounter in this book typically use relatively small sets of input data. But even so, you'll see in a few instances that our calculations can become tedious when done by hand. It's usually desirable, though, to follow the hand calculations discussed in the text and then use similar steps to solve a few additional problems. Once you've done that, you'll understand the uses and limitations of the procedures and be able to correctly interpret the results they produce. Then you can use a computer to carry out similar future work. The meaning of those computer results probably won't be explained in any detail, but you'll then have a background to understand the output.

A Computer Software Primer

Software is a general term that refers to the detailed sets of stored instructions—or *programs*—that control computer operations. Software, like television programs, turns a lifeless machine into something useful. There are several software categories.

Software Categories   Some software products—*applications programs*—are written to control the processing of particular tasks. Another software category—the *operating system*—is a collection of programs that allows a user's computing equipment or hardware to work with applications programs. Although *custom-made* applications programs must be written by professional programmers to process unique jobs, countless other common

---

[4] The story has often been told of a scientist who trained a flea to jump when a bell was sounded. Then, the scientist would ring the bell, and after each jump by the flea he would pull off one of the insect's legs. When the poor flea was down to its last leg, the scientist rang the bell, and the flea flopped over weakly. The scientist then removed the last leg, again rang the bell, and observed that the flea did not move at all. His conclusion: When you remove all of a flea's legs, it goes deaf.

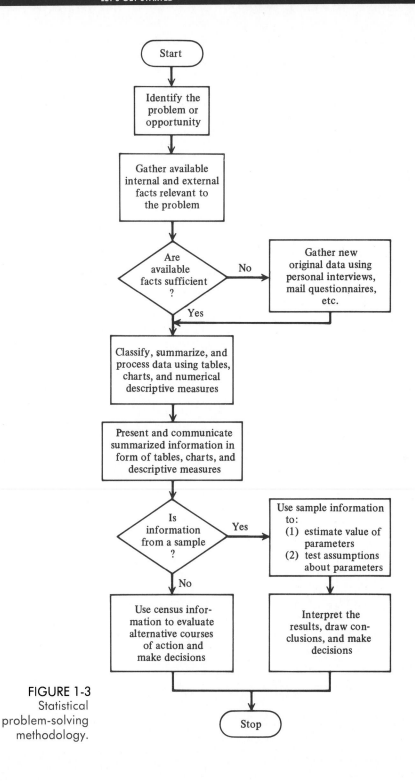

FIGURE 1-3
Statistical
problem-solving
methodology.

jobs found in workplaces and homes can be processed with *generalized* and *prewritten* applications programs that have been prepared by software vendors.

A prewritten applications package typically addresses the processing needs of many users. It might perform a *single* broad function such as processing words. Or it might analyze rows and columns of figures and then compute the statistical measures called for by the user. Then again, a single prewritten package might integrate *several* general functions such as processing words, managing files, manipulating data in rows and columns, and preparing graphic output.

Communicating with Prewritten Packages

In any computer environment there's a hardware/software connection or interface that permits communication between people and computers. This connection was made in the past when computer specialists keyed in the types of cryptic codes shown in Figure 1-4. But most people who work with computers today aren't computer specialists. They may have limited computer experience, or they may be knowledgeable users who are too busy to learn a complicated set of interface rules for every new software package. In either case, what people want is a "friendly" interface—one that has hardware that's easy and comfortable to use, and software that's easy to learn and apply.

Software developers have reduced (but not eliminated) user frustrations in recent years by producing packages that are easier to learn and simpler to use than those of just a few years ago. These packages typically employ menus, easier-to-learn commands, or a combination of these approaches.

A **menu** is a listing of the selection options available to program users. **Menu-driven programs** lead users through a series of data entry and process-

FIGURE 1-4   A few of the cryptic codes that might be used to tell a computer to process a billing program. Such communication between people and machines may be satisfactory when the people are computer specialists. But it hardly represents a "friendly" interface for most of today's computer users.

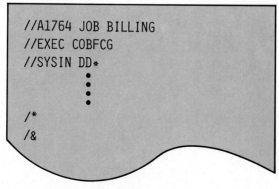

```
//A1764 JOB BILLING
//EXEC COBFCG
//SYSIN DD*
        •
        •
        •
/*
/&
```

ing procedures, giving and requesting information as necessary at each program level to complete a task. A main menu is typically followed by second-level menus, then third-level menus, and so on. As users move through these lower-level menus (called submenus), the choices become more detailed. Menus at all levels typically include a title, the possible options, and some type of **prompt**, or message telling the user how to respond. Or a *menu line* of key words may be displayed across the top or bottom of the screen. The words in the line represent available options. The menu line may remain displayed throughout the time the program is being used. The user can select an option from this line at any time, and when a choice is made a *pull-down* (sub)*menu* may then be shown. Sometimes a main menu is replaced with a *graphics interface* showing the options as drawings. These drawings, or *icons*, are pictures that represent options that can be picked. Users select an option by pressing a designated key or by pressing a button on a *mouse*—a hardware pointing device—to position a highlighted screen symbol (called a *cursor*) near the icon representing the desired choice.

A **command-driven program** is one that requires the user to learn and then enter the instructions needed to carry out program operations. Menus can speed the learning process and help people become productive users in a short time; learning commands generally takes a little more effort. But once a system has been learned, many users prefer to enter a few commands and get on with their work rather than page through several menu levels. For this reason, some software packages provide menus and on-screen help messages and tutorials for infrequent users but give others the option of bypassing those menus and messages.

The Electronic Spreadsheet

Data analysis involves separating a mass of related facts into constituent parts and then studying and manipulating those parts to achieve a desired result. And **data analysis packages** are the programs that people use to structure and rearrange data to accomplish their goals. Two software tools commonly used for data analysis are electronic spreadsheet and statistical analysis packages.

The **electronic spreadsheet** is a program that accepts data values and relationships in the columns and rows of its worksheet, then allows users to perform calculations and other operations on these facts to answer analysis questions. Since spreadsheets and statistical analysis packages have similarities, let's look at a spreadsheet example that can outline the financial results that might be expected if a new product is made and sold. Critical assumptions can be varied in the planning model to see if the product is likely to yield an acceptable profit (profit is revenue minus expenses).

The analyst first inserts a diskette with the command-driven spreadsheet program into the computer and presses a few keys. The display screen is divided as shown in Figure 1-5*a* into a series of blank rows (numbered) and blank columns (lettered). The intersection of a spreadsheet column and row is called a *cell* or *box*. The analyst decides if a cell is to hold a *label* (a set of characters that are used for descriptions, headings, and titles) a *value* (the numbers to be manipulated), or a *formula* (a relationship between values that determines what's displayed at a cell location). A special rectangular cursor or screen marker called a *cell pointer* is moved around with cursor control keys to locate cells where labels and other items are to be placed. In Figure 1-5*a*, the pointer is located at cell C1 (the intersection of column C and row 1). This location is displayed in an *edit area* shown in the upper left part of the screen.

After keying headings in the cells in rows 1 and 2, an analyst enters the row labels, beginning with "Unit Sales (Doz)" in box A3. The other label

FIGURE 1-5   (*a*) After a spreadsheet package is loaded, the display screen is divided into a series of blank numbered rows and lettered columns. The cell pointer is located in cell C1 in this figure. (*b*) The analyst has now entered the labels needed for the Product Planning Model.

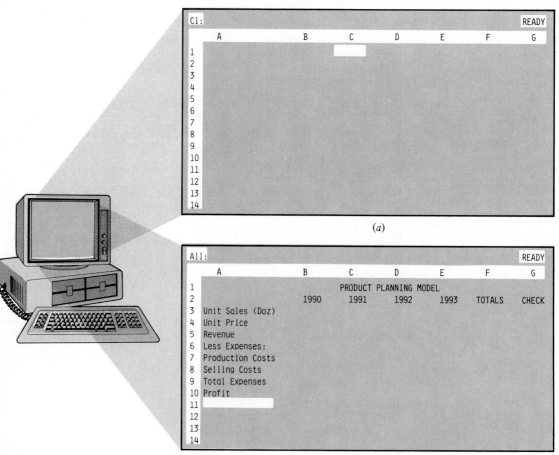

(a)

(b)

entries shown in Figure 1-5b complete the planning model format, and the analyst next enters numeric values into the spreadsheet. The first value keyed is the expected sales of the product in 1990. The pointer is positioned at cell B3 and the 6000 (dozen) estimate is entered there. Expected sales for 1991 through 1993 are similarly placed in cells C3, D3, and E3, and projected selling prices for the 4-year period are keyed in cells B4 through E4 (see Figure 1-6a).

The analyst next positions the cursor at cell B5 and enters a formula into the model. A formula produces a value in one cell that's dependent on values found in other cells. In this case, the formula in B5 causes the program to multiply the unit sales in B3 by the unit price in B4. Note that the formula *doesn't appear* in B5; rather, the formula result appears in this cell,

FIGURE 1-6 (a) Numeric values have now been keyed into rows 3 and 4 of the Product Planning Model. And the formula shown in the edit area at the top left part of the screen has been entered into cell B5. When this formula is executed, the result appears in the spreadsheet at location B5. (b) A listing of all the formulas used in the Product Planning Model.

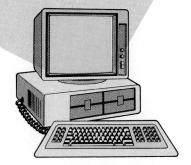

B5: +B3*B4   READY

	A	B	C	D	E	F	G
1			PRODUCT PLANNING MODEL				
2		1990	1991	1992	1993	TOTALS	CHECK
3	Unit Sales (Doz)	6000	7500	7800	6800		
4	Unit Price	5.50	6.25	6.50	6.75		
5	Revenue	33000					
6	Less Expenses:						
7	Production Costs						
8	Selling Costs						
9	Total Expenses						
10	Profit						
11							
12							
13							
14							

(a)

Cell	Formula
B5	+B3*B4
C5	+C3*C4
D5	+D3*D4
E5	+E3*E4
F3	+B3+C3+D3+E3
F5	+B5+C5+D5+E5
F7	+B7+C7+D7+E7
F8	+B8+C8+D8+E8
B9	+B7+B8
C9	+C7+C8
D9	+D7+D8
E9	+E7+E8
F9	+F7+F8
B10	+B5-B9
C10	+C5-C9
D10	+D5-D9
E10	+E5-E9
F10	+F5-F9

(b)

and the formula itself shows up in the edit area in Figure 1-6*a*. As you can see in the edit area, the formula for B5 is + B3*B4. (The + sign tells the package that a formula or a value is to be entered; the asterisk (*) symbol tells it to multiply; + , − , and / symbols tell it to add, subtract, and divide.)

Similar formulas are entered for cells C5, D5, and E5 (see Figure 1-6*b*), and other formulas are placed in boxes F3 and F5 so that the package will total the values found in the Unit Sales and Revenue rows. As you can see in Figure 1-6*b*, the formula for cell F3 is + B3 + C3 + D3 + E3. Or the analyst could use a built-in program function, write the formula as @SUM(B3..E3), and get the same results. The assumed costs to produce and sell the product for the 4-year period are entered in rows 7 and 8, and formulas are entered in the B9 through F9 cells to total these expenses. Finally, the assumed profit results of introducing the product are found in row 10 for the 4-year period. For example, the profit forecast in 1990 is found by subtracting the total expenses figure in B9 from the revenue figure in B5—that is, B10 = + B5 − B9. Formulas for all other cells are shown in Figure 1-6*b*.

The output produced by the data values and formulas entered into the planning model is shown in Figure 1-7. Of course, the sales, price, and cost figures are expected values (prepared, perhaps, with statistical techniques we'll discuss in Part 3), and unexpected events can change these values. But one of the reasons a spreadsheet package is such a powerful *decision support tool* is that decision makers can ask "what if" questions, insert changed values, and then watch as the values of dependent cells are recalculated and the effects of these changes ripple through their models. For example, an analyst can instantly see what may happen to yearly and total

FIGURE 1-7   The output of the Product Planning Model. The total profit figure in cell F10 is computed with the formula +F5 − F9. As a check, the analyst instructs the package to add the values in B10 through E10. This result is placed in G10. The values in F10 and G10 are then compared with another instruction to detect any discrepancy. If the figures in F10 and G10 had been different, the zero in G11 would have been replaced with a string of 9s to alert the user.

A2:							READY
	A	B	C	D	E	F	G
1			PRODUCT	PLANNING	MODEL		
2		1990	1991	1992	1993	TOTALS	CHECK
3	Unit Sales (Doz)	6000	7500	7800	6800	28100	
4	Unit Price	5.50	6.25	6.50	6.75		
5	Revenue	33000	46875	50700	45900	176475	
6	Less Expenses						
7	Production Costs	13200	16400	17500	16450	63550	
8	Selling Costs	16100	21750	23110	18325	79285	
9	Total Expenses	29300	38150	40610	34775	142835	
10	Profit	3700	8725	10090	11125	33640	33640
11							0
12							
13							
14							

profits if a new competitive product lowers sales or pricing estimates. In large models, literally thousands of calculations may occur between the time a few values are changed and the time their output effects are known.

We've already seen that spreadsheets have built-in functions such as the @SUM function that totals the values in all cells that fall in a specified range. There are also other built-in functions useful in statistical work, such as @AVG (used to compute and display a measure of central tendency—the average or arithmetic mean—of the values held in a group of cells), and @STD (used to compute the standard deviation—a measure of dispersion—of the values held in specified cells).

A *template* is a spreadsheet that contains all the labels and formulas needed to produce a specific type of model, but values haven't been entered into the data cells. For example, if the analyst creates the format shown in Figure 1-5*b*, but doesn't enter any data into rows 3, 4, 7, or 8, a template has been created. Template users can then insert data into those rows when they want to run the model.

Millions of people use spreadsheets to deal with common data analysis problems. But many of these people don't key in the labels, or prepare the formulas needed to create their models. Rather, they buy pretested commercial template packages that are designed to help them analyze problems in such predefined areas as budgeting, real estate investment, and statistical analysis. After the template and spreadsheet software is loaded into the computer, users merely enter the data (often with the help of menus and special data-entry screens) and then study the output.

Statistical Analysis Packages

Statistical packages are similar to spreadsheets in that they deal with columns and rows of numeric data. We've seen that a few statistical functions such as @AVG are built into spreadsheet packages, and other statistical formulas can be entered into spreadsheet cells to perform statistical studies. But **statistical analysis packages** are preprogrammed with all the specialized formulas and built-in procedures a user may need to carry out a range of statistical studies. Like spreadsheets, statistical programs can:

- Accept data from other sources.
- Copy and move data to duplicate the contents of one cell (or group of cells) into other locations, or to erase the contents of one or more cells in one place and place them in another.
- Add or remove data items, columns, or rows.
- Format the way cells, rows, and columns are laid out, save this format and the data it contains on a disk for future updates and analyses, and place all or part of the output on paper.
- Sort, merge, and manipulate facts in numerous ways.
- Perform analyses on single and multiple sets of data.
- Convert numeric data into charts and graphs that people can use to grasp relationships, spot patterns, and make more informed decisions.
- Print summary values and analysis results.

DOES ANYONE REALLY USE STATISTICAL PACKAGES?

You bet they do. Especially now, since micro-computer technology has drastically reduced the pain involved in learning, using, and inter-preting computerized statistical tools.

With the advent of the microcomputer, the application of statistics has become much more widespread.

No longer do users have to access a main-frame computer and pay enormous connection charges; nor do they have to spend months climbing up steep learning curves. Now a sim-ple IBM PC equipped with hard disk and suf-ficient memory can be coupled with a serious but flexible statistical package such as Systat by Systat Inc. of Evanston, Illinois, that can bring sophisticated statistical analysis to a vari-ety of situations—sometimes with remarkable results.

At MGM/United Artists Entertainment Co., in Culver City, California, Systat has recently become a critical tool in trying to determine whether a film will be a box office success or not. Information collected from people view-ing "sneak" previews of a film or advertise-ments is filtered and analyzed using Systat; MGM then tries to more closely predict—using Systat's statistical results—the ultimate eco-nomic success of any one film.

"In the past, the marketing of films has been predominantly based on hunch," says Greg Foster, manager of market research at Systat. "Now Systat helps validate our intuitions and provides a very effective quantitative base."

Systat is also being used at the United Na-tions, in New York. In the Department of Tech-nical Cooperation for Development, Michael Lackner, project coordinator, oversees a soft-ware department that provides software train-ing on Systat—as well as other custom-designed statistical packages—to demographic analysts from developing countries. China, El Salvador, and Honduras have sent analysts fa-miliar with mainframe-based statistical tools to Lackner's department and have been "fa-vorably impressed" with Systat and other microcomputer-based packages, says Lackner.

Yet another example of microcomputer-based statistical analysis is taking place at Bionetics Corp. at the Kennedy Space Center in Florida. William Alford, a biostatistician for the company, says he uses three different pack-ages—including Systat and Statgraphics—on a number of the company's NASA contracts.

Alford says he uses Systat to investigate the relationship between aerobic fitness and astro-naut tolerance levels when returning from a 0g space flight to the 1g earth environment. A cor-ollary project for the company involves analyz-ing the effectiveness of a specialized technique for the reduction of muscle atrophy after a space flight.

A final example of microcomputer-based sta-tistical applications circles back into the aca-demic realm, but with a new twist. At the East-ern Correctional Facility, a maximum security prison in Mapanoch, New York, 19 inmates of the facility are working toward master's de-grees in sociology; nine of the students are writing theses using statistical data analyzed with the Systat package.

The inmates' instructor, Susan Philliber, Ph.D., from the State University of New York at New Paltz, anticipates that a number of the theses will be eligible for publishing in aca-demic journals in the field of corrections.

The nine students using the Systat package are analyzing data taken from a survey per-formed within the prison. Out of a total prison population of some 1,000 men, 346 were ran-domly chosen for participation in the survey; a series of approximately 50 questions asked sur-vey participants to support or refute a wide va-riety of hypotheses.

One master's thesis is being written on the function of religious group membership in prison; another on what kind of behavioral changes occur among men approaching parole; and a third on a comparison of ethnic group identification inside and outside of prison.

According to Philliber, none of the men have had any prior experience with statistics, and none had worked with microcomputers before.

She says that after a month of instruction about statistics in general and Systat in particular (a package known for its thoroughness, but not particularly for its ease of use), the men could sit down at the IBM PCs in the laboratory and run basic statistical analyses using their own raw data. The men will be graduating in early October.

—From Hank Bannister, "Micro Applications Broaden the Impact of Statistical Analysis," Infoworld, Oct. 14, 1985, pp. 28ff. Reprinted with permission.

The most powerful statistical packages in use today tend to be command-driven, or they combine menus with commands. But very competent menu-driven programs are also available. Many of the most able (and most expensive) general-purpose programs such as *Minitab, SAS, SPSS,* and *BMDPC* were first written for large computer systems, and they've been around in various forms for well over a decade. Each of these command-driven products is available in a version that runs on personal computers. An equally potent product—*Systat*—was designed first for personal computers and is now also available on larger systems. And many other general-purpose packages may be found. *Statgraphics*, for example, is menu-driven, performs many statistical procedures, and offers users over 50 types of graphic output. More specialized programs are also around to carry out a specific type of statistical study. For example, *RATS, ESP,* and *SORITEC* are programs that appeal most to economists. (Some thoughts on selecting a statistical package that will fit *your* needs are presented in the Closer Look reading at the end of this chapter.)

Data entry procedures vary among packages. We'll look at these procedures for just two packages—*Mystat* and *Minitab*—since these programs are used throughout this book to demonstrate statistical concepts. (The procedures for other packages that you may have access to are easy to learn and are available from manuals and local sources.)

Figure 1-8 shows the data entry approach used in *Mystat,* a condensed educational version of *Systat.* After the *Mystat* program is loaded into a personal computer, the main menu shown in Figure 1-8*a* is displayed on the computer's screen. DEMO or HELP commands may be used to get assistance in using the program. To enter new data, the user types the word EDIT at the command prompt symbol (the > character shown below the ANOVA command) and then presses the [Enter] key. A data editor screen that resembles a blank spreadsheet then appears (Figure 1-8*b*). The user follows program procedures to (1) key the name(s) of the variable(s) to be processed in the Case row, and (2) enter the data values in the following 1, 2, 3, . . . rows. In Figure 1-8*c*, for example, a data set that you'll see again in Chapters 3 and 4 is named GALSOLD (for "gallons sold"), and the first 15 of the 50 values in this set are entered. After all 50 values are entered and saved on a magnetic disk, the user returns program control to the *Mystat* main menu to begin analyzing the data.

Figure 1-9 shows another data entry approach used by the *Minitab* package. After the program is loaded into the computer, a simple command prompt (MTB >) appears on the screen. The user elects to enter the same

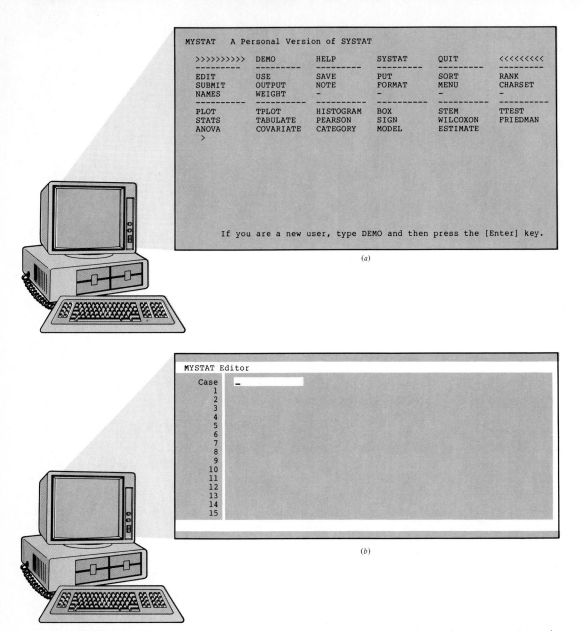

```
MYSTAT   A Personal Version of SYSTAT

  >>>>>>>>>   DEMO        HELP        SYSTAT      QUIT        <<<<<<<<<
  ---------   ---------   ---------   ---------   ---------   ---------
  EDIT        USE         SAVE        PUT         SORT        RANK
  SUBMIT      OUTPUT      NOTE        FORMAT      MENU        CHARSET
  NAMES       WEIGHT      -           -           -           -
  ---------   ---------   ---------   ---------   ---------   ---------
  PLOT        TPLOT       HISTOGRAM   BOX         STEM        TTEST
  STATS       TABULATE    PEARSON     SIGN        WILCOXON    FRIEDMAN
  ANOVA       COVARIATE   CATEGORY    MODEL       ESTIMATE
  >



      If you are a new user, type DEMO and then press the [Enter] key.
```

(a)

```
MYSTAT Editor
  Case   ▁
     1
     2
     3
     4
     5
     6
     7
     8
     9
    10
    11
    12
    13
    14
    15
```

(b)

FIGURE 1-8   (a) The *Mystat* main menu screen. (b) The *Mystat* data editor screen. (c) A data set named GALSOLD is entered into the *Mystat* data editor.

GALSOLD data set used above, and keys the SET C1 command at the prompt. This SET command tells the program that it's to place a data set in the first column of its worksheet. (Another data entry command—READ C1–C3—tells the program to put data into columns 1, 2, and 3.) The program next responds with a DATA > prompt, and the user enters a value and presses the [Enter] key. This procedure continues until the user types

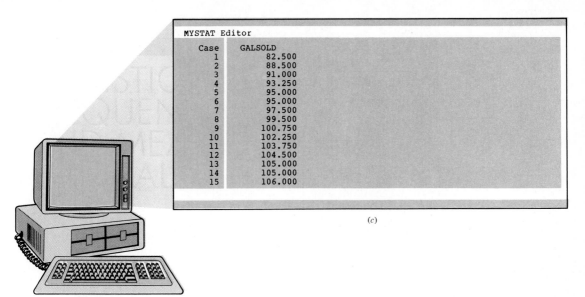

```
MYSTAT Editor

   Case      GALSOLD
     1         82.500
     2         88.500
     3         91.000
     4         93.250
     5         95.000
     6         95.000
     7         97.500
     8         99.500
     9        100.750
    10        102.250
    11        103.750
    12        104.500
    13        105.000
    14        105.000
    15        106.000
```

(c)

FIGURE 1-8   (Continued)

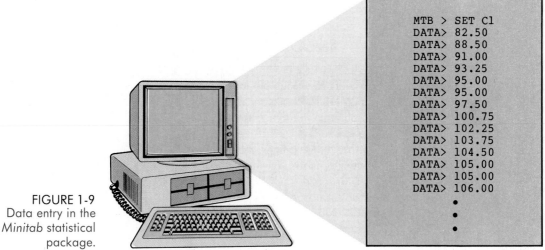

```
MTB >  SET C1
DATA>  82.50
DATA>  88.50
DATA>  91.00
DATA>  93.25
DATA>  95.00
DATA>  95.00
DATA>  97.50
DATA>  100.75
DATA>  102.25
DATA>  103.75
DATA>  104.50
DATA>  105.00
DATA>  105.00
DATA>  106.00
        •
        •
        •
```

FIGURE 1-9
Data entry in the
Minitab statistical
package.

END. The data set is then named—MTB > NAME C1 'GALSOLD'—and saved before analysis begins.

The analyses carried out by *Mystat* and *Minitab* result in numeric and graphic output. For example, a study of the GALSOLD data set produces the results shown in Figures 1-10 and 1-11. These results are explained in Chapters 3 and 4. You'll recall that a spreadsheet program requires users to supply the formulas needed to analyze the data, but the formulas and procedures to produce the outputs shown in Figures 1-10 and 1-11 are built into *Mystat* and *Minitab*.

TOTAL OBSERVATIONS: 50

GALSOLD

N OF CASES	50
MINIMUM	82.500
MAXIMUM	148.000
RANGE	65.500
MEAN	115.400
VARIANCE	222.278
STANDARD DEV	14.909
STD. ERROR	2.108
SUM	5770.000

(a)

STEM AND LEAF PLOT OF VARIABLE: GALSOLD   , N =   50

MINIMUM IS:	82.500
LOWER HINGE IS:	105.000
MEDIAN IS:	115.250
UPPER HINGE IS:	125.250
MAXIMUM IS:	148.000

```
 8    2
 8    8
 9    13
 9    5579
10    0234
10 H  556789
11    02344
11 M  555667889
12    1244
12 H  55688
13    234
13    5567
14    4
14    8
```

(b)

FIGURE 1-10   Some *Mystat* output results using the GALSOLD data set. (a) Summary descriptive measures produced with the STATS command. (b) A stem-and-leaf analysis using the STEM command. (c) A chart produced with the HISTOGRAM command. (d) A box plot created with the BOX command.

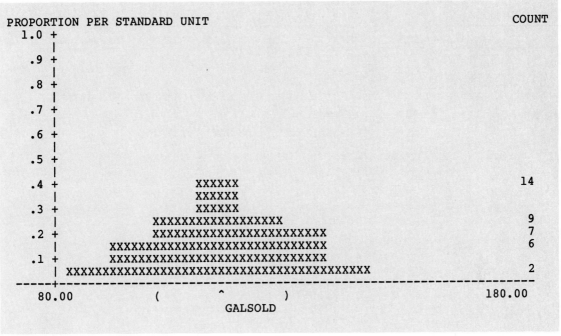

```
PROPORTION PER STANDARD UNIT                                    COUNT
 1.0 +
     |
  .9 +
     |
  .8 +
     |
  .7 +
     |
  .6 +
     |
  .5 +
     |
  .4 +             XXXXXX                                          14
     |             XXXXX
  .3 +             XXXXX
     |          XXXXXXXXXXXXXXXX                                    9
  .2 +          XXXXXXXXXXXXXXXXXXXXXX                              7
     |       XXXXXXXXXXXXXXXXXXXXXXXXX                              6
  .1 +       XXXXXXXXXXXXXXXXXXXXXXXXX                              2
     |    XXXXXXXXXXXXXXXXXXXXXXXXXXXXXXXXXXXXXXX
-----+-------------------------------------------------------------------
     80.00          (         ^          )                       180.00
                         GALSOLD
```

(c)

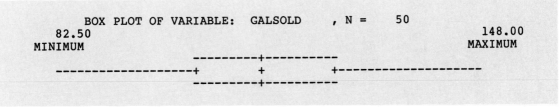

```
         BOX PLOT OF VARIABLE:   GALSOLD     , N =    50
     82.50                                              148.00
   MINIMUM                                             MAXIMUM
                       ---------+----------
     ------------------+        +        +----------------------
                       ---------+----------
```

(d)

```
MTB > DESCRIBE 'GALSOLD'

                N      MEAN    MEDIAN    TRMEAN     STDEV    SEMEAN
GALSOLD        50    115.40    115.25    115.43     14.91      2.11

               MIN       MAX        Q1        Q3
GALSOLD      82.50    148.00    104.87    125.44
```

(a)

```
MTB > STEM-AND-LEAF 'GALSOLD'

Stem-and-leaf of GALSOLD    N = 50
Leaf Unit = 1.0

      1      8  2
      2      8  8
      4      9  13
      8      9  5579
     12     10  0234
     18     10  556789
     23     11  02344
     (9)    11  555667889
     18     12  1244
     14     12  55688
      9     13  234
      6     13  5567
      2     14  4
      1     14  8
```

(b)

```
MTB > HISTOGRAM 'GALSOLD'

Histogram of GALSOLD    N = 50

Midpoint    Count
      80        1   *
      90        3   ***
     100        8   ********
     110       11   ***********
     120       13   *************
     130        8   ********
     140        5   *****
     150        1   *
```

(c)

FIGURE 1-11   Some *Minitab* output results for the GALSOLD data set. (a) Summary descriptive
measures produced with the DESCRIBE command. Note that although *Minitab* and
Mystat produce several of the same descriptive measures, they also differ in the
measures they compute. (b) A stem-and-leaf analysis using the STEM-AND-LEAF
command. (c) A chart produced with the HISTOGRAM command. (d) A box plot
created with the BOXPLOT command. (e) A plot of individual values in the
GALSOLD data set produced with the DOTPLOT command.

```
MTB > BOXPLOT 'GALSOLD'


                                  -------------------
                 ------------------I        +         I--------------------
                                  -------------------
             --+---------+---------+---------+---------+---------+-----GALSOLD
               84        96       108       120       132       144
```

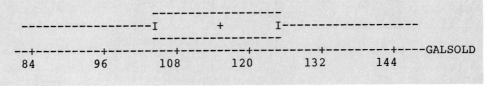

(d)

```
MTB > DOTPLOT 'GALSOLD'

                                       .           .         .
       .     . . .: .  .....:........:::: ....:. :  ..:::.   .   .
     --+---------+---------+---------+---------+---------+---------+-----GALSOLD
       84        96       108       120       132       144
```

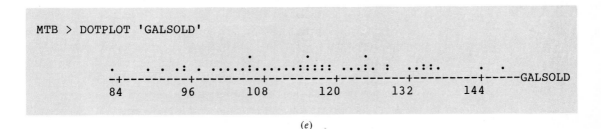

(e)

LOOKING BACK

1 The purpose of this book is to help you become an intelligent consumer of statistics, and the emphasis here is on explaining statistical procedures and interpreting the results. The book is organized into four parts and 16 chapters to achieve its purpose. Part 1 and much of Part 3 are devoted to a study of descriptive statistics (the procedures of data collection, classification, summarization, and presentation). Parts 2 and 4 consider statistical inference (the process of arriving at conclusions about the population under study on the basis of data obtained from a sample taken from that population).

2 The word "statistics" (plural) means numerical facts or data, while "statistics" (singular) refers to the body of principles and procedures developed for the collection, classification, summarization, interpretation, and communication of numerical data, and for the use of such data. Thus, statistics (singular) refers to a subject of study. The word statistics (either singular or plural) shouldn't be confused with the word "statistic" which means a measure taken from a sample. (The term parameter refers to a similar measure taken from a population.)

3 A knowledge of both descriptive statistics and statistical inference is needed by an educated citizen to (1) describe and understand relationships, (2) make better decisions, and (3) cope with change.

4 Several basic steps in a statistical problem-solving procedure are listed and discussed in this chapter. The first of these steps is to identify the problem or opportunity. The second step is to gather available facts from internal or external sources. Internal data are found in the departments of an organization while external data are available from government and other publications. If the data needed by decision makers can't be gleaned from primary or secondary sources, the third step in the problem-solving procedure is to gather new original data through the use of such tools as personal interviews and mail questionnaires. After data are collected,

the next step is to classify and summarize the facts. Finally, the data are presented and analyzed, and a decision is made.

5   Computers are efficiently used in statistical work. Software is a general term that refers to the detailed sets of stored instructions—or programs—that control computer operations. Generalized prewritten software packages are commonly used to carry out statistical studies. People communicate with these packages by using commands that the programs can recognize or by responding to the prompts displayed in a series of menus supplied by the programs. Two software tools commonly used for data analysis are electronic spreadsheet and statistical analysis packages. The electronic spreadsheet is a program that accepts data values and relationships in the columns and rows of its worksheet, and then allows users to perform calculations and other operations on these facts to answer analysis questions. An example of a spreadsheet application is presented in the chapter. A template is a spreadsheet that contains all the labels and formulas needed to produce a specific type of analysis, and many templates are available for statistical work. Statistical analysis packages are similar in many ways to spreadsheets, but they are preprogrammed with all the specialized formulas and built-in procedures a user may need to carry out a range of studies. The most powerful statistical packages in use today tend to be command-driven, or they combine menus with commands. Data entry procedures and the output results produced by two statistical packages—*Mystat* and *Minitab*—are presented in the chapter.

KEY TERMS AND CONCEPTS

You should now be able to define and use the following terms and concepts (the numbers shown indicate the pages where the terms and concepts are first mentioned):

Statistics (plural) 7

Statistics (singular) 7

Descriptive statistics 7

Statistical inference 7

Population (universe) 8

Sample 8

Census 8

Parameter 8

Statistic 8

Internal data 14

External data 14

Primary sources 14

Secondary sources 14

Personal interview 16

Mail questionnaire 17

Software 18

Menu 20

Menu-driven programs 20

Prompt 21

Command-driven program 21

Data analysis 21

Data analysis packages 21

Electronic spreadsheet 21

Statistical analysis packages 25

TOPICS FOR REVIEW AND DISCUSSION

1   Discuss the ways that the word "statistics" may be used.

2   What's the difference between descriptive statistics and inferential statistics?

3 "Statistical description is still an important part of the subject of statistics." Discuss this statement.

4 How can sample data be used in decision making?

5 "A statistic may be used to estimate a parameter." Discuss this statement.

6 How can a knowledge of statistics help you:
   a Describe relationships between variables?
   b Make better decisions?
   c Cope with change?

7 "Since nobody can know exactly what the future will bring, it's a waste of time to try to develop forecasts." Discuss this statement.

8 Identify and discuss the steps in the statistical problem-solving methodology.

9 What's the difference between primary and secondary sources of data?

10 "Some households either don't have telephones or have unlisted numbers, and this may bias the results of a survey." Discuss this statement.

11 Compare and contrast the personal interview and mail questionnaire approaches to gathering new original data.

12 a What is software?
   b What is an applications program?

13 "Software packages typically use menus, commands, or a combination of these approaches." Discuss this statement.

14 How does a menu-driven program differ from a command-driven program?

15 a What is data analysis?
   b Identify and discuss two types of data-analysis software packages.

16 a How are electronic spreadsheets similar to statistical analysis packages?
   b How do these packages differ?

17 a What is a spreadsheet template?
   b How could a template help a person carry out statistical studies?

18 Identify and discuss three capabilities of statistical software.

PROJECTS/ISSUES TO CONSIDER

1 Collect some questionnaires that you've received in the mail or located in the library. Bring these questionnaires to class and be prepared to discuss their purpose and design. How effective do you think these questionnaires are likely to be?

2 Go to your library and locate articles that incorporate the personal interview or mail questionnaire techniques used by researchers. Write a report outlining your findings.

3 While you're in the library, locate examples of primary and secondary data sources. Bring your examples to class and be prepared to discuss them.

4 Reviews of statistical software packages often appear in personal computing magazines and in *The American Statistician*. Find such a review of statistical packages and prepare a class presentation outlining the characteristics, strengths, and weaknesses of the packages being reviewed.

A CLOSER LOOK

SELECTING AND BUYING A STATISTICAL SOFTWARE PACKAGE

You've seen that statistical software packages offer many benefits to users, and perhaps you've decided to acquire such a tool. If you expect that your primary computer application will be to run such a program, and if you don't yet have a *personal computer* (pc), then you should follow a software-first approach. That is, you should first find the package that does exactly what you want to do and then get the hardware needed to run that package. But if you've already invested in (or have access to) a particular computer system, you'll probably want to limit your statistical software search to programs that are compatible with your equipment. Remember that all pc's won't run all packages, so you'll have to carefully match the selected software with your hardware and operating system.

Buying the wrong statistical software can produce expensive and embarrassing results: you can waste money and time, and you can wind up with a package that doesn't fit your needs or work habits. So how do you go about identifying prospective packages, deciding on the "right" selection, and then shopping for the "best" buy? That's not a simple question, but some general guidelines are given in the following paragraphs to help you identify a list of appropriate packages, narrow the field to a few suitable choices, and then make the final purchase decision. You should remember, though, that even the pros have trouble picking software for themselves because programs are complicated products.

IDENTIFYING SUITABLE PACKAGES

Some sources of information you can use in this first step are outlined below:

◗ The advertisements published in pc magazines give general descriptions of packages, and the members of local computer clubs can give specific accounts of the performance of competitive products. (There are hundreds of clubs and many knowledgeable members.)

◗ The reviews of software products published in many pc magazines go beyond advertising descriptions and give more information on the functions and features of selected packages.

◗ People in retail stores that carry software packages may be able to demonstrate products, answer questions, and furnish promotional literature.

◗ Software vendors sometimes supply demonstration disks and literature for their products. The demo disks may cost a few dollars. Some offer little more than advertising hype, but the better ones give you a realistic idea of the capabilities of the programs they represent. You can spend as much time with a demo disk as you want to get a feel for how the program will respond to your needs. It may also be possible to buy the manual that explains the use of a package.

◗ Computer consultants and acquaintances with statistical needs similar to yours can answer questions and make package recommendations.

◗ Don't overlook the possibility that your needs can be met with *shareware* (or *user-supported software*). Leads on these packages can be obtained from electronic bulletin boards or the software libraries of computer-using groups. The Boston Computer Society, for example, has a library of about 3,000 programs in all categories. Creators of these programs obtain distribution for their products by supplying free copies to user-group libraries. You're encouraged to copy these programs and try them out. If you find one you like, you're then asked to register with the author for a modest fee that may entitle you to a bound manual, the latest program updates, a newsletter, and a telephone number you can call for information if you run into trouble.

NARROWING THE FIELD

Once you've identified suitable package candidates, use this selection checklist to further narrow the field:

- Is the output produced by the package accurate and suitable in form and content? Can it do what must be done?
- Can the package accept the volume of input data you need to process in the format that you need to use?
- Can the package support an acceptable range of input/output devices?
- Does the package have an adequate response time, or must you wait long periods while the program digests your last data entry? Can the package support the use of a numeric coprocessor chip to speed response? If so, can such a chip be installed in a socket in your pc?
- Is the package user-friendly—is it easy to learn and easy to use? What's the quality of the documentation? That is, are the manuals and other documents in the package complete, clearly written, and well organized?
- Does the package have good error-handling controls? Are system errors reported to the user in clear messages or cryptic codes?

MAKING THE FINAL DECISIONS

The following questions can now help you screen the packages that are still in the running:

- Can it be demonstrated to *your* satisfaction that the package works with *your* data, in the format *you* use, to solve the problems that *you* specify? What's the level of support you can expect to receive from the vendor and/or the vendor's representative? Is some knowledgeable person able and willing to help you with any problems that may develop as you use the package?
- Does the package represent a good value for the money? Can an alternative, but acceptable, package be obtained for a significantly lower outlay?

SHOPPING FOR THE BEST BUY

Once you've decided on the package, where do you go to buy it? We've already seen where shareware may be obtained, but for other programs there are likely to be three outlets. The first source is a retail store (or an outlet maintained by a company for its employees). Stores *should* be familiar with the packages they sell. When that's true, they should be able to answer questions, provide support, and give quick delivery. The price charged by a retailer is likely to be close to the list price, but this is reasonable *if* you can depend on the retailer for help before and after the sale.

A second package source is a mail-order distributor. If you're buying a relatively simple package and don't need any local support, then you can usually save money by ordering it from this source. And a third way to buy the selected package is to order it directly from the producer. This may be about the only way to get a copy of a program supplied by a small vendor. Prices may be negotiated, and the firm may not be around later if you need help.

LIARS, #$%E*& LIARS, AND A FEW STATISTICIANS

Why Is This Man Blinking?

Colin Clark, the resource economist, is a mischievous fellow. He is the type of man who delights in examining the accepted wisdom in a disinterested, zero-based way. He is the type of man who, when ushered into the presence of the grand scenario writer, nods quietly, reaches for pad and pen, jots down copious notes, and then stashes them in a tickler file dated twenty years hence. . . .

In 1950, Clark has noted, our known oil reserves amounted to 75 billion barrels. Over the next two decades, our profligate society managed to consume 180 billion barrels, at the end of which period the official scorecard revealed known oil reserves of . . . 455 billion barrels. Nor was oil the only resource where the numbers seemed to move around on the page. World reserves of iron were recorded at 19,000 million metric tons in 1950. By 1970 we had taken 9,355 million metric tons out of the ground, leaving . . . what? In 1970, according to the accepted numbers, world reserves amounted to 251,000 million metric tons—an increase of more than 1,000 percent. Then there is the case of bauxite, a critically important mineral to the energy-conscious industrial world. In 1950, world reserves were recorded at 1,400 million metric tons. Twenty years later, after production of 505 million metric tons, reserves had soared to 5,300 million metric tons. And so on. Clark adduces the numbers for manganese and copper and zinc and lead and phosphate and chromite. In each case, the official numbers, reflecting a static view of an essentially dynamic process, were undone by the industry, wit, and "good luck" of those willing to suspend their belief in the accepted wisdom.

The calculation of "known reserves," much like the survey of public opinion, is an analytical tool of very limited utility. Literal-minded statisticians are proficient at recording the economic process in freeze-frame. At any given moment, they can align the matrices of time and data in perfect registration. The bottom-line figures for production and known reserves are thus almost always correct, and where they are incorrect the discrepancies are within tolerable allowances. Those numbers can tell us where we have been and, at least asymptotically, where we are. Where they cease to be useful is when they are adapted as the base for free-form extrapolation. . . .

—From Neal B. Freeman, "The Current Wisdom," *National Review*, March 20, 1981, p. 270. © 1981 by National Review, Inc., 150 East 35 Street, New York, NY 10016. Reprinted with permission.

LOOKING AHEAD

The preceding vignette shows some earlier errors made in estimating the "known reserves" of selected commodities. It's possible that these earlier statistics misled some people and caused them to take inappropriate actions. In this chapter, you'll see several other ways that statistics may be misapplied. The examples used are classified under nine headings. You'll also learn about tables and charts that legitimately help people understand and communicate numeric data, and you'll see how computer graphics software can help prepare these aids. The chapter then concludes with a series of questions you may ask yourself to reduce your chances of being misled.

Thus, after studying this chapter, you should be able to:

- Point out at least six ways in which statistics may be misused.
- Give examples of how statistics have been misapplied.
- Discuss the types of tables and charts that are used to honestly analyze and present numeric facts, and explain the purpose of computer graphics packages.
- Recall several questions you can ask yourself during your evaluation of quantitative information to reduce your chances of being misled.

CHAPTER OUTLINE

SOME UNFAVORABLE OPINIONS

Consider the following comments:

> *"There are three kinds of lies: lies, damned lies, and statistics."*—Benjamin Disraeli

> *"Get the facts first and then you can distort them as much as you please."*—Mark Twain

> *"In earlier times they had no statistics, and so they had to fall back on lies. Hence the huge exaggerations of primitive literature—giants or miracles or wonders! They did it with lies and we do it with statistics; but it is all the same."*—Stephen Leacock

> *"He uses statistics as a drunken man uses lampposts—for support rather than illumination."*—Andrew Lang

> *"It ain't so much the things we don't know that get us in trouble. It's the things we know that ain't so."*—Artemus Ward

These comments deal mostly with statistics, so it's not surprising that someone has also said that "if all the statisticians in the world were laid end to end—it would be a good thing!"

These unfavorable views of statistics are held by some who uncritically accepted statistical conclusions only to learn later that they had been misled. For it's true that statistical tools have been improperly used by politicians, writers, advertising agencies, lawyers, bureaucrats, statisticians—the list is virtually endless—to confuse people or deliberately mislead them. It's unnecessary to dwell on the motives of those who improperly use statistical tools. Technical errors may be made innocently, or valid statistical facts may be deliberately twisted, oversimplified, or selectively distorted, but the results are the same—people are misinformed and misled.

In this chapter we'll point out a few ways that statistical procedures have been misused. As a consumer of numerical facts, you should be alert to the possibility of misleading statistical conclusions.[1] It's no exaggeration to say that this course in statistics is worth your time if it succeeds only in helping you do a better job of distinguishing between valid and invalid uses of quantitative techniques.

THE BIAS OBSTACLE

You know that an early step in statistical problem solving is gathering relevant data. Your dictionary may define **bias** as the preference or inclination that inhibits impartial judgment, and the use of poorly worded and/or biased questions during the data-gathering phase may lead to worthless conclusions. The wording of questionnaires sent out by members of Congress to their constituents, for example, is often haphazard and often fails to produce focused answers. Even worse, questions are sometimes slanted to elicit answers that reinforce the congressperson's own biases. For example, in considering the military spending issue, New York Representative Frederick Richmond asked constituents if they favored "elimination of waste in the defense budget," and 95 percent naturally replied "yes." Sam Stratton, another New York Representative and a strong Pentagon champion, put the question of military spending to his constituents in this way.[2]

> *This year's defense budget represents the smallest portion of our national budget devoted to defense since Pearl Harbor. Any substantial cuts . . . will mean the U.S.A. is no longer number one in military strength. Which one of the following do you believe?*
>
> *"A. We must maintain our number-one status." [Nearly 63 percent voted for that.]*
>
> *"B. I don't mind if we do become number two behind Russia." [Only 27 percent checked that box.]*

Some people collecting or analyzing data may be tempted to use the "finagle factor" to give more emphasis to those facts that *support* their preconceived opinions than to those which *conflict* with their opinions. According to Thomas L. Martin, Jr.:[3]

---

[1] For more information on the misuses of statistics, the following sources are recommended: Stephen K. Campbell, *Flaws and Fallacies in Statistical Thinking*, Prentice-Hall, Inc., Englewood Cliffs, N.J., 1974; Darrell Huff, *How to Lie with Statistics*, W. W. Norton & Company, Inc., New York, 1954; and W. Allen Wallis and Harry V. Roberts, *Statistics: A New Approach*, The Free Press, Glencoe, Ill., 1956, pp. 64–89.

[2] "(Crab) Grass Roots: Questionnaires Sent by Congress Tap Voter Vitriol," *The Wall Street Journal*, Aug. 21, 1975, p. 13. Reprinted with permission of The Wall Street Journal.

[3] Thomas L. Martin, Jr., *Malice in Blunderland*, McGraw-Hill Book Company, New York, 1973, p. 6. Martin's Finagle Factor has been referred to by others as the Fudge Factor, the Fake Factor, or the B. S. Factor.

The Finagle Factor allows one to bring actual results into immediate agreement with desired results easily and without the necessity of having to repeat messy experiments, calculations or designs. [When discovered, the Finagle Factor] was instantly and immensely popular with engineers and scientists, but found its greatest use in statistics and in the social sciences where actual results so often greatly differ from those desired by the investigator. . . . Thus:

$$\begin{pmatrix} Desired\ results \\ on\ paper \end{pmatrix} = \begin{pmatrix} Finagle \\ Factor \end{pmatrix} \times \begin{pmatrix} Actual \\ results \end{pmatrix}$$

Unlike lies that quickly wear thin, the finagle factor is a very durable weave of logical fallacies and sophistries that often appears and is constantly being recycled into new applications. How can the finagle factor be employed? Thank you for asking. It can be (and often is) used by advertisers. Suppose that you see on television a professional-looking actor saying (with great sincerity) that "8 out of 10 doctors recommend the ingredients found in Gastro-Dismal elixir." Does this message convince you to rush out and buy a bottle? Even assuming that the "8 out of 10" figure is correct, it's likely that the doctors were merely giving their stamp of approval to a number of common ingredients found in many nonprescription products. They probably were not specifically recommending Gastro-Dismal as being any better than other similar brands. (In fact, it is even quite possible that the separate ingredients recommended by most doctors could be put together in some preposterous way by Gastro-Dismal employees so as to be injurious to health.)

One more example of the advertiser's art should be sufficient here. The California Raisin Advisory Board ran the following ad in at least one ladies' magazine a few years ago: "Your husband could dance with you for 11 minutes on the energy he'd get from 49 raisins. Think what would happen if he never stopped eating them." As Stephen Campbell observed, "I can think of a lot of things that might happen to a man if he never stopped eating raisins—and they are all very painful."[4]

Although bias in the form of the finagle factor is generally consciously applied, bias can also appear unintentionally. Sometimes, for example, a researcher is led to faulty conclusions because of unintentionally biased input data. The polls which predicted that Thomas Dewey would defeat Harry Truman in the 1948 election for president, the conclusion of the psychiatrist that most people are mentally unbalanced (based on the input of those with whom he came in contact), the conclusion of the *Literary Digest* that Alf Landon would defeat Franklin Roosevelt for president in 1936—all these examples have in common the fact that bias entered into the picture. In the *Literary Digest* fiasco, for example, the prediction was based on a sample of about 2 million ballots returned (out of a total of 10 million mailed out). Unfortunately for the *Digest* (which ceased to exist in 1937), the ballots had been sent to persons listed in telephone directories

---

[4] Stephen K. Campbell, *Flaws and Fallacies in Statistical Thinking*, Prentice-Hall, Inc., Englewood Cliffs, N.J., 1974, p. 174.

and automobile registration records. As any student of history will tell you, those who could afford a telephone and an automobile in 1936 were hardly a cross section of the electorate. Instead, they were among the more prosperous of the voting population, and as a class in 1936 they were predominantly in support of the Republican Landon.

AGGRAVATING AVERAGES

Let's assume that Sandy Loam decides to buy a small farm in the Ozark Mountains where the air and water are clean, where she can putter about and grow cucumbers, and where she'll have the peace and quiet to write her novel exposing the inhumanity and ribaldry of poets. Sandy finds a farm that suits her needs, and is surprised when the realtor tells her that the 100 farmers in the area have an average annual income of nearly $33,000. Six months later a rally is called to protest a proposed increase in property taxes. It's pointed out during the rally that the average income in the area is only $13,000 and that no new taxes can be afforded. Sandy is naturally confused. She hasn't made a dime growing cucumbers, though, so she's willing to go along with the argument.

How could there be such a drop in average income in just 6 months? The answer is that nothing has really changed. *Both $33,000 and $13,000 are correct and legitimate averages!* The 100 farmers in the area include 99 whose net income is about $13,000, and one farmer who has invested millions of dollars in a showplace cattle operation spreading over thousands of acres. This one farmer nets approximately $2 million annually. Thus, one average—the arithmetic mean—is found by first figuring the total income: The 99 farmers times $13,000 equals $1,287,000, and this figure is added to the $2 million of the one-hundredth farmer to give a total income of $3,287,000. This total is then divided by 100 farmers to arrive at the arithmetic mean of $32,870, or nearly $33,000. The **arithmetic mean,** then, is the sum of the values of a group of items divided by the number of such items. The average of $13,000 is the **median**—the value that occupies the middle position when all the values are placed in an ascending or descending order. Thus, the median value of $13,000 represents the earnings of the middle farmer in the group of 100.

In this example, the average of nearly $33,000 is misleading because it distorts the general situation. Yet it's not a lie; it has been correctly computed. The problem of the aggravating average is that the word **average** is a broad term that applies to several measures of central tendency such as the arithmetic mean, the median, or the mode. The **mode** is the most commonly occurring value in a data set. In our example, it is also $13,000, but in other examples it could be yet a third figure. It's not uncommon for one of these averages to be used where it isn't appropriate, and where it is selected to deliberately mislead the consumer of the information.

STUDY FINDS ARREST STATISTICS PUBLISHED BY FBI INACCURATE

WASHINGTON—A study has found that the local arrest statistics published each year by the Federal Bureau of Investigation, one of the most widely used measures of the nation's effort to fight crime, are highly inaccurate.

The study said that the inaccuracies seemed to have been largely caused by the failure of law enforcement officials to follow the bureau's complex and sometimes ambiguous instructions about what constitutes an arrest.

One issue is whether an arrest occurs when a person is stopped by a police officer, when the person is booked, or when the suspect is arraigned in court.

The study was conducted by the Police Foundation, a Washington-based law enforcement research group, under a Justice Department grant. An FBI spokesman said the bureau, in conjunction with the Justice Department's Bureau of Criminal Justice Statistics, "is conducting an in-depth study of the Uniform Crime Report program seeking ways to improve upon this important law-enforcement management tool." The arrest figures are published as part of the crime reports.

Criminal justice experts have acknowledged for many years that police agencies have periodically manipulated, for political purposes, their reports about the number of crimes reported to them by individual citizens. Very few studies, however, have examined the accuracy of arrest statistics.

The Police Foundation's examination of how arrests are reported to the FBI included a detailed audit of the 1980, 1981 and 1982 reports from the police departments of four unidentified cities and a survey of the procedures of 196 other law enforcement agencies.

The study said the major implication of the research for individual citizens was that the probability of ending up with an arrest record depended in large part on the jurisdiction where the arrest was made because of reporting differences.

In addition to its own research, the Police Foundation study cited several recent audits by Capt. John Sura, division commander of the Michigan State Police Central Records Division.

Police chiefs and mayors, as well as others responsible for controlling crime and understanding criminal behavior, use arrest statistics for a variety of purposes, such as measuring police productivity, determining the characteristics of suspected criminals and testing various strategies to apprehend criminals.

But one key finding of the study was that the summaries of the arrests provided by various police departments around the nation cannot be compared with each other "because there is widespread violation of the rules and procedures established by the FBI for compiling arrest statistics."

DISREGARDED DISPERSIONS

Suppose that Karl Tell, an economist specializing in nineteenth-century German antitrust matters, is being pressured to coach the track team by the president of the small college where he teaches. Karl isn't too enthused about this prospect, since it will distract him from his study of the robber barons of Düsseldorf, but the president reminds him that he doesn't yet have tenure and that his classes are not in enormous demand. Therefore, Karl does a little checking and finds that the four high jumpers can clear an

MARK TWAIN, THE STATISTICIAN

Mark Twain appreciated the concepts of central tendency and variability as this passage from *Sketches Old and New* shows:

I went to a watchmaker again. He took the watch all to pieces while I waited, and then said the barrel was "swelled." He said he could reduce it in three days. After this the watch averaged well, but nothing more. For half a day it would go like the very mischief, and keep up such a barking and wheezing and whooping and sneezing and snorting, that I could not hear myself think for the disturbance; and so long as it held out there was not a watch in the land that stood any chance against it. But the rest of the day it would keep on slowing down and fooling along until all the clocks it had left behind caught up again. So at last, at the end of twenty-four hours, it would trot up to the judges' stand all right and just in time. It would show a fair and square average, and no man could say it had done more or less than its duty. But a correct average is only a mild virtue in a watch, and I took this instrument to another watchmaker.

And perhaps we could be more forgiving of trend projectors who would mislead us if their messages were as entertaining as the following passage from *Life on the Mississippi:*

Now if I wanted to be one of those ponderous scientific people, and "let on" to prove what had occurred in the remote past by what had occurred in a given time in the recent past, or what will occur in the far future by what has occurred in late years, what an opportunity is here!

Please observe:

In the space of one hundred and seventy-six years the Lower Mississippi has shortened itself two hundred and forty-two miles. This is an average of a trifle over one mile and a third per year. Therefore, any calm person, who is not blind or idiotic, can see that in the Old Oölitic Silurian Period, just a million years ago next November, the Lower Mississippi River was upward of one million three hundred thousand miles long, and stuck out over the Gulf of Mexico like a fishing rod. And by the same token any person can see that seven hundred and forty-two years from now the Lower Mississippi will be only a mile and three-quarters long, and Cairo [Illinois] and New Orleans will have joined their streets together, and be plodding comfortably along under a single mayor and mutual board of aldermen. There is something fascinating about science. One gets such wholesale returns on conjecture out of such a trifling investment of fact.

average of only 4 feet and that the three pole vaulters can manage an average height of only 10 feet. Karl concludes that his first venture into athletic management is likely to result in considerable verbal abuse from both alumni and faculty colleagues. Is Karl correct? Probably, but not because of the data he has gathered. Karl has been the victim of aggravating averages (arithmetic means in this example). Had he checked further, he would have found that one of his four high jumpers consistently clears 7 feet—good enough to win every time in the competition he will face—while the other three can each only manage to stumble over 3-foot heights. Likewise, in the pole vault there is one athlete who vaults 16 feet (with a bamboo pole) and two others who can each manage to explode for only 7 feet.

The moral here is really the same as that of the preceding section: *Averages alone often don't adequately describe the true picture.* **Dispersion—** the amount of spread or scatter that occurs in the data—must also be considered. And we are simply making the further distinction here that *disregarded dispersion* exists when the spread or scatter of the values about the central measure is such that the average tends to mislead. Of course, disregarded dispersions and aggravating averages are usually found acting in concert to confuse and mislead. To summarize, the story is often told of the Chinese warlord who was leading his troops into battle with a rival when he came to a river. Since there were no boats, and since the warlord remembered that he had read somewhere that the average depth of the water in the river was only 2 feet at that time of year, he ordered his men to wade across. After the crossing, the warlord was surprised to learn that a number of his soldiers had drowned. Although the average depth was indeed just 2 feet, in some places it was only a few inches, while in other places it was over the heads of many who became the unfortunate victims of disregarded dispersion.

THE PERSUASIVE ARTIST

Statistical tables and charts are prepared to summarize data, uncover relationships, and interpret and communicate numerical facts to those who can use them.

Statistical Tables

Statistical tables efficiently organize classified data into columns and rows so users can quickly find the facts needed. Figure 2-1 shows examples of tables that classify different data sets. In Figure 2-1a, for example, the data are classified by the names of large companies included in *Datamation* magazine's list of the world's top information system firms, and by the percentage net return on sales of these companies in a 2-year period. How are the data classified in Figures 2-1b and 2-1c?

Line Charts

A **line chart** is one in which data points on a grid are connected by a continuous line to convey information. The vertical axis in a line chart is usually measured in quantities (dollars, bushels) or percentages, while the horizontal axis is often measured in units of time (and thus a line chart becomes a *time-series* chart). Line charts don't present specific data as well as tables, but they're usually able to show relationships more clearly. Both a table and a line chart are frequently used together in a presentation, with the chart employed to clarify or reinforce facts presented in a table (see Figure 2-2).

The *single-line* charts in Figure 2-3 show a snapshot of the French econ-

Top Performers (the companies with the highest return on sales)*

DTM 100 RANK		COMPANY	1987% NET RETURN ON SALES	1986% NET RETURN ON SALES*
1	61	Cray Research Inc.	21.4	20.9
2	78	Microsoft Corp.	20.4	22.1
3	88	Lotus Development Corp.	18.2	17.1
4	75	Amstrad plc	17.9	18.6
5	2	Digital Equipment Corp.	12.4	10.2
6	68	Bell Atlantic Corp.	12.0	11.8
7	89	Shared Medical Systems Corp.	11.6	8.5
8	37	Compaq Computer Corp.	11.1	6.8
9	67	Intergraph Corp.	10.9	11.6
10	44	Seagate Technology	10.7	12.7
11	30	ADP Inc.	10.2	8.9
12†	80	3M	9.7	9.1
12†	29	Amdahl Corp.	9.7	4.3
12†	1	IBM	9.7	9.3
15	43	Tandem Computers Inc.	9.4	9.4

FIGURE 2-1
Statistical tables.
(a) The companies
with the highest
return on sales. (b)
Banking operations
of foreign banks in
the United States.

*Cray and Microsoft take top honors for the second year in a row. Compaq and Amdahl upped their performances nicely.
†Before rounding off, the figure for 3M was 9.735; for Amdahl, 9.699; and for IBM, 9.698.
Source: Excerpted from DATAMATION (June 15, 1988, p. 20). © 1988 Cahners Publishing Company.

(a)

Banking Operations of Foreign Banks in the United States (total U.S. banking assets in billions* of major foreign countries as of December 31, expressed as a percentage of total U.S. banking assets)

COUNTRIES	1982		1983		1984		1985		1986	
	DOLLARS	%	DOLLARS	%	DOLLARS	%	DOLLARS	%	DOLLARS	%
Japan	113.0	5.0	126.0	5.0	151.3	6.1	181.3	6.1	245.4	8.7
Canada	22.1	1.0	27.8	1.2	38.1	1.5	42.3	1.7	42.4	1.5
United Kingdom	52.2	2.5	53.0	2.3	51.4	2.0	61.2	2.4	40.6	1.5
Italy	14.3	0.7	17.5	0.8	23.9	0.9	29.1	1.1	36.4	1.4
Switzerland	13.0	0.6	13.1	0.6	15.3	0.6	18.3	0.7	24.5	0.9
France	16.6	0.8	16.2	0.7	18.3	0.7	20.7	0.8	22.4	0.8
West Germany	8.9	0.4	7.4	0.3	7.6	0.3	8.8	0.4	11.0	0.4
All other countries	60.5	3.0	70.9	3.1	72.4	2.9	97.2	3.8	103.9	3.8
Total U.S. banking assets of foreign banks	300.6	14	331.9	14	378.3	15	458.9	18	526.6	19
Total assets of domestic banking institutions†	1,821.1	86	1,986.5	86	2,076.8	85	2,098.7	82	2,285.9	81
Total U.S. banking assets†	2,121.7	100	2,318.4	100	2,455.1	100	2,557.6	100	2,812.5	100

*Amounts for each country include the total U.S. banking assets of all banks from that country, namely the aggregate of the assets of their U.S. branches, agencies, bank subsidiaries, Edge Act and Agreement corporations and New York State–chartered investment companies (called Article XII corporations).
†Includes the total consolidated assets (domestic and international) of all U.S. banks.
Source: Quarterly Review, Federal Reserve Bank of New York, Spring, 1987, p. 4. Reprinted with permission.

(b)

Occupations with the Largest Job Growth, 1986–2000   (numbers in thousands)

OCCUPATION	EMPLOYMENT PROJECTED		CHANGE IN EMPLOYMENT, 1986–2000		PERCENT OF PROJECTED JOB GROWTH, 1986–2000
	1986	2000	NUMBER	%	
Salespersons, retail	3,579	4,780	1,201	33.5	5.6
Waiters and waitresses	1,702	2,454	752	44.2	3.5
Registered nurses	1,406	2,018	612	43.6	2.9
Janitors and cleaners	2,676	3,280	604	22.6	2.8
General managers and top executives	2,383	2,965	582	24.2	2.7
Cashiers	2,165	2,740	575	26.5	2.7
Truckdrivers	2,211	2,736	525	23.8	2.5
General office clerks	2,361	2,824	462	19.6	2.2
Food counter and related workers	1,500	1,949	449	29.9	2.1
Nursing aides, orderlies, and attendants	1,224	1,658	433	35.4	2.0
Secretaries	3,234	3,658	424	13.1	2.0
Guards	794	1,177	383	48.3	1.8
Accountants and auditors	945	1,322	376	39.8	1.8
Computer programmers	479	813	335	69.9	1.6
Food preparation workers	949	1,273	324	34.2	1.5
Teachers, kindergarten and elementary	1,527	1,826	299	19.6	1.4
Receptionists and information clerks	682	964	282	41.4	1.3
Computer systems analysts	331	582	251	75.6	1.2
Cooks, restaurant	520	759	240	46.2	1.1
Licensed practical nurses	631	869	238	37.7	1.1
Gardeners and groundskeepers	767	1,005	238	31.1	1.1
Maintenance repairers	1,039	1,270	232	22.3	1.1
Stock clerks	1,087	1,312	225	20.7	1.0
First-line clerical supervisors and managers	956	1,161	205	21.4	1.0
Dining room and cafeteria attendants	433	631	197	45.6	0.9
Electrical and electronics engineers	401	592	192	47.8	0.9
Lawyers	527	718	191	36.3	0.9

Source: Occupational Outlook Quarterly, Spring, 1988, p. 8

(c)

FIGURE 2-1
(Continued)
(c) Occupations
with the largest job
growth, 1986–2000.

omy (Figure 2-3a), and the ups and downs of income in one sector of the insurance business (Figure 2-3b). Of course, *multiple series* can also be depicted on one line chart, as shown in Figures 2-4 and 2-5. The lines in Figures 2-4a and b are all plotted against the same baseline. Note, though, that in the *component-part* (or area) line drawings in Figures 2-5a and b, the charts are built up in layers. Thus, in Figure 2-5a the "nonhousing stock" component is added to the "housing stock" element to get the "total capital stock" line at the top of the chart.

Bar Charts

A **bar** (or **column**) **chart** is one that uses the length of horizontal bars or vertical columns to represent quantities or percentages. As in the case of line

Farm Sector Income Statement (billions of dollars)

	1980	1981	1982	1983	1984	1985	1986	1987*
Farm receipts	$142.0	$144.1	$147.1	$141.1	$146.7	$149.2	$140.2	$138
Government payments	1.3	1.9	3.5	9.3	8.4	7.7	11.8	17
Total farm income†	149.3	166.3	163.5	153.1	174.7	166.0	159.5	163
Total expenses	133.1	139.4	140.0	140.4	142.7	133.7	122.1	119
Net farm income	16.1	26.9	23.5	12.7	32.0	32.3	37.5	45

*Values for 1987 are forecasts.
†Total net farm income includes the value of inventory changes. Net farm income totals may not add due to rounding. Data are not adjusted for inflation.
Data and source: *Agricultural Outlook,* March 1988, table 32, p. 54.

FIGURE 2-2
A table and a chart used together to analyze farm income. (Chart source: *Review,* Federal Reserve Bank of St. Louis, March–April 1988, p. 31. Reprinted with permission.)

U.S. Real Net Farm Income

charts, one scale on the bar chart measures values while the other may represent time. The bars typically start from a zero point, and are frequently used to show multiple comparisons (see Figure 2-6*a*). Bars may be *clustered* together to show how identified categories of interest change over time (Figure 2-6*b*). Or *component-part* (or *stacked*) bars can be used (Figure 2-6*c*). Stacked bars can compare many quantities in a single graph, but they can also be complex and difficult to interpret.

Pie Charts

Pie charts are simply one or more circles divided into sectors, usually to show the component parts of a whole. Single circles can be used (Figure 2-7*a*), or several pie charts can be drawn to compare changes in the com-

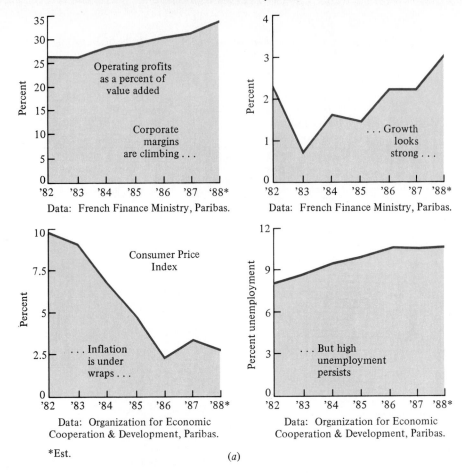

Behind France's Sunny Mood

Operating profits as a percent of value added

Corporate margins are climbing . . .

Data: French Finance Ministry, Paribas.

. . . Growth looks strong . . .

Data: French Finance Ministry, Paribas.

Consumer Price Index

. . . Inflation is under wraps . . .

Data: Organization for Economic Cooperation & Development, Paribas.

. . . But high unemployment persists

Data: Organization for Economic Cooperation & Development, Paribas.

*Est.

(a)

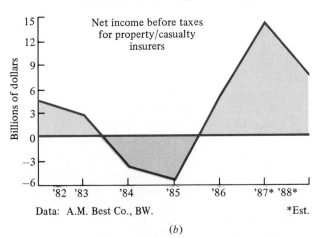

The Insurance Industry's Vicious Cycle

Net income before taxes for property/casualty insurers

Data: A.M. Best Co., BW.   *Est.

(b)

FIGURE 2-3   Single-line charts. (a) (Source: Reprinted from April 25, 1988 issue of *Business Week*, p. 77, by special permission, copyright © 1988 by McGraw-Hill, Inc.)
(b) (Source: Reprinted from April 11, 1988 issue of *Business Week*, p. 60, by special permission, copyright © 1988 by McGraw-Hill, Inc.)

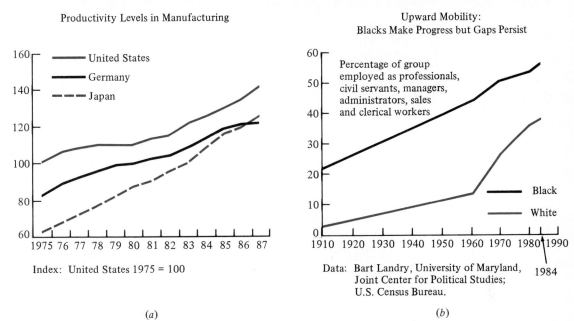

Productivity Levels in Manufacturing

Upward Mobility:
Blacks Make Progress but Gaps Persist

Index: United States 1975 = 100

Data: Bart Landry, University of Maryland, 1984
Joint Center for Political Studies;
U.S. Census Bureau.

(*a*)

(*b*)

FIGURE 2-4   Multiple series shown on single line charts. (*a*) (Source: *Quarterly Review,* Federal Reserve Bank of New York, Spring 1988, p. 10. Reprinted with permission.) (*b*) (Source: Reprinted from March 14, 1988 issue of *Business Week,* p. 68, by special permission, copyright © 1988 by McGraw-Hill, Inc.)

ponent parts over time (Figure 2-7*b*). A technique that's often used is to separate a segment of the drawing from the rest of the pie to emphasize an important piece of information.

Combination and Other Charts

As you can see in Figure 2-8, it's possible to include a mix of the charts we've now examined in a graphics presentation. An early chart—a combination data map and time-series plot—was drawn in 1861 by Joseph Minard, a French engineer. Minard's chart starkly shows (1) Napoleon's army of 422,000 men invading Russia in June 1812, (2) only 100,000 remaining when Moscow was reached in September, (3) the subzero weather conditions existing during Napoleon's retreat, and (4) the dwindling of his forces to a scant 10,000 troops by the time they returned to their starting point. As you'll see in the Closer Look reading at the end of the chapter, graphics expert Edward Tufte believes that Minard's effort remains one of the best statistical graphs ever drawn.

Other types of charts that can be used to display quantitative facts are **statistical maps** that present data on a geographical basis (Figure 2-9), and **pictographs** or pictorial charts that use picture symbols to convey meaning. Pictographs must be used with caution as we'll see in a few paragraphs.

The Allocation of Capital between
Housing and Nonhousing Assests

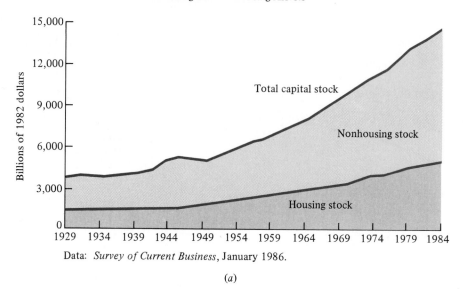

Data: *Survey of Current Business*, January 1986.

(*a*)

CD ROM Drive Unit Sales, 1987–1992

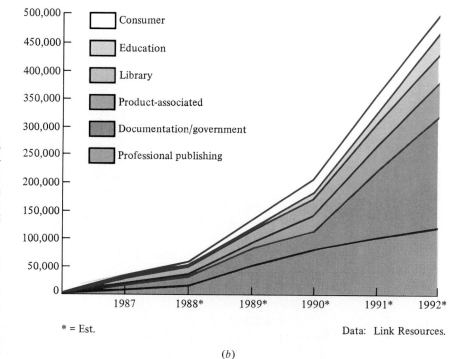

FIGURE 2-5
Component-part
li ne drawings. (*a*)
(Source: *Business
Review,* Federal
Reserve Bank of
Philadelphia,
March–April 1987,
p. 16. Reprinted
with permission.)
(*b*) (Source:
Infoworld, March
14, 1988, p. 27.
Reprinted with
permission.)

* = Est.

Data: Link Resources.

(*b*)

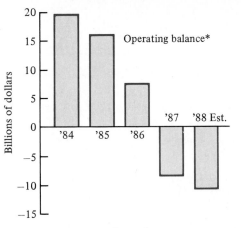

State and Local Budgets
Slide Deeper into the Red

*Includes capital spending
  but excludes social insurance accounts.

Data: Commerce Dept., Data Resources estimate.

(a)

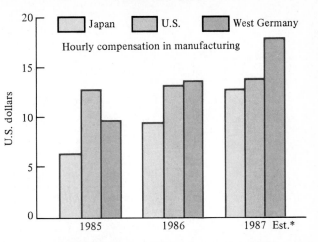

U.S. Wages Look
a Lot More Competitive

*Japanese and German 1987 wages at current (1988)
  exchange rates.

Data: Bureau of Labor Statistics, Data Resources Inc.

(b)

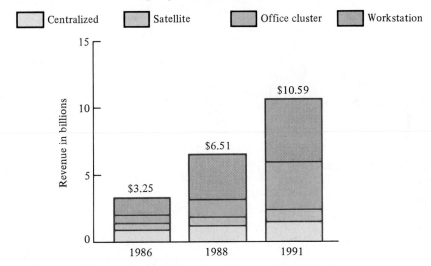

Low–end Growth
(workstations and office clusters will fuel growth in use of
intelligent printers and copiers)

Data: Information provided by Cap International, Inc.

(c)

FIGURE 2-6   (a) State operating balance, 1984–1988. (Source: Reprinted from May 2, 1988 issue
             of *Business Week*, p. 24, by special permission, copyright © 1988 by McGraw-Hill,
             Inc.) (b) An example of a clustered bar chart. (Source: Reprinted from March 7,
             1988 issue of *Business Week*, p. 20, by special permission, copyright © 1988 by
             McGraw-Hill, Inc.) (c) An example of a stacked bar chart. (Source: *Computer-
             world*, April 4, 1988, p. 43. Copyright 1989 by CW Publishing Inc., Framingham,
             MA 01701. Reprinted from *Computerworld*.)

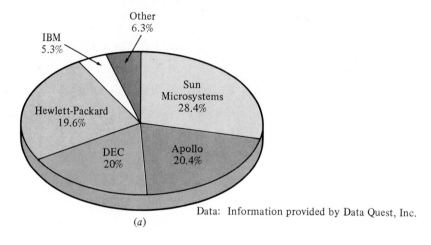

Sun Steals the Lead
Market share (based on unit shipment) of technical workstations

Data: Information provided by Data Quest, Inc.

(a)

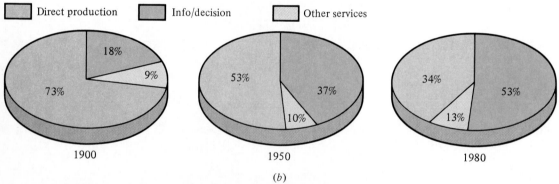

U.S. Employment Distribution by Occupation, 1900–1980

Direct production   Info/decision   Other services

1900   1950   1980

(b)

FIGURE 2-7   Pie charts. (a) (Source: *Computerworld*, March 7, 1988, p. 118. CW chart, Frank C. O'Connel. Copyright 1989 by CW Publishing Inc., Framingham, MA 01701. Reprinted from *Computerworld*.) (b) (Source: *Review*, Federal Reserve Bank of St. Louis, June/July, 1987, p. 18. Reprinted with permission.)

Computer Graphics Packages

You saw in Chapter 1 that statistical packages produce graphic output, but the **graphics packages** we'll consider here are separate software products that convert the numeric data used by computers into the visual images that people often prefer to use to communicate ideas. These packages may be classified into design, paint, analysis, and presentation categories. *Design packages* are the software tools used to create, edit, store, and make permanent prints of the designs of people such as engineers and architects. *Computer-aided design* (CAD) is a term that refers to work carried out by

Managers must make decisions based on quantitative data. Statistical tables can efficiently organize relevant data into columns and rows. But managers often prefer to use pictures to examine the relationships — contrasts, trends, comparisons — that exist in the data. Computer-generated graphics help managers produce better decisions by helping them visualize and understand these relationships. As you'll see in this photo gallery, computer graphics can show relationships through line charts, bar charts, pie charts, maps and three-dimensional images, and combinations of these presentation techniques.

Photo Courtesy of Hewlett-Packard Company.

Computer graphics programs enable these and other personal computers to produce high-quality, high-resolution graphics for decision-making purposes.

Courtesy of International Business Machines Corporation.

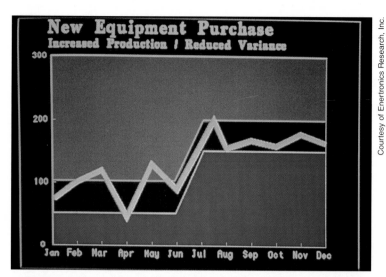

New Equipment Purchase
Increased Production / Reduced Variance

Courtesy of Enertronics Research, Inc.

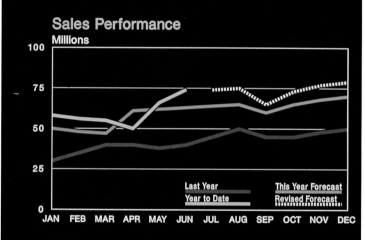

Sales Performance
Millions

Last Year
Year to Date
This Year Forecast
Revised Forecast

Courtesy of Computer Associates International, Inc.

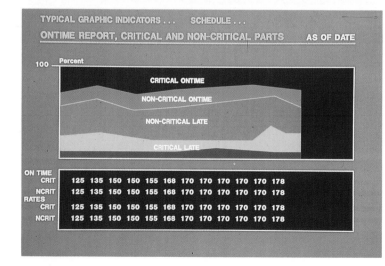

TYPICAL GRAPHIC INDICATORS . . .   SCHEDULE . . .

ONTIME REPORT, CRITICAL AND NON-CRITICAL PARTS   AS OF DATE

100 Percent

CRITICAL ONTIME

NON-CRITICAL ONTIME

NON-CRITICAL LATE

CRITICAL LATE

ON TIME												
CRIT	125	135	150	150	155	168	170	170	170	170	170	178
NCRIT	125	135	150	150	155	168	170	170	170	170	170	178
RATES CRIT	125	135	150	150	155	168	170	170	170	170	170	178
NCRIT	125	135	150	150	155	168	170	170	170	170	170	178

Courtesy of Computer Associates International, Inc.

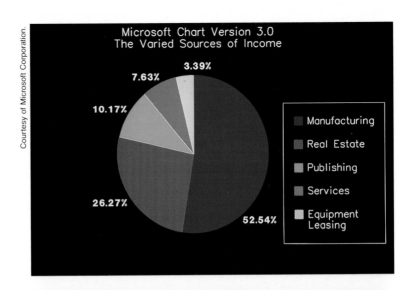

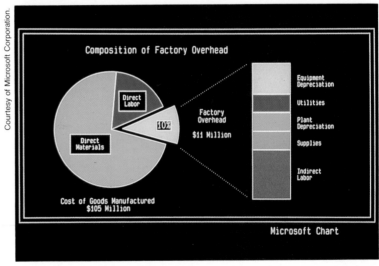

Pie charts may be used to show the component parts of a whole *(top)*.
A segment of a pie chart can be further analyzed through the use of
a component part bar chart *(bottom)*.

Line charts can be used to depict a single
series *(opposite top),* or a multiple
series can be shown in a line
chart *(opposite middle and bottom).*

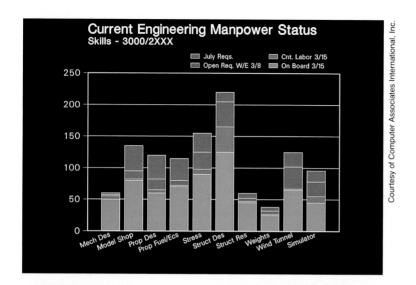

Courtesy of Computer Associates International, Inc.

Courtesy of International Business Machines Corporation.

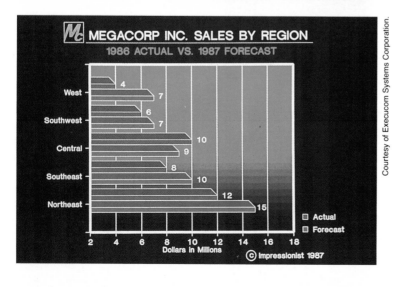

Courtesy of Execucom Systems Corporation.

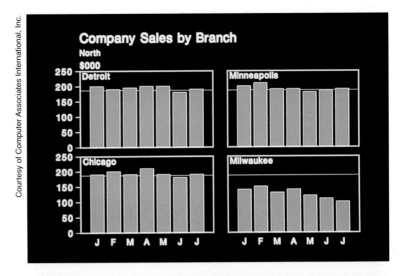

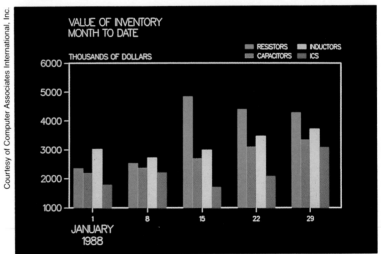

Of course, component part bar charts need not be combined with pie charts *(opposite top).*
And bar charts can also be used for projections *(opposite middle and bottom)* and to show multiple series *(above top and bottom).*

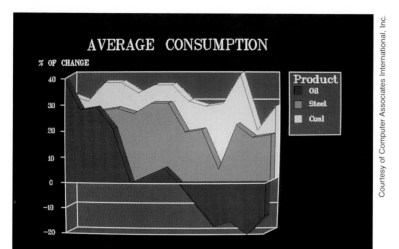

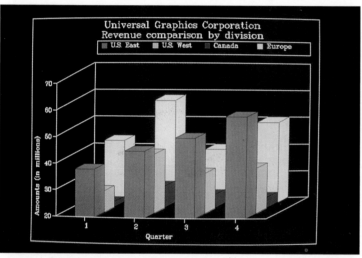

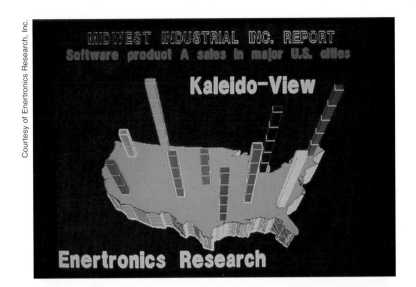

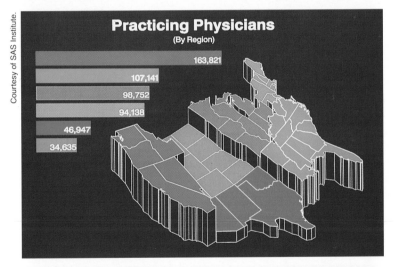

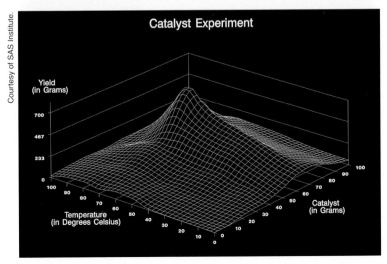

Three-dimensional images and maps are popular. Three-dimensional images may be created by using a line chart *(opposite top)* or a bar chart *(opposite middle)*. And two- and three-dimensional maps are used to show the geographic distribution of quantitative data *(opposite bottom and right top, middle, and bottom)*.

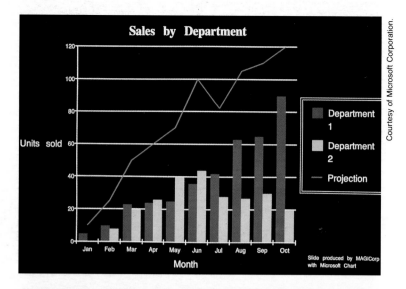

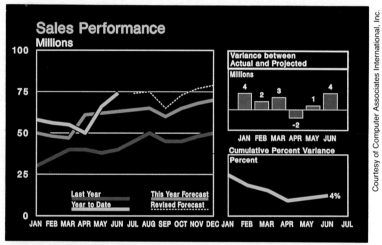

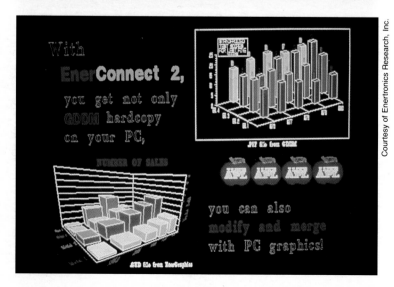

Combinations of graphic presentation techniques are often used to give managers better insights into underlying relationships.

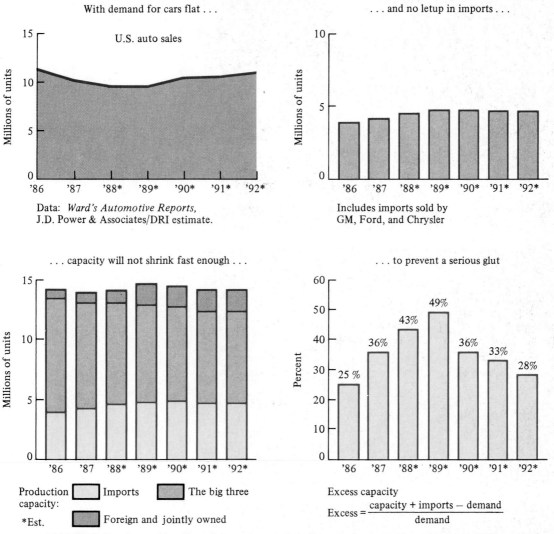

With demand for cars flat . . .

U.S. auto sales

. . . and no letup in imports . . .

Data: *Ward's Automotive Reports,*
J.D. Power & Associates/DRI estimate.

Includes imports sold by
GM, Ford, and Chrysler

. . . capacity will not shrink fast enough . . .

. . . to prevent a serious glut

Production capacity: Imports   The big three

*Est.   Foreign and jointly owned

Excess capacity

$$\text{Excess} = \frac{\text{capacity} + \text{imports} - \text{demand}}{\text{demand}}$$

FIGURE 2-8   A combination of charts in a series. (Source: Reprinted from March 7, 1988 issue of *Business Week*, pp. 54–55, by special permission, copyright © 1988 by McGraw-Hill, Inc.)

people working with these packages. Hobbyists and others with an artistic bent can use *paint packages* to create drawings on their display screens. Their "brush" may be a hand-held mouse that can produce fine lines or broad strokes. The artist can begin with a picture supplied from the package's clip art library of stored images, and then choose from dozens of color choices and patterns to add texture and interest to the selected image. Or the user can start with a clean screen "canvas" and begin painting with the mouse.

Design and paint packages are exciting graphics tools. But statistical information is usually manipulated with analysis and presentation packages.

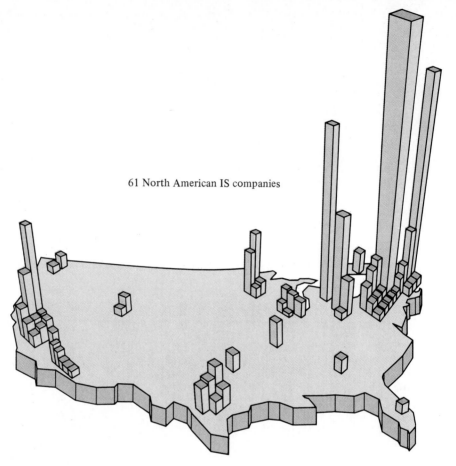

61 North American IS companies

An **analysis package** converts numeric data into a visual summary so that people can grasp relationships, spot patterns, and make more informed decisions. After supplying an analysis package with the data to be analyzed, the user can make a preliminary chart selection from a menu of chart formats supplied by the program (Figure 2-10*a*). The program quickly displays the data in the selected format (Figure 2-10*b*). The user can vary colors, add and delete lines and headings, change scales, and edit the appearance of the chart in other ways. If not pleased with the first choice, the user can try other formats (Figure 2-10*c*). After satisfactory results are obtained, the user can then print these results and store the chart format for future use.

Decision makers use analysis packages to gain insight into the relationships, changes, and trends that are buried in their data. But they use **presentation packages** to communicate messages to an audience. A presentation package has all the features found in an analysis package. Bar, pie, line, and other charts may still be used. But a presentation package can produce

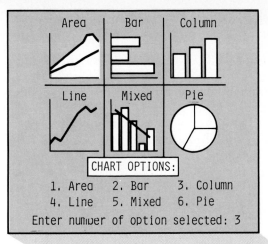

(a)

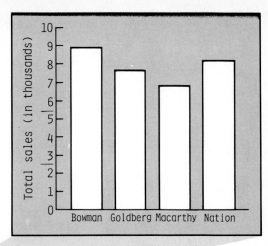

(b)

FIGURE 2-10
An analysis graphics package can generate charts and graphs from numeric input. Although data presented in graphic form are easier to absorb and remember, these visual aids were seldom used in the past because their preparation was time-consuming and expensive. Now, however, many inexpensive and easy-to-use analysis packages are available for personal computers.

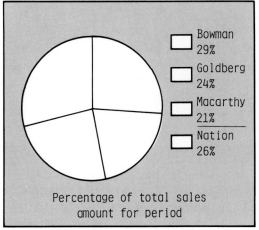

(c)

multiple three-dimensional images, and it allows the user to dress up charts in other ways. For example, it may include features found in paint packages to allow the user to spruce up charts with clip art library pictures or original drawings.

A presentation program makes better use of color than an analysis package, and it's likely to have built-in guidelines that use proven design rules to help users with format specifications. It may also support the use of maps and other graphic tools, and it may allow users to pick from an overwhelming number of type styles. Some presentation packages also have animation capabilities. For example, a user can make bars grow on a screen to dramatize increases. The best presentation graphics are produced on larger systems, but powerful and inexpensive presentation packages are also available for personal computers. Some examples of computer-generated graphics are shown in the nearby color photo gallery.

Misusing Graphics

So much for the ways that data can be honestly presented. But the purpose of some persuasive artists is to take honest facts and create misleading impressions. How is this done? There are numerous tricks, but we'll limit our discussion here to just a few examples.

Suppose you're running for reelection to a legislative body and during your past 2-year term appropriations have increased in your district from $8 million to $9 million. Now, as your fellow politicians know, this isn't a particularly good record, but the voters don't need to know that. In fact, you can perhaps turn this possible liability into an asset with the help of a persuasive artist. Figure 2-11a shows one way to present the information honestly. But since your objective is to mislead without actually lying, you prefer instead to distribute Figure 2-11b during your campaign. The difference between Figure 2-11a and Figure 2-11b, of course, lies in the changing of the vertical scale in the latter figure. (The wavy line correctly indicates a break in the vertical scale, but this is often not considered by unwary consumers of this type of information.) By breaking the vertical scale and by then changing the proportion between the vertical and horizontal scales, you've given the impression that you've certainly been doing a good job of getting appropriations.[5]

Having received favorable comments on your appropriations chart, you decide to employ another trick. New industry has come into your district during the past 2 years. There has been some increase in air and water pollution as a result, but there has also been an increase in average weekly wages of unskilled workers from $120 per week to $180 per week. Of course, you had little to do with bringing in the new industry, and there's also some disturbing evidence that the fact that wage increases have *followed* the new plants doesn't necessarily mean that they were *caused* by

[5] We could make Figure 2-11b even more impressive-looking by keeping the vertical scale and *compressing the horizontal* (time) *scale.*

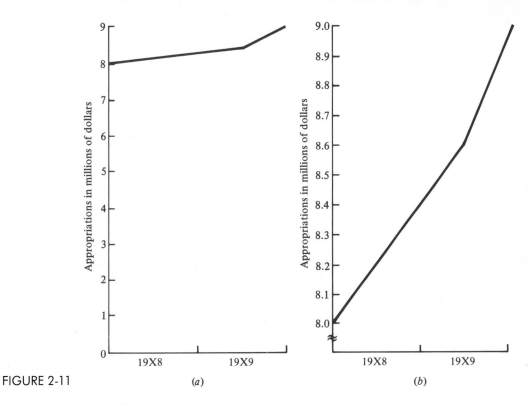

FIGURE 2-11   (a)   (b)

the new industry. But you see no reason to complicate matters with additional confusing facts.[6]

How can you best communicate this wage increase information to your constituents? After trying several approaches, you decide to use the pictograph in Figure 2-12a. The *height* of the small money bag represents $120, and the *height* of the large bag is correctly proportioned to represent $180. What's wrong and misleading, though, is the *area* covered by each figure. The space occupied by the larger bag creates a misleading visual impression. But that was the intent, wasn't it? If you think this example is farfetched, consider Figures 2-12b and c.

Let's now assume that in spite of your persuasive artwork the voters have seen fit to throw you out of office in favor of a write-in candidate. However, you're able to find work in your father's manufacturing company. One of the first jobs you're given is to prepare reports for stockholders and the union explaining company progress over the past year. The company has done well, and profits have amounted to 25 cents of every sales dollar. This can be accurately presented in picture form, as shown in Figure 2-13a. But a stronger impact can be made on stockholders if the coin is shown from the perspective of Figure 2-13b. Since you don't want

---

[6] We will discuss this matter in the next section.

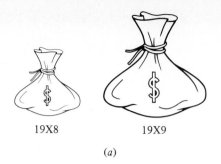

19X8   19X9

(a)

Drop in U.S. Unemployment

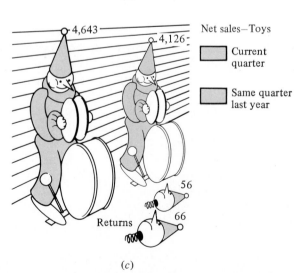

Percent of work force unemployed.
Percentages are seasonally adjusted.

10.5
10.0
9.5
9.0
8.5
8.0
7.5

10.4   10.3   10.2   10.1   10.0   9.5   9.5   9.2   8.8   8.4   8.2   8.0

Feb   March   April   May   June   July   Aug   Sept   Oct   Nov   Dec   Jan

|⟵————————————————1983————————————————⟶|1984

Data: Department of Labor.

(b)

Net sales—Toys

4,643
4,126

☐ Current quarter

☐ Same quarter last year

Returns   56   66

(c)

FIGURE 2-12   (a) The first of our perfidious pictographs. (b) Note the absence of any break in the vertical scale that compounds the area misrepresentation. (Source: *Fort Worth Star Telegram*, February 4, 1984, p. 9B. Reprinted with permission.) (c) And note the absence of any scale at all in this figure. (Source: *PC World*, April, 1986, p. 272. Reprinted with permission.)

Profits 25%
(a)

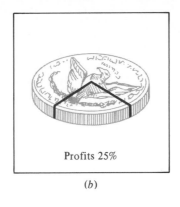
Profits 25%
(b)

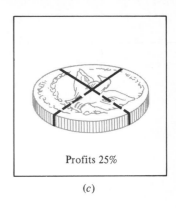
Profits 25%
(c)

FIGURE 2-13

the union members to become restless, though, you can show them the profit situation from the perspective of Figure 2-13c.

THE *POST HOC ERGO PROPTER HOC* TRAP

The Latin phrase for the logical reasoning fallacy that states that because B *follows* A, B was *caused* by A is *post hoc ergo propter hoc* which means "after this, therefore because of this." Erroneous cause-and-effect conclusions are often drawn because of the misuse of quantitative facts. Of course, as the following examples show, some errors are easy to spot:

> *Increased shipments of bananas into the port of Houston have been followed by increases in the national birthrate. Therefore, bananas were the cause of the increase in births.*

> *Ninety-five percent of all those who regularly use marijuana consumed large amounts of milk earlier in their lives. Thus, the early introduction of milk into the diet leads to the use of marijuana.*

> *The average human life-span has doubled in the world since the discovery of the tobacco plant. Therefore, tobacco . . . (this is just too gross to complete).*

But not all examples of the *post hoc* trap are so obvious. Some, in fact, can be subtle. To illustrate, you may recall that a few pages earlier (in the discussion of bias and the *Literary Digest* poll) the following sentence appeared: "Unfortunately for the *Digest* (which ceased to exist in 1937), the ballots had been sent to persons listed in telephone directories and automobile registration records." Did you perhaps conclude from this example that since the *Digest* folded in 1937, the cause was the poor forecast about the 1936 election? Obviously, the poll didn't help the magazine's reputation, but did it *cause* the publication to go out of business? Isn't it likely that a number of factors combined to bring about this demise?

Let's now conclude our discussion of the *post hoc* trap with a return to the antiseptic world of politics. We've already seen in the preceding section how a politician can claim credit for good events that occur after he or she is elected. Another technique—and one that is often just as misleading—is for a politician to accuse an opponent of being the cause of something bad. For example, *Time* magazine reported that Richard Nixon made the following remarks during the 1968 presidential campaign:

> *Hubert Humphrey defends the policies under which we have seen crime rising ten times as fast as the population. If you want your President to continue a do-nothing policy toward crime, vote for Humphrey. Hubert Humphrey sat on his hands and watched the U.S. become a nation where 50% of the American women are frightened to walk the streets at night.*

These remarks are ironic because of later events and misleading because of the great power attributed to the person occupying the position of vice president.

ANTICS WITH SEMANTICS

Failing to define terms and concepts that are important to a clear understanding of the message, *making improper or illogical comparisons* between unlike or unidentified things, *using an alleged statistical fact or statement to jump to a conclusion* which ignores other possibilities or which is quite illogical, *using jargon and lengthy words* to cloud the message when simple words and phrases are suitable—all these antics with semantics are used to confuse and mislead. The following examples should be adequate to illustrate this unfortunate fact:

- The Chapter 1 example of different unemployment figures produced by different groups shows what a failure to define terms can do to understanding. In fact, the Bureau of Labor Statistics has published no fewer than eight versions of the unemployment rate, using different labor-force definitions. Terms such as "poverty," "population," and "living standard," to name just a few, are also subject to different definitions, and the consumer of the information should be told which is being used.
- The Federal Trade Commission (FTC) took exception to the advertised claim made by the manufacturers of Hollywood Bread that their product had fewer calories per slice than other breads. According to the FTC, the claim was misleading, since the Hollywood slice was thinner than normal. Actually there was no significant difference in the number of calories when equal amounts of bread were compared.
- "One of four persons in the world is a Chinese communist. ...Think about that the next time you listen to Lawrence Welk." What in the world does this mean? "The Egress carburetor is up to 10% less polluting and up to 50% more efficient." Less polluting than what, a steel mill? And more efficient than what, a Boeing 747?
- "One in ten births is illegitimate. Thus, your estimate of your fellow

man is correct 10% of the time." Figures on many activities (including illegitimate births, rapes, marijuana smoking, etc.) are just not reliable, because many cases remain unreported.

- Representative Ben Grant, in arguing for plain speaking in a proposed new Texas constitution, points out how jargon can be used. If, noted Grant, a man were to give another an orange, he would simply say, "Have an orange." But if a lawyer were the donor, the gift might be accompanied with these words: "I hereby give and convey to you, all and singular, my estate and interests, right, title, claim, and advantages of and in said orange, together with its rind, juice, pulp, and pits, and all rights and advantages therein, with full power to bite, cut, suck, and otherwise eat same. . . ." Alas, the same kind of language may also be used to convey quantitative information.

THE TREND MUST GO ON!

Another way that a person may misuse statistical facts is to assume that because a pattern has developed in the past in a category, that pattern will certainly continue into the future. Such uncritical extrapolation, of course, is foolish. Changes in technology, population, and lifestyles all produce economic and social changes that may quickly produce an upturn or a downtrend in an existing social pattern or economic category. The invention of the automobile, for example, brought about significant growth in the petroleum and steel industries and a rapid decline in the production of buggies and buggy whips. Yet, as British editor Norman MacRae has noted, an extrapolation of the trends of the 1880s would show today's cities buried under horse manure.

Of course, the *judicious* projection of past patterns or trends into the future can be a very useful tool for the planner and decision maker, as we'll see in Chapter 13 in our study of time-series analysis. But the failure to apply a generous measure of common sense to extrapolations of past quantitative patterns can lead to faulty conclusions that people are seriously asked to accept.

FOLLOW THE BOUNCING BASE (IF YOU CAN)

In an editorial on minimum competency in English and Math published on April 7, 1978 the editors of the Pensacola Journal *actually stated that "After all, if you give the test to four students and four flunk, that's a 50 percent failure rate."*
—Ron McCuiston in The Matyc Journal, *Winter 1979, p. 59. From "Standard Deviations of the Square Root of Infinity." Reprinted with permission of* The Matyc Journal, *Inc.*

People today are often confused because of their failure to follow the bouncing base—that is, *the base period used in computing percentages.* A few examples will show how failure to clarify the base may lead to misunderstanding:

- A worker is asked to take a 20 percent pay cut from his weekly salary of $300 during a recessionary period. Later, a 20 percent increase is given to the worker. Is he happy? The answer may depend on what has happened to the base. If the *cut* is computed using the earlier period (and the salary of $300) as the base, the reduced pay amounts to $240 ($300 × 0.80). But if the pay *increase* of 20 percent is figured on a base that *has been shifted* from $300 to $240, the worker winds up with a restored pay of $288 (1.20 × $240). Thus, the bouncing base has cost him $12 each week, and this isn't likely to please him.

- In Chapter 12 we'll study the subject of index numbers, a valuable tool for measuring changes in such economic variables as prices and quantities produced. Unfortunately, index numbers are frequently misused by reporters. Let's assume, for example, that a price index uses 1982–1984 as the base period and assigns to prices in that period an index number value of 100. Later, in 1986, the index value has risen to 110, and in 1988 the price index is 117. These numbers mean that there was a 10 percent increase in prices between 1982–1984 and 1986 and a 17 percent increase in prices between 1982–1984 and 1988. So far, so good. But now a reporter uses these figures during an article and notes that there has been a 7 *percent increase* in prices between 1986 and 1988. It's true that the numbers 110 and 117 represent percentages, and it's also true that there's a difference of 7 *percentage points* between 1986 and 1988. But the *percentage increase* was actually 6.36 percent: $(117 − 110)/110 = 6.36$. In this case, the reporter failed to shift the base to 1986.

- Percentage *increases* can easily exceed 100 percent. For example, a company whose sales have increased from $1 million in 1987 to $4 million in 1989 has had a percentage increase of 300 percent: ($4 million − $1 million)/$1 million. [Of course, the sales in 1989 *relative* to the sales in 1987 were 400 percent—($4 million/$1 million) × 100— and this *percentage relative* figure is sometimes confused with the percentage increase value.] Remember, though, that percentage *decreases* exceeding 100 percent aren't possible if the original data are positive values. For example, *Newsweek* magazine reported in 1967 that Mao-Tse Tung had cut the salaries of certain Chinese government officials by 300 percent. Of course, once 100 percent is gone, there isn't anything left. Embarrassed editors later admitted that the cut was 66.67 percent rather than 300 percent. It isn't hard to find more recent examples. An article in the February 15, 1988 issue of *Datamation* began with these words: "We certainly understood the new industry economics. Oil prices had plunged more than 100% in the past two years. . . ."

The above examples show only a few types of abuses that may be associated with the use of percentages. But they do give you an idea of the importance of following the bouncing base.

SPURIOUS (AND CURIOUS) ACCURACY

Statistical data based on sample results are often reported in precise numbers. It's not unusual for several decimal places to be used, and the apparent precision lends an air of infallibility to the information reported—particularly when this information appears on a computer printout. Yet the accuracy image may be false. To illustrate, W. E. Urban, a statistician for the New Hampshire Agricultural Experiment Station, wrote a letter to the editor of *Infosystems* magazine taking issue with a previously published article. "Your magazine," wrote Urban, "has provided me with an excellent example of impeccable numerical accuracy and ludicrous interpretation which I will save for my statistics classes. With a total sample size of 55, reporting percentages with two decimal places is utter nonsense." The first sample percentage quoted in the article was 31.25, but, as Urban noted, the corresponding estimate of the population was likely to have been anywhere between 12 and 62 percent! As Urban concluded: "I realize that it is painful to throw away all the nice decimals the computer has given us. . . . but who are we kidding?" The reply of the editor: "No one. You're right."

Spurious accuracy isn't limited to sample results. In 1950 the *Information Please Almanac* listed the number of Hungarian-speaking people at 13,000,000, while in the same year the *World Almanac* placed the number at 8,001,112. Thus, there was a difference of about 5 million people in the estimates. This isn't particularly surprising. But isn't it curious that the *World Almanac* figure could be so precise? Does it stand to reason that the accuracy could be so great when the figures are well up into the millions? Albert Sukoff has observed that "Huge numbers are commonplace in our culture, but oddly enough the larger the number the less meaningful it seems to be. . . . Anthropologists have reported on the primitive number systems of some aboriginal tribes. The Yancos in the Brazilian Amazon stop counting at three. Since their word for 'three' is '*poettarrarorincoaroac*,' this is understandable."

Oskar Morgenstern summarizes the issue of spurious accuracy with these words:

> *It is pointless to treat material in an "accurate" manner at a level exceeding that of the basic errors. The classical case is, of course, that of the story in which a man, asked about the age of a river, states that it is 3,000,021 years old. Asked how he could give such accurate information, the answer was that 21 years ago its age was given as 3 million years.*

NUMBERS DON'T LIE—DO THEY?

I for one, am no longer unquestioningly in awe of high-powered number crunching. What does it mean, after all, when the eleventh canonical root is significant? What does it mean that the age of your grandparents, your preference in frozen desserts, and your hat size can be related to the number of clocks you own, the weight of the family dog, and your blood pressure. Yet, this kind of multivariate, multilayered analysis, whose input is rarely examined initially, whose assumptions are infrequently explicit, and whose technique is often obfuscating has become the basis for modern strategy. In fact, Peters and Waterman in their *In Search of Excellence* note that the word "strategy," which used to mean a damn good idea for knocking the socks off the competition, has come to be synonymous with "the quantitative breakthrough, the analytic coup, market share numbers, learning curve theory, positioning a business on a 4- or 9- or 24-box matrix and putting all of it on a computer."

It is quite evident that our ability to perform these feats of computational gymnastics has changed greatly since the 1940s. Our data-gathering capabilities have increased enormously. We are able to compile a vast number of minute observations through panel studies, WATS line telephone interviewing, and even cable TV interactive systems. The Simmons Market Research reports, for example, are based on a national sample of over 30,000 households. An even more impressive change has occurred in our ability to store and analyze such information. Years ago it would have taken weeks to perform the most rudimentary statistical analysis using a mechanical, desk top calculator. Today's mainframe computers can crank out very sophisticated statistical analyses in a matter of a few minutes.

But have the people who provide the grist for the analyst's numerical mill changed as well? Are their responses to pollsters' questions more accurate, more reasoned—better somehow?

WHAT'S BEHIND THE NUMBERS?

In 1947, a researcher named Sam Gill conducted a poll which solicited people's opinions of the Metallic Metals Act. Of those questioned, 70 percent expressed a definite opinion about the Act, ranging from "a good move" through "should be left to the individual states" and "all right for foreign countries but should not be required here" to "it is of no value at all." The pattern of responses reflected the public mood about government, foreign policy, and trade.

Recently, the poll was replicated. This time 64 percent of those interviewed had a definite opinion. The pattern of responses changed dramatically, probably because of the conservative shift of the last few years and because of uncertainties and anxieties about the deficit and foreign affairs. But in both surveys, people were giving their opinions about an Act which never existed. This phenomenon is not unusual.

In this age of Donkey Kong research, it may be worthwhile to remember several classic studies which tell us something about ourselves:

Richard LaPierre, in 1934, recorded the treatment that a Chinese couple received in hotels and restaurants. Of the 251 establishments they approached on a trip across the country, only one place refused to accommodate them. Later a questionnaire was sent to each establishment asking, "Would you accept members of the Chinese race as guests in your establishment?" Only one positive response was received.

A small rural community in Kansas was the setting for a 1958 study by Charles Warriner concerning the drinking of alcoholic beverages. Interviews with the citizenry of this "dry" town revealed a consistent personal belief that drinking alcoholic beverages was wrong. However, an unobtrusive garbage search showed that the public expressions and private behavior were not the same.

More recently, Dannick (1969) studied pedestrians crossing at an intersection. Those who crossed against a "don't walk" light were questioned. Most of the respondents insisted that they never engaged in such behavior and, even more frequently, expressed the sentiment that it was wrong for anyone to do it.

Other studies have shown that [employers'] attitudes toward hiring the handicapped are not reflected in their hiring practices, that responses to questions about income and education do not correspond to actual levels of income and education, and that there is a lack of consistency between students' attitudes toward cheating and actual behavior.

The fact is that people exaggerate, try to please, try to subvert, try to impress, forget, misinterpret, cheat, and lie. And yet when we are surrounded by such marvelous tools and techniques for processing numbers, it is easy to forget where those numbers originated. A modern-day LaPierre, without the benefit of his observational data and with the benefit of quantitative methods and computer knowledge, could turn out a most enlightening study—perhaps establishing a relationship between attitudes towards Orientals and restaurant seating capacity, geographic location, and the number of cars in the parking lot.

Our generation, reared on Pac Man, 48-station cable TV, and 256k-bit microchips, seems to be stuffing more things in between the strategist or decision maker and the individual. And the gap between how we think people feel and act and their actual beliefs and behaviors appears to be growing steadily wider. There has to be more emphasis on developing simple understandings of our research input—the data generated from interacting with people. Modern-day terminology would label this interaction "garbage in—garbage out," but I prefer the more eloquent description offered by Sir Josiah Stamp, a nineteenth-century official of the British Inland Revenue Department:

"The government are very keen on amassing statistics. They collect them, add them, raise them to the Nth power, take the cube root and prepare wonderful diagrams. But you must never forget that every one of these figures comes in the first instance from the village watchman, who just puts down what he damn pleases."

—Daniel Seymour, "Numbers Don't Lie—Do They?" Reprinted from *Business Horizons*, November–December 1984, pp. 36–37. Copyright 1984 by the Foundation for the School of Business at Indiana University. Used with permission.

AVOIDING PITFALLS

Harass them, harass them,
Make them relinquish the ball!
    —*Cheer at small but illustrious liberal arts college*

An important function of any statistics course is to help educated citizens do a better job of distinguishing between valid and invalid uses of quantitative techniques. Thus, information found in most of the chapters that follow should help you to avoid many of the pitfalls discussed in this chapter. Several later chapters, in fact, contain entire sections that point out the pitfalls and limitations associated with various statistical procedures. But even after you finish this book, you'll find that it isn't always easy to recognize or cope with statistical fallacies. You must, like the team being encouraged with the cheer printed above, remain on the defense to avoid serious statistical blunders. *To avoid pitfalls, you might ask yourself questions when evaluating quantitative information—questions such as:*

◗ *Who is the source of the information you are asked to accept?* Special interests have a way of using statistics to support preconceived positions. Using essentially the same raw data, labor unions might show that corporate profits are very high and thus higher wage demands are reasonable, while the company might make a case to show that profit margins are low and labor productivity is not keeping up with productivity in other industries and in other countries. Also, politicians of opposing parties can use the same government statistics relating to employment, taxation, national debt, welfare spending, budgets, and defense appropriations to draw surprisingly different conclusions to present to voters.

◗ *What evidence is offered by the source in support of the information?* Suspicious methods of data collection and/or presentation should put you on guard. And, of course, you should determine the relevancy of the supporting information to the issue or problem being considered.

◗ *What evidence or information is missing.* What *isn't* made available by the source may be more important than what is supplied. If assumptions about trends, methods of computing percentages or making comparisons, definitions of terms, measures of central tendency and dispersion used, sizes of samples employed, and other important facts are missing, there may be ample cause for skepticism.

◗ *Is the conclusion reasonable?* Have valid statistical facts or statements been used to support the jump to a conclusion that ignores other plausible possibilities? Does the conclusion seem logical and sensible?

LOOKING BACK

1   Many people have uncritically accepted statistical conclusions only to discover later that they've been misled. The aim of this chapter has *not* been to show you how to misapply statistics so that you may better con fellow humans. Rather, the purpose has been to alert you to the possibility of misleading statistical information so that you can do a better job of distinguishing between valid and invalid uses of statistical techniques.

2   Poorly worded and/or slanted questions may be used to gather data, and a "finagle factor" may be used to emphasize facts that support preconceived opinions. You should be aware of the bias that may exist in the information you're asked to accept.

3   The word "average" is a broad term that applies to several measures of central tendency such as the arithmetic mean, the median, or the mode. It's not uncommon for one of these averages to be used where it isn't appropriate, and where it's selected to deliberately mislead.

4   Averages alone usually don't adequately describe a data set. Dispersion—the amount of spread or scatter that occurs in the data—must also be considered. Disregarded dispersions and aggravating averages are usually found acting in concert to confuse and mislead.

5   Statistical tables and charts are prepared to summarize data, uncover relationships, and interpret and communicate numerical facts to those who can use them.

We've considered the legitimate uses of statistical tables, line charts, bar charts, pie charts, combination or mixed charts, and statistical maps in this chapter. The features of computer graphics packages that are used to analyze and present visual images have also been noted. But persuasive artists can easily take honest facts and create misleading impressions, as you've seen in this chapter.

6   Consumers of statistical information must be alert to the *post hoc ergo propter hoc* trap—the argument that because B follows A, B was caused by A. Erroneous cause-and-effect conclusions are often drawn because of the misuse of quantitative facts.

7   Failing to define terms, making improper or illogical comparisons, using an alleged statistical fact to jump to a conclusion, using jargon and lengthy words to cloud a message—all these antics with semantics are used to confuse and mislead.

8   Another way that a person may misuse statistical facts is to assume that because a pattern has developed in the past, that pattern will certainly continue into the future. Such uncritical extrapolation is foolish, but it often occurs.

9   People are often misled or confused because of a bouncing base—that is, the base period used in computing percentages. Several examples of the uses and misuses of percentages are noted in the chapter.

10 Statistical data are often reported in precise numbers, and it's not unusual for several decimal places to be used. This apparent precision lends an air of infallibility to the information reported—particularly when it appears on a computer printout. But as we've seen, the accuracy may be spurious (and curious).

11 To avoid the pitfalls presented in this chapter, you might ask yourself these questions: (1) Who is the source of the information you're asked to accept?, (2) What evidence is offered in support of the information?, (3) What evidence is missing?, and (4) Is the conclusion reasonable?

KEY TERMS AND CONCEPTS

Bias *41*	Bar (column) chart *48*
Arithmetic mean *43*	Pie chart *49*
Median *43*	Statistical maps *51*
Average *43*	Pictographs *51*
Mode *43*	Graphics packages *54*
Dispersion *46*	Analysis package *56*
Statistical tables *46*	Presentation package *56*
Line chart *46*	

TOPICS FOR REVIEW AND DISCUSSION

1   "According to the alumni office, the average Prestige U. graduate, Class of '65, makes $86,123 a year." Comment on this press release.

2   "An independent laboratory test showed that Krinkle Gum toothpaste users report 36 percent fewer cavities." Discuss this advertisement.

3   "Gastro-Dismal elixir has been used by a quarter of a million customers to cure baldness. We have a double-your-money-back guarantee, and only 2 percent of those who used Gastro-Dismal were not helped and asked for a refund." Discuss this advertisement.

4 "There are as many people whose intelligence is above the average as there are people with below-average intelligence." Discuss this statement.

5 "Our firm's income has gone from $5 million to $10 million in just 2 years—an increase of 200 percent." Discuss this statement.

6 "The number of aspirin poisonings makes birth control pills look like one of the safest drugs on the market." Discuss this statement.

7 "Since 66 percent of all rape and murder victims were one-time friends or relatives of their assailants, you are safer at night in a public park with strangers than you are at home." Comment on this remark.

8 "A group of Texas schoolteachers took a history test and failed with an average grade of 60. Thus, Texas schoolteachers are deficient in history." Do you agree?

9 "Last year 760.67 million marijuana cigarettes were smoked in the United States—a flaunting of the law unequaled since Prohibition." Discuss this statement.

10 Why were the polls wrong in 1948 when they predicted that Dewey would defeat Truman for president? (You'll have to do outside research to answer this question.)

11 "The word 'average' is a broad term that applies to several measures of central tendency." Identify and define three such measures.

12 "Averages alone don't adequately describe the true picture in a set of data." Why is this statement true?

13 a What's a statistical table?
    b What's a line chart?
    c What's a bar chart?
    d A pie chart?

14 a How are statistical maps used?
    b What is a pictograph?

15 "Computer graphics packages may be classified into design, paint, analysis, and presentation categories." Discuss this sentence.

16 Give two examples of how persuasive artists can take honest facts and create misleading impressions.

17 "Erroneous cause-and-effect conclusions are often drawn because of the misuse of quantitative methods." Discuss this statement.

18 How may antics with semantics be used to confuse and mislead?

19 Is there any distinction between percentage points and percentage increase? Explain.

20 "Percentage decreases exceeding 100 percent aren't possible if the original data are positive values." Explain why this is true.

21 What questions might you ask during the evaluation of quantitative information to avoid being misled?

PROJECTS/ISSUES TO CONSIDER

1 Robert Miller's column in the business section of the February 3, 1988 issue of *The Dallas Morning News* dealt with the average starting salaries for persons

with MBA degrees. Marketing analyst John K. Reagan was quoted in this column as follows:

Generally, MBA salary reports do not list MBA grads who join entrepreneurial companies. Why? These figures tend to lower the "average starting salary" figure that so many of the schools now judge their contemporaries upon. . . . I speak from personal experience in this matter, being a 1987 graduate of Duke University's Fuqua School of Business, where salaries average $42,000 for the first year out from B-school. I am involved in an import/export start-up here in Dallas, but I was nowhere to be found in Duke's report. . . . For Duke to include a low start-up salary that would depress those figures led them to drop me and a half-dozen other grads in favor of "better numbers" and better PR. Apparently this is the case at many of the Top 20 business schools.

Conduct interviews with business executives and/or business-school officials, get their reaction to Reagan's statement, and present your findings to the class.

2 Students in Professor Larry Schuetz's classes at Linn Benton Community College in Albany, Oregon, conduct "Prove that Claim" projects each term. A student first identifies an advertisement that contains a claim of superiority for a product. The student then writes a letter to a company official (names are obtained from library reference books) asking that person to supply information that will substantiate the claim. A copy of this letter goes to the instructor, and when a reply is received the student discusses the ad and the response in a class presentation. Follow these same steps to carry out your own Prove that Claim research project.

3 Find library (or other) examples of misleading graphics or questionable statistical usage and prepare a report of your findings.

A CLOSER LOOK

THE POWER OF GRAPHICS

Described by federal prosecutors as the leader of the nation's largest crime family, John Gotti was found innocent of conspiracy and racketeering charges after a long trial. The jury had deliberated for a week, reaching no decision. They then asked to see a chart previously introduced by the defense lawyers—a chart that totaled up the extensive criminal records of the seven prosecution witnesses against Mr. Gotti. The jury briefly contemplated the chart and then voted for acquittal on all charges. Such is the power of graphics.

The chart invites reading both horizontally and vertically; neither direction enhances the reputations of those testifying against Mr. Gotti

and his colleagues, as the eye detects patterns and unbroken runs of X's. Mr. Polisi, for example, has something of a streak going. The marks that indicate crimes committed by each witness are not modest or shy, and they dominate the spreadsheet grid (although only 37 percent of all possible combinations are marked). Placement of particularly obnoxious activities at the top (murder, drugs) and near the bottom of the list (pistol-whipping a priest) exploits the visual prominence of those positions.

Such displays are particularly effective and memorable in situations where most information communicated is verbal—a trial, a lecture, a business presentation. Courtroom graphics

Criminal Activity of Government Informants

CRIME	CARDINALE	LOFARO	MALONEY	POLISI	SENATORE	FORONJY	CURRO
Murder	X	X					
Attempted murder		X	X				
Heroin possession and sale	X	X		X			X
Cocaine possession and sale	X		X	X			
Marijuana possession and sale							X
Gambling business		X		X		X	
Armed robberies	X		X	X	X		X
Loansharking		X		X			
Kidnapping			X	X			
Extortion			X	X			
Assault	X		X	X			X
Possession of dangerous weapons	X	X	X	X	X		X
Perjury		X				X	
Counterfeiting					X	X	
Bank robbery			X	X			
Armed hijacking				X	X		
Stolen financial documents			X	X	X		
Tax evasion				X		X	
Burglaries	X	X		X	X		
Bribery		X		X			
Theft: auto, money, other			X	X	X	X	X
Bail jumping and escape			X	X			
Insurance frauds					X	X	
Forgeries				X	X		
Pistol whipping a priest	X						
Sexual assault on minor							X
Reckless endangerment							X

can overcome the linear, nonreversible, one-dimensional sequencing of talk talk talk, by allowing jury members to reason about an array of data at their own pace and in their own manner. Visual displays encourage different individual viewer styles and rates of understanding, editing, personalizing, and reasoning. Unlike speech, *visual displays are both a wide band and a perceiver-controllable channel.*

Note that the display—and Mr. Gotti's fate—stands or falls on the basis of the information itself. Adding color, or other sorts of interior decoration, or even sending the graphic to the chartroom of *USA Today* would not help; indeed, boutique graphics "styling" would reduce the presentation's credibility. After all, a serious decision has to be made on the basis of the evidence; the display must give full attention to evidence, not decoration. Data can only be compromised by the dreaded chartjunk. Worse, chartjunk has often come these days to replace information—just as Jonathan Swift indicted 17th-century cartographers who substituted drawings of animals for geographic knowledge:

With savage pictures fill their gaps
And O'er unhabitable downs
Place elephants for want of towns.

Clearing out chartjunk and turning our focus to the substance of the data brings out the deep and subtle paradox of graphics; an inevitable tension between the complexity of the world and the poverty of our methods for revealing that complexity. Even though we navigate daily through a perceptual world of three spatial dimensions and reason occasionally about still higher-dimensional arenas with mathematical ease, the world portrayed by our information displays is caught up in the two-dimensional poverty of endless flatlands of paper and screen. *Escaping this flatland is the major task of envisioning information,* for all the interesting worlds (imaginary, human, physical, biological) that we seek to understand are inevitably and happily multivariate worlds existing in hyperspace. Not flatlands.

When the toad (*Bufo americanus Le Conte*) sheds its skin during a quarterly molting, the skin leaves life's spaceland and collapses into flatland, much like our information displays.

All sorts of techniques for doing better than flattened-out toad suits have evolved during some 500 years of information design. Since the 15th-century Italian Renaissance, when Florentine architects perfected the necessary geometry, conventional techniques of perspective have enriched representations of physical objects.

For more abstract and richer information not snugly residing in three-space reality, several powerful techniques have evolved, often nearly silently, to be found in the workaday diagrams of those confronted with an overwhelming quantity of data. Some recently perfected statistical graphics enrich flatland with the color dynamics of rotating point clouds on a computer screen—a delight for exploring data from any point of view. . . .

The classic escape from flatland is the map of Charles Joseph Minard (1781–1870), the French engineer, depicting the tragic fate of Napoleon's army in Russia. Seeming to defy the pen of the historian by its brutal eloquence, this combination of data map and time-series plot, drawn in 1861, portrays the devastating losses suffered in Napoleon's Russian campaign of 1812. This is *War and Peace* told by a visual Tolstoy. Beginning at the left, on the Polish-Russian border near the Niemen River, the thick band shows the size of the army (422,000 men) when it invaded Russia in June 1812. The width of the band indicates the size of the army at each place on the map. In September the army reached Moscow, already sacked and deserted, with only 100,000 soldiers surviving.

The path of Napoleon's retreat from Moscow is depicted by the darker, lower band, which is linked to a temperature scale at the bottom of the chart. The winter was bitterly cold, and many froze on the march out of Russia. Crossing the Berezina River was a disaster, and the army finally struggled into Poland with only 10,000 men remaining—one soldier in 42 lived.

Minard's classic tells a rich, coherent story with multivariate data. Six variables are put down onto flatland: size of the army (1), its location on a two-dimensional surface (2, 3), direction of movement (4), and temperature (5) on various dates (6) during the retreat from

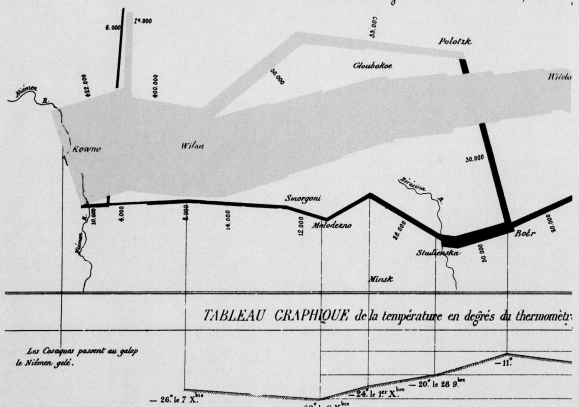

Photograph from Edward R. Tufte, *The Visual Display of Quantitative Information* (Cheshire, Connecticut: Graphics Press, 1983).

Moscow. This may well be among the best statistical graphics ever drawn. Minard did the drawing because *he hated war*; his map was meant as an antiwar poster. Thus, like all good information design, it was driven by an unyielding commitment to the content and substance, not method or technology of display. . . .

In excellent displays such as Napoleon's march . . . the information itself pays no atten-tion to those production or technological distinctions that segregate graphs from text from pictures from maps. It is all information, and our tools should reflect this. Albert Biderman, writing in the *Information Design Journal*, demonstrated that illustrations were once well integrated with text in scientific manuscripts, such as those of Leonardo da Vinci and Sir Isaac Newton, but that graphics became segre-

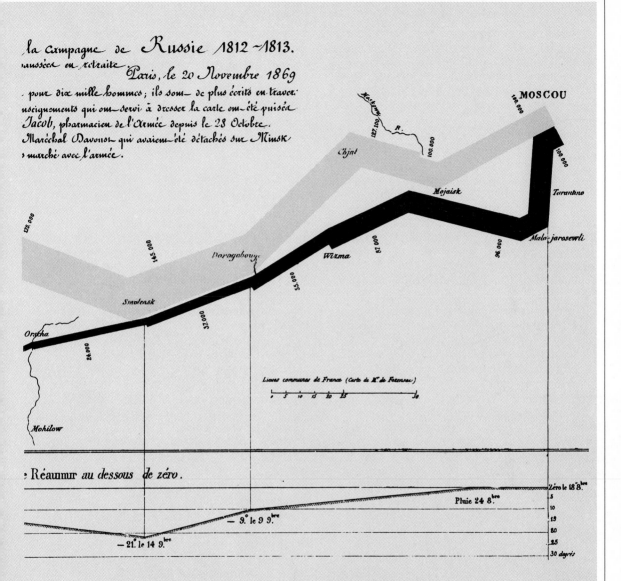

la Campagne de *Russie 1812-1813.*

Paris, le 20 Novembre 1869

MOSCOU

gated from text and table as printing technology developed. Biderman's deep insight is not compromised by the sociological jargon:

> The evolution of graphic methods as an element of the scientific enterprise has been handicapped by their adjunctive, segregated, and marginal position. The exigencies of typography that moved graphics to a segregated position in the printed work have in the past contributed to their intellectual seg-

regation and marginality as well. There was a corresponding organizational segregation, with decisions on graphics often passing out of the hands of the original analyst and communicator into those of graphic specialists— the commercial artists and designers of graphic departments and audio-visual aids shops, for example, whose predilections and skills are usually more those of cosmeticians and merchandisers than of scientific analysts and communicators.

Remedies for these problems—lack of text-figure integration and segregation of the graphics department from the content department—come from the personal computer and from desktop publishing. First of all, such systems allow users to combine different informational elements, including text, graphics, and photographs; second, they bring graphics power to the desks of those who understand the content and substance of the data.

Now it is obvious (although no less important for being obvious) that we are in for a lot of awful amateur design as a consequence of computer graphics. Also, awful professional design: in "makeover" examples in desktop publishing magazines, too often "After" looks worse than "Before"!

Nonetheless, bringing the computer to design and writing is a glorious achievement, a miracle. Nearly everyone should now be writing, drawing, and printing on the computer. Probably the best evidence for making the transition is that few people have gone back to the typewriter and ruling pen once they get a competent personal computer system. And some use language rarely found outside religion to describe the impact of personal computers on their lives.

At the same time, we celebrate classic graphics work, like that of Minard, in order to learn design strategies that effectively exploit the personal computer. Traditional design skill, care, craft, good judgment, and the ability to see wisely are all the more to be valued in the face of powerful—and sometimes empowering—technology.

The major ongoing graphical damage inflicted by computers is *chartjunk*—paraphernalia routinely added to every display that passes by: overbusy grid lines, garish color unrelated to the information, tarted-up three-dimensional representations of one-dimensional data, the debris of computer plotting, vibrating optical art, and ghastly little cartoons. The dithered texture and pattern fills found in drawing programs generate instant chartjunk, vibrating visual activity that has nothing to do with information.

The excuse for decoration is often, "The data is boring and we need to make it come alive." *Well, if the numbers are boring, you've got the wrong numbers.* Decoration won't save the day. Note the hidden assumption here that the audience for graphics is probably somewhat thick-headed, requiring a visual trick to induce them to look at some numbers. We should reject once and for all the quality-corrupting doctrines that numbers are boring and that graphics are for those with short attention spans. These doctrines blame the victims (the data and the audience) rather than the perpetrators. Finally, garish but data-starved graphics reduce the credibility of presentations, for most audiences will have a natural suspicion of hyped graphics. . . .

Chartjunk does not achieve the goals of its propagators. The overwhelming fact of data graphics is that they stand or fall on their content, gracefully displayed. Graphics do not become masterpieces or even attractive or interesting by the addition of ornamental hatching and false perspective to a few bars. Chartjunk can turn bores into disasters, but it can never rescue a thin data set. The best designs (as for Napoleon in Russia) are intriguing and curiosity-provoking, drawing the viewer into the wonder of the information, sometimes by narrative power, sometimes by immense detail, and sometimes by elegant presentation of simple but interesting data. But no information, no sense of discovery, no wonder, no substance is ever generated by chartjunk.

When a graphic is taken over by color-fill patterns, when all the data becomes Design Elements, when it is all style and no information, then that graphic may be called a *duck* in honor of the famous duck-shaped roadside stand, The Big Duck (following the critique of architectural theorists Robert Venturi, Denise Scott Brown, and Steven Izenour). For this building, the whole structure is itself decoration, just as in the duck data graphic.

Many ducks grow from the addition of a fake perspective to data. This variety of chartjunk, now in high fashion, abounds in corporate annual reports, mass media, and muddled academic research. . . .

Still, this is a highly competitive field, and worse ducks may be lurking on hard disks even now.

3

STATISTICAL DESCRIPTION: FREQUENCY DISTRIBUTIONS AND MEASURES OF CENTRAL TENDENCY

Is Upward Mobility Shrinking the Middle Class?

The widespread rumors of the decline of the great American middle class are not exaggerated, conclude economists Michael W. Horrigan and Steven E. Haugen of the Bureau of Labor Statistics in the May issue of the *Monthly Labor Review*. Their analysis indicates that the middle class has indeed been shrinking for two decades. But their most controversial finding is that almost all migration out of the middle class has been upward—toward greater affluence.

That's markedly different from the conclusions of earlier studies that generally showed either that the decline of the middle class reflected a widening of both the upper and lower classes, or that most of the movement was down the income ladder. The researchers note, however, that such findings are highly dependent on the ways in which the studies were designed, and they argue that their own study has the virtue of charting different definitions of the middle class over a number of years.

After evaluating various approaches, Horrigan and Haugen choose to define the middle class as consisting of families with incomes of $20,000 to $56,000 in 1986—that is, any family whose income was roughly between 68% and 190% of the median family income of $29,460. Surveying the 1969–86 period, they then analyze changes in: (1) the percentage of families in that specific real income range (equivalent to $6,680 to $20,100 in 1969 dollars), and (2) the proportion of families between 68% and 190% of annual median family incomes.

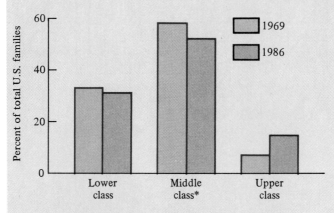

*Middle class is defined as consisting of families with
total pretax incomes of $20,000 to $55,999 in 1986
dollars. Lower and upper classes are families with
incomes below and above this range.

Data: Bureau of Labor Statistics.

More families
graduate to the
upper class.

The striking aspect of the BLS analysis is that both approaches yield similar results. Using the 68%-to-190% interval around the median income to define the middle class, for example, the researchers find that the middle class declined from about 60% of all families in 1969 to 53% in 1986, with 40% more families moving up than down. Using a fixed dollar range, they find that the middle class has shrunk from about 59% to 53% of families, with all of the consequent growth in the upper class (chart above). "The data," says Haugen, "clearly indicate that most of the decline in the middle class has gone to the upper class, while the lower class has remained relatively stable in size."

Although it's welcome news that more families have achieved upper-class status, the picture painted by the BLS study isn't entirely positive. For one thing, it shows that the share of income held by the lower class has declined and that income inequality in America has increased significantly. Moreover, much of the upward mobility may be the result of more women working full time to maintain or augment family living standards. Indeed, the study notes that at the same time the upper class was expanding, the median income of American families rose only slightly in real terms—just 12% over 17 years.

—From Gene Koretz, "Is Upward Mobility Shrinking the Middle Class?" Reprinted from August 15, 1988 issue of *Business Week* by special permission, copyright © 1988 by McGraw-Hill, Inc.

LOOKING AHEAD

The median discussed in the opening vignette is a popular measure of central tendency. It describes and summarizes in a single value a central characteristic of a set of values. You'll recall from Chapter 1 that "descriptive statistics" is a term applied to the procedures of data collection, classification, summarization, and presentation. In this chapter and in Chapter 4, we're concerned with all these aspects of statistical description. Thus, in this chapter we'll (1) briefly consider the collection and organization of raw (ungrouped) data, (2) examine ways to classify (and graphically present) data in a frequency distribution format, (3) survey the popular types of summary measures used to describe frequency distributions, and (4) show the procedures used to compute popular central tendency measures.

Thus, after studying this chapter, you should be able to:

- Explain how to organize raw data into an array and how to construct and interpret a frequency distribution.
- Graphically present frequency distribution data in the form of a histogram, frequency polygon, or stem-and-leaf display.
- Present an overview of the types of measures that summarize and describe the basic properties of frequency distribution data.
- Compute such measures of central tendency as the arithmetic mean, median, and mode for ungrouped data and for data in a frequency distribution.
- Compute a measure used to summarize qualitative data.

CHAPTER OUTLINE

LOOKING AHEAD
INTRODUCTION TO DATA COLLECTION AND ORGANIZATION
    The Raw Data
    The Data Array

FREQUENCY DISTRIBUTIONS
    Classification Considerations

GRAPHIC PRESENTATIONS OF FREQUENCY DISTRIBUTIONS
    The Histogram
    The Frequency Polygon
    A Comparison
    Stem-and-Leaf Displays
    Other Considerations
    Self-Testing Review 3-1

SUMMARY MEASURES OF FREQUENCY DISTRIBUTIONS: AN OVERVIEW
    Measures of Central Tendency
    Measures of Dispersion
    Measure of Skewness
    Measure of Kurtosis

INTRODUCTION TO DATA COLLECTION AND ORGANIZATION

We've already seen in Chapter 1 that existing data may be gathered from internal and external sources, and new original data may be obtained through the use of personal interviews and mail questionnaires. Regardless of the data-gathering methods used, though, all the items in a set of data are collected either by *counting* or by using some *measuring* instrument. The number of postal service mailboxes in a geographic area, for example, is found by counting, while such instruments as the automobile odometer, the service station gas pump, and the bathroom scale provide measurements of distances traveled, gallons of gasoline pumped, and weights of individuals.

The data collected for statistical analysis do not consist of items that are all identical, since there's little reason to study such a situation. Rather, variables are the focus of analytical attention. A **variable** is a characteristic that can vary from item to item in a data set. A variable with a countable or limited number of values is called a **discrete variable.** (The number of children per family in a study is an example of a discrete variable.) And a **continuous variable** is one with an unlimited number of values that may be measured and recorded to some predetermined degree of accuracy. Since measured quantities are continuously variable, they give only approximate results. The pointer on a bathroom scale, for example, might give a reading of 145 pounds. But if the pointer is lengthened and sharpened, and if the scale is calibrated more precisely, the reading might be 144.5 pounds. Further refinements and better instruments might give readings of 144.42 pounds, 144.4234 pounds, and so on.

Assuming that the data have been collected, let's now see how these facts may be organized in a meaningful way. (These steps, along with a summary of the topics covered in this chapter and in Chapter 4, are shown in Figure 3-1.)

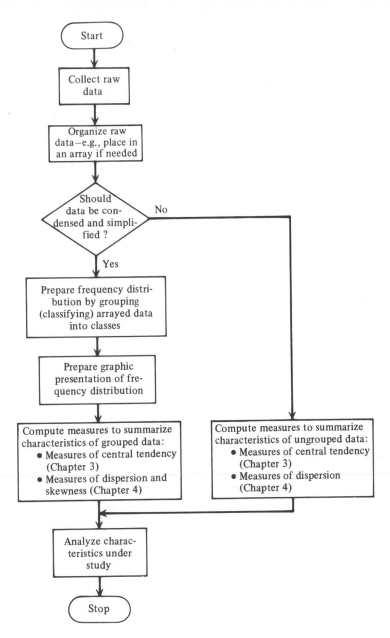

FIGURE 3-1

The Raw Data

A listing of the units produced by each plant worker would likely be of little value to a production manager who's trying to determine overall worker productivity. And a sales manager probably wouldn't learn much about general sales patterns by combing through each sales invoice issued for a particular period. To be meaningful, such unorganized raw data should be arranged in some systematic order.

An example of raw sales data is presented in Figure 3-2. Since we'll be computing various measures to summarize and describe these sales facts later in this chapter and again in Chapter 4, let's make sure we understand this data set. The Slimline Beverage Company makes and sells a line of dietetic soft drink products. These products are sold in bottles and cans. In addition, soft drink syrups are sold in restaurants, theaters, and other outlets that mix small amounts of the syrup with carbonated water and sell the result in paper cups. The sales manager wants to see how a new Fizzy Cola syrup is selling, and so the raw sales data on gallons of syrup sold were gathered as shown in Figure 3-2. This is the GALSOLD data set that was entered into the *Mystat* and *Minitab* statistical packages in Chapter 1 to produce some interesting output, but in its present form this unorganized mass of numbers probably isn't of much value to the manager. And what if there had been 500 values rather than just 50?

FIGURE 3-2 Raw Data (gallons of Fizzy Cola syrup sold by 50 employees of Slimline Beverage Company in 1 month)

EMPLOYEE	GALLONS SOLD	EMPLOYEE	GALLONS SOLD
P.P.	95.00	R.N.	148.00
S.M.	100.75	S.G.	125.25
P.T.	126.00	A.D.	88.50
P.U.	114.00	R.O.	133.25
M.S.	134.25	E.Y.	95.00
F.K.	116.75	Y.O.	104.50
L.Z.	97.50	O.U.	135.00
F.E.	102.25	U.S.	108.25
A.N.	110.00	L.T.	122.50
R.J.	125.00	E.A.	107.25
O.O.	144.00	A.T.	137.00
U.Y.	112.00	R.I.	114.00
T.T.	82.50	N.S.	124.50
G.H.	135.50	I.T.	118.00
R.I.	115.25	N.I.	119.00
O.S.	128.75	G.C.	117.25
U.S.	113.25	A.S.	93.25
P.O.	132.00	N.C.	115.00
O.R.	105.00	Y.A.	116.50
F.T.	118.25	T.N.	99.50
W.O.	121.75	H.B.	106.00
O.F.	109.25	I.E.	103.75
R.T.	136.00	N.F.	115.25
K.H.	124.00	G.U.	128.50
E.I.	91.00	X.N.	105.00

THE NUMERICAL MEASUREMENT SCALES

There are four levels of data measurement: nominal, ordinal, interval, and ratio. Nominal measurement means that the data (a number or symbol) is used for identification purposes only. An example of the nominal level of measurement is a social security number. It is used to identify a specific account; no one number is any better or worse than any other number.

The ordinal level of measurement is used when the data values can be ranked in some order of preference. An example of ordinal measurement is the horse race. Horses crossing an arbitrary finish line can be ranked in the order of their finishing. However, no standard scale of measurement is used. It would be impossible to say, for instance, that the horse that placed first was twice as good as the horse that finished second. Also, it would not matter if the horse coming in second was behind by a nose or by a furlong; it would still be the second-place horse.

The interval level of measurement is reached when the data values can be ranked in order on a standard scale. An example is the measurement of temperature on a thermometer. Each degree on the scale represents a uniform unit of measurement. For example, 5 degrees C is warmer than 1 degree C by four equal degrees. However, there is no absolute scale of measurement used—it is not possible to say that 5 degrees C is five times warmer than 1 degree C.

The ratio level of measurement is attained when the data can be ranked on a standard scale with a true zero point. An example is the measurement of mass in pounds. An item can not weigh less than zero pounds. An item weighing 5 pounds is heavier than an item weighing 1 pound by 4 equal pounds. The first item is also five times heavier than the second item.

—Richard T. Dué, "Predicting Results with Statistics," excerpted from DATAMATION (May 1980, p. 227). © 1980 Cahners Publishing Company.

The Data Array

Perhaps the simplest device for systematically organizing raw data is the **array**—an arrangement of data items in either an ascending (from lowest to highest value) or descending (from highest to lowest value) order. An array of the Fizzy Cola sales data given in Figure 3-2 is presented in Figure 3-3. This array, of course, is in an ascending order. Statistical packages such as *Mystat* and *Minitab* have built-in SORT commands and are easily able to sort an unordered raw data set into an array.

There are several *advantages in arraying raw data:*

- We see in Figure 3-3 that the sales vary from 82.50 to 148.00 gallons, a *range* of 65.50 gallons.
- It's obvious that the lower one-half of the values are distributed between 82.50 and 115.25 gallons, and the upper 50 percent of the values vary between 115.25 and 148.00 gallons.
- An array can show the presence of large concentrations of items at particular values. In Figure 3-3, no single value appears more than twice, but in other arrays there may be a pronounced concentration.

FIGURE 3-3   Data Array (gallons of Fizzy Cola syrup sold by 50 employees of Slimline Beverage Company in 1 month)

GALLONS SOLD	GALLONS SOLD
82.50	115.25
88.50	116.50
91.00	116.75
93.25	117.25
95.00	118.00
95.00	118.25
97.50	119.00
99.50	121.75
100.75	122.50
102.25	124.00
103.75	124.50
104.50	125.00
105.00	125.25
105.00	126.00
106.00	128.50
107.25	128.75
108.25	132.00
109.25	133.25
110.00	134.25
112.00	135.00
113.25	135.50
114.00	136.00
114.00	137.00
115.00	144.00
115.25	148.00

Source: Fig. 3-2.

In spite of these advantages, though, the array is still a rather awkward data organization tool, especially when the number of data items is large. Thus, there often exists the need to arrange the data into a more compact form for analysis and communication purposes.

FREQUENCY DISTRIBUTIONS

The purpose of a frequency distribution is to organize the data items into a more compact form without obscuring the essential information contained in those values. This purpose is achieved by grouping the arrayed data into a relatively small number of classes. Thus, a **frequency distribution** (or **frequency table**) is a tool for grouping data items into classes and then recording the number of items that appear in each class. In Figure 3-4, for example, we've grouped the gallons sold into seven classes and then indicated the number of employees whose sales have turned up in each of the seven

FIGURE 3-4   Frequency Distribution (gallons of Fizzy Cola syrup sold by 50 employees of Slimline Beverage Company in 1 month)

GALLONS SOLD	NUMBER OF EMPLOYEES (FREQUENCIES)
80 and less than 90	2
90 and less than 100	6
100 and less than 110	10
110 and less than 120	14
120 and less than 130	9
130 and less than 140	7
140 and less than 150	2
	50

Source: Fig. 3-3.

classes. (The term "frequency distribution" comes from this frequency of occurrence of values in the various classes.)

You'll notice in Figure 3-4 that the data are now arranged in a more compact form. A quick glance at the frequency distribution shows, for example, that the sales of about two-thirds of the employees ranged from 100 to 130 gallons (the sales of 33 of the 50 employees are distributed in the middle three classes). In short, Figure 3-4 gives us a reasonably good view of the overall sales *pattern* of Fizzy Cola syrup. Of course, the reduction or compression of the data has resulted in some loss of detailed information. We no longer know, for example, exactly how many gallons each employee sold. And we don't know from Figure 3-4 that the values have a range or spread of exactly 65.50 gallons. All we know about these matters is that there are two employees in the first class, for example, whose sales were somewhere between 80 and less than 90 gallons and that the range of values is going be somewhere between 50 and 70 gallons. On balance, though, the advantage of gaining new insight into the data patterns that may exist through the use of a frequency distribution can often outweigh this inevitable loss of detail.

To construct a frequency distribution, it's necessary to determine (1) the number of classes that will be used to group the data, (2) the width of these classes, and (3) the number of observations—or the **frequency**—in each class. In the next section we'll look at the first two interrelated considerations. The last step is a routine transfer of information from an array to a distribution, and so we'll not consider it here.

Classification Considerations

It's usually desirable to consider the following basic rules or criteria when creating a frequency distribution:

1   In formal presentations, the *number of classes* used to group the data generally varies from a minimum of 5 to a maximum of 18. The actual number of classes used depends on such factors as the number of observations being grouped, the purpose of the distribution, and the arbitrary preferences of the analyst. One could group the data in Figure 3-4 into many classes, with each class having a small width. Such a distribution can be useful for preliminary analysis. In fact the Closer Look reading at the end of this chapter gives an example of how a computer package can quickly create distributions with many classes and then show dotplots of those distributions to help people grasp the characteristics of the data. A **dotplot** shows the scale of the many classes used, and a dot is placed on the chart to show each item in the distribution. Of course, a distribution with many small classes is likely to contain too much detail to be used in a formal data presentation. And at the other extreme, a grouping of the data in Figure 3-4 into only three classes with intervals of 22 gallons each would result in the loss of needed detail.

2   Classes must be selected to *conform to two rules:* (*a*) both the smallest and largest data items must be included, and (*b*) each item must be assigned to one *and only one* class. Possible gaps and/or overlaps between successive classes that could cause this second rule to be violated must be avoided.

3   Whenever possible, the *width* of each class—that is, the **class interval**—should be *equal.* (It's also often desirable to use class intervals that are multiples of numbers such as 5, 10, 100, 1,000, and so on.) Although unequal class intervals may be needed in frequency distributions where large gaps exist in the data, such intervals may cause difficulties. For example, if frequencies in a distribution with unequal intervals are compared, the observed variations may merely be related to interval sizes rather than to some underlying pattern. Other difficulties of using unequal intervals can arise during the preparation of graphs. Our Figure 3-4 has arbitrarily been prepared with seven classes of equal size. How was the interval width of 10 gallons determined? You ask very perceptive questions. The following simple formula was used to estimate the necessary interval:

$$i = \frac{L - S}{c}$$

where i = width of the class intervals
L = value of the largest item
S = value of the smallest item
c = number of classes

Of course, as we've seen in Figure 3-3, the Fizzy Cola sales data range from a low of 82.50 gallons to a high of 148.00 gallons. Thus,

$$i = \frac{148.00 - 82.50}{7}$$

$$= \frac{65.60}{7}$$

= 9.36, a value close to the convenient class interval size of 10 gallons used in Figure 3-4

4 Whenever possible an **open-ended class interval**—one with an unspecified upper or lower class limit—should be avoided. Figure 3-5 has such an interval, and so it's an example of an **open-ended distribution.** An open-ended class may be needed when a few values are extremely large or small in comparison with the remainder of the more concentrated observations, or when confidential information might be revealed by stating an upper limit. For example, placing an upper limit on the data in Figure 3-5 might tend to reveal the income of an easily identifiable family in a small community. But open-ended classes should be used as sparingly as possible because of graphing problems and because (as we'll soon see) it's impossible to compute such important descriptive measures as the arithmetic mean and the standard deviation from an open-ended distribution.

5 When there's a concentration of raw data around certain values, it's desirable to construct the distribution in such a way that these points of concentration fall at the **class midpoint** or middle of a class interval. (The reason for this will become apparent later when we compute the arithmetic mean for data found in a frequency table.) In Figure 3-4, the midpoint of the class "110 and less than 120" is 115 gallons, the lower limit of that class is 110 gallons, and the upper limit is 119.999 gallons. Of course, another analyst could gather additional raw sales data for Fizzy Cola syrup and could then round the sales to the *nearest* gallon. This analyst might then set up a frequency distribution similar to Figure 3-4 with class intervals of 80 to 89, 90 to 99, 100 to 109, 110 to 119, and so

FIGURE 3-5   Open-Ended Distribution (total income reported by selected families)

TOTAL INCOME	NUMBER OF FAMILIES
Under $10,000	6
$10,000 and under $20,000	14
20,000 and under 30,000	18
30,000 and under 40,000	10
40,000 and under 50,000	5
50,000 and under 60,000	4
60,000 and over	3
	60

on. In this case, the *stated* limits are only 9 gallons apart, but the size of these class intervals is still 10 gallons. Why? Because the class "110 to 119" has a real lower limit or *lower boundary* of 109.5 and a real upper limit or *upper boundary* of 119.5. A **class boundary** is thus a number that doesn't appear in the stated class limits but is rather a value that falls midway between the upper limit of one class and the lower limit of the next larger one. In our example, the class interval still has a width of 10 gallons, but the class midpoint in this case is 114.5 gallons.

Once the data are grouped into a more compact form, the frequency distribution can be used for analysis, interpretation, and communication purposes. It's often possible to prepare a graphic presentation of a frequency distribution to achieve one or more of these aims. How can graphic presentations of frequency distributions be prepared, you eagerly ask? Well it just so happens. . . .

GRAPHIC PRESENTATIONS OF FREQUENCY DISTRIBUTIONS

In the absence of the raw data, a frequency distribution such as the one shown in Figure 3-4 is needed for reference purposes and to compute descriptive measures that summarize certain characteristics of the values. But a graphic presentation of the data found in a frequency table is more likely to get the attention of the casual observer and show trends or relationships that might be overlooked in a table. Let's look now at several graphic presentation techniques.

The Histogram

A **histogram** is a bar graph of a frequency distribution. Figure 3-6 is a histogram of the Fizzy Cola syrup sales data found in the table in Figure 3-4. As you can see, this histogram simply consists of a set of vertical bars. Values of the variable being studied—in this case gallons of syrup sold—are

FIGURE 3-6
Histogram of frequency distribution of gallons of Fizzy Cola syrup sold by 50 employees of Slimline Beverage Company in 1 month.

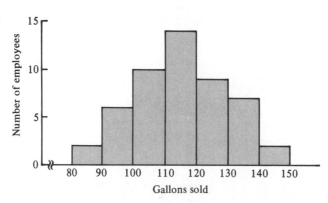

measured on an arithmetic scale on the horizontal axis. The bars in Figure 3-6 are of equal width and correspond to the equal class intervals in Figure 3-4; the height of each bar in Figure 3-6 corresponds to the frequency of the class it represents. Thus, the area of a bar above each class interval is proportional to the frequencies represented in that class.[1]

Most statistical software packages are programmed to produce histograms. A user merely enters a data set and then keys an appropriate command or menu response to tell the program what it should do. You saw in Chapter 1 how the gallons of Fizzy Cola syrup sold by the 50 Slimline employees was entered into the *Mystat* and *Minitab* packages and stored under the GALSOLD name. Now we can use a HISTOGRAM command with each of these packages to produce graphic results.

Figure 3-7*a* shows the *Mystat* **default graph**—one that's automatically produced for a data set by the software without any formatting instructions from the user. You'll notice that the *program has created* more classes than the seven shown in Figure 3-6. But a chart that resembles Figure 3-6 can be produced by *Mystat* with some format instructions from the user (Figure 3-7*b*). You can see that the *Mystat* histograms have three marks [(, ^ , and)] on the horizontal "gallons sold" scale. The ^ mark shows the location of the arithmetic mean in the distribution. Note that the mean is a "balance point"—that is, the histogram is in balance when supported at the mean. And about the middle two-thirds of the values in the histogram fall between the parentheses [( )] on the base line.

Figure 3-8*a* shows the *Minitab* default chart. You'll notice that the "bars" of the *Minitab* graph are asterisks that run horizontally rather than vertically. You can also see that the program has created eight rather than seven classes and has used class midpoints of 80, 90, . . . rather than 85, 95, ... To produce the same results as those shown in Figure 3-6, the *Minitab* user can modify the default graph with the two format subcommands shown in Figure 3-8*b*.

The Frequency Polygon

A **frequency polygon** is a line chart of a frequency distribution and is thus an alternative form of graphic presentation. Figure 3-9 is a frequency polygon using the same data and plotted on the same scales as the histogram in Figure 3-6. (In fact, Figure 3-6 has been lightly reproduced as background in Figure 3-9.) As you can see, points are placed at the midpoints of each class interval. The height of each plotted point in Figure 3-9, of course, represents the frequency of the particular class. These points are then connected by a series of straight lines. It's customary to close the polygon at both ends by (1) placing points on the baseline half a class interval to the left of the first class and half a class interval to the right of the last class,

---

[1] If unequal class intervals were used in a frequency distribution, the *areas of the bars above the various class intervals would still have to be proportional to the frequencies represented in the classes*—e.g., if the third interval is twice as wide as each of the first two, the frequency of the third interval must be divided by two to get the appropriate height for the bar.

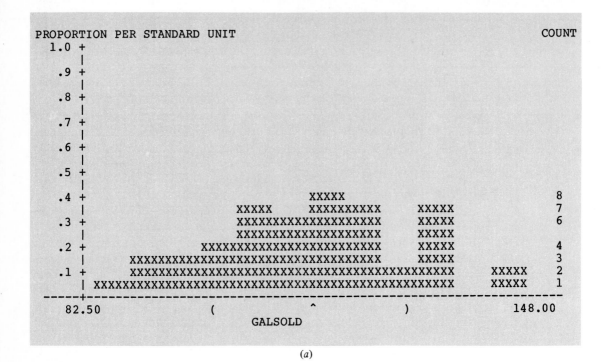

```
PROPORTION PER STANDARD UNIT                                      COUNT
 1.0 +
     |
  .9 +
     |
  .8 +
     |
  .7 +
     |
  .6 +
     |
  .5 +
     |
  .4 +                            XXXXX                               8
     |                XXXXX       XXXXXXXXX        XXXXX               7
  .3 +                XXXXXXXXXXXXXXXXXXXX         XXXXX               6
     |                XXXXXXXXXXXXXXXXXXXX         XXXXX
  .2 +            XXXXXXXXXXXXXXXXXXXXXXXX         XXXXX               4
     |        XXXXXXXXXXXXXXXXXXXXXXXXXXXXXX       XXXXX               3
  .1 +        XXXXXXXXXXXXXXXXXXXXXXXXXXXXXXXXXXXX        XXXXX        2
     |    XXXXXXXXXXXXXXXXXXXXXXXXXXXXXXXXXXXXXXXXXXXX    XXXXX        1
-----+---------------------------------------------------------------
    82.50               (            ^            )          148.00
                              GALSOLD
```

(a)

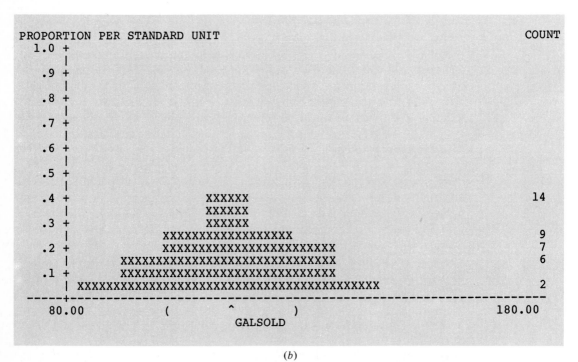

```
PROPORTION PER STANDARD UNIT                                      COUNT
 1.0 +
     |
  .9 +
     |
  .8 +
     |
  .7 +
     |
  .6 +
     |
  .5 +
     |
  .4 +              XXXXXX                                           14
     |              XXXXXX
  .3 +              XXXXXX
     |          XXXXXXXXXXXXXXXX                                      9
  .2 +          XXXXXXXXXXXXXXXXXXXXXXXX                              7
     |      XXXXXXXXXXXXXXXXXXXXXXXXXXXXXX                            6
  .1 +      XXXXXXXXXXXXXXXXXXXXXXXXXXXX                              2
     |  XXXXXXXXXXXXXXXXXXXXXXXXXXXXXXXXXXXXXX
-----+---------------------------------------------------------------
    80.00               (          ^          )             180.00
                              GALSOLD
```

(b)

FIGURE 3-7   Histograms produced by the *Mystat* statistical software package.

```
MTB > HISTOGRAM 'GALSOLD'

Histogram of GALSOLD   N = 50

Midpoint    Count
      80      1   *
      90      3   ***
     100      8   ********
     110     11   **********
     120     13   *************
     130      8   ********
     140      5   *****
     150      1   *
```

(a)

```
MTB > HISTOGRAM 'GALSOLD';
SUBC> INCREMENT=10;
SUBC> START=85.

Histogram of GALSOLD   N = 50

Midpoint    Count
    85.0      2   **
    95.0      6   ******
   105.0     10   **********
   115.0     14   **************
   125.0      9   *********
   135.0      7   *******
   145.0      2   **
```

FIGURE 3-8
Histograms
produced by the
Minitab statistical
software package.

(b)

then (2) drawing lines from the points representing the frequencies in the first and last classes to these baseline points (see Figure 3-9).

A Comparison

The *advantages of histograms over frequency polygons* are:

▸ Each individual class is represented by a bar which stands out clearly.
▸ The area of a bar represents the exact number of frequencies in a class interval.

However, the *frequency polygon also possesses certain advantages:*

▸ It's simpler and has fewer lines than a histogram, so it's especially suitable for making comparisons of two or more frequency distributions.

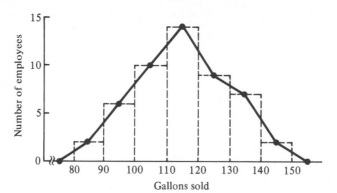

FIGURE 3-9
Frequency polygon
of the distribution
of gallons of Fizzy
Cola syrup sold by
50 employees of
Slimline Beverage
Company in 1
month.

◆ If the class intervals in a distribution are continuously reduced in size, and if the number of items in the distribution is continually increased, the frequency polygon will resemble a smooth curve more and more closely. Thus, if the frequency polygon in Figure 3-9 represented only a small *sample* of all the available data on Fizzy Cola syrup sales made by hundreds of employees, and if the frequency distribution—that is, the *population* frequency distribution—that could be prepared to account for all these data were made up of very narrow class intervals, the resulting population distribution curve might resemble the one shown in Figure 3-10. This bell-shaped or **normal curve**, which describes the distribution of many kinds of variables in the physical sciences, social sciences, medicine, agriculture, business, and engineering, is very important in statistics and will be reintroduced in many later chapters.

Stem-and-Leaf Displays

Exploratory data analysis (EDA) is a term that refers to several relatively new techniques that analysts can quickly use to get a feel for the data being studied. Developed by John Tukey, EDA techniques can be used be-

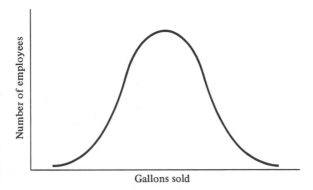

FIGURE 3-10
Generalized
normal population
distribution.

fore (or in place of) more traditional analysis approaches. A **stem-and-leaf display** is an example of an EDA tool; it lists the items in a data set to create a plot that looks like a histogram, but this plot includes the actual data and takes less time to prepare. The data array of the Fizzy Cola sales figures found in Figure 3-3 is used in Figure 3-11 to produce a stem-and-leaf display. A **stem** is a number to the left of the vertical line in Figure 3-11, and it represents the leading digit(s) of all data items that are listed on the same row. A **leaf** is a *single number* to the right of the vertical line that represents the trailing digit of a value. Thus, in Figure 3-11 there are 50 leaf values representing the 50 items in our data set. The first stem value in the top row of Figure 3-11 is 8, and the first leaf value is 2. These numbers are combined to give a sales figure of 82 gallons. (The lowest sales value in Figure 3-3 is 82.50 gallons, but fractional units are dropped in Figure 3-11.) The second leaf figure in the top row is used with its stem to produce a sales figure of 88 gallons. In the second row, the stem of 9 is combined with each of the leaves in that row to get six sales figures ranging from 91 to 99 gallons. The third-row stem of 10, when combined with leaves of 0, 2, 3, . . . produces data items of 100, 102, 103, . . . , and so it goes throughout the display. The appearance of our stem-and-leaf plot would have looked just the same if we had used the raw sales data from Figure 3-2 rather than the arrayed values from Figure 3-3. But, of course, our leaf values wouldn't have been in sequence.

Stem-and-leaf plots are easy to prepare, but it's even easier to turn the task over to a computer. Figure 3-12*a* shows how the *Mystat* program responds to a STEM command to process our GALSOLD data, and Figure 3-12*b* shows the output produced by *Minitab* when it executes its STEM-AND-LEAF command. You'll notice that both software packages go beyond the hand-drawn display shown in Figure 3-11. Both packages list each stem on two lines. Leaf values of 0 through 4 are printed on the first of the two lines, and leaves 5 through 9 appear on the second. If raw data (rather than arrayed values) had been supplied, the packages would have sorted the leaves on each line in an ascending order prior to printing.

The *Mystat* package also accompanies its display with several descrip-

FIGURE 3-11   A Stem-and-Leaf Display of the Slimline Beverage Company Sales Figures Found in Figure 3-3

STEM	LEAF VALUES	NUMBER OF DATA ITEMS
8	28	2
9	135579	6
10	0234556789	10
11	02344555667889	14
12	124455688	9
13	2345567	7
14	48	2

```
        STEM AND LEAF PLOT OF VARIABLE:   GALSOLD     , N =     50


MINIMUM IS:         82.500
LOWER HINGE IS:        105.000
MEDIAN IS:         115.250
UPPER HINGE IS:        125.250
MAXIMUM IS:        148.000


              8     2
              8     8
              9    13
              9    5579
             10    0234
             10 H 556789
             11    02344
             11 M 555667889
             12    1244
             12 H 55688
             13    234
             13    5567
             14    4
             14    8
```

(a)

```
        MTB > STEM-AND-LEAF 'GALSOLD'

        Stem-and-leaf of GALSOLD    N  = 50
        Leaf Unit = 1.0


              1      8 2
              2      8 8
              4      9 13
              8      9 5579
             12     10 0234
             18     10 556789
             23     11 02344
             (9)    11 555667889
             18     12 1244
             14     12 55688
              9     13 234
              6     13 5567
              2     14 4
              1     14 8
```

(b)

FIGURE 3-12   (a) Stem-and-leaf display produced by *Mystat* package. (b) Similar plot produced by *Minitab*.

tive measures (see Figure 3-12a). The *minimum* and *maximum* values in the data set are listed, the value of the middle item in the array—the *median*—is printed, and the lower and upper "hinges" are given. One-fourth of the data items fall below the *lower hinge* of 105.00 gallons, and one-fourth fall above the *upper hinge* of 125.25 gallons. The H, M, and H symbols between the stems and leaves in Figure 3-12a indicate the rows in the display that contain the lower and upper hinge and median values.

The first column (to the left of the stems) in the *Minitab* program in Figure 3-12b is also added to supply supporting information. This column begins at the top by accumulating the number of data values that have been accounted for in each line. There's one item in row 1, a *total* of two items in rows 1 and 2, four values through row 3, and 23 items are accounted for in the top 7 rows. The parentheses [( )] in the next row indicate that we've arrived at the line that holds the median value, and the number in the parentheses gives a count of the leaves on the line. Since the program tells us that there are 50 items in the data set, and since we've accounted for 23 of them in the first 7 rows, then the median is one of the early values in row 8 (about 115 in this case). After the median row is reached, the figures in column 1 of Figure 3-12b then show how many items remain on that line and the ones below it in the data set.

Other Considerations

It's sometimes useful to determine the number of data items that fall above or below a certain value rather than within a given interval. In such cases, a regular frequency distribution may be converted to a **cumulative frequency distribution**—one that adds the number of frequencies as shown in Figure 3-13. As you can see, we've merely arranged the data from Figure 3-4 in a different form. The eight employees who sold less than 100 gal-

FIGURE 3-13   Cumulative Frequency Distribution (gallons of Fizzy Cola syrup sold by 50 employees of Slimline Beverage Company in 1 month)

GALLONS SOLD	NUMBER OF EMPLOYEES
Less than 80	0
Less than 90	2
Less than 100	8
Less than 110	18
Less than 120	32
Less than 130	41
Less than 140	48
Less than 150	50

lons, for example, are the two who sold less than 90 plus the six in the class of "90 and less than 100" gallons. An **ogive** (pronounced "oh jive") is a graphic presentation of a cumulative frequency distribution. The ogive for Figure 3-13 is shown in Figure 3-14, and each point represents the number of employees having sales of less than the gallons indicated on the horizontal scale. By adding a percentage scale to the right of the ogive,[2] it's possible to graphically obtain the summary measures given in Figure 3-12*a*. For example, if, as shown in Figure 3-14, we draw a line from the 50 percent point on the percentage scale over to where it intersects with the ogive line, and if we then draw a perpendicular line from this intersection to the horizontal scale, we're able to read the median, the approximate amount of syrup sold by the middle employee in the arrayed group of 50. Drawing similar lines from the 25 and 75 percent points locates the lower and upper hinges.

Self-Testing Review 3-1

This is the first in a series of self-testing review sections that appear from time to time throughout most of the rest of the book. You're encouraged to pause here to test your understanding of the concepts that have just been presented. *The answers to self-testing review questions are found at the end of the chapter in which they appear.*

---

[2] Since each employee represents 2 percent of the total ($\frac{1}{50} \times 100$), we simply double the scale on the vertical axis to obtain our percentage scale.

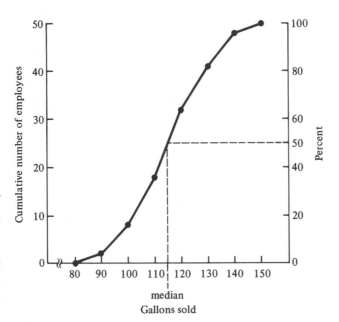

FIGURE 3-14
Ogive for the distribution of gallons of Fizzy Cola syrup sold by 50 employees of Slimline Beverage Company in 1 month.

Brock and Parse Lee, owners of the Lee Produce Company, are studying the size of the orders placed by customers in an outlying county. In the past week, the following 30 orders have been received:

$42.50	$45.00	$47.75	$52.10	$29.00	$31.25
21.50	56.30	55.60	49.80	35.55	42.30
43.50	34.60	65.50	45.10	40.25	58.00
30.30	44.80	36.50	55.00	59.20	36.60
38.50	41.10	46.00	39.95	25.35	49.50

1 a Arrange the above data in an ascending array.
   b What is the range of values?
2 Organize the data items according to order size into a frequency distribution having the classes "$20 and under $30," "$30 and under $40," . . . , and "$60 and under $70."
3 a Would it have been possible to have used six or seven classes rather than five classes in the above frequency distribution?
   b What would have been a reasonable class interval or width if you had prepared a frequency distribution using eight classes rather than five?
4 Draw a histogram of the frequency distribution prepared in problem 2 above.
5 Draw a frequency polygon of the frequency distribution prepared in problem 2 above.
6 Prepare a stem-and-leaf display using the arrayed data developed for part 1a above.

SUMMARY MEASURES OF FREQUENCY DISTRIBUTIONS: AN OVERVIEW

We'll soon be computing values that summarize and describe the basic characteristics of frequency distribution data. Before we begin these computations, though, let's pause here for a brief discussion and graphic overview (using Figure 3-15) of the summary measures we'll encounter.

Measures of Central Tendency

You know that the word "average" applies to several measures of central tendency. The purpose of these averages is to summarize in a single value the typical size, middle property, or central location of a set of values. The most familiar average is, of course, the *arithmetic mean,* which is simply the sum of the values of a group of items divided by the number of such items. But you also saw in Chapter 2 that the *median* and *mode* are other measures of central tendency that are commonly used. Figure 3-15 shows some possibilities that could exist in different data sets. Suppose in Figure 3-15*a* that we have the monthly sales distributions of two Slimline products—Opulent Orange (OO) and Ribald Root Beer (RR). Although the spread of the sales data in each distribution looks the same, it's obvious that the average sales of root beer are greater than the average sales of the

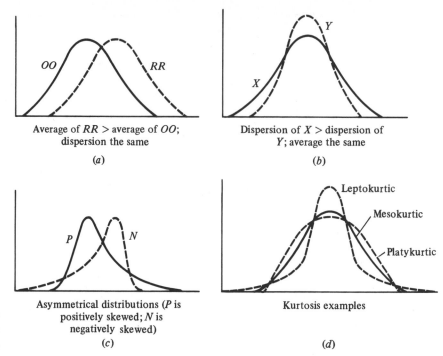

Average of RR > average of OO;
dispersion the same

(a)

Dispersion of X > dispersion of
Y; average the same

(b)

FIGURE 3-15
Summary measures
of frequency
distributions.

Asymmetrical distributions (P is
positively skewed; N is
negatively skewed)

(c)

Kurtosis examples

(d)

orange beverage; that is, the root beer sales are concentrated around a higher value than the orange sales.

Measures of Dispersion

What if two data sets have the same average? Does this mean that there's no difference in these sets? Perhaps, but then again, perhaps not. In Figure 3-15b, distributions X and Y would have the same average, but they're certainly not identical. The difference lies in the amount of spread, scatter, or dispersion of the values in each distribution as measured along the horizontal axis. Obviously, the dispersion in distribution X is greater than the spread of the values in distribution Y. One simple measure of dispersion is the *range*—the difference between the highest and lowest values. Other measures that we'll consider in Chapter 4 are the *average deviation, standard deviation,* and *quartile deviation*. But they all do the same thing: They measure the extent of the dispersion in the data.

Measure of Skewness

The data sets in Figures 3-15a and b have **symmetrical distributions:** If you draw a perpendicular line from the peaks of these curves to the baseline, you'll divide the area of the curves into two *equal* parts. As you can see in Figure 3-15c, however, curves may be skewed rather than symmetrical. A skewed (or asymmetrical) **distribution** occurs when a few values are much larger or smaller than the typical values found in the data set. For example, distribution *P* in Figure 3-15*c* might be the curve resulting from Professor Nastie's first statistics test. Most of the test scores are concentrated around the lower values, but a few curve breakers made extremely high grades. When the extreme values tail off to the *right* (as in distribution *P*), the curve represents a **positively skewed distribution.** Distribution *N* in Figure 3-15*c*, on the other hand, might be a curve of the test scores obtained by Professor Sweet's statistics students. As you can see, most of the students made high scores (although a few unfortunates had extremely low grades). When extreme values tail off to the *left* (as in distribution *N*), the curve shows a **negatively skewed distribution.** A measure of the extent to which a distribution departs from the symmetrical is presented in Chapter 4.

Measure of Kurtosis

He learns. He becomes educated. . . .
He instigorates knowledge.
                —The House at Pooh Corner

It's possible that three distributions (see Figure 3-15*d*) may be symmetrical, may have the same average value, and may have the same dispersion value (standard deviation). Yet the distributions may still possess different degrees of peakedness or **kurtosis.** It's beyond the scope of this book to consider the matter of kurtosis any further because little attention is paid to the subject in ordinary statistical analysis. Suffice it to say that a *mesokurtic* curve is generally normally peaked, with moderate-length tails; a *leptokurtic* curve is usually more peaked than normal, with lengthy tails; and a *platykurtic* curve is likely to be squat, with short tails. On the very remote chance that you should wish to remember these outrageous names, you might use the system of William S. Gosset, a famous early statistician who wrote: "I myself bear in mind the meaning of the words by [a] *memoria technica*, where the first figure represents platypus, and the second kangaroos, noted for 'lepping,' though, perhaps, with equal reason they should be hares!" (Figure 3-16 is a reproduction of a sketch supplied by Gosset.)

COMPUTING MEASURES OF CENTRAL TENDENCY

Data often have a tendency to congregate about some central value, and this central value may then be used as a summary measure to describe the general data pattern. If the collected facts are to be processed by noncomputer methods and are limited in number, an analyst may prefer to work directly with the **ungrouped data**—facts that haven't been put in a distribution format—rather than group them together in a frequency table. Likewise, if a computer is used, very large lists of ungrouped data can be

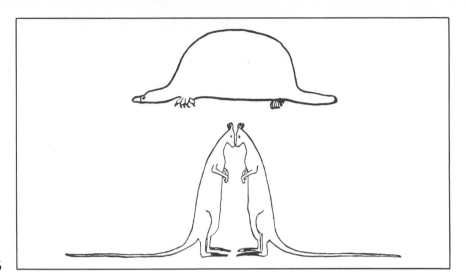

FIGURE 3-16

easily processed in a few seconds or minutes without any need for a frequency distribution. But if we're using secondary data that have been compressed into a frequency table to make them more easily understood, we should be able to compute the desired measures from the data in that format.

In the following sections we'll first compute measures of central tendency using *ungrouped data*, and then we'll look at methods of computing the same measures when the *data are grouped* into a frequency table.

Ungrouped Data

The Arithmetic Mean   When most people use the word "average," they're referring to the arithmetic mean. And when you have totaled the test grades you've made in a subject during a school term and divided by the number of tests taken, you've computed the *arithmetic mean*. The arithmetic mean is the most commonly used average.

Let's review the computation of the mean by considering the statistics grades made by Peter Parker[3] during one agonizing semester. (The grades have been *arrayed* in a descending order.)

75
75
61
50

---

[3] A name selected in memory of another loser, Sir Peter Parker, the British naval commander during the Revolutionary Battle of Sullivan's Island outside Charleston, South Carolina. During this battle, while giving orders aboard *HMS Bristol*, Sir Peter had the "unspeakable mortification" to have a cannon ball carry away the seat of his pants. (According to an old ballad, it "propelled him along on his bumpus.")

```
 40
 25
 10
  5
  1
───
342 Total of all grades
```

It's customary to let the capital letter X represent the values of variables (such as Peter's grades). Thus, the tough formula to compute the *mean for ungrouped data* is:

$$\mu = \frac{\Sigma X}{N} \qquad \text{(formula 3-1)}$$

where μ (the Greek letter mu) = arithmetic mean
Σ (the Greek letter sigma) = "the sum of"
N = number of X items in the listing

Since, in the case of Peter's grades, ΣX is 342, Peter's mean semester grade is 38 ($342/N$ or $342/9 = 38$).

A note about the symbols used in our formulas is in order here. The symbols used in this arithmetic mean formula, and in the other formulas presented in this chapter and Chapter 4, are appropriate when the data represent *all* the values in a *population.* Thus, the symbol μ refers to a population mean, and the symbol N refers to the total number of items in a population. But as you'll see in the first section of Chapter 6 (and in Figure 6-1), if the measures are being computed from *sample* data, different symbols must be used to maintain a distinction between population values and sample values.

The Median   You've seem that the *median* is a measure of central tendency that occupies the *middle position* in an *array* of values. Note that the word "array" has been emphasized; it's necessary to put the data into an ascending or descending order before selecting the median value. In the example of Peter's grades, the middle value in the array, and thus the median grade, is 40. [The median *position*—not the median value—is found by using the formula $(N + 1)/2$, where N in this example is 9. Thus, $(9 + 1)/2$ is 5, which is the median position in Peter's grade array.] Although the median in this example deviates by a small amount from the mean, the ultimate result of using either the mean or the median as the semester average grade is the same—Peter doesn't have a clue about the general subject of statistics and has flagged the course. As we saw in the example of farm income in the last chapter, however, one or a few extremely high (or extremely low) values in a series can cause a substantial difference between the mean and the median.

What if Peter's instructor had dropped Peter's lowest grade before computing the median? In that event, the middle position in Peter's grade array would have been midway between 40 and 50—i.e., the median value would then have been 45 (a change that doesn't do a thing for Peter's final grade).

The Mode   The *mode*, by definition, is the *most commonly occurring value* in a series. Thus, in the example of Peter's grades, the mode is 75—an average that appeals to Peter but not to his professor. Although not of much use in our grade example, the mode may be an important measure to a clothing manufacturer who must decide how many dresses of each size should be made. Obviously, the manufacturer will want to produce more dresses in the most commonly purchased size than in the other sizes.

Self-Testing Review 3-2

1   The parts in this question all pertain to the following ungrouped data:

(i)	(ii)
8	5
10	4
10	7
10	5
12	2
16	12
18	11
84	6
	52

   a   Determine the mean, median, and mode for the data in column *i*.
   b   Determine the mean, median, and mode for the data in column *ii*.
   c   If the data in column *i* are representative of the sizes of dresses sold during a typical day in a store, which measure of central tendency might be of most interest to the store's dress buyer?

2   Using the data array produced in Self-Testing Review 3-1, question 1a, determine the following measures:
   a   Arithmetic mean
   b   Median
   c   Mode

3   Peter Parker's instructor uses a computer to average student grades. One of Peter's classmates has a semester arithmetic mean grade of 84. The instructor has inadvertently deleted one of the student's grades from the computer storage medium. The remaining test scores for this student are 92, 76, 66, 95, 80, 79, 89, and 85. What's the value of the missing test score?

4   The number of faulty assemblies detected by a quality control inspector during a 15-day period was 57, 34, 39, 32, 35, 40, 27, 34, 29, 18, 20, 32, 25, 60, and 38.
   a   What's the mean number of faulty assemblies for the period?
   b   Construct a stem-and-leaf display of the data set.
   c   Locate the median and mode from your stem-and-leaf plot.

Grouped Data

Keep two points in mind as you study this section on how to compute averages for data grouped in a frequency distribution. The *first* point concerns the effort needed to process large data sets. Before computers became commonplace, it took less computational effort to approximate descriptive measures from frequency distributions than to compute the same measures from ungrouped data. Now, of course, computers can easily process huge lists of ungrouped data, a fact that has eliminated the computational advantages of frequency distributions. The *second* point to remember is that the values of the mean, median, and mode, when computed from a frequency distribution, *are not exact but are only approximations.* Certain assumptions (outlined later) are required in the computations, and the validity of these assumptions in any given problem determines the accuracy of the results. Of course, if you're using grouped data supplied by others (such as employees of a government agency), and the raw data are unavailable, you have no alternative but to compute approximations from a frequency table.

Let's now use the Slimline Company data found in Figure 3-4 to demonstrate computational procedures for grouped data.

The Arithmetic Mean Computing the arithmetic mean for a frequency distribution is similar to computing the mean for ungrouped data. But since the compression of data in a frequency table results in the loss of the actual values of the observations in each class in the frequency column, it's necessary to make an assumption about these values. The assumption (or estimate) is that *every observation in a class has a value equal to the class midpoint.* Thus, in Figure 3-17, it's assumed that the two employees (f) in the first class each sold 85 gallons (m) of Fizzy Cola syrup, giving a total of 170 gallons (fm) sold. Of course, we have the advantage of knowing from

FIGURE 3-17   Computation of Arithmetic Mean (gallons of Fizzy Cola syrup sold by 50 employees of Slimline Beverage Company in 1 month)

GALLONS SOLD	NUMBER OF EMPLOYEES (f)	CLASS MIDPOINTS (m)	fm
80 and less than 90	2	85	170
90 and less than 100	6	95	570
100 and less than 110	10	105	1,050
110 and less than 120	14	115	1,610
120 and less than 130	9	125	1,125
130 and less than 140	7	135	945
140 and less than 150	2	145	290
	$N = \Sigma f = 50$		5,760

$$\mu = \frac{\Sigma fm}{N} = \frac{5,760}{50} = \underline{\underline{115.2}} \text{ gallons sold}$$

Figure 3-3 that neither employee sold 85 gallons, but their actual total sales of 171 gallons is only 1 gallon over our estimate. And although our assumption in this first class has caused us to *underestimate* the true figure, it's quite possible that a similar error occurring in another class may cause us to *overestimate* that amount. Therefore, throughout a properly constructed distribution, the effect may be that most of these errors will cancel out. For example, we've slightly overstated the sales of the seven employees in the sixth class because it's assumed that they sold a total of 945 gallons (see the *fm* column in Figure 3-17). In fact, their sales in Figure 3-3 amounted to 943 gallons.

The computation of the mean of 115.2 gallons is shown in Figure 3-17. As you'll notice, the *formula for computing the mean for grouped data is:*

$$\mu = \frac{\Sigma fm}{N} \qquad\qquad\qquad \text{(formula 3-2)}$$

where f = frequency or number of observations in a class
  m = midpoint of a class and the assumed value of every observation in the class
  N = total number of frequencies or observations in the distribution

How does our estimate of 115.2 gallons compare to the true mean sales of the 50 Slimline employees? The *Mystat* program gives us the answer in Figure 3-18. Although the *Mystat* STATS command can generate a number of descriptive values, we've only asked for a few here. You can see that our estimate of 115.2 is close to the true mean of 115.4 gallons (5,770 gallons/50 = 115.4 gallons).

In the discussion of classification considerations a few pages earlier, you saw that (1) open-ended classes should be avoided if possible, and (2) points of data concentration should fall at the midpoint of a class interval. Perhaps the reasons for these comments may now be clarified. *First*, the uses of open-ended distributions are limited because it's impossible to compute the arithmetic mean from such distributions. Why is this true?

FIGURE 3-18
The *Mystat* program shows us that the true mean of the Fizzy Cola sales data is 115.4 gallons.

```
            TOTAL OBSERVATIONS:      50

                                 GALSOLD

            N OF CASES                     50
            MINIMUM                    82.500
            MAXIMUM                   148.000
            MEAN                      115.400
            SUM                      5770.000
```

AMAZING AMERICAN AVERAGES

"Average," when you stop to think of it, is a funny concept. Although it describes all of us, it describes none of us.

No one considers himself ordinary or run-of-the-mill. But while none of us wants to *be* the average American, we all want to know about him, or her. Averages define the national character, etch the routines of everyday life. They give us a benchmark to measure ourselves against. Do we earn, owe or work more than the average? Do we jog, eat, loaf or sweat less?

The average American man is 5 feet, 9 inches tall; the average woman, 5 feet, 3.6 inches. . . .

The average American is sick in bed seven days a year, missing five days of work.

The average woman is more inclined to be knock-kneed than the average man.

The average American high-school graduate recognizes about 15,000 words when reading or listening; the college graduate recognizes about 30,000.

The average body has over two million sweat glands and two square yards of skin.

On the average, the nonsmoking wives of smokers die four years younger than the nonsmoking wives of nonsmokers.

The average American female is 28-percent fat. The average American male is 15-percent fat.

The average Southerner has higher blood pressure than the average Yankee or Westerner.

Every generation of Americans has been taller than its parents by about an inch—except for the present one, which shows no change.

The average American man washes his hair more often than the average woman does. . . .

The average American nudist is 35 years old and married.

For every white-collar worker on the payroll, the typical business has 18,000 pieces of paper on file.

By the time he or she graduates from high school, the average American has watched 350,000 commercials and 18,000 murders on television. . . .

Thirty-two million Americans—nearly 15 percent—believe in astrology.

The average American worker gets about $4000 worth of fringe benefits a year. And of every 100 families, 47 get some form of monthly payment from Uncle Sam.

Nine of every ten shoppers enter a supermarket in a good mood, but only three of four are in a good mood when they leave.

On an average day, 24 mailmen receive animal bites.

Given a choice, four men in five take showers rather than baths. Among women, half would rather bathe, half shower.

The average husband in America is just a shade over 45 years old, and the average wife is just shy of 42. . . .

Three of every four working Americans get to work in less than half an hour, and only one in 20 commutes longer than an hour.

One of every three American workers feels overeducated for his or her job.

By his or her 70th birthday, the average American will have eaten 14 steers, 1050 chickens, 3.5 lambs and 25.2 hogs.

And finally, did you know that about half the people in America are below average?

Not *you*. Not *us*.

Them.

Because, as you can see in Figure 3-5, we cannot make any assumption about the income of each of the three families in the "$60,000 and over" class. Since there's no upper limit in this class, there's no midpoint value that we can assign to represent the total income for each of the three families. And *second*, if the raw data values are concentrated at the lower or upper limits of several classes rather than at the class midpoints, the assumption that we've made to compute the approximate value of the mean is incorrect and can lead to distorted results. For example, if the raw data are concentrated around the lower limits of several classes, the computed mean can overstate the true mean by a significant amount.

The Median   As we've seen, the median is the value that occupies the middle position in an array of values. However, since the actual values of a data set are lost when a distribution is constructed, it's only possible to approximate the median value from grouped data. We can illustrate this approximation process by referring to Figure 3-19. Figure 3-19*a* shows what we are looking for when we compute the median for our example problem. We might assume from the data given in the distribution in Figure 3-19*b* that the employee whose sales quantity was lowest (call him employee 1) sold approximately 80 gallons, and we might guess that the highest-selling employee (call her employee 50) sold approximately 149.9 gallons. What we are looking for, however, is the approximate quantity sold by the middle (or twenty-fifth) employee.

The *first* step in computing this median value is to locate the **median class**—the class in our example which contains the twenty-fifth or median worker in the group of 50 employees. Although this is a simple matter, the optional *cumulative frequencies* column in Figure 3-15*b* may be useful. As you can see, the 18 employees in the first three classes sold less than 110 gallons, and the 32 employees in the first four classes sold less than 120 gallons. Therefore, the twenty-fifth employee must be one of the 14 in the fourth or median class. If this approach seems familiar to you, that's not surprising. The *Minitab* stem-and-leaf display that we've examined in Figure 3-12*b* has a column that does the same thing.

The *next* step is to see which of the 14 is the median employee. (You've probably already figured it out, but for the sake of others let's explain this step.) If 18 have been accounted for in the first three classes, and if we're looking for number 25, that employee must be the *seventh* one in the group of 14 in the fourth class (that is, $25 - 14 = 7$). In other words, in our example it just happens that the median employee is found seven-fourteenths or one-half of the way through the median class. Of course, this is just a coincidence. The median observation could have been anywhere in the median class.

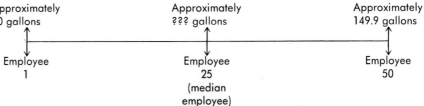

FIGURE 3-19b   Computation of the Median (gallons of Fizzy Cola syrup sold by 50 employees of Slimline Beverage Company in 1 month)

GALLONS SOLD	NUMBER OF EMPLOYEES (f)	CUMULATIVE FREQUENCIES (CF)
80 and less than 90	2	2
90 and less than 100	6	8
100 and less than 110	10	18
110 and less than 120*	14	32
120 and less than 130	9	41
130 and less than 140	7	48
140 and less than 150	2	50
	$\overline{50}$	

$$Md = L_{Md} + \left(\frac{N/2 - CF}{f_{Md}} \right)(i)$$

$$= 110 + \left(\frac{50/2 - 18}{14} \right)(10)$$

$$= 110 + \left(\frac{7}{14} \right)(10)$$

$$= 110 + 5$$

$$= \underline{\underline{115.0}} \text{ gallons}$$

*Median class.

The *final* step is to compute the median value by interpolating within the median class. To perform this step, it's assumed that the sales of the 14 employees in the median class are *evenly distributed throughout the class*. This assumption of an even distribution of values throughout the median class is seldom likely to be exactly correct, and so the computed median is only an approximation. But given this assumption, it's easy to see that if our median employee happens to be the middle one in the class, his or her sales should be halfway through the class interval. In short, the median value should be 115 gallons sold by the twenty-fifth employee.

The formula for computing the median given below is simply a formal presentation of the preceding paragraphs.

$$Md = L_{Md} + \left(\frac{N/2 - CF}{f_{Md}} \right)(i) \qquad \text{(formula 3-3)}$$

where Md = median
   L_{Md} = lower limit of the median class
   N = total number of frequencies in the distribution
   CF = cumulative frequencies in all classes up to, but not including, the median class
   f_{Md} = frequency of the median class
   i = size of the interval of the median class

The computation using formula 3-3 is shown in Figure 3-19*b*. You can verify in Figure 3-3, and in the output of the *Mystat* stem-and-leaf display in Figure 3-12*a*, that the true median is 115.25; therefore, our approximation of 115.0 gallons is quite close.

The Mode   The mode is, by definition, the most commonly occurring value. But since actual data values are unknown in frequency tables, the mode must also be approximated. It's assumed that the most commonly occurring value in a frequency distribution is found in the largest class and directly under the peak of a frequency polygon. Thus, the class in the distribution with the largest number of frequencies is the **modal class.** Figure 3-20 demonstrates how a mode may be *graphically* approximated from a histogram, and the following formula shows how a mode may be approximated in mathematical terms:

$$Mo = L_{Mo} + \left(\frac{d_1}{d_1 + d_2} \right)(i) \qquad \text{(formula 3-4)}$$

where Mo = mode
     L_{Mo} = lower limit of the modal class
     d_1 = difference between the frequency of the modal class and the frequency of the class immediately preceding it in the distribution
     d_2 = difference between the frequency of the modal class and the frequency of the class immediately following it in the distribution
     i = size of the interval of the modal class

Using the data from the example problem in Figure 3-19*b*, we can calculate the mode as follows:

$$Mo = 110 + \left(\frac{4}{4 + 5} \right)(10)$$

$$= 110 + \frac{40}{9}$$

$$= 110 + 4.44$$

$$= 114.44 \text{ gallons sold}$$

Other Measures   In addition to the mean, median, and mode, there are other specialized measures of central tendency that are occasionally used. The **weighted arithmetic mean,** for example, is a modification of the measure we have been computing that assigns *weights* or indications of *relative importance* to the values to be averaged. Thus, if you get grades of 83 and 87 on hourly statistics exams and a grade of 95 on the final, if the hourly exams each carry a weight of 25 percent of your semester grade, and if the final counts 50 percent of your grade, your weighted mean or semester average will be

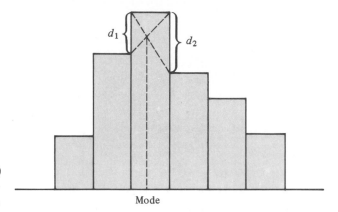

FIGURE 3-20
Graphic represen-
tation of the mode.

$$\frac{83\,(25) + 87\,(25) + 95\,(50)}{100} = 90 \qquad \text{(congratulations!)}$$

In addition to the weighted arithmetic mean, there are two other lesser means. Although it's beyond the scope of this book to go into the computation of these measures, they should be used in special situations. The *geometric mean,* for example, is the measure that should be employed to average ratios or rates of change expressed in positive numbers, and the *harmonic mean* is a measure that should be used to average time rates under certain conditions.

Summary of Comparative Characteristics

There's no general rule that will always identify the proper measure of central tendency to use. In a perfectly *symmetrical distribution,* the issue of which average to use is simplified because the arithmetic mean, median, and mode have the *same value* (see Figure 3-21*a*). But if the data produce a skewed distribution, the values of the three measures are different. In a *positively skewed* distribution, for example, the *mode* remains under the peak of the curve and has the smallest value; the *mean,* influenced by the extremely large values, is pulled out from under the peak of the distribution in the direction of those extreme values, and has the largest value; and the *median* lies between the mode and the mean (see Figure 3-21*b*). In a *negatively skewed* distribution, the *mode* has the largest value and is still found under the peak of the curve; the *mean* has the smallest value because extremely small data items have been used in its computation; and, as always, the *median* lies between the mode and the mean (see Figure 3-21*c*).[4]

---

[4] In fact, it has been observed that in moderately skewed continuous distributions, the median will be located approximately two-thirds of the distance toward the mean from the mode. This approximation can be useful in checking the reasonableness of computed values of the three measures. In our Slimline example problem, the distance between *(Cont. on p. 110)*

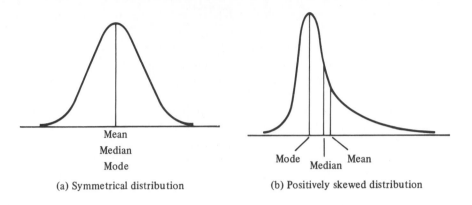

Mean
Median
Mode

(a) Symmetrical distribution

Mode   Mean
   Median

(b) Positively skewed distribution

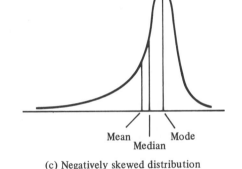

Mean   Mode
     Median

(c) Negatively skewed distribution

FIGURE 3-21
Typical locations of
the principal
measures of central
tendency.

In selecting the proper average to use, the characteristics of each measure must be considered, and the type of data available must be evaluated.
A summary of comparative characteristics for each measure follow.

The Arithmetic Mean   Some of the more important characteristics of the
mean are:

1 *It's the most familiar and most widely used measure.* Long explanations of its meaning are thus not usually required.
2 *It's a computed measure whose value is affected by the value of every
   observation.* A change in the value of any observation will change the
   mean value; however, the mean value may not be the same as any of
   the observation values.
3 *Its value may be distorted too much by a relatively few extreme
   values.* Because it is affected by all the values of the variable, the mean

(Cont. from p. 109) the mean of 115.2 gallons and the mode of 114.44 gallons is 0.76 gallon.
If we add two-thirds of this 0.76 gallon to the mode of 114.44, we arrive at an approximation
of the median of 114.95, which verifies the reasonableness of the median computation of
115.0 gallons.

(as we saw in Chapter 2) can lose its representative quality in badly skewed distributions.

4   *It cannot be computed from an open-ended distribution in the absence of additional information.* This point has been discussed earlier.

5   *It's the most reliable average to use when sample data are being used to make inferences about populations.* As we will see in Chapters 6 and 7, the mean of a sample of observations taken from a population may be used to estimate the value of the population mean.

6   *It possesses two mathematical properties that will prove to be important in subsequent chapters.* The *first* of these properties is that the sum of the differences between data items and the mean of those items will be zero—that is, $\Sigma\,(X - \mu) = 0$. And the *second* property is that if these differences between data items and the mean are *squared*, the sum of the squared deviations will be a *minimum value*—or $\Sigma\,(X - \mu)^2 =$ minimum value. To illustrate these properties suppose we have the following observations: 2, 3, 4, 7, and 9. The mean of these items is 25/5 or 5. These two properties of the mean may now be demonstrated as follows:

X	$(X - \mu)$	$(X - \mu)^2$
2	−3	9
3	−2	4
4	−1	1
7	+2	4
9	+4	16
25	0	34

As you can see, $\Sigma\,(X - \mu)$ *must* equal zero (this is really a definition of the mean); and $\Sigma\,(X - \mu)^2$ in this case is 34. If we were to use any other value in place of the true mean of 5 in our example and follow through with the same procedure of determining and then squaring the deviations of the actual values from this other value, the result would be a total *greater than* our minimum value of 34. If, for example, we substituted the number 4 for our true mean of 5 and follow through with the procedure of determining and then squaring differences, $\Sigma\,(X - 4)$ would be 39. Of course, there isn't anything profound to note about the number 34 in our original example; a different data set would almost certainly produce a minimum value of some other number.

The Median   Some of the important characteristics of the median are:

1   *It's easy to define and easy to understand.* The computation and interpretation of the median, as we have seen, is not difficult.

2   *It's affected by the number of observations, but not by the value of these observations.* Thus, extremely high or low values will not distort the median.

3 *It's frequently used in badly skewed distributions.* The median will not be affected by the size of the values of extreme items, and so it's a better choice than the mean when a distribution is badly skewed.

4 *It may be computed in an open-ended distribution.* Since the median value is located in the median class interval, and since that interval is virtually certain of not being open-ended, the median may be determined.

5 *It's generally less reliable than the mean for statistical inference purposes.* In the statistical inference chapters in Part 2 we will use the mean exclusively as the measure of central tendency.

The Mode   Some of the characteristics of the mode are:

1 *It's generally a less popular measure than the mean or median.*

2 *It may not exist in some sets of data, or there may be more than one mode in other sets of grouped data.* A distribution with two peaks—a **bimodal distribution**—should probably be reclassified into more than one distribution.

3 *It can be located in an open-ended distribution.*

4 *It's not affected by extreme values in a distribution.*

Self-Testing Review 3-3

1 Using the frequency distribution for Lee Produce Company orders that you constructed in Self-Testing Review 3-1, question 2, determine the following measures:
   a   The arithmetic mean
   b   The median
   c   The mode

2 Compare the mean and median for the Lee Produce Company data computed in question 2a and b of Self-Testing Review 3-2 with the same measures computed in problem 1a and b above. How do you account for the differences?

3   a   From the information obtained in question 1 above, is the distribution for the Lee Produce Company negatively or positively skewed?
   b   Why?

SUMMARIZING QUALITATIVE DATA

Our purpose in this chapter has been to look at the *quantitative* measures of central tendency that are used to summarize data. Before concluding this chapter, though, we should briefly mention the key measure used to summarize the relative frequency with which a particular characteristic occurs. This measure, used to summarize *qualitative* data, is the *percentage* (or *proportion*).

If, for example, a machine produces 250 parts and a quality control check shows that 21 of the parts are defective, the percentage of defective parts is (21/250) (100) or 8.4 percent. This percentage can be used in an analytical approach (to be discussed in Part 2) to give a production manager help in deciding if corrective action is needed. And if a candidate receives 540 votes in a poll of 1,200 voters before an election—that is, the candidate gets (540/1,200) (100) or 45 percent of the poll votes—the candidate's campaign manager can use this qualitative summary measure for analysis and planning purposes, as you'll see in Part 2. Similarly, a television executive whose program is watched in 252 of the 900 homes surveyed (that is, in 28 percent of the homes) at the time the program is aired can also use this percentage result to make future programming decisions. We'll summarize qualitative data as percentages rather than proportions in this book because percentages are more frequently used in everyday discussion. A proportion, of course, is obtained simply by moving the percentage decimal point two places to the left. Thus 28 percent is .28 in proportion terms.

LOOKING BACK

1   The data collected for statistical analysis are discrete or continuous variables. A discrete variable has a countable or limited number of values, while a continuous variable is one that may be measured to some predetermined degree of accuracy.

2   Statistical description begins with the collection of raw data. These unorganized facts may then be arranged in some systematic order. One simple device for organizing raw data is the data array—an arrangement of data items in either an ascending or descending order.

3   A frequency distribution is a tool for grouping data items into classes and then recording the number of items that appear in each class. The compression of data into a frequency table results in some loss of detailed information. For example, the individual values in a data set can no longer be identified. But the advantage of gaining new insight into the data patterns that may exist through the use of a distribution may outweigh this loss of detail.

4   It's usually desirable to consider several basic rules when creating a frequency distribution. For example, the number of classes used to group the data generally varies from a minimum of 5 to a maximum (for formal presentations) of 18. Classes must be selected so that the smallest and largest data items are included in the table, and each item must be assigned to only one class. Whenever possible, class intervals should have equal widths, and open-ended classes should be avoided. The distribution should be designed so that any data concentration points fall at the class midpoints.

5   A graphic presentation of the data found in a frequency distribution can show trends or relationships that might be overlooked in a table, and such graphs are more likely to get the attention of casual observers. A histogram is a bar graph of a frequency distribution. Statistical software packages are programmed to produce histograms of user-supplied data sets, and default and modified histograms produced by the *Mystat* and *Minitab* programs are shown in this chapter. A frequency polygon is a line chart of a frequency distribution, and is thus an alternative

to the histogram. The stem-and-leaf display, an exploratory data analysis tool, looks like a histogram, but the plot includes the actual data and takes less time to prepare. Examples of the stem-and-leaf displays created by *Mystat* and *Minitab* are shown in the chapter. A standard frequency table can be used to produce a cumulative frequency distribution.

6 The basic properties or characteristics of frequency distribution data may be summarized and described by measures of central tendency, dispersion, skewness, and kurtosis. In this chapter we've concentrated on the computation of measures of central tendency such as the mean, median, and mode for both ungrouped and grouped data. Before computers became commonplace, it took less effort to approximate descriptive measures from frequency distributions than to compute the same measures from ungrouped data. Now, of course, that's no longer the case. And since the values of the mean, median, and mode, when computed from a distribution, are only estimates, many people now prefer to work with the raw data and a computer. Still, there are times when frequency tables have been supplied by others and the raw data are unavailable. In that case, the analyst has no alternative but to compute approximations from the tables.

7 In selecting the proper average to use, the characteristics of each measure must be considered, and the type of data available must be evaluated. The mean, for example, is the most familiar measure, is affected by the value of every observation, can be distorted by extreme values, cannot be found from an open-ended distribution, and possesses important mathematical properties. The median is easy to understand, isn't affected by extreme values, can be used in badly skewed distributions, and may be computed in open-ended distributions. The mode can also be located in open-ended tables, and it's not affected by extreme values, but it's not as popular as the mean or median, and it may not exist in some data sets.

8 A measure used to summarize the relative frequency with which a particular characteristic occurs is the percentage or proportion. This qualitative measure is often used for statistical inference purposes, as we'll see in Part 2.

KEY TERMS AND CONCEPTS

Median class   *106*

$$Md = L_{Md} + \left(\frac{N/2 - CF}{f_{Md}}\right) (i) \quad \text{(formula 3-3)} \quad 107$$

Modal class   *108*

$$Mo = L_{Mo} + \left(\frac{d_1}{d_1 + d_2}\right) (i) \quad \text{(formula 3-4)} \quad 108$$

Weighted arithmetic mean   *108*

Biomodal distribution   *112*

PROBLEMS

1   The following scores were made by Professor Shirley A. Meany's accounting students on a test:

68	52	49	56	69
74	41	59	79	81
42	57	60	88	87
47	65	55	68	65
50	78	61	90	85
65	66	72	63	95

    a   Arrange the above grades in an ascending array.
    b   What is the range of the values?
    c   Compute the arithmetic mean, median, and mode for these grade values.

2   a   Organize the data items from problem 1 above into a frequency distribution having the classes "40 and under 50," "50 and under 60," . . . , and "90 and under 100."
    b   Now place the data in a stem-and-leaf display.
    c   Compute the mean, the median, and the mode for this frequency distribution.
    d   Compare the values obtained in problem 2b with the values found in problem 1c and explain any discrepancies.
    e   Is the distribution of test scores positively or negatively skewed? Explain your answer.

3   The following distribution gives the miles traveled by 100 Lee Produce Company route trucks for the year 198x.

MILES TRAVELED	NUMBER OF TRUCKS
5,000 and under 7,000	5
7,000 and under 9,000	10
9,000 and under 11,000	12
11,000 and under 13,000	20
13,000 and under 15,000	24
15,000 and under 17,000	14
17,000 and under 19,000	11
19,000 and under 21,000	4

    a   Construct a histogram and a frequency polygon of this mileage distribution.
    b   Construct an ogive, and graphically locate the value of the median.
    c   Compute the mean.
    d   Compute the median and interpret its meaning.
    e   Compute the mode and interpret its meaning.
    f   Is the distribution skewed? Explain your answer.

4 The following data represent the annual earnings of families in two small New England communities, Languor and Friskyville:

	NUMBER OF FAMILIES	
ANNUAL EARNINGS	LANGUOR	FRISKYVILLE
$5,000 and under $8,000	5	2
8,000 and under 11,000	40	20
11,000 and under 14,000	73	32
14,000 and under 17,000	52	58
17,000 and under 20,000	22	35
20,000 and under 23,000	8	30
23,000 and under 26,000	—	15
26,000 and over	—	8
	200	200

a Compute the arithmetic mean annual earnings for each community.
b Compute the median annual earnings for each community.
c Compute the modal annual earnings for each community.
d Now that you have summarized the distributions, compare and analyze the economic situations in these two communities.

Note: If it is not possible to perform all the operations in any or all of the above parts, please indicate your reasons for omitting the operation(s).

5 In order to prepare a government report, a university must determine the percentage of men and women faculty members in its several schools and colleges. The faculty data are as follows:

SCHOOL / COLLEGE	A	B	C	D
Men	148	64	12	102
Women	32	42	26	48

What is the percentage of women faculty members in each school or college?

6 In his last game, the kicker for Gridiron University had punts of 42, 38, 51, 48, and 45 yards. Compute the arithmetic mean and median.

7 The following distributions show the salaries paid to male and female employees by Chauvinists, Inc.

	NUMBER OF EMPLOYEES	
ANNUAL SALARIES	MALE	FEMALE
$6,000 and under $9,000	2	10
9,000 and under 12,000	5	25
12,000 and under 15,000	20	9
15,000 and under 18,000	18	6
18,000 and under 21,000	15	3
21,000 and under 24,000	8	2
24,000 and under 27,000	5	0
27,000 and under 30,000	2	0
	75	55

a   Compute the arithmetic mean for each distribution.

b   Compute the median for each distribution.

c   Compute the mode for each distribution.

8   The following distribution gives the monthly number of warranty repair jobs per month handled by an automobile dealiership:

WARRANTY REPAIR JOBS PER MONTH	NUMBER OF MONTHS
95 and under 105	4
105 and under 115	14
115 and under 125	18
125 and under 135	15
135 and under 145	10
145 and under 155	7
155 and under 165	2

a   Compute the mean monthly number of repair jobs.

b   Compute the median monthly number of repair jobs.

9   The following scores were made by Professor Wendell Nastie's students on a recent statistics test:

TEST SCORES	NUMBER OF STUDENTS
40 and under 50	4
50 and under 60	6
60 and under 70	10
70 and under 80	4
80 and under 90	4
90 and under 100	2

a   Compute the mean test score.

b   Compute the median test score.

10 As a class project, a group of commuting students attending an evening course at a local college decided to approximate the average distance traveled to class. The following data were gathered:

MILES TRAVELED	NUMBER OF STUDENTS
0 and under 2	2
2 and under 4	5
4 and under 6	4
6 and under 8	8
8 and under 10	1

a   What was the mean distance traveled?

b   What was the median distance traveled?

11 The miles traveled by six of the students attending an evening class at a local college are given below:

STUDENT	MILES TRAVELED
A	1
B	4
C	9
D	8
E	5
F	6

a   What was the mean distance traveled?
b   What was the median distance traveled?
c   What was the modal distance traveled?

12   The following data have been gathered by a nurse at the headquarters of a retailing organization. The data show the weights of headquarters managers in 1979 and 1989.

MANAGER	WEIGHTS (IN POUNDS) 1979	1989
A	145	156
B	130	151
C	159	161
D	120	147
E	160	175
F	180	196
G	140	141
H	180	187
I	151	155
J	155	201

a   What's the arithmetic mean weight for both years?
b   What's the median weight for both years?
c   What's the modal weight for both years?

13   In five pro baseball seasons, "Doc" Oglesvee had the following batting averages:

.227   .294   .301   .303   .297

a   What was Doc's overall arithmetic mean batting average for the five seasons?
b   What was Doc's median batting average for the five seasons?

14   Greg's test scores for two semesters of calculus are listed below. The percentage of each semester's grade represented by each score is also given.

1st SEMESTER	2d SEMESTER	% OF GRADE
72	87	15
68	66	15
74	71	15
68	69	15
90	88	40

a   Compute the weighted arithmetic mean for each semester.

b   Is Greg improving? Explain your answer.

15 To set a size limit on bass, Krimpee Lake Wildlife Council took the following 50 specimens from the lake:

SIZE (IN INCHES)	NUMBER CAUGHT
10 but less than 12	6
12 but less than 14	14
14 but less than 16	17
16 but less than 18	8
18 but less than 20	5
	50

a   Compute the arithmetic mean size.

b   What is the median size?

c   Compute the mode.

16 Reverend Droner, wishing to learn if the men or women of his flock were more pious, kept track of individual attendances at each of the 50 sermons he gave last year. The attendance records revealed the following information:

NO. OF SERMONS ATTENDED	MEN	WOMEN
1 to 10	4	1
11 to 20	7	5
21 to 30	13	16
31 to 40	20	21
41 to 50	3	9
	47	52

a   Compute the arithmetic mean for each distribution.

b   Compute the median for each distribution.

c   Compute the mode for each distribution.

d   What conclusion can you reach from your analysis?

17 Below are the figures for absences in Mr. Dry's botany class:

DAYS ABSENT	NO. OF STUDENTS
1 to 3	9
4 to 6	6
7 to 9	13
10 to 12	9
13 to 15	1

a   Compute the mean numbers of days absent.

b   Compute the median number of days absent.

18 Figgy's Ice Cream Company operates a fleet of vending trucks. The sales figures of each truck for a recent day were:

SALES IN DOLLARS	NO. OF TRUCKS
300 < 350	1
350 < 400	2
400 < 450	7
450 < 500	12
500 < 550	13
550 < 600	9
600 < 650	2

a   Compute the mean sales for the day.
b   Compute the modal sales for the day.
c   What was the median sales figure for the day?

19 The traffic tickets issued by police in Crossville in a one-week period are shown below:

NO. OF TICKETS	POLICE OFFICERS
0 < 5	2
5 < 10	6
10 < 15	10
15 < 20	4
20 < 25	1

a   What is the mean number of tickets written weekly?
b   What's the median number of weekly tickets?
c   What's the modal number of weekly tickets?

20 The number of traffic tickets issued for a period by five Crossville police officers is given below:

NAME	NO. OF TICKETS
R. Oldman	16
A. Trapper	9
L. Perez	10
J. Ketchum	8
F. Wheeler	5

a   What is the median number of tickets written?
b   What is the mean number of tickets written?

21 After six holes, 25 golfers at the Duffers International Tournament had the following scores:

| 71 | 68 | 85 | 96 | 12 | 92 | 37 | 41 | 54 | 25 | 66 | 15 | 73 |
| 23 | 14 | 55 | 65 | 43 | 88 | 92 | 19 | 22 | 51 | 62 | 84 |

a   Array the scores in an ascending array.
b   What is the range of the scores?
c   What is the mean and median for the above scores?

22 Using the data from problem 21 above:
   a Organize the scores into a stem-and-leaf display using nine classes.
   b Determine the mean using the frequency distribution produced in the stem-and-leaf display.
   c How does the mean calculated in part b above compare with the true mean produced in problem 21c?

23 A study is conducted to determine the mean income of a population of salespersons. The study concludes that for the population of 100 salespersons the mean income is $33,000. It is discovered later that the income of the last person in the group was incorrectly reported to be $20,000 when it should have been $50,000. What's the true mean income of the population?

TOPICS FOR REVIEW AND DISCUSSION

1 Distinguish between a discrete variable and a continuous variable.
2 "Since measured quantities are continuously variable, they give only approximate results." Discuss this comment.
3 What are the possible advantages to be obtained from arraying raw data?
4 a What's the purpose of a frequency distribution?
   b Give one advantage of a frequency distribution.
   c Outline one disadvantage of a frequency table.
5 What basic rules or criteria should be observed in constructing a frequency distribution?
6 "Whenever possible, the width of the classes should be equal." Discuss this statement.
7 "Open-ended classes should be avoided whenever possible." Why is this statement true?
8 Why should data-concentration points be placed at the midpoint of a class interval?
9 a What is a histogram?
   b How can computers be used to produce these charts?
10 a What is a frequency polygon?
    b What are the advantages and disadvantages of frequency polygons when compared to histograms?
11 "A stem-and-leaf display is an example of an exploratory data analysis tool that lists the items in a data set to create a plot that looks like a histogram." Discuss this statement.
12 a What is a stem in a stem-and-leaf display?
    b What is a leaf?
    c How many leaves would there be in the display of a data set with 200 values?
13 a What's a cumulative frequency distribution?
    b How can an ogive be used to locate the median?
14 Discuss the types of summary measures that may be used to describe the characteristics of frequency distributions.
15 a What basic assumption is needed to approximate the arithmetic mean from grouped data?

    b What assumption is needed to approximate the median from grouped data?

    c What assumption is made to produce the mode from grouped data?

16 "If the raw data are concentrated around the lower limits of several classes, the computed mean could overstate the true mean by a significant amount." Discuss this statement.

17 a Discuss the typical locations of the mean, median, and mode in a negatively skewed distribution.

    b Discuss the locations of these three measures in a positively skewed distribution.

18 Summarize the comparative characteristics of the mean, median, and mode.

19 Identify and discuss the two mathematical properties of the arithmetic mean.

20 "A quantitative measure of central tendency tells us how much; a qualitative summary measure tells us how many." Discuss this statement.

PROJECTS/ISSUES TO CONSIDER

1 Team up with several of your classmates, locate a data set (in your library or elsewhere) that's of interest to the group, and perform the following steps as a group:

    a Array the data.

    b Experiment with different class intervals until you've settled on the "best" frequency distribution format for your data.

    c Create a graphic presentation to show the overall pattern of your data. Use two of the tools discussed in this chapter (histogram, frequency polygon, stem-and-leaf display, ogive) for your presentation.

    d Compute the mean, median, and mode for your data set, and select the one that best describes your data.

    e Present an analysis of the data to the class.

    f Keep copies of the work you've done for additional projects in later chapters.

2 Let's duplicate the class project presented in problem 10, page 117, in this chapter. Have a class recorder list the miles each student has traveled to get to class. After the data have been gathered, work as a class to group the data into a frequency distribution. Once that's done, divide into three teams. The assignment of *team 1* is to prepare an appropriate graphic presentation of the data (histogram, stem-and-leaf plot, frequency polygon). *Team 2* is responsible for computing the descriptive measures (mean, median, mode). The task of *team 3* is then to (1) analyze the graphic presentation and descriptive measures prepared by teams 1 and 2 and (2) present their analysis to the class. All teams should keep copies of their work for future projects.

3 Using the suggestions in project 2 (above) as a guide, carry out similar projects to record and analyze student-generated data such as miles-per-gallon scores of student automobiles, amount of change (in cents) in student pockets, student pulse rates (per minute), and so on.

ANSWERS TO SELF-TESTING REVIEW QUESTIONS

3-1
1   a   The data array is:

$21.50	$43.50
25.35	44.80
29.00	45.00
30.30	45.10
31.25	46.00
34.60	47.75
35.55	49.50
36.50	49.80
36.60	52.10
38.50	55.00
39.95	55.60
40.25	56.30
41.10	58.00
42.30	59.20
42.50	65.50

   b   The range is $44.00 ($65.50–$21.50).

2

SIZE OF ORDER	NUMBER OF ORDERS
$20 and under $30	3
30 and under 40	8
40 and under 50	12
50 and under 60	6
60 and under 70	1
	30

3   a   Yes, the distribution could have had more classes, but because of the limited number of frequencies, five classes adequately present the data.

   b $i = \dfrac{L - S}{c} = \dfrac{\$65.50 - \$21.50}{8} = \dfrac{\$44.00}{8} = \$5.50$, or an interval of $6

4   See the figure presented below.

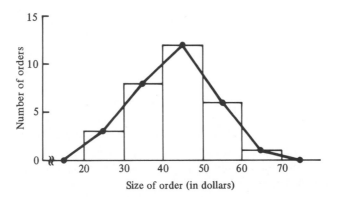

5   See the frequency polygon superimposed on the preceding figure.
6   The stem-and-leaf display should look something like this (the appearance of your plot may differ if you've used a computer program).

```
2 ‖ 159
3 ‖ 01456689
4 ‖ 012234556799
5 ‖ 255689
6 ‖ 5
```

3-2

1   a $\mu = 84/7 = 12$; $Md = 10$; $Mo = 10$
    b $\mu = 52/8 = 6.5$; $Md = 5.5$ (You didn't forget to array the data, did you?); $Mo = 5$
    c   The mode

2   a $\mu = \$1{,}298.40/30 = \$43.28$
    b $Md = \$43.00$ (the middle position in the array)
    c   There is no mode, since no value occurs more than once.

3   With a mean of 84 for 9 grades, the total points scored is 756. Since the grades remaining for the student total 662, the missing grade needed to bring the semester total to 756 is 94.

4   a $520/15 = 34.67$ or 35 assemblies.
    b
```
1 ‖ 8
2 ‖ 0579
3 ‖ 2244589
4 ‖ 0
5 ‖ 7
6 ‖ 0
```

    c   The median *position* is $(15+1)/2$ or 8 in the display. Counting down from the top of the display, we see that the eighth leaf is 4. Thus, the median value is 34 assemblies. The two modal values in this example are 32 and 34. Both values occur twice.

3-3

1   a $\mu = \$1{,}290/30 = \$43.00$

    b $Md = \$40 + \left(\dfrac{15 - 11}{12}\right)(10)$

        $= \$40 + \left(\dfrac{4}{12}\right)(10)$

        $= \$43.33$

    c $Mo = \$40 + \left(\dfrac{4}{4 + 6}\right)(10)$

        $= \$40 + \left(\dfrac{4}{10}\right)(10)$

        $= \$44.00$

2   The differences are due to the fact that the above measures computed from a frequency distribution are approximations of the true values. The true mean, for example, is the $43.28 found in question 2a of Self-Testing Review 3-2. But since we did not have the actual values when we computed the mean from the frequency distribution, our computed value of $43.00 is only approximately correct.

3   a   Negatively skewed.
    b   Because the mean is the smallest value and the mode is the largest value.

A CLOSER LOOK

DOTPLOTTING THE SLIMLINE DATA

The array of sales of Fizzy Cola syrup by 50 employees of the Slimline Beverage Company are shown in Figure 3-3 in this chapter. Let's assume now that the sales manager wants to compare the cola syrup sales with the syrup sales of Plum Natural—a carbonated diet drink with a fruit juice component. The array of sales of Plum Natural syrup by 50 Slimline employees is shown in Figure CL3-1.

FIGURE CL3-1   Array of Gallons of Plum Natural Syrup (sold by 50 employees of Slimline Beverage Company in 1 month)

GALLONS SOLD			
58.25	75.00	94.50	111.25
63.50	77.75	96.00	114.00
69.75	78.75	97.00	114.75
69.75	80.00	99.50	116.25
70.00	82.50	100.25	119.50
71.25	84.00	102.50	121.00
71.25	84.50	102.50	125.75
72.50	85.75	102.50	128.00
72.50	87.85	104.00	130.75
72.50	90.75	104.50	130.75
73.75	91.50	106.75	135.00
73.75	91.50	108.00	
74.00	91.50	110.50	

A built-in charting function in the *Minitab* software package uses a DOTPLOT command. A *dotplot* is a useful tool for preliminary data analysis; it groups the study data into many classes or intervals, and shows each data item as a dot on the chart. Dotplots for the Fizzy Cola data (labeled GALSOLD1) and the Plum Natural data (labeled GALSOLD2) are shown in Figure CL3-2.

The *Minitab* package has responded to the command DOTPLOT 'GALSOLD1' 'GAL-SOLD2' to produce two charts with different horizontal scales (in gallons). These scales are assigned by the program. In the GALSOLD1 (cola) plot, the first markings on the scale—the tick marks with the +'s—show values of 84 and 96 gallons. Since there are 10 spaces from mark to mark, each interval represents 1.2 gallons. And each interval in the GALSOLD2 plot covers 1.5 gallons. It's difficult to compare these two dotplots and analyze the two data sets because of the different interval sizes and because each plot starts and ends at different values. This difficulty is easily overcome, though, with another *Minitab* command that produces the result shown in Figure CL3-3.

FIGURE CL3-2

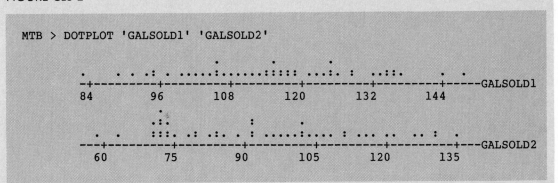

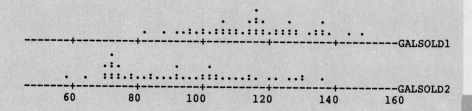

```
MTB > DOTPLOT 'GALSOLD1' 'GALSOLD2';
SUBC> SAME.
```

FIGURE CL3-3

You'll notice that the subcommand SAME is used to produce these revised plots. This subcommand tells the software to devise a scale that can be used for both plots. The Slimline sales manager can now see at a glance that cola sales are concentrated just below the 120 gallon mark, while the heaviest sales of Plum Natural syrup fall below 80 gallons. It's also obvious at a glance that: (1) the mean sale of cola syrup exceeds the mean sale of the Plum Natural product, (2) the median sales figure is larger for the cola drink, and (3) there's a larger spread or scatter of the dots in the Plum Natural plot than in the cola chart.

This preliminary data analysis may be enough to give the sales manager the information needed to support sales promotion and other marketing decisions. But to confirm the visual image supplied by the dot plots, the manager can quickly have the package compute important descriptive measures with a DESCRIBE 'GALSOLD1' 'GALSOLD2' command. The results obtained are shown in Figure CL3-4.

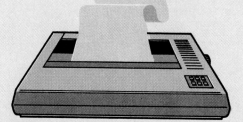

FIGURE CL3-4

```
MTB > DESCRIBE 'GALSOLD1' 'GALSOLD2'
```

	N	MEAN	MEDIAN	TRMEAN	STDEV	SEMEAN
GALSOLD1	50	115.40	115.25	115.43	14.91	2.11
GALSOLD2	50	93.79	91.50	93.22	20.10	2.84

	MIN	MAX	Q1	Q3
GALSOLD1	82.50	148.00	104.87	125.44
GALSOLD2	58.25	135.00	73.94	108.62

As you can see, the minimum, maximum, mean, and median sales figures are given for each data set. The TRMEAN figure shown is a *modified mean*. That is, the program trims or deletes the smallest 5 percent and the largest 5 percent of the values from the data set and then computes the mean of the middle 90 percent of the items. The STDEV heading refers to the standard deviation—a measure of the spread or scatter of the items in a series. We'll examine this measure at length in Chapter 4, but you'll note that our dotplot impression is correct: The scatter of the values in the Plum Natural data set is greater then the spread of the cola values. The Q_1 and Q_3 measures are also explained in Chapter 4, and a discussion of the SEMEAN is left to a later chapter. Don't worry about these values now.

4

STATISTICAL DESCRIPTION: MEASURES OF DISPERSION AND SKEWNESS

Living and Dying by the Numbers

Any nation that uses as many numbers as we do will invariably make mistakes. An egregious one was made this summer in a loudly trumpeted study by Congress's Joint Economic Committee. A staff analysis showed that from 1963 to 1983, the fraction of the nation's wealth held by the top 0.5 percent of the population went from 25 percent to 35 percent. Egalitarian outcry ensued.

A short time later, however, it was shown that there was a coding error in the Federal Reserve data upon which the study was based. A single household with $2 million in business assets was misreported as possessing $200 million. Because it happened to be weighted heavily in the sample, that error badly distorted the entire result. In fact, no statistically significant change in the distribution of wealth has occurred. But this discovery received distressingly little press coverage compared with the attention directed at the original report, demonstrating once again how vulnerable we have become to statistical manipulation.

On another issue, though, in another part of the Government, the statistically honorable thing was done. For some time it had been clear that the Government's quarterly GNP "flash" estimate, based on preliminary figures, was unreliable. Final figures differed substantially from the flash. Recognizing that the inaccurate early numbers were causing stock market swerves, managerial complications, and confidence problems, the Bureau of Economic Analysis decided in January 1986 to stop issuing flash estimates altogether. Good riddance.

Some of the year's most original and interesting numbers concerned various aspects of illegal activity. Figures for the President's Commission on Organized Crime suggest that the American Mafia will net $30 billion in profits this year on gross receipts of $50 billion. By comparison, after-tax profits of all domestic American corporations totaled $99.5 billion in 1985.

All of this activity is organized by a tiny band of managers. The FBI says that there are fewer than 2,000 sworn Mafia members, and perhaps 10 times that many "associates." They are well rewarded for their energetic efforts: Wharton Econometrics estimates that members average $250,000 in annual income.

This ill-gotten lucre costs the rest of us. *Fortune* magazine reports that by stifling competition and siphoning off capital, organized crime results in the loss of 40,000 jobs, increases consumer prices by 0.3 percent, and will cost every American $77 this year in disposable income.

As it happens, in organized crime as in so many other things, it appears that Japan is now number one. In Japan there are some 94,000 *yakuza*—gangsters—sporting vintage 1930s suits, vibrant body tattoos, and partly amputated fingers. And *they* manage a total annual take of $46 billion on $58 billion in revenues.

—From Karl Zinsmeister, "1986 By the Numbers," *Public Opinion*, January/February 1987, p. 17. Reprinted with the permission of the American Enterprise Institute for Public Policy Research.

LOOKING AHEAD

The opening vignette shows how a single faulty data item badly distorted a Congressional report. But considerable variability may exist in a data set, even when the facts are accurate. The purpose of this chapter is to show you how to measure this variability. The measures of central tendency discussed in Chapter 3 are generally not, by themselves, sufficient to summarize adequately the data being studied. Also needed are the measures shown in this chapter that describe the dispersion in the data set. Measures that summarize dispersion show to what extent the individual data items are scattered about an average size. In addition to learning the characteristics of various measures of dispersion, you'll also see how to compute these measures for ungrouped and grouped data. Finally, you'll see how to compute measures of relative dispersion and skewness.

Thus, after studying this chapter, you should be able to:

- Explain the reasons for measuring absolute dispersion, relative dispersion, and skewness.
- Compute such measures of absolute dispersion as the range, the average deviation, and the standard deviation for ungrouped data.

♦ Compute the standard deviation and the quartile deviation for data organized in a frequency distribution.
♦ Explain the meaning of, and some of the characteristics of, the measures of absolute dispersion discussed in this chapter.
♦ Compute (and explain the purpose of) the coefficient of variation and the coefficient of skewness.

CHAPTER OUTLINE

MEASURES OF ABSOLUTE DISPERSION

A measure of absolute dispersion is one that gauges the variability that exists in a data set, and it is expressed in the units of the original observations—

such as gallons sold, dollars earned, or miles driven. Measures of absolute dispersion may be computed for ungrouped data and for data grouped into frequency distributions.

Reasons for Measuring Absolute Dispersion

There are at least two reasons for measuring absolute dispersion. The *first* reason is to form a judgment about how well the average value depicts the data. For example, if a large amount of scatter exists among the items in a series, the average size used to summarize those values may not be representative of the data being studied. This isn't a new thought; we saw in Chapter 2 the dangers of disregarded dispersions.

A *second* reason for measuring dispersion is to learn the extent of the scatter so that steps may be taken to control the existing variation. For example, a tire manufacturer tries to produce a product that has a long average mileage life. But the manufacturer should also want to build tires of a uniform high quality so that there isn't a wide spread in tire mileage results to alienate customers. (You're pleased with the 40,000 miles you got from a set of these tires, but I'm really chapped by the 12,000 miles I got with the same tires.) By measuring the existing variation, the manufacturer may see a need to improve the uniformity of the product through better inspection and other quality control procedures.

COMPUTING MEASURES OF ABSOLUTE DISPERSION FOR UNGROUPED DATA

Three common measures of absolute dispersion are often computed from ungrouped data. These measures are the range, the average deviation, and the standard deviation.

The Range

The *range* is the simplest and crudest measure of dispersion, and we've seen that it's merely the difference between the highest and lowest values in an array. The range is used to report the movement of stock prices over a time period, and weather reports typically state the high and low temperature readings for a 24-hour period. Since we discussed the range in Chapter 3, we need not consider it further here.

The Average Deviation

Let's assume that after his recent academic ordeal (discussed in the last chapter) our friend Peter Parker decides to get away and go to a seaside resort where he can romp, play, and drown his sorrows (but, it is hoped, not himself). A fellow student who works part-time at a travel agency tells Peter that her agency is making group travel arrangements for two resorts. We've seen that a population is the total collection of values being studied by a decision maker. Our decision maker now is Peter, and he has defined two populations—all the single females going to resort A and all those headed for resort B. At Peter's urging, his friend reports that the population mean age of all single females signed to go to resort A is 19, while the population mean age of all those going to resort B is 31. Peter quickly signs up to go to Resort A and goes home to pack. The actual ages of the single females going to each resort are as follows:

AGES OF SINGLE FEMALES GOING TO RESORT A	AGES OF SINGLE FEMALES GOING TO RESORT B
2	18
2 triplets	19
2	19
4	19
5	19
7	19
10	20
11	20
11	45
34	45
35	46
35	47
50	48
58	50
266 Total age	434 Total age

$$\mu = \frac{266}{14} = 19 \qquad \mu = \frac{434}{14} = 31$$

If Peter had looked beyond the mean age, it's likely that he would have made a different decision. What Peter apparently wanted was a population mean age of 19 and very little spread or scatter of the individual ages about the mean. In short, Peter would have preferred a small measure of dispersion to go along with the mean of 19. (Of course, Peter made a grade of 5 on the test covering measures of dispersion.) One measure of dispersion that would have alerted Peter to the spread of ages is the **average** (or **mean**) **deviation**—an average of the absolute deviations of the individual items about their mean.

To calculate the average deviation it's necessary to (1) compute the mean of the items being studied; (2) determine the **absolute deviation**, which is the numeric difference of each item from the mean *without regard to the algebraic sign;* and (3) compute the average (mean) of these absolute deviations. The appropriate formula is:

$$AD = \frac{\Sigma |X - \mu|}{N} \qquad \text{(formula 4-1)}$$

THE IMPORTANCE OF STATISTICIANS

Mankind has been doing applied statistics for thousands of years. When faced with a problem—almost any problem—people can be observed deciding what data are needed and are practical to collect; then obtaining the data, analyzing the data, developing a conceptual framework that is consistent with the data (a predictive model, if you will); and then choosing a course of action based—at least in part—upon the data. Statistics has been a recognizable discipline for several centuries. That is, there has been some generalized theoretical basis and some form of documentation. In fact, the discipline of statistics is older than many of the other major disciplines we know today.

Statistics is now widely employed in all areas of society including industry and government. . . . The statistical data supplied by government agencies is but one example of the dependence of corporations, and the whole of society, upon statistical services. Statistics and statisticians also have vital internal roles in production, research, marketing, and support functions of the modern corporation.

The American Statistical Association was founded in 1839. Only the American Philosophical Society, the forerunner umbrella organization for all scientific disciplines, is older among American professional associations. Anyone who has followed the growth of the ASA during the last several decades must be impressed with the breadth of its activities, which mirror the growth of statistics as a discipline. As the ASA approaches its 150th anniversary, we can take great pride in the accomplishments of our discipline and our association. . . .

—From Donald W. Marguardt, "The Importance of Statisticians," *Journal of the American Statistical Association*, March, 1987, p. 1. This article was presented as the Presidential Address at the 1986 Annual Meetings of the American Statistical Association in Chicago. Reprinted with permission.

where AD = average deviation
  X = values of the observations
  μ = mean of the observations
  $|\ |$ = algebraic signs of the deviations are to be ignored
  N = total number of observations

Figure 4-1 shows the use of formula 4-1 in computing the average deviation for the ages of the single females going to resort B. Note that if the signs of the deviations about the mean aren't ignored, the sum of these deviations will always equal zero. Why? Because you'll recall that the first mathematical property of the mean is $\Sigma(X - \mu) = 0$. Thus, it's impossible to compute the average deviation unless absolute values are used. The average deviation or spread in our example is 13.57 years. (Would the average deviation have been larger or smaller if we had computed it for those going to resort A?)

Unlike the range, the average deviation takes every observation into account and shows the average scatter of the data items about the mean; however, it's still relatively simple to understand and compute. Unfortunately, the procedure of ignoring the algebraic signs limits its use in further calculations.

FIGURE 4-1   Computation of the Average (Mean) Deviation (Peter Parker's pick lacked perspicacity)

AGES OF FEMALES GOING TO RESORT B (1)	MEAN AGE (2)	$\|X - \mu\|$ (1) − (2)
18	31	13
19	31	12
19	31	12
19	31	12
19	31	12
19	31	12
20	31	11
20	31	11
45	31	14
45	31	14
46	31	15
47	31	16
48	31	17
50	31	19
434		190

$$\mu = \frac{434}{14} = 31$$

$$AD = \frac{\Sigma |X - \mu|}{N} = \frac{190}{14} = \underline{\underline{13.57}} \text{ years}$$

The Standard Deviation

The standard deviation is also used with the mean and is the most popular measure of dispersion. In a precise sense, the **standard deviation** is the square root of the average of the squared deviations of the individual data items about their mean. What a tongue twister! In easier-to-understand terms, though, the standard deviation is a measure of how far away items in a data set are from their mean. As we'll see later, a majority of the values in the data set will fall no more than 1 standard deviation away from their mean, and only a few will lie more than 2 standard deviations from the mean.

Like the calculation of the average deviation, the computation of the standard deviation is based on, and is representative of, the deviations of the individual data items about the mean of those values. And another similarity with the average deviation is that, as the actual observations become more widely scattered about their mean, the standard deviation becomes larger and larger. Of course, if all the items in a series are identical in value—that is, if there is no spread or scatter of values about the mean— the standard deviation is zero. We ignored algebraic signs to calculate the average deviation, but we don't do that to compute the standard deviation. Let's now work through an example problem that shows you how to compute the standard deviation for ungrouped data.

Figure 4-2 shows the calculation of the standard deviation for the single females going to resort B. The steps in the computation are:

1 The arithmetic mean of the data is computed. (We have seen that it is 31.)
2 The mean is subtracted from each of the individual ages in column 1. (See column 3.)
3 The deviations of the individual ages about the mean (column 3) are squared (see column 4) and totaled. This total, $\Sigma(X - \mu)^2$, is the second mathematical property of the mean discussed in Chapter 3; it's a minimum value. Appendix 5, at the back of the book, has been provided to help you quickly look up the squares of numbers from 1 through 1,000.
4 The mean of the deviations in column 4 is computed. This average of the squared deviations about the mean is called the **variance**. Take another look at the steps needed to compute the variance, for it's an important statistical measure in its own right. In Chapter 10, we'll compute the variances of several samples as a part of an *analysis of variance* procedure that's used to see if the arithmetic means of several populations are likely to be equal. And note that although the variance measures the average amount of variability that exists about the mean

FIGURE 4-2 Computation of the Standard Deviation; Ungrouped Data (Peter Parker's pick lacked perspicacity)

AGES OF FEMALES GOING TO RESORT B (X) (1)	MEAN AGE (μ) (2)	$(X-\mu)$ (1) − (2) (3)	$(X-\mu)^2$ $[(1)-(2)]^2$ (4)
18	31	−13	169
19	31	−12	144
19	31	−12	144
19	31	−12	144
19	31	−12	144
19	31	−12	144
20	31	−11	121
20	31	−11	121
45	31	14	196
45	31	14	196
46	31	15	225
47	31	16	256
48	31	17	289
50	31	19	361
434		0	2,654

$$\mu = \frac{\Sigma X}{N} \qquad \sigma = \sqrt{\frac{\Sigma(X - \mu)^2}{N}}$$

$$= \frac{434}{14} = 31 \qquad = \sqrt{\frac{2,654}{14}}$$

$$= \sqrt{189.57}$$

$$= \underline{\underline{13.8 \text{ years}}}$$

of a data set, it's not expressed in the units of the original data. That is, the variance in our example is 189.57, but this value represents the average variability of *squared* ages. Thus, to obtain a measure of dispersion expressed in terms of the original values, the following final step is needed.

5 *The standard deviation is computed by taking the square root of the variance.* (Appendix 5, at the back of the book, will also help you quickly look up square root values.) As you can see in Figure 4-2, the standard deviation for our example is 13.8 years of age. The standard deviation is always larger than the average deviation for the same data set because the squaring of deviations puts more emphasis on extreme values. Would the standard deviation figure have been larger or smaller if we had computed it for the single females going to resort A?

The appropriate formula for the above steps is:

$$\sigma = \sqrt{\frac{\Sigma(X - \mu)^2}{N}} \qquad \text{(formula 4-2)}$$

where σ (small Greek letter sigma) = standard deviation of the population
X = values of the observations
μ = mean of the observations
N = total number of observations in the population

It's not too difficult to use formula 4-2 with a small data set. But when many more items are included, the procedure becomes tedious. In that case, you might prefer to use the following "shortcut" variation of formula 4-2 (the symbols haven't changed):

$$\sigma = \sqrt{\frac{\Sigma X^2}{N} - \left(\frac{\Sigma X}{N}\right)^2} \qquad \text{(formula 4-3)}$$

Figure 4-3 shows that this alternative approach must yield the same results achieved in Figure 4-2.

"But the Computer Can't Be Wrong . . . "

The symbols used in formula 4-3 and in the other standard deviation formulas presented in this chapter refer to values in a *population*. But as you'll see in Chapter 6, if the standard deviation is being computed from *sample* data, different symbols must be used to differentiate between population and sample values. Furthermore, when you read Chapter 7, you'll learn that the formula used to compute a sample standard deviation differs slightly from the one for a population that we've just used.

FIGURE 4-3   An Alternative Computation of the Standard Deviation; Ungrouped Data

AGES OF FEMALES
GOING TO RESORT B

(X) (1)	(X^2) (2)
18	324
19	361
19	361
19	361
19	361
19	361
20	400
20	400
45	2,025
45	2,025
46	2,116
47	2,209
48	2,304
50	2,500
434	16,108

$$\sigma = \sqrt{\frac{\Sigma X^2}{N} - \left(\frac{\Sigma X}{N}\right)^2} = \sqrt{\frac{16,108}{14} - \left(\frac{434}{14}\right)^2}$$

$$= \sqrt{1,150.57 - 961} = \sqrt{189.57} = \underline{\underline{13.8 \text{ years}}}$$

That fact can lead to some unexpected results if you use a statistical software package such as *Mystat* or *Minitab* to compute the standard deviation for a data set. There's no problem if you're looking for the standard deviation for a *set of sample data*, because that's what most packages are programmed to deliver. But if you've entered all the items in a population and are trying to get the *population standard deviation*, the package doesn't know that it's *using the wrong formula*. What you'll receive is a standard deviation value that's slightly larger than it should be. You'll see why this is true in Chapter 7.

For now, if you have access to a statistical package (or an electronic calculator with a built-in standard deviation function), try this little experiment:

1   Enter the data set from column 1 of Figure 4-2 or 4-3 into your program.
2   Issue the command—such as STATS in *Mystat* or DESCRIBE in *Minitab*—that causes the program to calculate the standard deviation for the data set.
3   Compare the result with the answer of 13.8 years given in Figure 4-2 or 4-3.

Figure 4-4 shows how *Mystat* and *Minitab* have processed our example data. Both packages have given us a standard deviation value of 14.29 years—the correct figure *if* our data set had come from a sample selected from a population. How can we get a population standard deviation, then, if the package is programmed to supply only a sample value? It's simple: *just divide the package result by* $\sqrt{N/(N-1)}$, where N is the number of items in the population data set. In our example, this divisor is

$$\sqrt{\frac{14}{13}} \quad \text{or} \quad \sqrt{1.0769} \quad \text{or} \quad 1.038$$

Dividing the program results of 14.29 years by this divisor of 1.038 yields the same 13.76- or 13.8-year value found in Figures 4-2 and 4-3. (And dividing the variance of 204.15 in the *Mystat* output in Figure 4-4*a* by the 1.0769 figure given above yields the variance of 189.57 presented in Figures 4-2 and 4-3.)

FIGURE 4-4   Output produced by *Mystat* and *Minitab* statistical packages when supplied with Peter Parker's Resort B data set. (a) The values of selected descriptive statistics generated by *Mystat*. (b) *Minitab* output using the Resort B data set. *Minitab* is programmed to produce a "trimmed" or modified mean (TRMEAN) that averages the middle 90 percent of the values in a data set. The meanings of Q_1 and Q_3 are discussed later in this chapter. An explanation of SEMEAN must wait until Chapter 6.

```
              TOTAL OBSERVATIONS:      14


                                      AGES


                 N OF CASES                14
                 MINIMUM                18.000
                 MAXIMUM                50.000
                 RANGE                  32.000
                 MEAN                   31.000
                 VARIANCE              204.154
                 STANDARD DEV           14.288
```

(a)

```
MTB > DESCRIBE 'AGES'

                N      MEAN    MEDIAN    TRMEAN    STDEV    SEMEAN
   AGES        14     31.00     20.00     30.50    14.29      3.82

              MIN       MAX        Q1        Q3
   AGES     18.00     50.00     19.00     46.25
```

(b)

Self-Testing Review 4-1

1 Using the ages of the single females going to resort A in the Peter Parker fiasco, compute the following measures:
   a The range
   b The average deviation
   c The variance
   d The standard deviation

2 The following statistics grades were made by Peter Parker last semester and were presented in Chapter 3:

GRADES
75
75
61
50
40
25
10
5
1
342

Compute the following:
   a The average deviation
   b The variance
   c The standard deviation

COMPUTING MEASURES OF ABSOLUTE DISPERSION FOR GROUPED DATA

As you read this section, keep in mind the two points about grouped data computations raised in Chapter 3. *First,* the ability of computers to easily process lengthy lists of ungrouped data has eliminated the computational advantages of using frequency distributions. And *second,* only *approximate* values can be computed for the measures of dispersion we'll discuss in this section.

When data grouped in a frequency distribution must be processed, however, the average deviation is seldom used. Rather, the primary measure of dispersion is the standard deviation, which is used along with the mean for descriptive purposes. An alternative is the quartile deviation, which, together with the median, is used to describe those distributions whose characteristics either cannot or should not be represented by the mean and the standard deviation. In demonstrating the procedures for computing approximations of the standard and quartile deviations, we'll once again use the Slimline Beverage Company data found in Chapter 3, Figure 3-4. We'll also assume that we have a population data set—that is, our data items represent all those of interest to the sales manager.

The Standard Deviation

Computing the standard deviation from a frequency distribution is similar to calculating the measure from ungrouped data. The formula used to approximate the population standard deviation in Figure 4-5 is:

$$\sigma = \sqrt{\frac{\Sigma f (m - \mu)^2}{N}} \qquad \text{(formula 4-4)}$$

where f = frequency or number of observations in a class
$\quad\quad m$ = midpoint of a class and the assumed value of every observation in the class
$\quad\quad N$ = total number of frequencies or observations in the distribution

   As you can see in Figure 4-5, the standard deviation for the Slimline Company data is 14.63 gallons. The class intervals and the (f), (m), and (fm) columns of Figure 4-5 duplicate the same columns in Figure 3-17, page 103, that were used to compute the mean. But the last three columns in Figure 4-5 are new. The mean of the distribution (115.2 gallons) is subtracted from each of the class midpoints in the $(m - \mu)$ column to get a deviation amount. Each of these deviations is squared in the $(m - \mu)^2$ column. And each of these squared deviations is multiplied by the frequencies in each class in the last $f(m - \mu)^2$ column. The total of this last column   (10,698.00)

FIGURE 4-5   Computation of Standard Deviation (gallons of Fizzy Cola syrup sold by 50 employees of Slimline Beverage Company in 1 month)

GALLONS SOLD	NUMBER OF EMPLOYEES (f)	CLASS MIDPOINTS (m)	(fm)	DEVIATION ($m - \mu$)	($m - \mu)^2$	$f(m - \mu)^2$
80 and less than 90	2	85	170	−30.2	912.04	1,824.08
90 and less than 100	6	95	570	−20.2	408.04	2,448.24
100 and less than 110	10	105	1,050	−10.2	104.04	1,040.40
110 and less than 120	14	115	1,610	−0.2	0.04	0.56
120 and less than 130	9	125	1,125	9.8	96.04	864.36
130 and less than 140	7	135	945	19.8	392.04	2,744.28
140 and less than 150	2	145	290	29.8	888.04	1,776.08
	50		5,760			10,698.00

$$\mu = \frac{\Sigma fm}{N} = \frac{5,760}{50} = 115.2 \text{ gallons} \qquad \sigma = \sqrt{\frac{\Sigma f (m - \mu)^2}{N}}$$

$$= \sqrt{\frac{10,698.00}{50}}$$

$$= \sqrt{213.96}$$

$$= \underline{14.63} \text{ gallons}$$

is then divided by the total frequencies, the variance of 213.96 is obtained, and the square root of this value—the standard deviation of 14.63 gallons—finally emerges.

You've undoubtedly noticed that using formula 4-4 to compute the standard deviation requires several columns and a number of tedious calculations. You can reduce the work load by using the following shortcut variation of formula 4-4:

$$\sigma = \sqrt{\frac{\Sigma f(m)^2 - (\Sigma fm)^2/N}{N}}$$ (formula 4-5)

where $\Sigma f(m)^2$ = sum of fm times m for each class, and *not* Σf times $(\Sigma m)^2$.

Figure 4-6 shows the use of this shortcut method. The results of Figures 4-5 and 4-6 must agree, and they do, as you can verify. Only one additional column beyond those used to calculate the mean is needed in Figure 4-6. The figures in this column—labeled $f(m)^2$—are found by multiplying the

FIGURE 4-6   Computation of Standard Deviation Using Formula 4-5 (gallons of Fizzy Cola syrup sold by 50 employees of Slimline Beverage Company in 1 month)

GALLONS SOLD	NUMBER OF EMPLOYEES (f)	CLASS MIDPOINTS $(m) \times$	$(fm) =$	$f(m)^2$
80 and less than 90	2	85	170	14,450
90 and less than 100	6	95	570	54,150
100 and less than 110	10	105	1,050	110,250
110 and less than 120	14	115	1,610	185,150
120 and less than 130	9	125	1,125	140,625
130 and less than 140	7	135	945	127,575
140 and less than 150	2	145	290	42,050
	50		5,760	674,250

$$\mu = \frac{\Sigma fm}{N} = \frac{5,760}{50} = 115.2 \text{ gallons}$$

$$\sigma = \sqrt{\frac{\Sigma f(m)^2 - (\Sigma fm)^2/N)}{N}}$$

$$= \sqrt{\frac{674,250 - (5,760)^2/50}{50}}$$

$$= \sqrt{\frac{674,250 - 663,552}{50}}$$

$$= \sqrt{\frac{10,698}{50}}$$

$$= \sqrt{213.96}$$

$$= \underline{14.63} \text{ gallons}$$

(m) and (fm) values in each table row. For example, in the first row 85 is multiplied by 170 to get the $f(m)^2$ result of 14,450. (Squaring the m value of 85 and then multiplying by the f value of 2 will, of course, produce the same product.) Once the $f(m)^2$ column is completed and totaled, all figures needed for formula 4-5 are available.

Interpreting the Standard Deviation

As we saw in the last chapter, in Figure 3-10, when many values are analyzed, they are often found to be distributed or scattered about their arithmetic mean in a reasonably symmetrical way. The standard deviation is a particularly important measure of dispersion because of its relationship to the mean in such bell-shaped or normal distributions. Although this relationship is considered in detail in the next chapter, it's appropriate here to show how the standard deviation is used with the mean to indicate the proportions of the observations in a distribution that fall within specified distances from the mean.

Suppose, for example, that many people are given an IQ test, and the resulting raw scores are organized into a frequency distribution. A frequency polygon is prepared from the distribution and is found to be symmetrical or normal in shape. The arithmetic mean of this mound-shaped distribution is 100 points, and the standard deviation is 10 points. In such a situation, the mean IQ score is directly under the peak of the curve, and the following relationships exist: (1) *68.3 percent* of the test scores fall within *one* standard deviation of the mean—that is, 68.3 percent of the people have test scores between 90 and 110 points; (2) *95.4 percent* of the test scores fall within *two* standard deviations of the mean—that is, slightly over 95 percent of those taking the test have scores between 80 and 120 points; and (3) *virtually all* (99.7 percent) of the test scores fall within *three* standard deviations of the mean (scores between 70 and 130). Figure 4-7 shows these relationships. The precise percentage figures we've just used are obtained from Appendix 2 at the back of the book. The use of this table is explained in the next chapter.

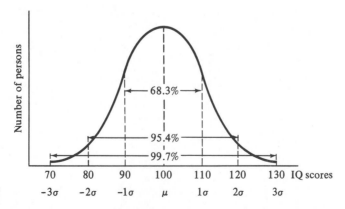

FIGURE 4-7

The exact relationships that exist between the mean and the standard deviation in a normal distribution may also be used for analysis purposes with distributions that are only approximately normal. Thus, we're now in a position to interpret the meaning of the standard deviation of 14.63 gallons in our Slimline example distribution, since that distribution is approximately normal. We can conclude that about the middle two-thirds of the 50 employees sold syrup quantities between $\mu \pm 1\sigma$, that is, between 115.20 gallons $\pm$ 14.63 gallons (or from 100.57 to 129.83 gallons). This result is graphically demonstrated in the *Mystat*-generated histogram of our example distribution in Figure 3-7*b*, page 90. The ^ mark on the horizontal scale locates the mean value, and the parenthesis marks [( )] to the right and left of the ^ mark show the positions found by adding and subtracting one standard deviation. Furthermore, about 95 percent of the employees sold syrup quantities between $\mu \pm 2\sigma$, or between 85.94 and 144.46 gallons. You can verify from the data array in Figure 3-3, page 84, that 66 percent of the employees sold between 100.57 and 129.83 gallons and that 96 percent of them sold between 85.94 and 144.46 gallons. Of course, all the employees sold syrup quantities between $\mu \pm 3\sigma$.

The Quartile Deviation

Like the range, the **quartile deviation** is a measure that describes the existing dispersion in terms of the *distance* between selected observation points. With the range, the observation points are simply the highest and lowest values. In determining the quartile deviation, though, we'll compute an **interquartile range** that includes approximately the *middle 50 percent* of the values in the distribution. (Since actual data items are lost in a distribution, it's only possible to approximate the interquartile range.) Thus, our observation points, as shown in Figure 4-8, are at the *first* (Q_1) and *third* (Q_3) quartile positions.

The interquartile range is simply the distance or difference between Q_3 and Q_1. The **first quartile** (Q_1) position is the point that separates the *lower* 25 percent of the values from the upper 75 percent. And the **third quartile**

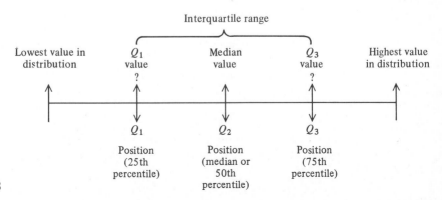

FIGURE 4-8

(Q_3) position is the point that separates the *upper* 25 percent of the values from the lower 75 percent. Thus, the lower and upper 25 percent of the values aren't considered in the computation of the quartile deviation (and open-ended distributions present no problem). Note in Figure 4-8 that the first quartile value is the same as the 25th percentile, the third quartile value is the same as the 75th percentile, and the **second quartile** (Q_2) is just another name for the median or 50th percentile.[1] As a matter of fact, in computing the values of Q_3 and Q_1 to find the interquartile range, we'll follow the same procedure used in the last chapter to compute the median. So let's get started on the quartile deviation by once again using the Slimline Company data.

The computation of the quartile deviation is shown in Figure 4-9. The *formula for computing the Q_1 value is*

$$Q_1 = L_{Q_1} + \left(\frac{N/4 - CF}{f_{Q_1}} \right)(i) \qquad \text{(formula 4-6)}$$

where Q_1 = first quartile value

       L_{Q_1} = lower limit of the first quartile class—i.e., the first class whose cumulative frequency exceeds $N/4$ observations

       CF = cumulative frequencies in all classes up to, but not including, the *first quartile class*

       f_{Q_1} = frequency of the first quartile class

       i = size of the interval of the first quartile class

And *the formula for computing the Q_3 value is*

$$Q_3 = L_{Q_3} + \left(\frac{3N/4 - CF}{f_{Q_3}} \right)(i) \qquad \text{(formula 4-7)}$$

where Q_3 = third quartile value

       L_{Q_3} = lower limit of the third quartile class—i.e., the first class whose cumulative frequency exceeds $3N/4$ observations

       CF = cumulative frequencies in all classes up to, but not including, the *first quartile class*

       f_{Q_1} = frequency of the third quartile class

       i = size of the interval of the first quartile class

As you'll notice in Figure 4-9, the computed values of Q_1 and Q_3 are 104.5 gallons and 126.11 gallons. Thus, approximately the middle 50 percent of the Slimline employees sold syrup quantities between 104.5 and

---

[1]Also, the median is the 5th decile . . . it's enough to make you cry.

FIGURE 4-9   Computation of the Quartile Deviation (gallons of Fizzy Cola syrup sold by 50 employees of Slimline Beverage Company in 1 month)

GALLONS SOLD	NUMBER OF EMPLOYEES (f)	CUMULATIVE FREQUENCIES (CF)
80 and less than 90	2	2
90 and less than 100	6	8
100 and less than 110	10	18
110 and less than 120	14	32
120 and less than 130	9	41
130 and less than 140	7	48
140 and less than 150	2	50
	50	

$$Q_1 = L_{Q_1} + \left(\frac{N/4 - CF}{f_{Q_1}}\right)(i) \qquad Q_3 = L_{Q_3} + \left(\frac{3N/4 - CF}{f_{Q_3}}\right)(i)$$

$$= 100 + \left(\frac{50/4 - 8}{10}\right)(10) \qquad = 120 + \left(\frac{150/4 - 32}{9}\right)(10)$$

$$= 100 + \left(\frac{12.5 - 8}{10}\right)(10) \qquad = 120 + \left(\frac{37.5 - 32}{9}\right)(10)$$

$$= 100 + 4.5 \qquad\qquad\qquad = 120 + 6.11$$

$$= 104.5 \text{ gallons} \qquad\qquad = 126.11 \text{ gallons}$$

Interquartile range $= Q_3 - Q_1$
$$= 126.11 - 104.5$$
$$= 21.61 \text{ gallons}$$

Quartile deviation $= \dfrac{Q_3 - Q_1}{2}$
$$= \frac{21.61}{2} \text{ gallons}$$
$$= \underline{\underline{10.81}} \text{ gallons}$$

126.11 gallons. Stated another way, the interquartile range $(Q_3 - Q_1)$ of 21.61 gallons shows that the sales of the central 50 percent of the employees varied within an approximate spread of 21.61 gallons.

The *quartile deviation (QD) is simply one-half the interquartile range* and is sometimes referred to as the *semiinterquartile range*. That is,

$$QD = \frac{Q_3 - Q_1}{2} \qquad\qquad \text{(formula 4-8)}$$

Obviously, the smaller the QD, the greater the concentration of the middle half of the observations in the distribution. If a distribution is normal in shape, exactly 50 percent of the values are found in the range of the median $\pm 1QD$ because the values of Q_1 and Q_3 are equal distances from the median. This relationship can also be used for analysis purposes with distributions that are only approximately normal. Thus, in our Slimline example, we can conclude that approximately the middle 50 percent of the employees sold syrup quantities between $Md \pm 1QD$, that is, between

115.0 gallons ± 10.81 gallons (or from 104.19 to 125.81 gallons). You can verify from the data array in Figure 3-3, page 84, that the middle 50 percent of the employees actually sold between 105.00 and 125.25 gallons.

Box-and-Whiskers Display

You saw in Chapter 3 that the stem-and-leaf plot is an exploratory data analysis tool that analysts use to get a feel for the data being studied. Another graphic technique used in exploratory analysis is the box-and-whiskers display. A **box-and-whiskers display** (also called a **boxplot**) shows the middle half of the values in a data set—the values that lie in the interquartile range—as a *box,* and then draws lines or *whiskers* extending to the left and right from the box to indicate the remaining 50 percent of the data items. Figure 4-10 shows the boxplots generated by the *Mystat* and *Minitab* software packages when supplied with our Slimline Company data set.

As you can see in Figure 4-10, the box is a rectangle. The + mark or marks inside the box show the *median* location. Each end of the box is called a **hinge** and is marked with some symbol such as + (*Mystat*) or I (*Minitab*). For our purposes, the left or **lower hinge** of a box is basically the same as Q_1, and the right or **upper hinge** is essentially the same as Q_3. (Different approaches are used in texts and software packages to define and calculate hinges and quartiles. The results are only slightly different, so we'll not go into these details.) To understand hinge terminology a little

FIGURE 4-10   Box-and-whiskers displays of the Slimline Company data set produced by the (a) *Mystat* and (b) *Minitab* statistical software packages.

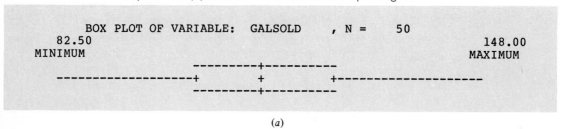

(a)

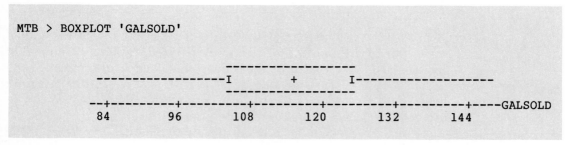

(b)

better, try this experiment: (1) Put a column of evenly spaced and arrayed numbers on a sheet of paper, with the lowest value placed at the top edge, and the largest value placed at the bottom edge of the sheet. (2) Fold the paper in the middle so it's half as long. (3) Now fold the paper in the middle again, crease it, and then unfold it to its original shape. You'll find that the two outer creases—or hinges—should be about where the Q_1 and Q_3 data values are located, and the middle crease should locate the median.

The box-and-whiskers display gives analysts a quick pictorial representation of the median, Q_1, Q_3, and the smallest and largest values in a data set. They can also tell at a glance if a data set is reasonably symmetrical, or if it's skewed. How? A median mark that's about in the center of the box, as it is in the plots in Figure 4-10, tells the analyst that the data set is reasonably symmetrical. But a median mark that's to the right side of the box indicates negative skewness, while a mark that's placed on the left side of the box suggests positive skewness. Also, skewness is indicated if one whisker line is appreciably longer than the other. For example, a longer right whisker suggests positive skewness.

Self-Testing Review 4-2

1   The data for the following frequency distribution were first presented in Self-Testing Review 3-1, and the mean, median, and mode for this distribution of orders placed by customers of the Lee Produce Company were computed in Self-Testing Review 3-3.

SIZE OF ORDER	NUMBER OF ORDERS
$20 and under $30	3
30 and under 40	8
40 and under 50	12
50 and under 60	6
60 and under 70	1
	30

Determine the following measures:
a   The standard deviation
b   The quartile deviation

2   Analyze the above distribution by the use of
a   The mean and the standard deviation
b   The median and the quartile deviation

3   Three computer printouts are shown on the next page. These are: Output 1: Descriptive measures prepared by *Mystat*. Output 2: Descriptive measures prepared by *Minitab*. Output 3: The top part of the Mystat stem-and-leaf display shown in Figure 3-12a. Each of these printouts describes GALSOLD, the Slimline data set we've just about worked to exhaustion.
a   Compare the standard deviation figures in outputs 1 and 2 to the computed standard deviation value found in Figures 4-5 and 4-6. How do you account for the small difference?

```
              TOTAL OBSERVATIONS:      50


                            GALSOLD


              N OF CASES                   50
              MINIMUM                  82.500
              MAXIMUM                 148.000
              RANGE                    65.500
              MEAN                    115.400
              VARIANCE                222.278
              STANDARD DEV             14.909
              SUM                    5770.000
```

Output 1

```
MTB > DESCRIBE 'GALSOLD'

                N      MEAN    MEDIAN    TRMEAN     STDEV    SEMEAN
   GALSOLD     50    115.40    115.25    115.43     14.91      2.11

              MIN       MAX        Q1        Q3
   GALSOLD  82.50    148.00    104.87    125.44
```

Output 2

```
       STEM AND LEAF PLOT OF VARIABLE:   GALSOLD     , N =    50


MINIMUM IS:       82.500
LOWER HINGE IS:      105.000
MEDIAN IS:       115.250
UPPER HINGE IS:      125.250
MAXIMUM IS:      148.000
```

Output 3

b   Compare three sets of figures: the values of Q_1 and Q_3 in output 2, the upper and lower hinge figures in output 3, and the quartile figures computed in Figure 4-9. How do you account for the differences?

SUMMARY OF COMPARATIVE CHARACTERISTICS

There's no general rule that will always identify the proper measure of absolute dispersion to use. In selecting the appropriate measure, the characteristics of each must be considered, and the type of data available must be evaluated. A summary of comparative characteristics for each measure discussed in the preceding pages is given below.

The Range

Some of the characteristics of the range are:

1 *It's the easiest measure to compute.* Since its calculation involves only one subtraction, it is also the easiest measure to understand.
2 *It emphasizes only the extreme values.* Because the more typical items are completely ignored, the range may give a very distorted picture of the true dispersion pattern.

The Average Deviation

Some of the characteristics of the average deviation are:

1 *It gives equal weight to the deviation of every observation.* Thus, it's more sensitive than measures such as the range or quartile deviation that are based on only two values.
2 *It may be computed from the median as well as from the mean.* Although formula 4-1 called for the computation of the deviations about the mean of the values, the median could also have been used to compute an average deviation.
3 *It's an easy measure to compute.* It is also not difficult to understand.
4 *It's not influenced as much by extreme values as the standard deviation.* The squaring of deviations in the calculation of the standard deviation places more emphasis on the extreme values.
5 *Its use is limited in further calculations.* Because the algebraic signs are ignored, the average deviation is not as well suited as the standard deviation for further computations.

The Standard Deviation

Included among the characteristics of the standard deviation are:

1 *It's the most frequently encountered measure of dispersion.* Because of the mathematical properties it possesses, it's more suitable than any other measure of dispersion for further analysis involving statistical inference procedures. We shall use the standard deviation extensively in the chapters in Part 2.
2 *It's a computed measure whose value is affected by the value of every observation in a series.* A change in the value of any observation will change the standard deviation value.

3 *Its value may be distorted too much by a relatively few extreme values.* Like the mean, the standard deviation can lose its representative quality in badly skewed distributions.

4 *It cannot be computed from an open-ended distribution in the absence of additional information.* As formula 4-2 shows, if the mean cannot be computed, neither can the standard deviation.

The Quartile Deviation

Some of the characteristics of the quartile deviation are:

1 *It's similar to the range in that it's based on only two values.* As we've seen, these two values identify the range of the middle 50 percent of the values.

2 *It's easy to define and easy to understand.* (All right, so the computations were a bit tedious. But the concept of the quartile deviation wasn't hard to understand, was it?)

3 *It's frequently used in badly skewed distributions.* The quartile deviation will not be affected by the size of the values of extreme items, and so it may be preferable to the average or standard deviation when a distribution is badly skewed.

4 *It may be computed in an open-ended distribution.* Since the upper and lower 25 percent of the values are not considered in the computation of the quartile deviation, an open-ended distribution presents no problem.

MEASURE OF RELATIVE DISPERSION

The standard deviation and the other gauges of dispersion we've studied in the preceding pages are measures of *absolute dispersion*. That is, they are expressed in the units of the original observations. Thus, one distribution of the annual earnings of a group of fire fighters may have a standard deviation of $800 and another distribution of the annual earnings of a group of master plumbers may have a standard deviation of $1,600. Or, to use another example, one distribution may have a standard deviation of 14.63 gallons, while another has a standard deviation of $9.80. If we desire to *compare* the dispersions of the first two distributions, can we conclude that the distribution with the $1,600 standard deviation has twice the variability of the one with the $800 standard deviation? Or can we conclude that the distribution with the 14.63-gallon standard deviation is more widely dispersed than the one with the $9.80 measure? (Can we even logically compare gallons and dollars?) The answer to these questions is that we can't conclude anything by simply comparing the measures of absolute dispersion. Rather, what is needed for *comparison* purposes is a measure of the degree of *relative dispersion* that exists in the distributions being studied.

The most popular measure of relative dispersion is the ***coefficient of variation (CV)***, which is simply the standard deviation of a distribution expressed as a percentage of the mean of the distribution. That is,

$$CV = \frac{\sigma}{\mu}(100) \hspace{5cm} \text{(formula 4-9)}$$

Thus, if the mean for the distribution of annual earnings of the fire fighters is $20,000, the coefficient of variation is computed as follows:

$$CV = \frac{\$800}{\$20,000}(100) = 4 \text{ percent}$$

And let's assume that the mean annual earnings for the group of plumbers is $44,000. The coefficient of variation for this distribution is then 3.64 percent: $CV = (\$1,600/\$44,000)(100)$. Therefore, the distribution with the standard deviation of $1,600 not only *does not* have twice the variability of the one with the $800 standard deviation; it actually has *less* relative dispersion. In other words, the annual earnings received by the group of plumbers are slightly more uniform than the earnings received by the fire fighters.

Self-Testing Review 4-3

1 Compute the coefficient of variation for the Slimline example problem.
2 Compute the coefficient of variation for the Lee Produce Company data discussed in Self-Testing Review 4-2.
3 Which of these distributions has the greater relative dispersion?

MEASURE OF SKEWNESS

You'll remember from Figure 3-21 that the mean, median, and mode are all located directly under the peak of a symmetrical frequency polygon. You'll also recall that as a distribution departs from the symmetrical—that is, as it becomes skewed—these measures of central tendency are separated (see Figures 3-21*b* and *c*), with the mode remaining under the peak of the curve and the mean moving the greatest distance out in the direction of the tail of the distribution. These relationships between measures of central tendency are utilized in the following *coefficient of skewness* formula, which gives the direction (negative or positive) as well as an indication of the degree of skewness (*Sk*):

$$Sk = \frac{3(\mu - Md)}{\sigma} \qquad \text{(formula 4-10)}$$

When a distribution is symmetrical, formula 4-10 gives a value of zero, because the mean and median are equal, and the numerator of the formula is thus zero. As the mean and median become separated, however, the co-efficient of skewness moves from zero toward either a negative or positive value of one. (It will seldom exceed ±1.) Obviously, the closer the value is to zero, the less skewed the distribution. The value is not, of course, expressed in terms of any units of measure. The coefficient of skewness for our Slimline example distribution is computed as follows:

$$Sk = \frac{3(115.2 - 115.0)}{14.63} = +.041$$

This value of + .041 represents a very small degree of positive skewness. (If the distribution had been skewed to the left, the value of the mean would have been smaller than the median and the sign of the numerator of formula 4-10 would have been negative.)

Self-Testing Review 4-4

1 Compute the coefficient of skewness for the Lee Produce Company distribution. (The values of the measures needed are found in earlier self-testing review sections.)
2 Interpret the value computed in question 1.

LOOKING BACK

1 A measure of absolute dispersion gauges the variability that exists in a data set, and it's expressed in the units of the original observations. Such a measure is needed to form a judgment about the reliability of the average value and to learn the extent of the scatter of the observations so that steps may be taken to control the existing variation.
2 Three measures of absolute dispersion computed for ungrouped data are the range, the average deviation, and the standard deviation. The range is simply the difference between the highest and lowest values in an array. The average deviation is an average of the absolute deviations of the individual items about their mean. To calculate this measure you find the mean of the items being studied, determine the absolute deviation of each item from the mean, and then compute the mean of these absolute deviations.
3 The standard deviation is the square root of the average of the squared deviations of the individual data items about their mean. In terms that don't have to be interpreted by a rocket scientist, the standard deviation is simply a measure of how far away items in a data set are from their mean. A majority of the values in the data

set will fall no more than one standard deviation away from their mean, and only a few will lie more than two standard deviations from the mean. To calculate the standard deviation for ungrouped data, you find the mean of the data set, subtract this value from each of the data items, square and total the deviations thus obtained, divide this total by the number of data items to get the variance, and then take the square root of the variance to get the standard deviation. If you use a statistical software package to bypass these steps, though, remember that the package probably uses a formula to find a sample standard deviation. Much of the time that's what package users are seeking. But if you're working with a population data set, you'll have to make a simple adjustment to the computer output to get the correct population standard deviation.

4   Only approximate values can be computed for measures of dispersion when data are grouped in a frequency distribution. The primary measure of dispersion for grouped (or ungrouped) data is the standard deviation. Computing this measure from a frequency table is similar to calculating it from ungrouped data, as you saw in the chapter. If a frequency distribution is symmetrical or approximately so, about two-thirds of the items in the distribution lie within one standard deviation of the mean, about 95 percent of the items fall within two standard deviations of the mean, and virtually all of the observations are within three standard deviations of the mean.

5   The quartile deviation measures the existing dispersion in terms of the distance between selected observation points, and these points are the first and third quartile values. Subtracting Q_1 from Q_3 produces the interquartile range, and the quartile deviation is one-half of this value. If a distribution is symmetrical, 50 percent of the values are found in the range of the median plus and minus the quartile deviation. A box-and-whiskers display shows the values that lie in the interquartile range as a box, and it then draws lines or whiskers extending to the left and right from the ends (or hinges) of the box to indicate the remaining 50 percent of the data items. A boxplot is an exploratory data analysis tool that can be quickly generated by a software package to give a quick picture of the median, Q_1, Q_3, and the smallest and largest values in a data set. It can also show if a data set is symmetrical or skewed.

6   The chapter gives a summary of the comparative characteristics for each of the measures of absolute dispersion discussed. The range is the easiest measure to compute, but it emphasizes only extreme values. The average deviation gives equal weight to the deviation of every data item, and is easy to compute, but its use is limited because it ignores algebraic signs. The standard deviation is by far the most popular measure we've considered because it's the most suitable for further analysis, and its value is affected by the value of every data item, but its value may be distorted by extreme values, and it cannot be computed from an open-ended distribution in the absence of additional information. The quartile deviation is based on two quartile values that identify the range of the middle 50 percent of the values, it's easy to understand, it may be used in badly skewed distributions, and it may be computed in an open-ended distribution, but it ignores 50 percent of the data set.

7   If there's a need to compare the dispersions of two or more distributions, a measure of the degree of relative dispersion that exists should be used. The coefficient

of variation is well suited for comparison purposes. And in addition to measuring dispersion, it may be desirable to measure the extent to which a data set departs from the symmetrical to form an opinion about the suitability of various descriptive measures. The coefficient of skewness may be used for this purpose.

KEY TERMS AND CONCEPTS

Measure of absolute dispersion   *130*
Average (mean) deviation   *132*
Absolute deviation   *132*

$$AD = \frac{\Sigma |X - \mu|}{N} \quad \text{(formula 4-1)} \quad 132$$

Standard deviation   *134*
Variance   *135*

$$\sigma = \sqrt{\frac{\Sigma (X - \mu)^2}{N}} \quad \text{(formula 4-2)} \quad 136$$

$$\sigma = \sqrt{\frac{\Sigma X^2}{N} - \left(\frac{\Sigma X}{N}\right)^2} \quad \text{(formula 4-3)} \quad 136$$

$$\sigma = \sqrt{\frac{\Sigma f (m - \mu)^2}{N}} \quad \text{(formula 4-4)} \quad 140$$

$$\sigma = \sqrt{\frac{\Sigma f (m)^2 - (\Sigma fm)^2/N}{N}} \quad \text{(formula 4-5)} \quad 141$$

$\mu \pm 1\sigma$ *143*
$\mu \pm 2\sigma$ *143*
$\mu \pm 3\sigma$ *143*
Quartile deviation   *143*

Interquartile range   *143*
First quartile (Q_1) *143*
Third quartile (Q_3) *143*
Second quartile (Q_2) *144*

$$Q_1 = L_{Q_1} + \left(\frac{N/4 - CF}{f_{Q_1}}\right)(i) \quad \text{(formula 4-6)} \quad 144$$

$$Q_3 = L_{Q_3} + \left(\frac{3N/4 - CF}{f_{Q_3}}\right)(i) \quad \text{(formula 4-7)} \quad 144$$

$$QD = \frac{Q_3 - Q_1}{2} \quad \text{(formula 4-8)} \quad 145$$

Box-and-whiskers display (boxplot)   *146*
Hinge   *146*
Lower hinge   *146*
Upper hinge   *146*
Coefficient of variation   *151*

$$CV = \frac{\sigma}{\mu}(100) \quad \text{(formula 4-9)} \quad 151$$

Coefficient of skewness   *151*

$$SK = \frac{3(\mu - Md)}{\sigma} \quad \text{(formula 4-10)} \quad 152$$

PROBLEMS

1   The following scores were made by a few of Professor Shirley A. Meany's accounting students who dared to take an optional makeup test:

72
65
43
50
68
62

Compute the following measures for these test scores:
a   The range
b   The average deviation
c   The variance
d   The standard deviation

2 The following distribution represents the test scores made by Professor Meany's students in problem 1, Chapter 3:

TEST SCORES	NUMBER OF STUDENTS
40 and under 50	4
50 and under 60	6
60 and under 70	10
70 and under 80	4
80 and under 90	4
90 and under 100	2

Compute the following measures for this distribution:
a The standard deviation
b The quartile deviation
c The coefficient of variation
d The coefficient of skewness
Note: The mean and median were to be computed in problem 2c, Chapter 3.

3 Analyze the distribution in problem 2 above by the use of
a The mean and the standard deviation
b The median and the quartile deviation

4 Using the mileage data in problem 3, Chapter 3, compute the following measures:
a The standard deviation
b The quartile deviation
c The coefficient of variation
d The coefficient of skewness

5 Compare the distributions in the preceding problems 2 and 4, and indicate which has the greater relative dispersion.

6 Using the annual earnings data for families in Languor and Friskyville found in problem 4, Chapter 3, perform the following operations:
a Compute the standard deviation for each community.
b Compute the quartile deviation for each community.
c Compute the coefficient of variation for each community.
d Compute the coefficient of skewness for each community.
e Compare and analyze the economic situations in these two communities.
Note: If it is not possible to perform all the operations in any or all of the above parts, please indicate your reasons for omitting the operation(s).

7 Compute the average deviation and the standard deviation for the yardage figures given in problem 6, Chapter 3.

8 Using the annual salary data of Chauvinists, Inc., found in problem 7, Chapter 3, perform the following operations:
a Compute the standard deviation for both male and female employees.
b Compute the quartile deviation for both male and female employees.
c Compute the coefficient of variation for each group.
d Compute the coefficient of skewness for each group.
e Compare and analyze the economic situations of each employee group.

9 Using the warranty repair job data found in problem 8, Chapter 3, perform the following operations:

    a　Compute the standard deviation.

    b　Compute the quartile deviation.

    c　Compute the coefficient of variation.

10　Using Professor Nastie's test score data from problem **9**, Chapter 3, compute the following measures:

    a　The standard deviation

    b　The coefficient of skewness

11　Using the mileage data from problem **10**, Chapter 3, compute the following measures:

    a　The standard deviation

    b　The quartile deviation

12　Using the mileage data from problem **11**, Chapter 3, compute the following measures:

    a　The average (mean) deviation

    b　The variance

    c　The standard deviation

13　The weights of headquarters' managers given in problem **12**, Chapter 3, may be used to compute the following measures:

    a　The standard deviation for 1979

    b　The standard deviation for 1989

14　Entrance exam scores for applicants to Parcheesi University are tabulated below:

| 990 | 1403 | 1059 | 1213 | 763 | 1352 | 898 | 999 | 1181 |
| 1264 | 269 | 428 | 582 | 381 | 1141 | 760 | 455 | 345 |

    a　Find the highest and lowest scores and compute the range of the scores.

    b　Compute the mean score.

    c　Compute the average deviation.

    d　What is the value of the variance?

    e　What is the value of the standard deviation?

15　Let's assume that a *population* data set of 100 items is entered into a computer program such as *Mystat* that is programmed to compute the standard deviation for a *sample*. If the program returns a standard deviation value of 25, what's the true standard deviation for the population? What is the true variance?

16　The entrance exam scores in problem **14** above are grouped into classes as shown below:

SCORE	APPLICANTS
200 and less than 500	5
500 and less than 800	3
800 and less than 1100	4
1100 and less than 1400	5
1400 and less than 1700	1

    a　What is the arithmetic mean for this distribution of scores?

b What is the standard deviation for this distribution?

c Compare the answers obtained in problems 14 and 16 and explain the discrepancies.

17 What is the value of the quartile deviation for the distribution of entrance exam scores given in problem 16 above?

18 The following table gives the monthly number of soldiers afflicted with dysentery in a Confederate infantry unit in 1863:

J	F	M	A	M	J	J	A	S	O	N	D
3	6	9	12	12	15	18	20	9	2	1	0

a What was the mean and standard deviation values for the dysentery cases in 1863?

b What was the median number of dysentery cases for the year?

c Using the results of parts a and b, compute the coefficient of skewness and explain the meaning of this value.

d Using the results found in part a, calculate the coefficient of variation.

TOPICS FOR REVIEW AND DISCUSSION

1 Why is it necessary to measure dispersion?

2 Why must the algebraic signs of the deviations about the mean be ignored when computing the average deviation?

3 a What is the standard deviation?

b How does the computation of the standard deviation for ungrouped data differ from the computation of the average deviation?

4 a What is the variance?

b Discuss the limitations of variance as a measure of absolute dispersion.

5 a How is it possible that a statistical software package can give an incorrect answer for a population standard deviation?

b What can be done to correct the answer supplied by the program?

6 Discuss the relationship that exists between the mean and the standard deviation in a normal distribution.

7 Discuss the relationship that exists between the median and the quartile deviation in a symmetrical distribution.

8 a What is the interquartile range?

b How is this range depicted in a boxplot?

9 a What is a hinge in a box-and-whiskers display?

b How is the median located in such a display? How can such a display tell an analyst if a data set is reasonably symmetrical?

10 Discuss the important characteristics of:

a The range

b The average deviation

c The standard deviation

d The quartile deviation

11   What is the purpose of the coefficient of variation?

12   What is the purpose of the coefficient of skewness?

PROJECTS/ISSUES TO CONSIDER

1   Retrieve the data set that your group located and analyzed for project 1 in Chapter 3. Using the same data, carry out these additional steps as a group:

    a   Compute the range, standard deviation, quartile deviation, and coefficient of skewness for your data set.

    b   Prepare a box-and-whiskers display of your data.

    c   Select the measure of absolute dispersion that best describes your data set, and show its relationship to one of the measures of central tendency you computed for the Chapter 3 project.

    d   Present an analysis of your data to the class.

    e   Keep copies of the work you've done for additional projects in later chapters.

2   Three teams were assigned to analyze student-generated data in projects 2 and 3 in Chapter 3. Teams 1 and 2 can now divide up the work of computing the range, standard deviation, quartile deviation, and coefficient of skewness for the data sets used in Chapter 3. Team 3 can then use this work to prepare a boxplot, analyze the data, and present their findings to the class. All teams should keep copies of their work for future projects.

ANSWERS TO SELF-TESTING REVIEW QUESTIONS

4-1

1   a   Range = 56 years

    b $AD = 234/14 = 16.71$ years

    c $\sigma^2 = 4,860/14 = 347.14$

    d $\sigma = \sqrt{347.14} = 18.63$ years

2   a $AD = 222/9 = 24.67$ points

    b $\sigma^2 = 6,826/9 = 758.4$

    c $\sigma = \sqrt{758.4} = 27.54$ points

4-2

1   a $\sigma \text{ (direct method)} = \sqrt{\dfrac{2,880}{30}} = \sqrt{96} = \$9.80$

     or

     $\sigma \text{ (shortcut method)} = \sqrt{\dfrac{58,350 - 1,290^2/30}{30}} = \sqrt{\dfrac{58,350 - 55,470}{30}}$

                     $= \sqrt{\dfrac{2,880}{30}} = \sqrt{96} = \$9.80$

b $Q_1 = 30 + \left(\dfrac{7.5 - 3}{8}\right)(10)$ $Q_3 = 40 + \left(\dfrac{22.5 - 11}{12}\right)(10)$

$\quad = \$35.62$ $\qquad\qquad\qquad\qquad = \$49.58$

$QD = \dfrac{\$49.58 - \$35.62}{2} = \$6.98$

2 a Since this distribution may be considered approximately normal, Brock and Parse Lee may conclude that about the middle two-thirds of their orders from the outlying county are within one standard deviation of the mean, that is, between $\$43.00 \pm \$9.80$ (or from $\$33.20$ to $\$52.80$).

   b And about the middle 50 percent of the orders are between $Md \pm 1\ QD$, that is, between $\$43.33 \pm \$6.98$ (or from $\$36.35$ to $\$50.31$).

3 a There are two reasons why the value of 14.91 gallons computed by *Mystat* and *Minitab* doesn't equal the 14.63 gallons found in Figures 4-5 and 4-6. *First*, the 14.63 gallon figure is an approximation computed from a frequency table, so it can only be an estimate. *Mystat* and *Minitab* both had the actual data items to work with, but, as noted earlier in the chapter, the programs treat the data set as a *sample* and use a formula that slightly overstates the population standard deviation. The net result is that both standard deviation figures are slightly off. The true population standard deviation, found by dividing 14.91 by $\sqrt{N/(N-1)}$ or $\sqrt{50/49}$, is 14.76 gallons.

   b Again, the Q_1 and Q_3 values in Figure 4-9 are estimates computed from a frequency distribution. We didn't expect that they would be exact. *Minitab* and *Mystat* use slightly different methods to compute their quartile and hinge results from the actual data set. A look at the array of the raw data in Figure 3-3 shows that the actual Q_1 and Q_3 values are, in this case, the hinges given in Output 3.

4-3

1 $CV = (14.63/115.2)\ (100) = 12.70$ percent

2 $CV = (\$9.80/\$43.00)\ (100) = 22.79$ percent

3 There is a much greater degree of dispersion in the Lee Produce Company distribution.

4-4

1 $Sk = \dfrac{3(\$43.00 - \$43.33)}{\$9.80} = -.10$

2 There is a slight degree of negative skewness in the Lee Produce Company data, but not enough to prevent the use of the mean and the standard deviation as representative descriptive measures.

A CLOSER LOOK

INTRODUCING STATISTICS TO GEOGRAPHY STUDENTS: THE CASE FOR EXPLORATORY DATA ANALYSIS

A grasp of data analysis and statistics enables students to analyze their own data and affords them access to the large body of geographical research involving the use of quantitative methods. Unfortunately, many first- and second-year college students are ill-prepared to take a course on elementary numerical techniques, and those who had limited success with mathematics at high school often are intimidated by the prospect of further study.

A reassuring introduction to data analysis can be provided by exposure to some techniques developed by John Tukey. Tukey designed exploratory data analysis (EDA) to give students or analysts a "feel" for the data being considered and to suggest avenues for further work. He devised techniques for simple arithmetical manipulation and visual portrayal of data sets to indicate salient features of their structure. Many of these techniques and ideas are presented in a readable and easily accessible text (Tukey 1977).

Although there are some exceptions (McNeill 1977; Erickson and Nosanchuk 1977), a majority of textbooks focus on confirmatory data analysis (CDA) techniques (Cox and Jones 1981), which include regression, analysis of variance, and the use of samples to estimate population characteristics. These parametric statistical tests demand a normal distribution of observations, but little geographical data follow an exactly normal distribution (Gould 1970). EDA provides students with the preliminary exploration necessary to understand the structure of the data they are investigating and to avert use of inapplicable or spurious procedures. This paper discusses . . . aspects of EDA [including] the use of displays [and] "resistant" statistics. . . .

DISPLAYS

Plotting data is an important, preliminary aspect of any statistical investigation. A rapid and efficient way of plotting data is the stem and leaf plot (Tukey 1977). This task involves constructing a plot of actual data values that looks like a histogram, but contains more information and takes less time to complete.

Consider the case of personal income in Canada (Table CL4-1). To construct the stem and leaf plot, each value is represented by a stem (thousands) that is recorded to the left of a vertical line, and a leaf (hundreds) recorded to the right of the line. The range of stems in the data set (6 to 11) and the leaves, rounded to the nearest hundred, should be listed. The leaves of the first four provinces are:

```
11 | 1 0   Alberta, British Columbia
10 |
 9 |
 8 | 9     Manitoba
 7 | 1     New Brunswick
 6 |
```

A full stem and leaf plot of the data set is:

```
11 | 1 0
10 | 6 2
 9 | 4 0
 8 | 9
 7 | 1 8 0
 6 | 3
```

TABLE CL4-1   Personal Income in Canada, 1980

PROVINCE OR TERRITORY	Can$
Alberta	11,067
British Columbia	11,027
Manitoba	8,876
New Brunswick	7,085
Newfoundland	6,343
Nova Scotia	7,845
Ontario	10,614
Prince Edward Island	7,048
Quebec	9,370
Saskatchewan	9,028
Yukon and Northwest Territories	10,231

Source: Ministry of Treasury and Economics (1982).

The stems are already in numerical order. The leaves on each stem can be arranged in ascending order. Thus:

```
11 | 0 1
10 | 2 6
 9 | 0 4
 8 | 9
 7 | 0 1 8
 6 | 3
```

From this display, students can obtain an impression of the distribution of the data. (If the display were turned on its side, the leaves would resemble the bars of a histogram. Unlike a histogram, however, the values decrease from left to right.) From the stem and leaf plot, the modal class (i.e., $7,000–$7,900 in our example) can be identified. By rearranging the leaves into ascending order, we can also identify the median value of the data ($9,000) by counting off one-half the values. This measure of central tendency can be placed in a box:

```
11 | 0 1
10 | 2 6
 9 | [0] 4
 8 | 9
 7 | 0 1 8
 6 | 3        Newfoundland
```

Similarly, the quartiles or "quarter-markers" can be identified by counting a quarter of the values from the top (to locate the upper quartile) and bottom (to locate the lower quartile) of the sequence. These values can be circled:

```
11 | 0 1
10 | 2 (6)
 9 | [0] 4
 8 | 9
 7 | 0 (1) 8
 6 | 3        Newfoundland
```

Then the distance between the upper and lower quartiles (midspread or interquartile range) can be calculated: $10,600 − $7,100 = $3,500. The midspread, which encompasses the middle fifty percent of values, is a measure of the dispersion of the data. Students find it easier to understand the midspread than the more commonly used standard deviation, because the midspread can be visualized from a stem and leaf plot. Of the measures derived from this plot, only the modal class can be obtained from a histogram.

The arithmetic mean cannot be readily identified from the plot of personal income. If the plot were symmetrical about the median, then the mean would be roughly equal to the median. If not, the mean would be skewed toward the tail of the distribution. We can, however, notice and label any extreme or outlying values in the data set. Once these have been identified, students can be asked to provide reasons for them. In the income example, Newfoundland has the lowest value, and we may ask why personal income in Newfoundland is lower than in the other Canadian provinces and territories. Stem and leaf displays are intended to prompt questions and ideas; they are not ends in themselves.

A box-plot can be easily constructed from a stem and leaf plot. Maximum and minimum values in a data set are marked against a vertical scale of measurement (Figure CL4-1), and the median and quartiles are marked by horizontal bars that form a box representing the midspread or interquartile range. For comparison of several data sets, box-plots drawn side by side provide an uncluttered image of the level and spread of each set. As with stem and leaf plots, box-plots can be used to generate questions, ideas, or hypotheses. For example, from the box-plot of river discharge data (Figure CL4-1), differences in discharge of the rivers can be identified. Although the streams recorded the same minimum value, Marching-

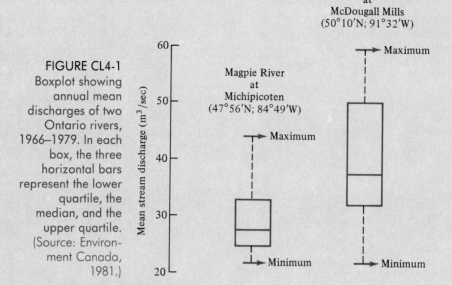

FIGURE CL4-1
Boxplot showing annual mean discharges of two Ontario rivers, 1966–1979. In each box, the three horizontal bars represent the lower quartile, the median, and the upper quartile. (Source: Environment Canada, 1981.)

ton River has a more variable discharge and a higher median discharge than the Magpie River. Students could be asked to discuss factors affecting stream discharge with the aid of precipitation data, and topographical and geological maps.

RESISTANT STATISTICS

Because of the desirability of using exploratory techniques in the preliminary study of data sets, investigators require measures that are resistant (i.e., measures that work well under a variety of circumstances). The median is an example of a resistant measure. As an estimate of the middle of the data set, the median is less strongly affected by one or two outlying (or extreme) values than the mean. It is also a more representative measure of the middle of a data set than the mean when the distribution is skewed or non-normal. Geographers frequently encounter skewed data sets in physical geography; for example, most precipitation data are positively skewed. In general, there tend to be many rainfall events with a low total precipitation and relatively few with a high total precipitation. In these cases, the median may be a better indicator of "average" rainfall than the arithmetic mean (Figure CL4-2). Similarly, the midspread is more resistant to outliers than the standard deviation. . . .

21	9
20	
19	
18	0
17	
16	5 8
15	0
14	
13	2 5
12	3 9 　Mean 120.2
11	
10	0 5 　Median 105
9	2 2 9 9
8	5
7	7 9 9
6	8

FIGURE CL4-2
Stem and leaf plot of rainfall totals in Sydney, Australia, 1934–1952. (Source: Hammond and McCullagh, 1978.)

CONCLUSION

Exploratory techniques are intended for preliminary data analysis. They are not exhaustive methods; although, depending on the purpose of the analysis, they may provide sufficient insight into the structure of the data so that no further work need be conducted. Their importance as educational tools lies in their potential to indicate rapidly, and in an easily understood way, several important features of a data set to the student without the need for excessive calculation or understanding of statistical theory.

ACKNOWLEDGMENT

The authors wish to thank Nicholas J. Cox, Durham University, for introducing them to Exploratory Data Analysis.

REFERENCES

Cox, N.J., and Jones, K. 1981. "Exploratory data analysis." In *Quantitative Geography—A British View*, eds. N. Wrigley and R.J. Bennett, pp. 135–43. London: Routledge and Kegan Paul.

Erickson, B.H., and Nosanchuk, T.A. 1977. *Understanding Data*. Toronto: McGraw-Hill, Ryerson.

Gould, P.R. 1970. "Is statistix inferens the geographical name for a wild goose?" *Economic Geography* 46:439–48.

McNeil, D.R. 1977. *Interactive Data Analysis*. New York: John Wiley.

Ministry of Treasury and Economics. 1982. *Ontario Statistics 1982*. Toronto: Government of Ontario.

Tukey, J.W. 1977. *Exploratory Data Analysis*. Reading, MA: Addison-Wesley.

—From Christopher R. Burn and Michael F. Fox, "Introducing Statistics to Geography Students: The Case for Exploratory Data Analysis," *Journal of Geography*, January-February 1986, pp. 28–31. Reprinted with permission.

PART TWO

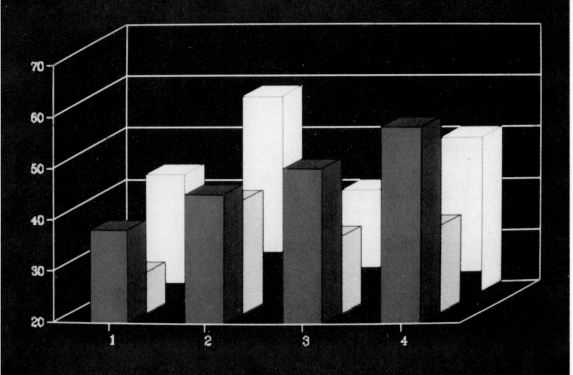

SAMPLING IN THEORY AND PRACTICE

Sometimes he thought sadly to himself, "Why?" and sometimes he thought, "Wherefore?" and sometimes he thought, "Inasmuch as which?"
— *Winnie-the-Pooh*

If complete knowledge were readily available, many world problems would be solved, since complete knowledge is synonymous with certainty. Instead of relying on complete knowledge, though, decision makers must often resign themselves to using inferences based on sample results. Alas, they are destined to make statements using a vocabulary of "perhaps," "likely," "maybe," "almost surely," and "probably."

Fortunately for decision makers, the science of statistics allows them to gather and analyze sample data in an objective way. Statistical sampling methods don't provide absolute certainty in decisions, but they can reduce the level of uncertainty which would have existed without any sample data. Thus, statistical techniques don't guarantee results; they merely allow a person to move from a position of complete ignorance to one of definable doubt. Statistical laws merely bestow upon the individual the privilege of saying "maybe."

In this part of the text, we'll cover an area formally known as *inferential statistics*. Specifically, in Chapter 7 we'll discuss methods of *objectively making estimates*. For example, these methods may be applicable in estimating the average time required to produce an item, or in estimating the percentage of voters who favor a political candidate. And in Chapters 8 through 11, we'll discuss *hypothesis-testing* techniques to be used in specified situations. Thus, you may want to test the assumption that the average time required to assemble a circuit board is 23 minutes, or you may want to determine the validity of a candidate's statement that 60 percent of the voters support her bid for office.

But before we walk, we must learn to crawl. Before covering the material in Chapters 7 through 11, we must understand why these methods are valid. Chapters 5 and 6, then, explain why there's justification for the methods you'll be using in the later chapters of this part. Thus, these early chapters will show you why you can place confidence in your estimates and conclusions. Chapter 5 covers the area of *probability*. Complete certainty is nonexistent in decision making, so everything must be placed in the context of likely outcomes. And Chapter 6 ties together the principles of probability and sampling techniques.

The chapters included in Part 2 are:

5

PROBABILITY AND PROBABILITY DISTRIBUTIONS

Quotable Quotations

Lest men suspect your tale untrue, keep probability in view.

—John Gay

It is truth very certain that, when it is not in our power to determine what is true, we ought to follow what is most probable.

—Rene Descartes

Probability must atone for the want of truth.

—Matthew Prior

A reasonable probability is the only certainty.

—E. W. Howe

A thousand probabilities do not make one fact.

—Italian Proverb

But to us, probability is the very guide of life.

—Bishop Joseph Butler (1736)

The laws of probability, so true in general, so fallacious in particular.

—Edward Gibbon: Autobiography

No priest or soothsayer that ever lived could hold his own against old probabilities.

—Oliver Wendell Holmes (1875)

Ignorance gives one a large range of probability.

—George Eliot

Games of chance are probably as old as the human desire to get something for nothing; but their mathematical implications were appreciated only after Fermat and Pascal in 1654 reduced chance to law.

—Eric T. Bell

It is probable that many things will happen contrary to probability.

—Anonymous

The principal means for ascertaining truth—induction and analogy—are based on probabilities; so that the entire system of human knowledge is connected with the theory of probability.

—Pierre Simon, Marquis de Laplace (1819)

—From Hardeo Sahai, "Some Quotable Quotations Usable in a Probability and Statistics Class," *School Science and Mathematics,* October 1979, pp. 486–492. Reprinted with permission of *School Science and Mathematics.*

LOOKING AHEAD

Probability—the subject of all the above quotations—is also the subject of this chapter. There are many instances when we must make decisions under conditions of uncertainty. A retailer, for example, must decide how much inventory to stock, and a manager must decide whether or not to market a new product.

In each of these cases, the individual makes the decision on the basis of what he or she thinks will happen. The retailer, for example, looks at past sales records to determine the likely demand, and then makes an inventory decision. This decision, then, is based on the retailer's probability judgment that a particular sales level will occur. Since probability is so important in many statistical analyses, we'll look in this chapter at the (1) meaning and types of probability, (2) methods of computing probabilities, (3) use of probabilities in computing expected values, (4) concept of the probability distribution, and (5) characteristics of the binomial, Poisson, and normal probability distributions.

Thus, after studying this chapter you should be able to:

- Define probability and explain how probabilities may be classified.
- Correctly use addition and multiplication rules to perform probability computations.
- Use probabilities in computing expected values.
- Explain what a probability distribution is and compute the probabilities in binomial and Poisson distributions.
- Use the table of areas for the standard normal probability curve (Appendix 2) to determine probabilities for a normal distribution.

CHAPTER OUTLINE

LOOKING AHEAD
MEANING AND TYPES OF PROBABILITY

MEANING AND TYPES OF PROBABILITY

Some Basic Concepts

For our purposes, **probability** may be defined as the chance or likelihood that an event will occur. Probability measurements are stated as fractions between 0 and 1. A probability of 0 means an event can *never* occur, and a probability of 1 means an event will *always* occur.

If you were given a penny to flip (which is about all a penny is good for these days), you would likely believe that the probability of getting a head is .5. This doesn't mean that if you flipped the penny twice you would have to get one head. It does mean, though, that if you flipped the penny a large number of times, the proportion of heads would approach .5.

You might try an experiment along these lines: Flip a penny 10 times and keep a record of the number of heads; then flip the penny 10 more times and keep a cumulative record of the number of heads obtained in the 20 trials. If this experiment is continued for a total of 100 or 200 flips, you'll likely find the proportions of heads getting ever closer to .5.

If you have access to a statistical software package, you may be able to save wear and tear on your flipping thumb by letting the computer take over this tedious task. Figure 5-1 shows a **computer simulation** or imitation of our penny-flipping experiment carried out with the *Minitab* program. The top lines in Figure 5-1 direct the program to simulate 200 coin tosses by a random process, and then store the results in a column. The probability of "success" in a single toss—that is, that the coin falls heads up—is set at .5. The simulated results of the 200 tosses are then printed. The first 15 "tosses" are shown in the first row of 1s and 0s. Each 1 represents a success in getting a head, and each 0 shows that a tail was produced in the trial. Notice that 10 of the first 15 trials were heads, but that the proportion of heads and tails moved toward balance as the trials continued. In fact, as you can see from the tally at the bottom of Figure 5-1, there were 98 heads and 102 tails in this simulated effort. Another simulation, of course, would likely produce a different total count.

Probability Categories

Probabilities may be classified in several ways. An *a priori* (before the fact) **probability** is one that can be determined in advance without experimentation. Assigning a figure of .5 to the chance of getting a head in a single flip of that penny is one example of an *a priori* probability. In rolling a single die (one-half of a pair of dice) there are six possible outcomes because there are six sides. Unless someone has slipped you a loaded die, each outcome is equally likely to occur. Therefore, the *a priori* probability of throwing any given number is 1/6. Figure 5-2 checks the validity of this last sentence with another computer simulation, this time an experiment in rolling a die 180 times. Since each of the six sides has an equal chance of facing up in a single trial, we should expect that each face would appear about 30 times in 180 trials. As you can see in the count column at the bottom of Figure 5-2, our expectation is on target, although, of course, each face didn't appear exactly 30 times.

A **relative frequency** (or **empirical**) **probability** is one that's determined after the fact from observation and experimentation. In this type of probability, the frequency with which an event of interest has actually occurred in a number of observed trials is noted. For example, let's assume that you're still flipping that penny and have now flipped it 10,000 times. If the relative frequency of heads, let's assume, continues to remain at about .65 of the total number of observations—that is, if about 6,500 of your 10,000 flips have produced heads—you may be justified in revising your *a priori* probability expectation and concluding, perhaps, that you've been using an unfair or loaded coin. Empirical probability is important to many types of de-

```
MTB > RANDOM 200, SET IN C1;
SUBC> BERNOULLI P = .5.
MTB > PRINT C1

C1
    0    1    1    0    0    1    1    1    1    1    1    0    1    0    1
    1    1    0    0    0    1    0    1    1    1    0    0    0    0    1
    0    1    0    0    1    0    1    0    1    1    0    1    1    0    0
    1    0    0    0    1    1    0    0    0    1    1    1    1    0    1
    0    1    0    1    0    1    1    1    0    0    1    1    1    1    0
    0    0    0    1    0    0    1    0    1    0    1    1    0    0    0
    0    0    0    0    1    1    1    1    0    1    0    1    0    0    0
    1    0    0    1    1    1    0    0    1    1    0    1    0    1    1
    1    1    1    1    0    1    0    0    1    0    0    1    0    0    1
    0    0    1    1    0    0    1    1    0    0    0    1    0    1    0
    1    1    1    1    0    0    1    1    0    0    1    0    0    0    1
    1    1    0    1    1    1    0    0    0    1    1    1    0    0    0
    0    1    1    1    0    1    0    0    0    0    0    1    0    1    0
    1    0    0    0    0


MTB > TALLY C1;
SUBC> ALL.


    C1    COUNT  CUMCNT  PERCENT   CUMPCT
    0      102     102    51.00    51.00
    1       98     200    49.00   100.00
    N=     200
```

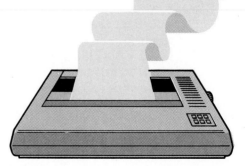

FIGURE 5-1   A *Minitab* simulation of 200 flips of a penny.

cisions. The quantity of inventory a retailer should stock, the number of nurses a hospital should employ, and the rates that should be charged by an insurance company may be determined on the basis of empirical probabilities.

Subjective probability—a third classification—occurs in situations where *a priori* or empirical values can't be determined, and so probabilities are as-

```
MTB > RANDOM 180, SET IN C1;
SUBC> INTEGERS 1 TO 6.
MTB > PRINT C1

C1
   4   1   3   1   4   2   5   3   1   6   5   2   6   6   4
   3   5   6   1   4   5   2   2   1   1   6   2   4   5   2
   4   6   3   2   6   6   5   5   1   1   3   4   1   1   4
   5   4   4   3   2   1   5   2   6   4   6   3   6   2   3
   3   4   4   6   1   6   2   6   6   5   4   3   5   4   2
   6   6   5   3   1   4   3   5   4   1   5   4   2   2   2
   1   4   6   4   5   5   5   2   2   5   3   2   4   3   2
   2   2   4   4   2   5   2   6   2   4   5   6   4   6   1
   4   1   1   2   4   5   1   3   3   2   5   4   2   3   6
   4   6   1   3   4   2   3   1   1   6   4   1   5   3   3
   3   6   3   1   6   6   5   4   6   3   4   1   6   4   3
   1   3   3   4   3   3   3   3   6   6   5   1   2   1   2

MTB > TALLY C1;
SUBC> ALL.


     C1  COUNT  CUMCNT  PERCENT   CUMPCT
      1    28     28    15.56    15.56
      2    30     58    16.67    32.22
      3    31     89    17.22    49.44
      4    35    124    19.44    68.89
      5    25    149    13.89    82.78
      6    31    180    17.22   100.00
     N=   180
```

FIGURE 5-2   A computer simulation of 180 rolls of a die.

signed on the basis of someone's personal judgment or intuition. For example, a plant manager may believe that there's a .6 probability that the union will call a strike next week. This probability is the manager's subjective estimate of the likelihood of a strike, and is not an *a priori* or empirical value. The accuracy of the strike estimate, of course, depends on the manager's experience and skill. In the following pages our focus is on *a priori* and empirical values, but subjective probability plays a vital role in advanced statistical decision theory.

PROBABILITY COMPUTATIONS

Anybody can win unless there happens to be a second entry.
—George Ade

Two basic rules are used during probability computations. The first is an **addition rule** that's used, for example, to calculate the probability that one or the other of two events will happen. That is, an analyst is trying to find the probability that *either* A *or* B occurs. This is written as $P(A \text{ or } B)$. The second is a **multiplication rule** that's used to determine the probability that *both* events will happen. This dual occurrence is written as $P(A \text{ and } B)$. Applications of these rules are shown in the following sections.

Addition Rule for Mutually Exclusive Events

Two events are said to be **mutually exclusive** if the occurrence of one event prevents the occurrence of the other event. For example, when you flip a penny a single time, if you get a head it's obviously impossible to get a tail, and vice versa. Therefore, these two possible outcomes of a single trial are mutually exclusive.

When two events are mutually exclusive, the probability that *one or the other* of the two events will occur is the *sum* of their separate probabilities. For two mutually exclusive events, A and B, this addition rule is:

$$P(A \text{ or } B) = P(A) + P(B)$$

These symbols mean that the probability that either A or B will occur is equal to the probability that A will happen plus the probability that B will happen. Thus, if you roll a single die, the probability of getting either a 1 or a 2 is computed as follows:

$$P(1 \text{ or } 2) = P(1) + P(2)$$
$$= \frac{1}{6} + \frac{1}{6}$$
$$= \frac{2}{6} \text{ or } \frac{1}{3}$$

Now let's suppose that Jane is shopping for new tires. The probability is .25, .30, .20, .15, or .10 that she will buy Michelin, Goodyear, General, Firestone, or Continental tires. What's the probability that she will buy either General or Continental tires? The answer here is:

$$P(\text{General or Continental}) = P(\text{General}) + P(\text{Continental})$$
$$= .20 + .10$$
$$= .30$$

Addition Rule When Events Are Not Mutually Exclusive

If two events are not mutually exclusive, it's possible for both events to occur. For example, if you draw a card from a deck of 52 cards, and if you want either an ace or a spade, it will be possible for you to draw the ace of spades. Therefore, these two events, ace and spade, are not mutually exclusive. In a case like this, our addition rule must be stated as:

$$P(A \text{ or } B) = P(A) + P(B) - P(A \text{ and } B)$$

The probability of getting either an ace or a spade from a deck of cards is:

$$P(\text{ace or spade}) = P(\text{ace}) + P(\text{spade}) - P(\text{ace and spade})$$
$$= \frac{4}{52} + \frac{13}{52} - \frac{1}{52}$$
$$= \frac{16}{52}$$

And if the probabilities are .37, .30, and .20 that a gardener will buy a lawn mower, edger, or lawn mower and edger on April 1, then the probability that the gardener will buy a mower or edger on that day is:

$$P(\text{mower or edger}) = P(\text{mower}) + P(\text{edger}) - P(\text{mower and edger})$$
$$= .37 + .30 - .20$$
$$= .47$$

Multiplication Rule for Independent Events

In determining a **joint probability**, that is, the probability that both events will occur, it's necessary to know if the two events are independent. We have **independent events** if the occurrence or nonoccurrence of one has no effect on the probability that the other happens. This may be shown by assuming that you have one red die and one green die and you wish to know the probability of throwing a 2 with this pair of dice. This means, of course, throwing a 1 on the red die and a 1 on the green die. The probability of throwing a 1 on the red die is 1/6 and will be 1/6 regardless of the result obtained by tossing the green die. Since the probabilities of getting a 1 on the green die or a 1 on the red die aren't affected by the result on the other die, these events are said to be independent.

If two events are *independent*, the probability that they will *both occur* is the product of their separate probabilities. This may be stated as:

$$P(\text{A and B}) = P(\text{A}) \times P(\text{B})$$

For the problem of throwing a 2 with the dice, the probability is:

$$P(\text{1 on red and 1 on green}) = P(\text{1 on red}) \times P(\text{1 on green})$$
$$= \frac{1}{6} \times \frac{1}{6}$$
$$= \frac{1}{36}$$

Another illustration of independent events is *sampling with replacement*. Let's consider a bowl containing 10 poker chips, 6 red and 4 white. A chip is drawn, its color is noted, and it is replaced in the bowl; then a second chip is drawn. The probability that the second chip will be red or the probability that it will be white isn't affected by the result of the first draw. Therefore, the probability that a sample of two with replacement will result in two red chips is:

$$P(\text{two red}) = P(\text{red on first draw}) \times P(\text{red on second draw})$$
$$= .6 \times .6$$
$$= .36$$

Multiplication Rule for Dependent Events

What if we want to find the probability that both events A and B will happen, but these events *aren't* independent? That is, what if we have a **dependent** (or **conditional**) **events** situation where the probability of occurrence of one event depends on whether or not the other happens? In that case, the probability that *both* of the dependent events will occur is:

$$P(A \text{ and } B) = P(A) \times P(B|A)$$

The term $P(B|A)$ is a dependent or conditional probability and is read "the probability of B given A."

Let's again consider our previous example of the bowl containing six red and four white poker chips. A chip is drawn, and then a second chip is drawn and the first chip isn't replaced. (We are thus *sampling without replacement*.) The probability that the second chip is red or the probability that it is white depends on the result of the first draw. The probability that a sample of two drawn in this fashion results in two red chips is:

$$P(\text{two red}) = P(\text{red on first draw}) \times P(\text{red on second draw|red on first draw})$$

$$= \frac{6}{10} \times \frac{5}{9}$$

$$= \frac{30}{90} \text{ or } \frac{1}{3}$$

For another example of the use of conditional probability concepts, let's assume that the Good Business Bureau conducts a survey of the quality of service offered by the 86 automobile repair shops in its city. The bureau's findings are shown below:

	GOOD SERVICE	QUESTIONABLE SERVICE	TOTAL
New-car dealers	18	6	24
Independent shops	34	28	62
Total	52	34	86

From this table you can determine a number of probabilities. For example, the probability that a shop selected at random from this group of 86 shops is one that provides good service is 52/86, or about .60 (the "good service" column divided by the total number of shops). Or the probability that the shop is an independent garage is 62/86 or about .72. And the probability that an independent shop is also one that gives good service is 34/62 or about .55. Now, if a motorist with a sputtering vehicle limps into one of these 86 shops, what's the probability first that it's an independent garage (IG), and then that it provides good service (GS)? Our formula for dependent events gives us the answer:

$$P(IG \text{ and } GS) = P(IG) \times P(GS|IG)$$

$$= \left(\frac{62}{86} \text{ or } .72 \right) \times \left(\frac{34}{62} \text{ or } .55 \right)$$

$$= .40$$

Of course, we could also have seen from the table that 34 of the 86 shops are independent garages that give good service, and 34/86 = .40.

Self-Testing Review 5-1

1 An urn contains eight red marbles, seven green marbles, and five white marbles. What's the probability that a marble selected at random will be:
   a Either red or green?
   b Either green or white?

2 If one card is drawn from a deck of playing cards, what's the probability that it will be either a face card or a heart?

3 An urn contains 14 red marbles and 6 green marbles. If sampling is done with replacement, what's the probability that a sample of two will contain:
   a Two red marbles?
   b Two green marbles?

4 An urn contains 12 red marbles and 8 green marbles. If sampling is done without replacement, what's the probability that a sample of two will contain:
   a Two red marbles?
   b Two green marbles?

5 Referring back to the Good Business Bureau survey of auto repair shops discussed above, what's the probability that the owner of the malfunctioning car will drive into a new car dealer's shop and then receive questionable service?

6 The members of a country club were surveyed to see how many read the club's newsletter. The table below summarizes the survey results:

	READERS	NONREADERS	TOTAL
Men	65	46	111
Women	38	52	90
Total	103	98	201

   a What's the probability that a member selected at random is a man who ignores the newsletter?
   b What's the probability that the person selected is a woman who reads the newsletter?

EXPECTED VALUE

Probabilities can be put to good use to see if the long-term likelihood of the monetary payoff(s) resulting from an activity exceeds the cost of participating in the activity. For example, if you buy one of the 1,000 raffle tickets that are printed and sold, your probability of winning is 1/1,000. Now let's suppose that one person will win the raffle prize of $10,000. In that case, the ticket you bought has a 1/1,000 chance of winning the $10,000. The expected value of your ticket, then, is [$10,000(1/1,000)] or $10.

This example of expected value is simple to understand, and you could

ORDERS OF MAGNITUDE

Quick: how fast does human hair grow in miles per hour? What is the volume of all the human blood in the world? If you don't know, it's no surprise; even math students sometimes don't, either. The answers are perhaps intriguing: hair grows a little faster than 10 to the minus-8th miles an hour (that is, a decimal point followed by 7 zeros and a 1); the totality of human blood will fill a cube 800 feet on a side, or cover New York's Central Park to a depth of about 20 feet.

What does surprise me, however, is how often I find adults who have no idea of easily imagined numbers: the population of the United States, say, or the approximate distance from the East Coast to the West. Many otherwise sophisticated people have no feel for magnitude, no grasp of large numbers like the federal deficit or small probabilities like the chances of ingesting cyanide-laced painkillers. This disability, which the computer scientist Douglas Hofstadter calls innumeracy, is so widespread that it can lead to bad public policies, poor personal decisions—even a susceptibility to pseudoscience.

Without some intuition for common numbers, it's hard to react with the proper skepticism to terrifying reports of the exotic disease of the month. It's impossible to respond with the proper sobriety to a warhead carrying a megaton of explosive power—the equivalent of a million tons of TNT. Without some feel for probability, car accidents appear to be a relatively minor problem of local travel while being killed by terrorists looms as a major risk of international travel. While 28 million Americans traveled abroad in 1985, 39 Americans were killed by terrorists that year, a bad year—1 chance in 700,000. Compare that with the annual rates for other modes of travel within the United States: 1 chance in 96,000 of dying in a bicycle crash; 1 chance in 37,000 of drowning and 1 chance in only 5,300 of dying in an automobile accident.

There's a joke I like that's marginally relevant. An old married couple in their 90s contact a divorce lawyer. He pleads with them to stay together. "Why get divorced now after 70 years of marriage? Why not last it out? Why now?" Finally, the little old lady responds in a creaky voice, "We wanted to wait until the children were dead." A feeling for which magnitudes are appropriate for various contexts is essential to getting the joke.

It might seem that one way to combat innumeracy is for newspapers to use scientific notation—9.3×10^7 instead of 93 million or 93,000,000; 2.2×10^{13} instead of 22 trillion or 22,000,000,000,000. In a sense we do this when we use the Richter scale for earthquakes, and decibels for sound intensity. But how many people remember that a 6 on the Richter scale indicates an earthquake 10 times as severe as a 5?

Reasonable guess: The media could also discuss probabilities in contexts besides weather forecasting—particularly in medical reporting. This would acclimate people to the concepts involved. Even more helpful when dealing with magnitude would be frequent comparisons that invoke everyday life. For example: knowing that there are approximately a million seconds in 12 days gives one a new grasp of that number. By contrast, it takes almost 32 years for a billion seconds to tick away. And the human species is not much more than a trillion seconds old.

Another quiz: how many trees were cut down last year to print books, articles and newspaper columns on astrology and psychic readings? Whatever the number, its sheer magnitude leads me to an unappreciated consequence of innumeracy: a predisposition to believe in pseudoscience. Consider first an intriguing result in probability—a subject, by the way, that should be taught to everyone. Since a year has 366 days (counting Feb. 29), there would have to be 367 people gathered together for one to be 100 percent certain that at least two people have the same birthday. How many people would be required for one to be just 50 percent certain? A reasonable guess might be 184, about half of 367. The unexpected answer: there need be only 23! For one to be 50 percent certain that somebody has a particular birthday, however—say, July 4—a crowd of 253 would be necessary.

The example may be odd, but the principle is quite general. That *some* unlikely event will come to pass is likely; that a *particular* one will is not. The pronouncements of psychics or astrologers are sufficiently vague so that the probability that *some* prediction will occur is very high. It's the *particular* incidents that are seldom true. Yet the overwhelming majority of incorrect predictions are conveniently forgotten and the correct ones are greatly magnified by publicity. Without a feel for number and chance, people can easily be misled.

The primary reason innumeracy is so pernicious is the ease with which numbers are invoked to bludgeon the innumerate into dumb acquiescence. Even though mathematics deals with certainties, its applications are only as good as the underlying assumptions, simplifi-

cations and estimations that go into them. Any bit of nonsense, for example, can be computerized—astrology, biorhythms, certain items in the military budget—but that doesn't make the nonsense more valid. Statistical projections are invoked so thoughtlessly that it wouldn't be surprising to see someday that the projected waiting period for an abortion is a year.

When I write of innumeracy, I'm not referring to ignorance of any abstract higher mathematics. I'm bemoaning a lack much more basic. Somehow, too many Americans escape education in mathematics with only the haziest feel for numbers and probability and for the ways in which these notions are essential to understanding a complex world.

—John Allen Paulos, "Orders of Magnitude." From NEWSWEEK, November 24, 1986, pp. 12–13, © 1986, Newsweek, Inc. All rights reserved. Reprinted by permission.

justify paying $10 or less for a raffle ticket under these conditions. (You might even pay a little more if the raffle proceeds go to your favorite charity.) Sometimes, though, expected value problems are more complex. Several monetary payoffs (or other nondollar amounts such as the number of items produced) may be called for when different outcomes are realized. Regardless of whether the problem is as simple as our raffle example, or is more detailed, though, an **expected value** (or **mathematical expectation**) is the quantity obtained by multiplying the probability of an event happening by the payoff or amount that will occur in that event, and by then adding all of these results for all possible events. Or stated another way, an expected value is simply a weighted mean of all the possible values or outcomes that a discrete random variable can assume in an experiment. (**A random variable** is one that takes on different values depending on the chance outcome of an experiment.)

Let's look at one more example. Suppose that Peter Parker of Chapters 3 and 4 fame offers to pay you a dollar for each spot that appears on the surface of a die in a single toss. The price charged by Peter for each roll of the die is $3.00. Formula 5-1 gives the expected value for simple or more involved problems, and it's used to compute the expected value of Peter's game:

$$E(X) = \Sigma[xP(x)]$$

(formula 5-1)

where $E(X)$ = expected value of X

x = each value that the X variable can assume

$P(x)$ = The probability of occurrence of that x value

In our raffle example there was only one value for x and it was $10,000. In Peter's game, though, there are six possible outcomes, and six different payout amounts. Figure 5-3 shows how to use formula 5-1 to find the expected value for Peter's game. If you played Peter's game *long enough*, you could expect an average win of $3.50 on each roll of the die. It would be impossible, of course, to win $3.50 on a single roll, but you ought to be able to eventually clean Peter out. If, for example, you play 1,000 games, your expected value will be (1,000) ($3.50) or $3,500, and your cost to play should be about (1,000) ($3.00) or $3,000. Peter again is victimized by statistical obfuscation.

Self-Testing Review 5-2

1   Suppose that Peter changes the rules of his game so that you win $3.00 if you roll a five or a six in a single trial, but if you roll a one, two, three, or four in the trial you must pay Peter $2.75. Should you play this game?
2   In the last year, a bakery has sold as many as nine of its special cakes in one day, but has never sold less than four. The sales records for the year show that only four cakes were sold on 5 percent of the days. Thus, the baker used .05 as the probability that four cakes would be sold (an empirical figure). The following probabilities were assigned in a similar relative frequency way: five cakes = .15; six cakes = .35; seven cakes = .25; eight cakes = .12; and nine cakes = .08. What's the expected amount of daily cake sales?

PROBABILITY DISTRIBUTIONS

A **probability distribution** is simply a complete listing of all possible outcomes of an experiment, together with their probabilities. This concept can be illustrated by the probabilities of throwing various numbers with a

FIGURE 5-3   Computing the Expected Value of Peter Parker's Game

DIE SURFACE (AND DOLLAR VALUE) x	PROBABILITY OF OCCURRENCE OF SURFACE $P(x)$	$xP(x)$
1	1/6	.17
2	1/6	.33
3	1/6	.50
4	1/6	.67
5	1/6	.83
6	1/6	1.00
	1.00	3.50

$E(X) = \Sigma[xP(x)]$
$3.50, the expected average win

pair of dice. This example will also allow us to further illustrate the application of the addition and multiplication rules discussed earlier.

In a previous section we computed the probability of throwing a 2 with a pair of colored dice to be 1/36. Let's now consider the probability of throwing a 3. We can obtain this result either by getting a 1 on the red die and a 2 on the green die or by getting a 2 on the red die and a 1 on the green die. The probability of a 1 on the red die and a 2 on the green die is computed as follows:

$$P(\text{1 on red and 2 on green}) = P(\text{1 on red}) \times P(\text{2 on green})$$
$$= \frac{1}{6} \times \frac{1}{6}$$
$$= \frac{1}{36}$$

The probability of 2 on red and 1 on green is:

$$P(\text{2 on red and 1 on green}) = P(\text{2 on red}) \times P(\text{1 on green})$$
$$= \frac{1}{6} \times \frac{1}{6}$$
$$= \frac{1}{36}$$

We wish to know the probability that *one or the other* of these events will occur. Therefore, we now use the addition rule to determine the probability of throwing a 3 with the pair of dice:

$$P(3) = P(\text{1 on red and 2 on green}) + P(\text{2 on red and 1 on green})$$
$$= \frac{1}{36} + \frac{1}{36}$$
$$= \frac{2}{36}$$

The probabilities of throwing a 4, 5, 6, 7, 8, 9, 10, 11, or 12 may be computed in the same way. The complete probability distribution is given in Figure 5-4.

Note that the total of the probabilities in Figure 5-4 is 1. This is always true for all probability distributions. Since the distribution is a complete listing of *all* outcomes of the experiment, the probability that one or the other of these outcomes will occur is a certainty.

There are several probability distributions that are important in understanding and applying statistical methods. We'll now examine three of these, the binomial distribution, the Poisson distribution, and the normal distribution.

THE BINOMIAL DISTRIBUTION

The **binomial distribution** describes the distribution of probabilities when there are only two possible outcomes for each trial of an experiment. For

FIGURE 5-4   Probability
               Distribution of
               Results of Throwing
               a Pair of Dice

RESULT	PROBABILITY
2	1/36
3	2/36
4	3/36
5	4/36
6	5/36
7	6/36
8	5/36
9	4/36
10	3/36
11	2/36
12	1/36
Total	36/36 = 1

example, when flipping a penny, there are only two possible outcomes, heads or tails. Therefore, the probability distribution showing the probabilities of zero, one, two, three, and four heads in four flips of the penny is a binomial distribution.

This distribution also describes the probabilities associated with many more practical problems. When a manufacturer is trying to determine the quality of a product, each item inspected is either good or defective. In an election poll, each person interviewed either plans to vote for a particular candidate or doesn't expect to vote for that office seeker. In a market survey, each person interviewed either plans to buy a new car this year or doesn't have such an intention.

Combinations—A Brief Digression

There was once a brainy baboon
Who always breathed down a bassoon,
For he said, "It appears
That in billions of years
I shall certainly hit on a tune."
                    —Sir Arthur Eddington

Before discussing how to compute binomial probabilities, it's desirable here to consider briefly the subject of combinations.[1] A **combination** is a selection of r items from a set of n distinct objects *without regard to the order* in which the r items are picked. For example, let's assume that you have one card identified by the letter A, another by the letter B, and a third

---

[1] The subject of *permutations* is often discussed in statistics texts at about this time. We have elected to delete the topic here; however, problem 9 at the end of the chapter briefly explains how permutations differ from combinations.

by the letter C. One combination of two cards picked from this set of three cards is AB, and another combination is AC. How many different combinations of two cards can be made without regard to the order in which the cards appear? That is, how many combinations of three cards taken two at a time are possible? We can answer this terribly difficult question by simply listing the possible combinations: AB, AC, and BC.

The general formula for the number of combinations of n things taken r at a time—or $_nC_r$—is:

$$_nC_r = \frac{n!}{r!(n-r)!}$$ (formula 5-2)

The symbol $n!$ (or n *factorial*) means the product of

$$n(n-1)(n-2)(n-3)\ldots[n-(n-1)]$$

(You may remember that $0!$ is, by definition, equal to 1.) Thus, $6!$ equals $6 \times 5 \times 4 \times 3 \times 2 \times 1$, or 720. We noted above that three combinations of two could be made of the three cards A, B, and C. Verifying this result by our formula gives:

$$_3C_2 = \frac{3!}{(2!)(3-2)!}$$
$$= \frac{3 \times 2 \times 1}{2 \times 1 \times 1}$$
$$= 3$$

And if we wish to know the number of combinations of seven things taken three at a time, we find:

$$_7C_3 = \frac{7!}{(3!)(7-3)!}$$
$$= \frac{7 \times 6 \times 5 \times 4 \times 3 \times 2 \times 1}{(3 \times 2 \times 1)(4 \times 3 \times 2 \times 1)} = \frac{7 \times \overset{1}{6} \times 5 \times \overset{2}{4} \times 3 \times 2 \times 1}{(3 \times 2 \times 1)(4 \times 3 \times 2 \times 1)}$$
$$= 35$$

Back to the Binomial

In dealing with binomial probabilities, one of the two possible outcomes will be regarded as a success and the other will be deemed a failure. A *success* is simply the outcome for which we wish to find the probability distribution, such as a head on a penny or a defective item from a production line. The probability of success in *any one trial* may be identified here as p, and the probability of failure in the *same single trial* may be labeled q. Thus, the sum of p and q is one in all cases. The probability of r successes in n trials is

$$P(r) = (_nC_r)\,(p)^r(q)^{n-r} \qquad\qquad \text{(formula 5-3)}$$

We can illustrate the computation of binomial probabilities with some examples.

Example 1 You're back to flipping that penny. If you toss it four times, what's the probability of getting *zero* heads? For this problem p is the probability of getting heads on any one flip of the penny and is equal to 1/2. The probability of not getting heads, q, is also 1/2. The number of trials, n, is 4, and the number of successes for which we want the probability, r, is 0. The probability of zero heads in four flips is:

$$P(0) = {_4C_0}\,(1/2)^0\,(1/2)^4$$

$$= \frac{4!}{0!\,4!}\,(1)\,(1/16)$$

$$= \frac{1}{16} \quad \text{or} \quad .0625$$

Of course, in addition to the possibility of no heads when a penny is flipped four times, there are also the possibilities of getting one head, two heads, three heads, or four heads. The probabilities for these other possible outcomes are shown below:

$$P(1) = {_4C_1}\,(1/2)^1\,(1/2)^3$$

$$= \frac{4!}{1!\,3!}\left(\frac{1}{2}\right)\left(\frac{1}{8}\right)$$

$$= (4)\,(1/16)$$

$$= \frac{4}{16} \quad \text{or} \quad .25$$

$$P(2) = {_4C_2}\,(1/2)^2\,(1/2)^2$$

$$= \frac{4!}{2!\,2!}\left(\frac{1}{4}\right)\left(\frac{1}{4}\right)$$

$$= \frac{4 \times 3 \times 2 \times 1}{2 \times 1 \times 2 \times 1}\left(\frac{1}{16}\right)$$

$$= \frac{6}{16} \quad \text{or} \quad .375$$

$$P(3) = {_4C_3}\,(1/2)^3\,(1/2)^1$$

$$= \frac{4!}{3!\,1!}\left(\frac{1}{8}\right)\left(\frac{1}{2}\right)$$

$$= \frac{4}{16} \quad \text{or} \quad .25$$

FIGURE 5-5   Probability Distribution
for the Number of
Heads in Four Flips of
a Penny

NUMBER OF HEADS	PROBABILITY
0	1/16
1	4/16
2	6/16
3	4/16
4	1/16
Total	16/16 = 1

$$P(4) = {}_4C_4 \, (1/2)^4 \, (1/2)^0$$

$$= \frac{4!}{4! \, 0!} \left(\frac{1}{16}\right)(1)$$

$$= \frac{1}{16} \quad \text{or} \quad .0625$$

These probabilities are summarized in Figure 5-5, and a histogram of the probability distribution is given in Figure 5-6.

We can also use a statistical software package (*Minitab* in this case) to duplicate our penny-flipping experiment. Figure 5-7*a* shows the "proba-

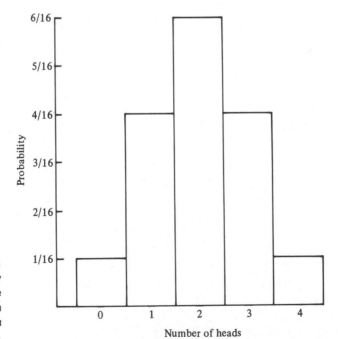

FIGURE 5-6
Probability
distribution of the
number of heads in
four flips of a
penny.

bility density function" or PDF computations for the probability of 0, 1, 2, 3, and 4 heads in n or 4 trials. And Figure 5-7b gives the probability distribution generated by the computer for our binomial experiment. You'll notice that this computer-generated probability distribution is the same as the one shown in Figure 5-5.

Example 2 The output of a production process is 10 percent defective.

```
MTB > PDF 0;
SUBC> BINOMIAL N = 4, P = .50.
        K              P( X = K)
      0.00                0.0625


MTB > PDF 1;
SUBC> BINOMIAL N = 4, P = .50.
        K              P( X = K)
      1.00                0.2500


MTB > PDF 2;
SUBC> BINOMIAL N = 4, P = .50.
        K              P( X = K)
      2.00                0.3750


MTB > PDF 3;
SUBC> BINOMIAL N = 4, P = .50.
        K              P( X = K)
      3.00                0.2500


MTB > PDF 4;
SUBC> BINOMIAL N = 4, P = .50.
        K              P( X = K)
      4.00                0.0625
```

(a)

FIGURE 5-7
(a) The probabilities of getting 0 through 4 heads in four coin tosses. (b) The probability distribution for the number of heads in four tosses.

```
MTB > PDF;
SUBC> BINOMIAL N = 4, P = .50.

      BINOMIAL WITH N =    4  P = 0.500000
        K              P( X = K)
        0                0.0625
        1                0.2500
        2                0.3750
        3                0.2500
        4                0.0625
```

What's the probability of selecting *exactly two* defectives in a sample of *five?* For this problem, $p = .1$ and $q = .9$. Thus, the solution is:

$$P(2) = {}_5C_2\,(.1)^2(.9)^3$$

$$= \frac{5!}{2!3!}\,(.01)\,(.729)$$

$$= \frac{5 \times 4 \times 3 \times 2 \times 1}{2 \times 1 \times 3 \times 2 \times 1}\,(.00729)$$

$$= (10)\,(.00729)$$

$$= .0729$$

The computer-generated probabilities for this event and for *all* the other possible outcomes in this sample of five are given in Figure 5-8, and this probability distribution is shown in Figure 5-9. A comparison of Figures 5-6 and 5-9 shows a marked difference in shape. The probability distribution for the number of heads obtained in four flips of a penny is symmetrical because p and q are equal. However, in Figure 5-9, where p and q aren't equal, we find a marked skewness.

You've seen now how binomial probabilities are calculated by hand and with a computer program. Not surprisingly, tables have been prepared that generate binomial values for a selected number of trials using predetermined probability levels. Such a table is presented in Appendix 1 in the back of the book, and you can look up in this table all the binomial probability values that we've now computed. But the computer comes in handy when the probability of occurrence of an event in a single trial isn't some

FIGURE 5-8
(a) The probability of selecting exactly two defectives in a sample of five. (b) The probability distribution for the number of defectives in a sample of five. The K column shows the number of possible defectives, and the P(X = K) column gives the corresponding probability.

```
MTB > PDF 2;
SUBC> BINOMIAL N = 5, P = .10.
        K              P( X = K)
      2.00                0.0729
```

(a)

```
MTB > PDF;
SUBC> BINOMIAL N = 5, P = .10.

      BINOMIAL WITH N =    5   P = 0.100000
        K              P( X = K)
        0                0.5905
        1                0.3281
        2                0.0729
        3                0.0081
        4                0.0005
        5                0.0000
```

(b)

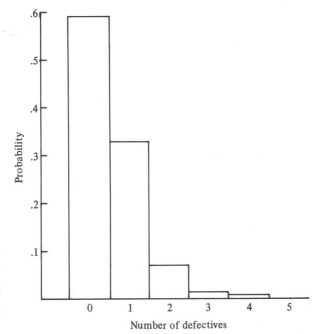

FIGURE 5-9
Probability distribution of the number of defectives in a sample of 5.

convenient value such as .10, .20, .25, ... What if the output of a production process is 16.7 percent defective and you want to find the probability of selecting exactly 2 defectives in a sample of 5? You couldn't use the Appendix 1 table in that case. You could use the binomial distribution formula we've been working with, but you might prefer to let the computer do this work, as it has done in Figure 5-10.

```
MTB > PDF;
SUBC> BINOMIAL N = 5, P = .167.

BINOMIAL WITH N =    5  P = 0.167000
    K              P( X = K)
    0               0.4011
    1               0.4020
    2               0.1612
    3               0.0323
    4               0.0032
    5               0.0001
```

FIGURE 5-10 The probability distribution for the number of defectives in a sample of five when 16.7 percent of all parts are defective. The K column shows the number of possible defectives, and the P(X = K) column gives the corresponding probability.

Self-Testing Review 5-3

1   In six flips of a penny, what is the probability of getting:
   a   Exactly three heads?
   b   Exactly four heads?
   c   At least four heads?

2   If 30 percent of a population own their own homes, what is the probability that a sample of seven from this population will contain:
   a   Exactly two homeowners?
   b   Exactly four homeowners?
   c   At least five homeowners?

THE POISSON DISTRIBUTION

We've just seen that if we know the probability of success in a single trial, the binomial probability distribution can tell us the probability of getting a specific number of occurrences in a given number of trials. But we'll need to use other probability distributions when conditions differ. If, for example, we're interested in the number of specific occurrences that take place per *unit of time or space* rather than during a given number of trials, then it's appropriate to use the **Poisson distribution.** This distribution is named after Simeon Denis Poisson, the Frenchman who developed it in the early 1800s. It can be used, for example, to calculate the probability that there will be a specified number of calls (0, 1, 2, 3, or some other discrete random variable) coming into a telephone switchboard during a given *period of time.* Or, it may be used to calculate the probability that there will be a specific number of flaws found on the *surface space* of a sheet-metal panel used in the production of truck trailers.

Further situations with a time-unit reference that call for the use of the Poisson distribution in probability computations include demand for a product, demand for service, number of accidents, and number of arrivals at toll booths, supermarket check stands, and airports. And another application that's space related is the number of typeset errors per newspaper page. In these and other situations, it's appropriate to use the Poisson distribution if the following conditions are met: (1) the average number of occurrences (μ) is constant for each unit of time or space, (2) the probability of more than one occurrence at any single point in time or space is practically zero, and (3) the number of occurrences in any interval of time or space is independent of the number of occurrences in other intervals.

If the use of the Poisson distribution is appropriate to a given situation, then the probability of observing exactly *x* number of occurrences per unit of measure (hour, minute, square meter, page) can be found using this formula:

$$P(x) = \frac{\mu^x e^{-\mu}}{x!} \qquad \text{(formula 5-4)}$$

where $P(x)$ = probability of exactly x number of occurrences

μ = mean number of occurrences per unit of time or space

e = a constant, the base of the natural logarithm system $(2.71828\dots)$

To see how this formula is used, let's suppose that an average of 3 cars arrive at a highway tollgate every minute. Assuming that this rate is approximated by a Poisson process, what's the probability that exactly 5 cars arrive in a 1-minute period? The answer is:

$$P(x) = \frac{\mu^x e^{-\mu}}{x!}$$

$$P(x = 5) = \frac{(3^5)(2.71828)^{-3}}{5!}$$

$$= \frac{(243)(.0498)}{120}$$

$$= \frac{12.10}{120}$$

$$= 0.1008$$

Thus, the probability that 5 cars arrive in 1 minute is .1008. Other results may be computed in the same way to show the probability of arrival of 0, 1, 2, 3, 4, 6, ... cars at the tollgate.

The table in Figure 5-11 simplifies the use of the Poisson formula. This table gives the values of $e^{-\mu}$ that correspond to selected values of μ.

FIGURE 5-11   Values of $e^{-\mu}$

μ	$e^{-\mu}$	μ	$e^{-\mu}$	μ	$e^{-\mu}$	μ	$e^{-\mu}$
.10	.9048	1.90	.1496	3.60	.0273	5.40	.0045
.20	.8187	2.00	.1353	3.70	.0247	5.50	.0041
.30	.7408			3.80	.0224		
.40	.6703	2.10	.1225	3.90	.0202	5.60	.0037
.50	.6065	2.20	.1108	4.00	.0183	5.70	.0033
		2.30	.1003			5.80	.0030
.60	.5488	2.40	.0907	4.10	.0166	5.90	.0027
.70	.4966	2.50	.0821	4.20	.0150	6.00	.0025
.80	.4493			4.30	.0136		
.90	.4066	2.60	.0743	4.40	.0123	6.10	.0022
1.00	.3679	2.70	.0672	4.50	.0111	6.20	.0020
		2.80	.0608			6.30	.0018
1.10	.3329	2.90	.0550	4.60	.0101	6.40	.0017
1.20	.3012	3.00	.0498	4.70	.0091	6.50	.0015
1.30	.2725			4.80	.0082		
1.40	.2466	3.10	.0450	4.90	.0074	6.60	.0014
1.50	.2231	3.20	.0408	5.00	.0067	6.70	.0012
		3.30	.0369			6.80	.0011
1.60	.2019	3.40	.0334	5.10	.0061	6.90	.0010
1.70	.1827	3.50	.0302	5.20	.0055	7.00	.0009
1.80	.1653			5.30	.0050		

Thus, the value of e^{-3} needed for our example can be found by locating the value of μ (3.00), and by then reading the corresponding value of $e^{-\mu}$ (.0498). However, it's possible to use another table of selected values of the Poisson distribution to look up the answer to our example problem without performing any computations at all. This table is found in Appendix 11 at the back of the book.

The columns in this table represent selected values of μ, so our first step is to locate the column with a μ value of 3.0. Having located the correct column, the next step is to identify the row with an x value of 5 that corresponds to the arrival of 5 cars in our example. The answer to our problem, then, is the value of .1008 that's found at the intersection of the identified column and row. The other entries in the μ column give the probability values for the arrival of 0, 1, 2, 3, etc. cars at the tollgate. Since the total of all the entries under a specific μ column is approximately 1.00, these entries collectively represent a probability distribution.

It probably comes as no surprise to you that a statistical program can easily produce all the results we've found by calculating and/or table-lookup efforts. In Figure 5-12a, the *Minitab* program has printed the Poisson probabilities that 0, 1, 2, 3, . . . cars arrive at the tollgate in a 1-minute period, when a mean of 3 cars stop at the gate every minute. As you can see, the computer-generated probability figure that exactly 5 cars will arrive at the gate in a 1-minute interval is .1008, and this agrees with the answer obtained in our computed example. The sum of all the possible probabilities shown in Figure 5-12a must equal 1.00, as it does when the "cumulative distribution function" or CDF command is used in Figure 5-12b. You can easily see, for example, that the probability is about .92 that from 0 to 5 cars will arrive at the gate during a 1-minute period.

The statistical package also allows an analyst to *simulate* the arrival of cars at the gate in a 60-minute period (assuming, again, that a mean of 3 cars arrive every minute). Figure 5-13 shows this simulation. You'll notice that no cars appeared in 4 of the 60 simulated minutes, while 2 cars and 3 cars arrived in each of 14 and 18 minutes in the hour. The actual counts at the gate in a real 60-minute period will likely differ from this simulated count, and, indeed, another simulation will also produce slightly different counts. But the overall pattern should persist, and this information can be used by highway officials for planning purposes.

Self-Testing Review 5-4

1 Calls at the switchboard of the Charleston Cab Company follow a Poisson process and occur at an average rate of eight calls per hour.
   a What's the probability that there will be exactly six calls in any one hour?
   b What's the probability that there will be exactly 10 calls in an hour?
2 In the preceding problem, what's the probability that there would be five or more—i.e., at least five—calls in an hour?

```
MTB > PDF;
SUBC> POISSON 3.

     POISSON WITH MEAN =   3.000
         K          P( X = K)
         0           0.0498
         1           0.1494
         2           0.2240
         3           0.2240
         4           0.1680
         5           0.1008
         6           0.0504
         7           0.0216
         8           0.0081
         9           0.0027
        10           0.0008
        11           0.0002
        12           0.0001
        13           0.0000
```

(a)

```
MTB > CDF;
SUBC> POISSON 3.

     POISSON WITH MEAN =   3.000
        K  P( X LESS OR = K)
        0           0.0498
        1           0.1991
        2           0.4232
        3           0.6472
        4           0.8153
        5           0.9161
        6           0.9665
        7           0.9881
        8           0.9962
        9           0.9989
       10           0.9997
       11           0.9999
       12           1.0000
```

FIGURE 5-12
Poisson probability
tables for the
highway tollgate
example. (a)
Probability that K
cars will arrive at
gate in one minute.
(b) Cumulative
probability table.

(b)

THE NORMAL DISTRIBUTION

You'll recall from Chapter 3 that a *discrete variable* is one with a countable number of values, while a *continuous variable* is one with an unlimited number of values that can be measured and recorded to some predetermined degree of accuracy. The binomial and Poisson distributions are examples of **discrete probability distributions**—distributions in which there's a

```
MTB > RANDOM 60, SET IN C1;
SUBC> POISSON 3.
MTB > PRINT C1

C1
    4    2    3    5    0    3    5    2    3    2    3    7    4    6    4
    6    3    3    3    3    4    2    2    4    2    4    5    3    1    1
    3    2    6    0    5    6    4    2    3    3    4    3    2    0    3
    2    3    0    3    2    2    2    8    5    3    4    3    4    6    2

MTB > TALLY C1;


SUBC> ALL.


      C1   COUNT  CUMCNT  PERCENT   CUMPCT
       0      4       4     6.67     6.67
       1      2       6     3.33    10.00
       2     14      20    23.33    33.33
       3     18      38    30.00    63.33
       4     10      48    16.67    80.00
       5      5      53     8.33    88.33
       6      5      58     8.33    96.67
       7      1      59     1.67    98.33
       8      1      60     1.67   100.00
      N=     60
```

FIGURE 5-13   Computer simulation of number of cars arriving at tollgate in 60-minute period.

finite number of values a discrete random variable can take. But when we're dealing with a continuous random variable (such as weight, length, or time) that can have an infinite number of values, we need a **continuous probability distribution**—one that can be used to study variables that can be measured to whatever degree of precision is needed.

By far the most important continuous probability distribution is the **normal distribution.** (An expression of its importance—and its shape—was presented by W. J. Youden of the National Bureau of Standards in the manner

TRACKING DOWN THE "NORMAL" DISTRIBUTION

In *The Small House at Allington*, Anthony Trollope, writing about his Plantagenet Palliser—later to become prime minister—as a young man, notes: "Statistics were becoming dry to him, and love was very sweet. Statistics, he thought, might be made as enchanting as ever, if only they could be mingled with love."

I digress for the sake of love to a brief discussion of one of the more fascinating words in statistics, the word "normal." We have in statistics the normal distribution and the normal equations. In other parts of life there are normal schools, normal hydrocarbons, normal ranges in medicine, and (some time ago) "back to normalcy" as a political slogan. . . .

To call the normal distribution normal is more than an arbitrary naming; a normative aspect is one that persists. Some statistics textbooks published even in this decade misleadingly say that the normal distribution is so called because "it is the distribution normally encountered."

The origin of the phrase "normal distribution" is unclear. Stephen Stigler, professor of statistics at the University of Wisconsin, and I have been trying to track it down, with partial success, by starting with the writings of Sir Francis Galton. He wrote *about* the normal distribution at great length, and with an amazing range of elegant variation. Among the terms he used were "the exponential distribution," "the law of error," "the law of deviation," "the Gaussian Law," "the law of statisti-

cal constancy," and, of course, "the normal law," the last first used by Galton in 1877 as far as we can find. Karl Pearson claimed at a later time that he himself had coined the term "normal" as a neutral word to avoid the chauvinistic, competing, nationalistic "Gaussian" and "Laplacian," but it seems to us that his memory was wrong.

Our current hunch is based on the observation that Galton and his contemporaries, when writing of the normal distribution, not only used terms like "the exponential distribution" but, to avoid monotony, brought in vaguer expressions: "the usual distribution," "the commonly encountered distribution," and, of course, "the normal distribution." (Indeed, "normal" in that sense was used still earlier, in 1873, by Charles Sanders Peirce, the great American philosopher and statistician.) We think that somehow "normal" won out among the synonyms. Why should it have won? I speculate that "normal" is a word with powerful, positive connotations because of the ambiguity between its two meanings: (1) something desirable and (2) something commonly found. A major theme in our culture, after all, is the desirability of what is commonly found, so the two senses reinforce each other. . . .

—From William Kruskal, "Formulas, Numbers, Words: Statistics in Prose." Reprinted from THE AMERICAN SCHOLAR, Volume 47, Number 2, Spring, 1978, p. 227. Copyright © 1978 by the author. By permission of the publisher.

shown in Figure 5-14*a*.) As indicated in Figure 5-14*a*, the normal curve is symmetrical and, because of its appearance, is sometimes called a bell-shaped curve. It's actually not a single curve but a family of curves. A particular normal curve is defined by its mean μ and its standard deviation σ. Figure 5-14*b* shows three normal curves with the same mean, but with different standard deviation values; Figure 5-14*c* shows three normal curves with the same standard deviation, but with different mean values.

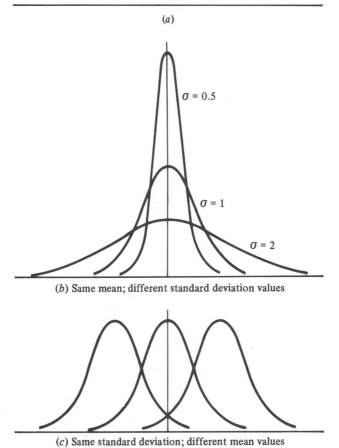

THE
NORMAL
LAW OF ERROR
STANDS OUT IN THE
EXPERIENCE OF MANKIND
AS ONE OF THE BROADEST
GENERALIZATIONS OF NATURAL
PHILOSOPHY ◆ IT SERVES AS THE
GUIDING INSTRUMENT IN RESEARCHES
IN THE PHYSICAL AND SOCIAL SCIENCES AND
IN MEDICINE AGRICULTURE AND ENGINEERING ◆
IT IS AN INDISPENSABLE TOOL FOR THE ANALYSIS AND THE
INTERPRETATION OF THE BASIC DATA OBTAINED BY OBSERVATION AND EXPERIMENT

(*a*)

$\sigma = 0.5$

$\sigma = 1$

$\sigma = 2$

(*b*) Same mean; different standard deviation values

(*c*) Same standard deviation; different mean values

FIGURE 5-14
Normal
distributions.

Areas under the Normal Curve

Probabilities for continuous distributions are represented by areas under the curve. That is, the *probability* that the variable will take on a value *between a and b is the area under the curve between* two vertical lines erected at *points a and b.* For example, if the breaking strength of a material is normally distributed with a mean of 110 pounds and a standard deviation of 25 pounds, the probability that a piece of this material has a breaking strength between 100 and 120 pounds is the area under the curve covered by this interval, as shown in Figure 5-15.

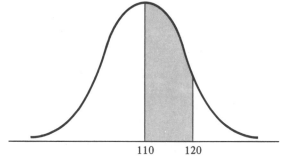

FIGURE 5-15
Probability of
breaking strength
between 110 and
120.

The area covered by an interval under the normal curve is determined by the use of a *table of areas* developed for this purpose. It's impossible, of course, to create tables measured in the actual units of every normal curve. But we can construct a table for what is known as a *standardized normal curve*. This table can then be used to determine probabilities for any normal distribution.

To understand the use of the table of areas, it's necessary to understand the relationship of the *standard deviation* to the normal curve. An interval of a given number of standard deviations from the mean covers the same area in any normal curve. Thus, the interval from 50 to 70 for a normal curve with a mean of 50 and a standard deviation of 20 covers the same area as the interval from 170 to 200 in a normal curve with a mean of 170 and a standard deviation of 30, since both of these intervals cover a distance of one standard deviation from the mean (see Figure 5-16).

Using the Table of Areas

The table of areas under the normal curve (see Appendix 2 at the back of the book) shows the area between the mean and a given number of standard deviations from the mean. The symbol Z is used to represent this number of standard deviations. Thus, a **Z value** (or **Z score** or **standard score**)

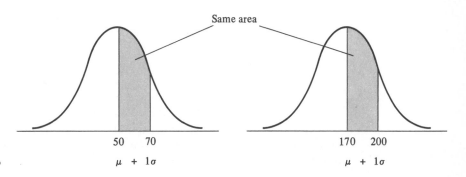

FIGURE 5-16

is the *difference* between any value in a data set (x) and the mean of all the x values (μ) divided by the standard deviation (σ). Or,

$$Z = \frac{x - \mu}{\sigma} \qquad \text{(formula 5-5)}$$

How is the table of areas used? Thank you for asking. Let's assume that we have a normal distribution with a mean of 50 and a standard deviation of 20. Let's also assume that one of the items in the data set used to compute the mean of 50 has a value of 75. How many Z values (or standard deviations) from the mean will 75 fall? The answer is:

$$Z = \frac{x - \mu}{\sigma}$$
$$= \frac{75 - 50}{20}$$
$$= 1.25$$

Thus, a value of 75 will lie 1.25 standard deviations to the *right* of the mean of 50. Similarly, the Z value for a figure of 25 in the data set is:

$$Z = \frac{x - \mu}{\sigma}$$
$$= \frac{25 - 50}{20}$$
$$= -1.25, \text{ or } 1.25 \text{ standard deviations to the left of the mean of 50}$$

The first column of the table of areas in Appendix 2 gives values of Z to one decimal place. The remaining columns give the second decimal place. To find the area for $Z = 1.25$, go down the column to 1.2 and across to the column headed .05. The area given by the table is .3944.

Since the normal curve is symmetrical, what's true for one-half of the curve is true for the other half. Therefore, the table gives areas for only one-half of the curve, and the area for $Z = -1.25$ is also .3944. All right, you may say, but what's the meaning of this .3944? This question is answered in the next section.

Computing Normal Curve Probabilities

Let's pause here just long enough to emphasize two precautions. *First*, remember that Z is the number of standard deviations *from the mean*. The interval from $Z = 1$ to $Z = 2$ is one standard deviation wide, but it *does not* cover the same area as the area between the mean and one Z value.

Second, remember that the normal curve is a probability distribution, and thus the total area under the curve is 1. Therefore, the area covered by one-half the curve is .5.

To determine the probability of getting a value within a particular interval, *the following procedure should be used:*

1 Determine the Z value for each limit of the interval.
2 From the table of areas, determine the *area* for each Z value.
3 If both limits of the interval are on *opposite sides* of the mean, *add* the areas determined in the previous step. If the limits are on the *same side* of the mean, *subtract* the smaller area from the larger one.

Let's illustrate this procedure with several examples—all of which deal with the life expectancy of light bulbs whose lifetimes are normally distributed with a mean life of 750 hours and with a standard deviation of 80 hours.

Example 1   What's the probability that a light bulb will last between 750 hours and 830 hours (see Figure 5-17)?

For 830 hours, Z $= \dfrac{830 - 750}{80}$
$= 1.00$

For Z = 1.00, the area = .3413

The probability that a light bulb will last between 750 and 830 hours is thus .3413 (the shaded area in Figure 5-17).

Example 2   What's the probability that a light bulb will last between 790 hours and 870 hours (see Figure 5-18)?

For 790 hours, Z $= \dfrac{790 - 750}{80}$
$= .50$

For 870 hours, Z $= \dfrac{870 - 750}{80}$
$= 1.50$

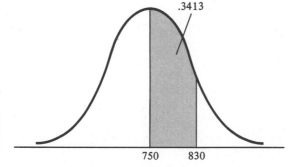

FIGURE 5-17
Probability of a
light bulb lasting
between 750 and
830 hours.

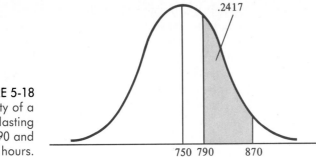

FIGURE 5-18
Probability of a
light bulb lasting
between 790 and
870 hours.

Then, for Z = 1.50, the area = .4332
and for Z = .50, the area   = .1915
and the probability is $\overline{.2417}$

The *difference* between the two areas is .2417, which is both the shaded area in Figure 5-18 and the probability that a bulb will last between 790 and 870 hours.

Example 3   What's the probability that a light bulb will last between 730 hours and 850 hours (see Figure 5-19)?

$$\text{For 730 hours, Z} = \frac{730 - 750}{80}$$
$$= -.25$$
$$\text{For 850 hours, Z} = \frac{850 - 750}{80}$$
$$= 1.25$$

Then for Z = −.25, the area = .0987
and for Z = 1.25, the area   = .3944
and the probability is $\overline{.4931}$

The *sum* of the two areas is .4931, which is both the shaded area of Figure 5-19 and the probability that a bulb will last between 730 and 850 hours.

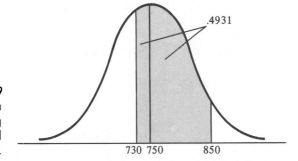

FIGURE 5-19
Probability of a
light bulb lasting
between 730 and
850 hours.

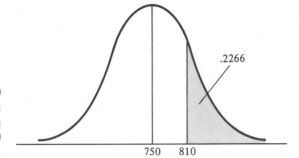

FIGURE 5-20
Probability of a
light bulb lasting
more than 810
hours.

.2266

750   810

Example 4   What is the probability that a light bulb will last more than 810 hours (see Figure 5-20)?

For 810 hours, Z $= \dfrac{810 - 750}{80}$

$= .75$

For Z = .75, the area = .2734

Thus, if .2734 is the area between 750 and 810 hours, the area *beyond* 810 hours must be the *difference* between .5000 (the total area greater than the mean of 750 hours) and .2734. In short, the probability that a bulb will last more than 810 hours (and the shaded area in Figure 5-20) is .2266.

Example 5   What's the probability of a light bulb lasting less than 770 hours (see Figure 5-21a)?

For 770 hours, Z $= \dfrac{770 - 750}{80}$

$= .25$

The area less than the mean of 750 = .5000
and for Z = .25, the area   = .0987
                                     ‾‾‾‾‾‾
                                     .5987

The answer is the *sum* of the areas (and probabilities) or .5987. This same probability figure is produced by the *Minitab* software package in Figure 5-21b. The user supplied the "cumulative distribution function" (CDF) command and specified an x value of 770 hours on line 1. The program was told that a normal probability distribution with a mean of 750 hours and a standard deviation of 80 hours was to be used (line 2). The program then supplied the correct answer in line 3.

Example 6   What's the probability that a light bulb will last more than 670 hours (see Figure 5-22)?

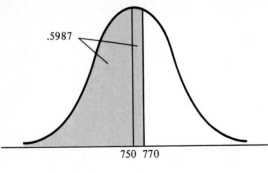

.5987

```
MTB > CDF 770;
SUBC> NORMAL 750, 80.
  770.0000      0.5987
```

750 770

(a)   (b)

FIGURE 5-21   (a) Probability of a light bulb lasting less than 770 hours. (b) Computer printout showing probability of light bulb lasting less than 770 hours.

FIGURE 5-22
Probability of a
light bulb lasting
more than 670
hours.

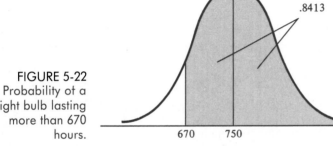

.8413

670   750

For 670 hours, Z $= \dfrac{670 - 750}{80}$

$= -1.00$

The area greater than the mean of 750 = .5000

and for Z $= -1.00$, the area $= \dfrac{.3413}{.8413}$

The *sum* of the areas (and probabilities) is .8413, and this is the answer.

Self-Testing Review 5-5

1 The breaking strength of a material is normally distributed with a mean of 90 pounds and a standard deviation of 20 pounds. What is the probability that a piece of this material will have a breaking strength:
   a Between 90 and 114 pounds?
   b Between 95 and 110 pounds?
   c Between 80 and 110 pounds?
   d Greater than 70 pounds?
   e Greater than 100 pounds?

2 Interpret the computer-generated results below for the light bulb data given previously in:

a Example 4.

b Example 6.

```
MTB > CDF 810;              MTB > CDF 670;
SUBC> NORMAL 750, 80.       SUBC> NORMAL 750, 80.
   810.0000     0.7734         670.0000     0.1587
```

(*a*) Printout   (*b*) Printout

LOOKING BACK

1 Probability is defined as the chance or likelihood that an event will occur. A probability of 0 means an event can never occur, and a probability of 1 means it will always occur. Probabilities of interest, of course, lie between these two extremes. Probabilities are assigned using *a priori,* relative frequency (empirical), and subjective approaches. An *a priori* probability is one that can be determined in advance without experimentation. The validity of the *a priori* probabilities assigned to coin flips and die tosses have been checked in the chapter with computer simulations. A relative frequency probability is one that's determined after the fact from observation and experimentation, and subjective probability is assigned on the basis of someone's personal judgment.

2 Two basic rules are used during probability computations. The first is an addition rule that's used, for example, to calculate the probability that one or the other of two events will happen. The second is a multiplication rule that's used to determine the probability that both events will happen. The addition rule is used with mutually exclusive events and in some other situations. The multiplication rule is used in joint probability cases where the events are either independent or dependent. An independent event is one where the occurrence or nonoccurrence of one outcome has no effect on the probability that the other happens. A dependent (or conditional) event is one where the probability of occurrence of one outcome depends on whether or not the other happens.

3 An expected value (or mathematical expectation) is the quantity obtained by multiplying the probability of an event happening by the payoff or amount that will occur in that event, and by then adding all of these results for all possible events. We've seen how to compute expected values in this chapter.

4 A probability distribution is a complete listing of all possible outcomes of an experiment, together with their probabilities. We've seen how to construct discrete probability distributions by computing the probabilities of throwing various numbers with a pair of dice, and by calculating the probabilities of getting zero to four heads in four flips of a coin. This coin-flipping example created a binomial distribution because there are only two possible outcomes for each trial of the experiment. The probability of exactly *r* successes in *n* trials in a binomial situation can be

computed with formula 5-3, and this formula requires that you be able to compute the number of combinations of *n* things taken *r* at a time. Formula 5-2 shows how to do this. Of course, binomial probability values can also be found in special tables such as the one in Appendix 1 in the back of the book, or a statistical software package can generate these values.

5 Another type of discrete probability distribution is the Poisson distribution—one that's used if an analyst is interested in the number of specific occurrences that take place per unit of time or space rather than during a given number of trials. We've seen an example of how to compute the probability that exactly 5 (or 2, 3, 4, . . .) cars arrive at a highway tollgate in a 1-minute period. And we've seen how statistical software can calculate Poisson probability values and simulate an experiment using Poisson values.

6 The binomial and Poisson distributions are examples of discrete probability distributions, but when we're dealing with a continuous random variable (such as the life expectancy of light bulbs) we need to use a continuous probability distribution. By far the most important continuous probability distribution is the normal distribution. The probability that a variable will take on a value between *a* and *b* is the area under the normal curve between two vertical lines erected at points *a* and *b*. And the area covered by an interval under the normal curve is found by using a table of areas such as the one found in Appendix 2 at the back of the book. This table shows the area between the mean of a normal curve and a given number of standard deviations from that mean. The symbol *Z* is used to represent this number of standard deviations. Several light bulb examples are discussed in the chapter to show how to compute normal curve probabilities.

KEY TERMS AND CONCEPTS

Probability   *169*
Computer simulation   *170*
a priori probability   *170*
Relative frequency (empirical) probability   *170*
Subjective probability   *171*
Addition rule   *172*
Multiplication rule   *173*
Mutually exclusive events   *173*
Joint probability   *174*
Independent events   *174*
Dependent (conditional) events   *175*
Expected value (mathematical expectation)   *178*
Random variable   *178*
$E(X) = \Sigma[xP(x)]$ (formula 5-1)   *178*
Probability distribution   *179*

Binomial distribution   *180*
Combination   *181*
$$_nC_r = \frac{n!}{r!(n-r)!} \text{ (formula 5-2)}\quad 182$$
$P(r) = (_nC_r)(p)^r(q)^{n-r}$ (formula 5-3)   *183*
Poisson distribution   *188*
$$P(x) = \frac{\mu^x e^{-\mu}}{x!} \text{ (formula 5-4)}\quad 188$$
Discrete probability distributions   *191*
Continuous probability distribution   *192*
Normal distribution   *192*
Z value (Z score or standard score)   *195*
$$Z = \frac{x - \mu}{\sigma} \text{ (formula 5-5)}\quad 196$$

PROBLEMS

1 What's the probability of throwing a 1 or 2 with a single die?
2 If a single card is drawn from a deck of playing cards, what's the probability that it will be either a black card or an ace?

3 A box of parts contains eight good items and two defective items.
 a What's the probability that in a sample of two, taken with replacement, both items will be good?
 b What's the probability that they will both be defective?

4 A box of parts contains seven good items and three defective items.
 a What's the probability that in a sample of two, taken without replacement, both items will be good?
 b What's the probability that they will both be defective?

5 If a penny is flipped seven times, what's the probability of getting:
 a Exactly four heads?
 b Exactly five heads?
 c Four or five heads?

6 If 60 percent of a population are Democrats, what's the probability that a sample of six from this population will contain:
 a Exactly four Democrats?
 b Exactly five Democrats?
 c Four or five Democrats?
 d At least four Democrats?

7 The lifetimes of batteries produced by a firm are normally distributed with a mean of 110 hours and a standard deviation of 10 hours. What's the probability that a battery will last:
 a Between 110 and 115 hours?
 b Between 107.5 and 120 hours?
 c Between 112 and 123 hours?
 d Between 90 and 102 hours?
 e More than 113 hours?

8 The scores of students on a standardized test are normally distributed with a mean of 300 and a standard deviation of 40. What's the probability that a student will score:
 a Between 310 and 330?
 b Between 280 and 340?
 c Less than 320?
 d More than 260?
 e More than 380?

9 *Permutations.* We saw in the chapter that *order was not important* in considering the number of possible combinations—i.e., the card combination AB was considered the same as BA. *Permutations differ from combinations in that order is important*—i.e., AB is one permutation, and BA is a second permutation. The formula for the number of permutations of *n* things taken *r* at a time is:

$$nPr = \frac{n!}{(n-r)!}$$

 a In how many different ways can a football fan enter a stadium by one gate and leave by a different gate if there are 15 gates in the stadium?
 b In how many different ways can a committee of 3 be chosen from a group of 10 deans? (Are you sure this is a permutation?)

10 There are 50 students in the M.B.A. program at Systems Tech. In this class, 20

students are taking statistics, 15 are taking finance, and 10 are taking both statistics and finance. If a student is chosen at random:

   a   What's the probability that he or she is taking either statistics or finance?

   b   What's the probability that he or she is not taking statistics?

11 The diameters of a particular group of parts are normally distributed with a mean of 2 inches and a standard deviation of 0.2 inches. If a part is chosen at random, what's the probability that it will have a diameter:

   a   Between 1.8 and 2.1 inches?

   b   Between 1.5 and 1.7 inches?

   c   Greater than 2.2 inches?

   d   Less than 2.3 inches?

12 An automobile dealership in Denver has compiled the following sales data over the past year:

NUMBER SOLD PER WORK DAY	RELATIVE FREQUENCY
0	.20
1	.30
2	.30
3	.15
4	.05
	1.00

   a   What's the expected value of daily automobile sales?

   b   Based on the expected value figure, how many cars would the dealer expect to sell in 200 days?

13 If an investor has a .60 probability of making a $20,000 profit, and a .40 probability of suffering a $25,000 loss, should she make the investment on the basis of her expected value?

14 If a life insurance company offers a $1,000 1-year term policy to a woman at age 48 for an annual premium of $15, and if insurance tables show that there are five deaths for every 1,000 women in this age group, what's the expected gain for the company, assuming that large numbers of such policies are issued?

15 If a keyboard operator averages two errors per page of newsprint, and if these errors follow a Poisson process, what's the probability that exactly four errors will be found on a given page?

16 In the preceding problem, what's the probability that at least two errors will be found?

17 Police records show that there has been an average of four accidents per week on the West Freeway. Assuming these accidents follow a Poisson process, what's the probability that the police must respond to exactly six accidents in a given week?

18 The Bobbs Trailer Manufacturing Company uses large panels of sheet metal in the truck trailers it produces. If there's an average of two blemishes per panel, and if the blemishes follow a Poisson process, what's the probability that there will be no blemishes found on a given panel?

19 If the proportion of families with two children in Suburbia is .30, and the proportion of families with three, four, and five or more children respectively is

.30, .10, and .05, what's the probability that a family selected at random has three or more children?

20 A shipment of 10 stepladders includes one that's defective. What's the probability that if two ladders are selected at random and inspected neither of the two will be defective?

21 Sue Ellen Cosmetic Company's average salesperson has sales of $200,000 per year. Company records indicate a normal distribution and a standard deviation of $25,000. What percentage of the salespersons sell $230,000 per year or more?

22 The scores made by students at Macbeth University on a standardized test are normally distributed with a mean of 300 and a standard deviation of 40. What's the probability that a student selected at random will score between 310 and 330?

23 If a box of transistors contains 18 good components and two defective ones, what's the probability that if a first transistor is selected, tested, and then replaced before a second one is drawn, both components will be good?

24 John and Margaret Malone have four children. What's the probability that they have at least two boys?

25 If the average verbal SAT score of all students attending McGuire College is 540, and if the population standard deviation is 30 SAT points, what's the probability that the verbal SAT score of a student selected at random is greater than 535?

26 A pair of dice is rolled in a game of chance. The payoff in dollars is equal to the combined number of spots on the dice. Thus, a single spot on each die results in a payoff of $2, a one-spot on one die and three spots on the second die yields a payoff of $4, and so on. How much should one be willing to pay to play this game? (Hint: Figure 5-4 can help you answer this question.)

27 A group of defense contractors estimates that they can build a missile defense system for New York City that will be 95 percent effective at stopping incoming missiles. A potential enemy is estimated to have 20 warheads targeted on New York.

    a What's the probability that New York will survive an attack (assume independent probabilities)?

    b What's the probability that New York will survive an attack if the potential enemy targets 40 warheads on New York?

28 A survey is conducted to determine eye and hair color in a population. The results are given below:

HAIR COLOR	BLUE EYES	BROWN EYES	TOTAL
Blond	10	30	40
Black	40	100	140
Red	5	25	30
Totals	55	155	210

    a What's the probability that a person selected at random from this survey group is a blue-eyed blond?

    b What's the probability that a person has red hair?

    c What's the probability that a person has blue eyes?

29 A drawer contains 6 black socks and 15 red socks.
    a What is the probability of drawing a black sock at random?
    b What's the probability of drawing two red socks at random assuming the first sock is put back in the drawer before the second is selected (that is, drawing with replacement)?
    c What's the probability of drawing two red socks at random without replacement?
    d What's the probability of drawing one red sock and one black sock without replacement?
    e What's the probability of drawing two black socks at random without replacement? (*Hint:* the probabilities in parts c, d, and e must total what value?)

30 A die is cast six times. What's the probability of obtaining
    a Exactly 1 six?
    b Exactly 2 sixes?
    c At least 3 sixes?

31 The typical applicant at Pristine U. has a 40 percent chance of passing the entrance exam. If the total number of applicants is 30 on a particular day,
    a What's the probability that exactly 9 applicants will pass the entrance exam? (*Hint:* the binomial probability tables in Appendix 1 at the back of the book can be used here and in the other parts of this problem.)
    b What's the probability that the number of applicants who pass the exam will lie between 10 and 14 inclusive?
    c What's the probability that exactly 12 students will pass?

32 Let's assume now that the typical applicant at Pristine U. has a 65 percent chance of passing the entrance exam. What's the probability that exactly 19 of 30 applicants will pass the exam?

33 The fire department in Crossville can put out a fire in 1 hour, and the average number of alarms per hour is 2.4.
    a What's the probability that no alarms are received for 1 hour? (*Hint:* The Poisson probability distribution table in Appendix 11 can be used here.)
    b What's the probability that no alarms are received for 2 hours?
    c What's the most probable number of alarms in an hour (use integer values), and what's the probability that exactly this integer number will be received in an hour?
    d If the department has three trucks, what's the probability that an alarm will go unanswered in a 1-hour period because all trucks are busy?
    e By how much is the probability in part d reduced if an extra truck is added?

34 Transmission failures on the XPS automobile are normally distributed, the mean lifetime of an XPS transmission is 70,000 miles, and the standard deviation is 10,000 miles.
    a What's the probability that an XPS transmission will last at least 70,000 miles?
    b What's the probability that a transmission will fail between 65,000 miles and 80,000 miles?
    c Suppose that the manufacturer of the XPS wants to limit warranty work on the transmission to no more than 20 percent of the cars sold. What would you advise about the transmission warranty period?

TOPICS FOR REVIEW AND DISCUSSION

1 a What is probability?
   b How can a computer program simulate a probability experiment?
2 Explain the difference between *a priori* probability and empirical probability.
3 "Two basic rules are used during probability computations. The first is an addition rule and the second is a multiplication rule." Explain when each of these rules should be used.
4 "In determining a joint probability, it's necessary to know if the two events are independent." Discuss this statement.
5 How does a conditional probability event differ from an independent probability event?
6 Define a probability distribution, and give an example of how one could be created.
7 Discuss the procedure used to find the expected value of a discrete random variable.
8 a What is a binomial probability distribution?
   b Why is the formula for combinations incorporated into the formula for computing binomial probabilities?
9 How may a Poisson distribution be used?
10 "The binomial and Poisson distributions are examples of discrete probability distributions." Discuss this sentence.
11 "Probabilities for continuous probability distributions are represented by areas under the curve." Discuss this statement.
12 Discuss the procedure for computing normal curve probabilities.

PROJECTS/ISSUES TO CONSIDER

1 Conduct a library search to locate an article that discusses probability concepts/applications or the use of probability distributions. Prepare a brief summary of the article for a class presentation.
2 Team up with several of your classmates and research the contributions of two of those who pioneered in the development of probability concepts, probability distributions, and other statistical topics. Some (but not all) of the names your team (and other class teams) might consider are: Thomas Bayes, Jacques Bernoulli (and other members of the Bernoulli family), Francis Galton, William S. Gossett, Karl Friedrich Gauss, Christian Huygens, Pierre-Simon de Laplace, Abraham De Moivre, Blaise Pascal, Karl Pearson, and Simeon Denis Poisson. Outline the contributions of those your team has selected in a brief class presentation.

ANSWERS TO SELF-TESTING REVIEW QUESTIONS

5-1
1 a .75
  b .6

2 $\dfrac{22}{52}$ or   .4231

3 a .49
   b .09

4 a $\dfrac{132}{380}$ or   .34737

   b $\dfrac{56}{380}$ or   .14737

5 Let's abbreviate new-car shop as NCS and questionable service as QS. Then:

$$P \text{ (NCS and QS)} = P \text{ (NCS)} \times P(\text{QS}|\text{NCS})$$
$$= \dfrac{24}{86} \times \dfrac{6}{24}$$
$$= .28 \times .25$$
$$= .07$$

6 a $P(\text{man and nonreader}) = P \text{ (man)} \times P \text{ (nonreader}|\text{man)}$
$$= \dfrac{111}{201} \times \dfrac{46}{111}$$
$$= .55 \times .41$$
$$= .23$$

   b $P(\text{woman and reader}) = P \text{ (woman)} \times P \text{ (reader}|\text{woman)}$
$$= \dfrac{90}{201} \times \dfrac{38}{90}$$
$$= .45 \times .42$$
$$= .19$$

5-2

1 $E(X) = -\$2.75(1 \text{ or } 2 \text{ or } 3 \text{ or } 4) + \$3(5 \text{ or } 6)$
$$= -\$2.75\left(\dfrac{1}{6} + \dfrac{1}{6} + \dfrac{1}{6} + \dfrac{1}{6}\right) + \$3\left(\dfrac{1}{6} + \dfrac{1}{6}\right)$$
$$= -\$2.75\left(\dfrac{4}{6}\right) + \$3\left(\dfrac{2}{6}\right)$$
$$= -\$1.83 + \$1.00$$
$$= -\$0.83$$

2 You shouldn't play; Peter's trying to recoup his losses.

CAKES SOLD PER DAY x	PROBABILITY FOR NUMBER SOLD P(x)	xP(x)
4	.05	.20
5	.15	.75
6	.35	2.10
7	.25	1.75 .
8	.12	.96
9	.08	.72
	1.00	6.48

The expected amount of daily cake sales = 6.48 cakes.

5-3

1 a .3125

 b .2344

 c The answer here is the *sum of* the probabilities of getting four heads, five heads, and six heads. That is, the probability of getting at least four heads is .2344 + .0938 + .0156, or .3438.

2 a .3177

 b .0972

 c The sum of the probabilities of five, six, and seven homeowners is .0250 + .0036 + .0002, or .0288.

5-4

1 a With $\mu = 8$ and $x = 6$, the probability from Appendix 11 is .1221.

 b With $\mu = 8$ and $x = 10$, the probability from Appendix 11 is .0993.

2 From Appendix 11, we see that the probabilities of 0, 1, 2, 3, and 4 calls totals .0996. Therefore, the probability of getting five or more calls is 1.000 − .0996 or .9004.

5-5

1 a .3849

 b .2426

 c .5328

 d .8413

 e .3085

2 a In line one the user asked for a cumulative distribution function (CDF) that gives the area under a normal curve *up to the specified x value*, which is 810 hours in this case. (The mean of 750 hours and the standard deviation of 80 hours are supplied by the user in line two of the printout.) The program responded with a probability value of .7734, which is the cumulative probability under the curve up to the 810-hour position. That is, the probability is .7734 that a bulb will last less than 810 hours. You saw in Example 4 that the probability is .2266 that a bulb will last more than 810 hours. Both probability figures, of course, add to 1.

 b In this printout, the computer supplies the probability that a bulb will last up to, but not more than, 670 hours. This .1587 value and the .8413 probability figure given in Example 6 must total 1.

A CLOSER LOOK

MISAPPLICATIONS REVIEWS: ICE FOLLIES

In mid-February, is there any place worse to reside than the Northeastern United States? Of course there is; the Arctic Circle comes to mind. But the cumulative effect of two months of Northeastern winter, coupled with the knowledge that any real spring weather is at least two months away, creates a deep freeze of the heart that hardens us all in great despair.

Now there is some evidence that the crisis of the spirit is accompanied by a crisis of the intellect. In a two-day period starting on February 16, 1986, *The New York Times* published [two] distinct pieces of statistical silliness. Lest this indication of our mental hibernation disappear with the melting snow, I decided to use this column to record it. That we lose brain cells in the cold the way deciduous trees lose their leaves is, I would think, a point of great biological interest.

A MIRACLE ON THE JERSEY SHORE
In its review of the previous week's news, the *Times* of February 16 included the following item:

> The odds were one in 3,200,000 when Evelyn Marie Adams won almost $4 million in the New Jersey state lottery in October. Last week, Mrs. Adams hit on a real longshot— one in 17,300,000,000,000, according to a Rutgers University professor—to win again, making the manager of a 7-Eleven store in Point Pleasant Beach the first two-time million-dollar winner in any of the nation's 22 state lotteries. Since winning in October, Mrs. Adams had increased from 25 to 100 the number of $1 tickets she purchased each week. After laying claim to her new $1.4 million prize, she reflected, "They say good things come in threes, so. . . ."

Actually, the *Times* had calmed down a bit by February 16; two days earlier, it had hailed Mrs. Adams for "defying odds so great as to be almost preposterous."

What is objectionable about the quoted paragraph? For one thing, it implies that one in 1.7×10^{13} is the probability of Mrs. Adams's second victory *in itself*. But her accomplishment on that occasion—identifying which six numbers between one and 42 would be drawn in the lottery—was an event with prior probability of one in 5.2 million. Evidently, the statistic from Rutgers was the product of this latter number and the chance of the original win (one in 3.2 million); hence it was actually some sort of joint probability. And precisely which joint probability was it? It was the chance that if Mrs. Adams had purchased exactly two lottery tickets, they would have yielded the two jackpots that brought her fame. Such an event, of course, bears no relationship to what really happened. Mrs. Adams's first victory (on something like the 3,000th ticket she bought) was not in itself extraordinary: so many people bet in these high-stakes lotteries that some jackpot winners are inevitable. (In Massachusetts, there are sometimes five times as many bets as there are possible lottery outcomes.) And Mrs. Adams's strategy after her first win of buying 100 tickets per week meant that, over the next year, the chance was roughly one in 800 that she would sooner or later hit the jackpot again.

Given that Mrs. Adams's second triumph— whenever it occurred during the year—would have provoked delirium in the newspapers, the chance that we would be informed of her "one-in-17-trillion" achievement was itself around one in 800. In other words, the press was preprogrammed to depict as miraculous that which was merely unusual.

And there is yet another point to bear in mind. Presumably many other jackpot winners also continued to play their state lotteries, which means that the chance of getting some double winner in 1986—whether in New Jersey or elsewhere—was probably substantial.

Indeed, we cannot even ascertain whether, given the betting patterns of former winners, the time until we witnessed the first double triumph fell below the median of its distribution. Against this backdrop, the astonished gasps of *The New York Times* seem ludicrous.

My hope now is that Mrs. Adams will again quadruple her betting (to $400 per week) and continue at that level for (say) 20 years. If she does, the probability is something like one in 10 that someday she will be a triple jackpot winner. Should that happen, we can all enjoy the efforts by excitable professors to describe the rarity of her feat ("less likely than having Halley's Comet take the Lanchester Prize in an off year"). Mrs. Adams could then offer the more rational explanation that good things come in threes—especially when one spends a few hundred thousand dollars to help them along. . . .

WASHINGTON BENEATH THE SURFACE

The *Times* of February 17 contained the following innocuous item:

> The other day on the Metro, a young woman was overheard telling a companion that six out of ten Washingtonians were lawyers. When this lopsided statistic was challenged, she interviewed fellow riders at random, and eight of the first ten people she asked said yes, they were indeed members of the legal profession. The District of Columbia Bar Association, however, cast doubt on this survey. The city has 44,000 lawyers, a bar official said; in a population of about 625,000 this comes to seven in 100. Which leads to the conclusion that either a lot of lawyers ride the subway or a lot of subway riders have professional pretensions.

It might be petty to quibble about this amiable anecdote, but I can see another resolution to the mystery. Imagine that at 8:30 on a Tuesday morning on the IRT subway near Wall Street, a survey was taken about which jobs the commuters held. An amazingly large fraction of respondents might identify themselves as stockbrokers, but that would reflect neither mendacity nor the fact [that] stockbrokers ride the subway a lot (that is, more than people with other occupations). It would merely demonstrate the eccentric conditions of the survey.

Perhaps what happened in Washington was that the woman, sensing from their appearance and demeanor that there were many lawyers around her, was prompted to make her exaggerated remark. The skeptic, treating the woman as something of a Poisson generator of demographic statistics, insisted on an immediate hypothesis test. The upshot was that the very conditions that provoked the peculiar conjecture also provided its apparent confirmation.

FINAL WORD

These items—which the *Times* clearly intended to be light-hearted—might reflect nothing more important than seasonal aberrations. But they also raise another possibility. Perhaps the most prestigious newspaper in the world has somehow been transformed into a high-sprouting geyser of superficial statistical statements. Should that be the case, then we must prepare to wear our mathematical raincoats every day of the year and to help dry a hapless populace that will be drenched by fallacious arguments.

6

SAMPLING CONCEPTS

Misuses of a Term

Another often misused statistical expression with a precise technical meaning is "random sample." I shall not present a one-paragraph introductory description here (although that would not be difficult), but shall content myself with illustrating the range of misuse.

One misuse is pejorative; for it, a random sample is an erratic grab bag. An example of that misuse occurs in a book review by Allan K. Wildman in the *Journal of Modern History:* "What we are given is a careless and incomplete random sampling to illustrate a passing point." At the other extreme, the term "random sample" may be used to provide a spuriously positive scientific effect, like the white jacket worn by an actor in a television toothpaste commercial. For example, in a 1973 *New York Times* article, Lawrence K. Altman reports a Swedish study on views of death. He paraphrases a Swedish cardiologist: "Dr. Biörk . . . emphasized that his analysis was . . . not a scientific study based on a random sample." In terms of the statistical meaning of "random sampling," it is neither necessary nor sufficient for ensuring that a study be scientific.

—From William Kruskal, "Formulas, Numbers, Words: Statistics in Prose." Reprinted from THE AMERICAN SCHOLAR, Volume 47, Number 2, Spring, 1978, p. 225. Copyright © 1978 by the author. By permission of the publisher.

LOOKING AHEAD

In the opening vignette, the author discusses the misuse of the term "random sample," and mentions that the term could be easily introduced in a one-paragraph description. Well, you'll find a one-paragraph introduction to the random or probability sample in Chapter 6. In later chapters, you'll be introduced to a number of statistical inference procedures that are based on the use of random samples. Before using these procedures, however, you should understand why it's valid to use probability sample results to make estimates and decisions about population characteristics. Chapter 6 gives you this understanding of the basic nature of inferential statistics by presenting the theoretical and intuitive bases for estimating population values and for testing hypotheses about those values.

More specifically, after briefly *reviewing* some population and sample *definitions* in Chapter 6, we'll examine (1) the *importance of and advantages of sampling,* (2) some methods of *sample selection,* and (3) the extremely important concepts associated with *sampling distributions of means and percentages.* These latter topics represent the cornerstone of statistical inference, and in these sections we will discuss *why* it's possible to infer a population characteristic with a sample characteristic.

Thus, after studying this chapter, you should be able to:

- Understand and appreciate *(a)* the purpose and importance of sampling and *(b)* the advantages made possible by sampling.
- Describe the types of samples that may be selected.
- Trace through the steps that are required to *(a)* produce a sampling distribution of sample means, *(b)* compute the mean of this sampling distribution, and *(c)* compute the standard deviation of this sampling distribution.
- Define the Central Limit Theorem and explain the relationship that exists between the standard error of the mean and the size of the sample.
- Trace through the steps necessary to *(a)* produce a sampling distribution of sample percentages, *(b)* compute the mean of this sampling distribution, and *(c)* compute the standard deviation of this sampling distribution.

CHAPTER OUTLINE

LOOKING AHEAD
POPULATION, SAMPLE, SYMBOLS: A REVIEW
  Self-Testing Review 6-1

IMPORTANCE OF SAMPLING
  Self-Testing Review 6-2

ADVANTAGES OF SAMPLING
  Cost
  Time
  Accuracy of Sample Results
  Other Advantages
  Self-Testing Review 6-3

POPULATION, SAMPLE, SYMBOLS: A REVIEW

On Tuesday, when it hails and snows,
The feeling on me grows and grows
That hardly anybody knows
If those are these and these are those.
        —*Winnie-the-Pooh*

Usually, the term "population" brings to mind a large mass of people who reside in a geographic area. In statistics the term has a broader meaning. A *population* is the total of any kind of unit under consideration by the analyst. These individual units may be items such as business firms, credit accounts, transistors, and even people. A **finite population** is one where the total is a limited, specific number. An **infinite population** is unlimited in size.

A *sample* is any portion of the population selected for study. Consider, for example, the American League as a population of baseball teams. If the Boston Red Sox and Kansas City Royals are selected for a statistical study of the American League, these two teams are considered a sample.

Whether a group of items is a population or whether the data set is a sample, a statistician usually describes the group with measures such as total number, an average, a standard deviation, and the like. For example, a group of students may be described by the total number of students, the average grade, and a standard deviation of grades. If a particular measure describes a population, it's a parameter. A *parameter* is a characteristic of a population. If a particular measure describes a group of items which is a sample, the measure is known as a statistic. A *statistic* is a characteristic of a sample.

You may think that we are "slicing the baloney awfully thin" with all these distinctions in terminology, but the terminology is so important that there is justification for any apparent overemphasis in presentation. As you'll see, sample results are generalized to describe the population; that is, *statistics are used to estimate parameters.* The distinctions made now between parameters and statistics minimize the danger of confusion in later chapters.

Statisticians maintain the distinction between parameters and statistics through the use of different symbols. *Greek letters are usually used to denote parameters, while lowercase italic letters denote sample statistics.* Figure 6-1 illustrates the common symbols. The population mean is designated by μ (mu), while the sample mean is denoted by $\bar{x}$. The standard deviation of the population is indicated by σ (sigma), while the sample standard deviation is represented by s. And the population percentage is represented by π (pi), while the sample percentage is designated by p.

Self-Testing Review 6-1

1 Suppose we are interested in performing a study of a university, and there are four possible groups from which we could collect data. Would each of the following groups be the population of the university or only a sample?
   a All students enrolled at the university
   b All students enrolled in a psychology course

FIGURE 6-1   Distinctions between a Population and a Sample

AREA OF DISTINCTION	POPULATION	SAMPLE
Definition	Defined as a total of the items under consideration by the researcher	Defined as a portion of the population selected for study
Characteristics	Characteristics of a population are parameters	Characteristics of a sample are statistics
Symbols	Greek letters or capitals μ = population mean σ = population standard deviation N = population size π = population percentage	Lowercase italic letters $\bar{x}$ = sample mean s = sample standard deviation n = sample size p = sample percentage

    c All students enrolled in the business school

    d All students in every division of the university

2 In an earlier example, the American League was considered a population. Is it possible to consider the American League a sample? If so, under what circumstances?

3 A workers' union has a membership of 300 persons. Data were collected from 25 of them, and their average age was 39. The average age of the entire union membership was therefore estimated to be approximately 39. A subsequent polling of all members indicated the true average age was 42.

    a What figure is a parameter?

    b What figure is a statistic?

    c The sample statistic 39 was used to estimate the _____ .

IMPORTANCE OF SAMPLING

Sampling occurs frequently in the course of daily events and should not be viewed as just a concept employed solely by statisticians. Although the samplings in daily life do not have the sophistication of formal statistical studies, they do serve a fundamental purpose of providing information for judgments. Here are a few examples.

1 A homemaker tastes a spoonful of the soup which she is preparing for supper. She wants to know if the soup has an acceptable flavor.

2 A prospective car buyer test-drives an automobile in order to judge whether or not it's a potential lemon.

3 Pieces of ore are analyzed to determine the potential of a new mine.

The examples could go on *ad infinitum* (and *ad nauseam*), but let's look at the rationale for sampling.

The purpose of sampling is to provide sufficient information so that inferences may be made about the characteristics of a population. Many times it's not feasible to study an entire population to determine its true character. The homemaker can't taste the whole pot of soup to see if it's really acceptable, and the test driver can't drive the car for 3 years to find out if it will eventually be a lemon. But the data produced by sampling can be used to support judgments about the population.

The goal in sampling is to select a portion of the population that displays all of the characteristics of the population. If we're to make a judgment about a population from sample results, we want those results to be as representative of the population as possible. Suppose we want to perform a sociological study of a town. Not only should the sample contain enough people to represent the upper, middle, and lower income groups, it should also be representative of such characteristics as age, education, and ethnic background. Unless a sample is similar to the population, there can be no reliability in estimates based on sample results.

Unfortunately, it's extremely difficult, if not nearly impossible, to have a sample that's completely representative of the population. And it would be unreasonable to expect a sample result to have *exactly* the same value as some population characteristic because sampling error is always present. But this fact doesn't mean that sample results are useless. Statisticians have learned to cope with error. If sampling error can be objectively assessed, the precision of estimates can also be objectively gauged. Statisticians are often willing to give up the benefits of a complete enumeration for the advantages (described in the next section) that can be gained from sampling.

Self-Testing Review 6-2

1   "Sampling is a concept which can be best illustrated by statisticians." Comment on this statement.
2   What is the purpose of sampling?
3   What is considered the goal of sampling?
4   If there are any errors in sampling, the results of the sample have little use. True or false? Why?
5   How can the precision of an estimate be determined?

ADVANTAGES OF SAMPLING

Complete information acquired through a census is generally desirable, and the Closer Look reading at the end of this chapter details the efforts made by the United States Census Bureau to carry out its decennial census responsibility. If every item in a population data set is examined, we would be quite confident in describing the population. But, as in many situations, what you want is not necessarily what you can get. Census data represent a luxury item in most situations and are thus often not available for studying a population. Data gathering by sampling rather than census taking is the rule rather than the exception because of the *sampling advantages* discussed below.

Cost

Any data-gathering effort incurs costs for such things as mailings, interviews, and data tabulations. The more data to be handled, the higher the costs will likely be. Consider a consumer survey of the United States: If an attempt were made to poll every citizen, the cost would easily run into many millions of dollars. Any benefits derived from such census data would likely be negated by the cost. For example, a national food company might want to improve a product to increase sales. The company could survey every potential customer, but it's very likely that the costs of a census would wipe out any additional revenues generated from an improved product. Any time a sample can be taken with less expenditure than that required for a census, cost becomes an acceptable (although not sufficient) reason for sampling.

Time

The oft-quoted maxim "Time is money" characterizes many important business decisions. Speed in decisions is often crucial, since many profitable opportunities pass quickly. Let's assume you're the owner of a company and you've got this innovative idea for a better mousetrap. Your top advisers tell you that rival firms are also racing to build a better mousetrap. Being the first company to have a better product on the market may lead to high sales revenues, but do you actually have a better mousetrap? Will the public beat a path to your company's door for your idea, or will the new trap fail to appeal to the public? Obviously a census to help answer these questions requires too much time, which is a valuable commodity. The answer lies in sampling, since it can produce adequate information in a shorter period.

Accuracy of Sample Results

Sometimes the results of a small sample provide information that's almost as accurate as that resulting from a complete census. How is this possible? Remember that the object in sampling is to achieve representation of the population characteristics. There are sampling methods which produce samples that are highly representative of the population. In these situations, larger samples will not produce results which are *significantly* more accurate. Let's again consider the situation of the homemaker. If she has stirred the soup well before sampling, two sips should be sufficient to tell her about the entire pot. Any additional sips will only serve to decrease the volume of soup available for supper. (It's fortunate that our homemaker isn't brewing whiskey.)

Other Advantages

Destructive tests are often employed in testing product quality. For example, a company may be interested in the tensile strength of a truckload of iron bars which it has received. To test their tensile strength, the iron bars are subjected to pressure until they break. All the bars can be tested only if the company wants a truckload of broken iron bars. Since that's certainly not the case, a sample of bars must be used.

Sometimes the resources may be available for a census, but the nature of the population requires a sample. Suppose we're interested in the number of humpback whales left in the world. Environmental organizations may be willing to sponsor our project, but migration movements, births, and deaths prevent complete enumeration. One approach to the problem is to sample a small area of the ocean and use the results to make a projection.

Self-Testing Review 6-3

1 "Complete information is always desirable, but sample results may sometimes be almost as accurate." Comment on this statement.
2 When will costs justify sampling over a census? Is cost alone a sufficient reason for a sample?

SAMPLE SELECTION

A sample should be representative of the population, but there are a variety of ways that samples can be selected. The selection method used in a particular situation is determined by the nature of the population and the skill of the researcher.

Judgment Samples

Sample selection is sometimes based on the opinion of one or more persons who feel sufficiently qualified to identify items for a sample as being characteristic of the population. Any sample based on someone's expertise about the population is known as a **judgment** (or **purposive**) **sample.** As an example, consider the seasoned political campaign manager who intuitively designates certain voting districts as reliable indicators or estimators of public opinion of his candidate. The sample of voting districts is based on the campaign manager's expertise and involves no complex statistical computations.

A judgment sample is convenient, but this convenience is also a disadvantage. Since the judgment sample is not determined by any statistical techniques, it's difficult to assess how closely it measures reality. This difficulty of objective assessment leaves an uncomfortable uncertainty in any estimation based on the sample results. This doesn't mean, though, that a judgment sample should never be used. The quality of a judgment sample depends on the researcher's expertise, but experience may serve as a valuable tool in surveys.

Probability Samples

Anyone who uses arithmetic methods of producing random digits is, of course, in a state of sin.
—John von Neumann

(Here's that paragraph you were warned about at the beginning of the chapter.) A **random** (or **probability**) **sample** is one in which the probability of selection of each element in the population is known prior to sample selection. Unlike the use of a judgment sample, the use of a probability sample results in evaluations of population characteristics that may be objectively assessed. Let's look now at three of the major types of probability

SURVEYS WASTE LOTS OF CASH TO FIND "BEST" PLACE TO LIVE

NEW YORK—Why do you live where you do?

You know the answer to that question far better than I, but I suspect that, for most Americans, the reason is not that you read a survey telling you to move there.

Nonetheless, in this age of the worship of pseudo-information, a vast amount of money is being spent on projects designed to determine the "best" place to make one's home. Some people, I suppose, even take this stuff seriously.

More to the point, the sponsors of such surveys hope to convince people with really big bucks to deploy, such as corporate executives seeking a new plant location, that they've been looking for love in all the wrong places.

And so, as you come home each night to the street where you unscientifically live, a heap of hefty interests with a heap of hefty computers are hard at work on "objective" new surveys designed to persuade your bosses that it's time for a change.

The current round of State Wars began with the latest edition of a widely publicized index published annually since 1979 by Grant Thornton, a Chicago accounting firm, which ranks the general manufacturing and business climates of the 48 contiguous states.

As Grant Thornton saw it, South Dakota— site of Mount Rushmore and home of the big bison herds—now has the No. 1 business climate in America. Utah ranked second, followed by Nebraska, Arizona and North Dakota. As for the top manufacturing states, New Jersey ranked 23rd; California, 26th; Massachusetts, 27th; Illinois, 34th; Pennsylvania, 40th; New York, 42nd; Ohio, 43rd; and Michigan, heartland of the automotive industry, at the very bottom.

This survey naturally infuriated—you guessed it—the losers. It wasn't just damaged civic pride, they insisted, or the sort of transient mass despondency engendered by the home team's losing the big game. This was a serious matter: manufacturers might be misled in deciding where to locate a new factory, and local businesses might be buffaloed into moving to Rapid City.

Now comes the losers' response: a new index called the AmeriTrust/SRI Indicators of Economic Capacity, commissioned by a big Rust Belt bank holding company, AmeriTrust Corp. of Cleveland, and put together by SRI International of Menlo Park, Calif., formerly Stanford Research Institute.

And this new index shows—surprise!—that the most attractive regions for business are the mid-Atlantic, Pacific, Northwest and Midwest, the very regions that (entirely coincidentally, of course) had the lowest four rankings in the Grant Thornton Index.

The authors of the new, counterattacking survey tell me their noble effort was essential because the Grant Thornton Index measures the wrong things—thereby presenting an "unnecessarily narrow vision of the American economy," and giving insufficient weight to the availability of an educated and skilled work force, technology and capital.

As it happens, the bruised titans of the industrial Midwest are not alone in having doubts about Grant Thornton. The Washington-based Corporation for Enterprise Development, for example, has criticized it for ignoring "quality of life" issues while emphasizing low public expenditures, low wages and low taxes. (Grant Thornton retorts that its index in fact ranks the states on 22 criteria.)

Predictably, in so subjective a game, still more players are rushing to get into the act. There's already another complete ranking of manufacturing climates from Inc. magazine, which focused on just three criteria—job generation, number of new businesses and growth of small firms—and concluded that the real No. 1 was Arizona, with Wyoming last. (South Dakota tumbled to 45th.)

What we have here on all sides, under the guise of science, is actually just proof of a simpler reality about every such statistic: if you want to get the right answers, just be sure to ask the right questions.

The only scary part of all this well-financed foolishness is that somebody might be deluded into making a key individual or corporate decision based only on somebody else's "authoritative" survey. But take heart: maybe it's just that we Americans love polls, yet are too smart to take them seriously. After all, one recent survey named Pittsburgh America's "most livable" city, which understandably spurred joy in the revivified canyons of the Golden Triangle—but, at last report, had not resulted in any mass evacuation from San Diego.

—Louis Rukeyser, "Surveys Waste Lots of Cash to Find 'Best' Place to Live," January 4, 1987, © 1987 Fort Worth Star-Telegram. Reprinted by permission: Tribune Media Services.

samples. (Unless otherwise indicated, the word "sample" will always refer to a probability sample in the pages that follow.)

Simple Random Sample   A **simple random sample** is one in which each possible combination has an equal probability of occurrence, and each item in the population has an equal chance of being included in the sample. Suppose we have a basketball team with 10 players and we want to estimate the average score of each player per game. Assume further that we would like to select a simple random sample of size 3. With 10 players and a sample of 3, the number of possible combinations of 3 players is

$$_{10}C_3 = \frac{10!}{3!\ 7!} = 120$$

Therefore, each combination must have a 1/120 chance of being selected, and each ball player must have a 3/10 chance of being in the sample.

A computer may be programmed to carry out a random selection process, as you saw in Chapter 5 (Figures 5-1, 5-2, and 5-13). We'll also use a statistical software package to select random samples later in this chapter. And to assure that sample selection is left to chance, a **table of random numbers** may also be used in the selection of a simple random sample. Each digit in such a table is determined by chance and has a 1/10 probability of appearing at any single-digit space in the table. Appendix 3 in the back of the book contains random numbers.

To show the use of random numbers in simple random sampling, suppose we have a list of 200 gardeners eligible for a consumer survey and we want a sample of 20. We could obtain a simple random sample in the following manner:

1   Assign each and every gardener a number from 000 to 199. Each gardener should have a unique number. The first gardener would have 000, the second 001, and so forth.
2   Next consult a table of random numbers. Figure 6-2 is a brief excerpt from such a table.
3   It's essential that we establish a systematic way of selecting a sequence of digits from the table so that no bias enters into the selection process.

FIGURE 6-2   An Example of a Table of
Random Numbers

5124	0746	6296	9279
5109	1971	5971	1264
4379	6296	8746	5899
8194	3721	4621	3634

In our case, we need a sequence of three digits. Let's say that our pattern of selection is the last three digits of each block of numbers and that we will work down a column.

4   Select a random number in the systematic manner and match the random number assigned to a gardener. For example, gardener number 124 is selected first, and then gardener number 109 is added to the sample. If we have a random number we cannot use, as in the case of 379, we proceed to the next random number, 194, and continue this process of selection until there's a sample of 20 gardeners.

Stratified Sample   If a population is divided into relatively homogeneous groups or strata and a sample is drawn from each group to produce an overall sample, this overall sample is known as a **stratified sample.** Stratified sampling is usually performed when there's a large variation within the population and the researcher has some prior knowledge of the structure of the population that can be used to establish the strata. The sample results from each stratum are weighted and calculated with the sample results of other strata to provide an overall estimate.

As an illustration, suppose our population is a university student body and we want to estimate the average annual expenditures of a college student for nonschool items. Assume we know that because of different lifestyles, the older students spend more than the younger students, but there are fewer older students than younger students because of some dropout factor. To account for this variation in life-style and group size, the population of students can easily be stratified into freshmen, sophomores, juniors, and seniors. A sample can be taken from each stratum and each result weighted to provide an overall estimate of average nonschool expenditures.

Cluster Sample   A **cluster sample** is one in which the individual units are groups or clusters of single items. It's always assumed that the individual items within each cluster are representative of the population. Consumer surveys of large cities often employ cluster sampling. The usual procedure is to divide a map of the city into small blocks, each block containing a cluster of households to be surveyed. *A number of clusters are selected for the sample,* and all the households in a cluster are surveyed. A distinct benefit of cluster sampling is savings in cost and time. Less energy and money are expended if an interviewer stays within a specific area rather than traveling across stretches of the city.

Self-Testing Review 6-4

1   Which, if any, sampling method is applicable to all situations?

2   Is a probability sample more representative of a population than a judgment sample? Why is it or isn't it?

3   Why is a probability sample more desirable than a judgment sample?

4   Assume we have a population of 10 and we wish to take a simple random sample of 2. What are the chances of selection for each sample combination? What are the chances of each individual item to be selected for the sample?

5   What's the basic assumption in cluster sampling?

6   What must we know about a population in order to obtain a stratified sample?

SAMPLING DISTRIBUTION OF MEANS—A PATTERN OF BEHAVIOR

The sample mean only approximates the population mean; it seldom has exactly the same value as the population mean. Suppose the average income of a probability sample of city residents is $16,251. We could venture to say that the approximate value of the population mean is $16,251. But we intuitively know that the chances are slim that the sample mean equals the population mean. A different sample of residents would most likely yield a different mean, such as $16,282, while another sample might produce a mean of $16,249. This variation in a sample statistic is known as **sampling variation.**

In stating that a sample mean is an approximation of the population mean, we've made an assumption that the sample mean is related in some manner to the population mean. We intuitively assume that the value of the sample mean *tends toward* the value of the population mean. As we shall see later in this chapter, our intuition is correct; the population characteristics determine the range of values a sample mean may take.

Let's assume that we have a *population* of 15 cards numbered from 0 through 14. And let's further assume that we want to select random samples of size 6 from this population. The number of *possible* samples that *could* be selected is a combinations problem. We've been using $_nC_r$ to represent the combination of n things taken r at a time, but in this case it might be clearer to use $_NC_n$ to represent the combination of N items in a population taken n at a time in a sample. Anyway:

$$_{15}C_6 = \frac{15!}{6! \; 9!} = 5,005 \text{ possible samples}$$

One of these 5,005 possible samples consists of the cards numbered 2, 4, 6, 8, 10, and 12; a *second* possible sample selection is made up of the cards numbered 1, 4, 3, 7, 8, and 13; and a *third* possible sample comprises the cards numbered 14, 0, 7, 10, 9, and 8. (You can figure out the other 5,002 possible samples next summer at your leisure.) The arithmetic *means* of these 3 possible samples, in the order presented, are 7, 6, and 8. If there are

5,005 possible samples, there are, of course, 5,005 possible sample means. And if we were to select all the 5,005 possible samples, compute the mean of each of these samples, and arrange the 5,005 sample means in a frequency distribution, this distribution would be called a *sampling distribution of means*. Thus, a **sampling distribution of means** is the distribution of the arithmetic means of all the possible random samples of size *n* that could be selected from a given population.

If we're not careful at this point, we can run into some semantic difficulties, so let's pause here to consider the three fundamental types of distributions in Figure 6-3. There's nothing new about Figure 6-3*a;* it's merely the frequency distribution of whatever population happens to be under study and could, of course, take many shapes. The mean and standard deviation of the *population distribution* (μ and σ) are quite familiar to us. And there's really nothing very new about the distributions in Figure 6-3*b;* they are simply the frequency distributions of some of the samples that could be selected from a population. As is true in any frequency distribution, the individual values (in a sample) are distributed about the mean of those values ($\bar{x}$). The sample standard deviation (*s*) is the measure of dispersion in these *sample distributions*, which, like the population distribution, can take many shapes. Theoretically, of course, there are as many sample distributions as there are possible samples that could be taken in a given situation. For example, if there are 5,005 possible samples, there are 5,005 possible sample distributions. Finally, in Figure 6-3*c*, we come to the *sampling distribution of the means*—a distribution that *is* new to us, and one that *should not be confused with a sample distribution* (even though the terms are confusingly similar). In the sampling distribution in Figure 6-3*c*, the possible sample mean values are distributed about the mean of the sampling distribution (sometimes called the grand mean). The mean of the sampling distribution, you'll see in Figure 6-3*c*, is identified by the symbol $\mu_{\bar{x}}$, and the standard deviation of the sampling distribution is identified by symbol $\sigma_{\bar{x}}$.

Mean of the Sampling Distribution of Means

My center is giving way, my right is falling back, the situation is excellent. I attack.
    —*Marshal Foch, Battle of the Marne*

Be alert and pay attention now, because we are about to attack you with a very important fact: *The mean of the sampling distribution of means is equal to the population mean*—that is, $\mu_{\bar{x}} = \mu$. What's that you said? "I have read some ridiculous statements in the past, and more than a few of them have come from this book, but. . . . " Anticipating just such a skeptical attitude, we've prepared an example to redeem our credibility.

Let's assume the population is a small group of five students enrolled in a statistics course and an instructor wants to estimate the average amount of time spent by each student preparing for classes each week. Figure 6-4

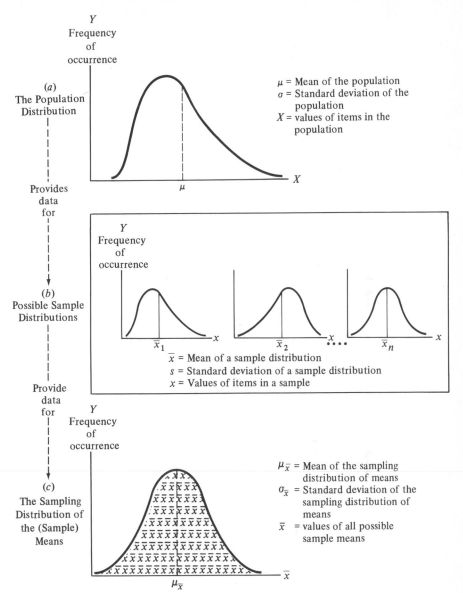

FIGURE 6-3
Educational
schematic of three
fundamental types
of distributions.

lists the amount of time each student spends per week preparing for class
(but the instructor does not have access to this information). As we can see
from Figure 6-4, the population mean preparation time is 6 hours.

If the instructor takes a sample of three students, what are the possible
values of the sample mean? How different from the true mean of 6 might a
sample mean be? Figure 6-5 provides the answer; it also provides us with
the data needed to compute the mean of the sampling distribution as follows:

FIGURE 6-4   Population of Students and Their
Weekly Preparation Time

STUDENT	PREPARATION TIME (HOURS)
A	7
B	3
C	6
D	10
E	4
	$\Sigma X = 30$

$$\mu = \frac{\Sigma X}{N} = \frac{30}{5} = 6$$

FIGURE 6-5   Sampling Distribution of Means

SAMPLE COMBINATIONS	SAMPLE DATA	SAMPLE MEANS $(\bar{x})$	$(\bar{x} - \mu_{\bar{x}})$	$(\bar{x} - \mu_{\bar{x}})^2$
1. A, B, C	7, 3, 6	5.33	−.67	.45
2. A, B, D	7, 3, 10	6.67	.67	.45
3. A, B, E	7, 3, 4	4.67	−1.33	1.77
4. A, C, D	7, 6, 10	7.67	1.67	2.79
5. A, C, E	7, 6, 4	5.67	−.33	.11
6. A, D, E	7, 10, 4	7.0	1.00	1.00
7. B, C, D	3, 6, 10	6.33	.33	.11
8. B, C, E	3, 6, 4	4.33	−1.67	2.79
9. B, D, E	3, 10, 4	5.67	−.33	.11
10. C, D, E	6, 10, 4	6.67	.67	.45
		60.0		10.0

$$\mu_{\bar{x}} = \frac{\Sigma(\bar{x}_1 + \bar{x}_2 + \bar{x}_3 + \cdots + \bar{x}_{N}C_n)}{{}_{N}C_n} = \frac{60}{10} = 6$$

where the numerator is the sum of all possible sample means, and the denominator is the number of possible samples.

$$\mu_{\bar{x}} = \frac{\Sigma(\bar{x}_1 + \bar{x}_2 + \bar{x}_3 + \cdots + \bar{x}_{N}C_n)}{{}_{N}C_n} \qquad \text{(formula 6-1)}$$

$$= \frac{\Sigma(5.33 + 6.67 + 4.67 + \cdots + 6.67)}{5!/(3!)(2!)}$$

$$= \frac{60}{10}$$

$$= 6$$

Thus, as you can see, the mean in Figure 6-4 equals the mean in Figure 6-5. That is, $\mu_{\bar{x}} = \mu$.

This isn't an isolated case. Let's go back to our earlier example of the population of 15 cards numbered 0 to 14. The sum of the integers on the 15 cards is 105, and the mean of this population of 15 numbers is 7 (105/15). We didn't take those 5,000 + different samples to prove that $\mu_{\bar{x}}$ equals this μ of 7. But we did take 150 samples of 6 cards from our population of 15, and we then computed the sample mean for each of these 150 samples. Well, actually we didn't do this tedious task at all. Rather, we turned it over to a computer running the *Minitab* statistical package and promptly received the output shown in Figure 6-6.

In the first two lines of Figure 6-6, the program was told to put 150 samples into 150 rows, with each row or sample having six data items ran-

FIGURE 6-6   A computer simulation of 150 samples of six cards each, selected from a population of 15 integers numbered 0 to 14. The μ is 7, and the mean of this data set of 150 simulated sample means is 7.0233.

```
MTB > RANDOM 150 SAMPLES INTO C1-C6;
SUBC> INTEGERS 0 TO 14.
MTB > RMEAN C1-C6 INTO C7
MTB > MEAN C7
   MEAN    =       7.0233
MTB > STEM-AND-LEAF C7

Stem-and-leaf of C7          N  = 150
Leaf Unit = 0.10


      1      3 1
      5      3 5688
      8      4 011
     16      4 55556688
     23      5 0111333
     44      5 5555556668888888888888
     55      6 00001333333
     71      6 5555555666888888
    (20)     7 00000000000111111333
     59      7 556666688888
     47      8 00000111113333
     33      8 55556668888
     22      9 00011113333
     11      9 55556
      6     10 013
      3     10 558

MTB > BOXPLOT C7
```

```
                              -----------------
           ---------------------I     +     I--------------------
                              -----------------
         --+---------+---------+---------+---------+---------+----C7
          3.0       4.5       6.0       7.5       9.0      10.5
```

domly selected from the integers 0 to 14. Line 3 of Figure 6-6 then instructed the program to (1) compute the mean of the six data items in each of the 150 rows and (2) store these 150 sample means in a separate column (C7). The next instruction calls for the program to compute the mean from the data set of 150 sample means. This value is 7.0233, very close to the population mean of 7.00. A stem-and-leaf display is then prepared to show the values of the 150 sample means. As you can see, the smallest of the 150 sample means has a value of about 3.1, and the largest sample mean is about 10.8. Finally, a box-and-whiskers display of our data set of 150 sample means is presented. You'll notice that the box representing the middle 50 percent of the sample means has a lower hinge of about 5.8 and an upper hinge of about 8.1.

A dozen more simulations, each with 150 random samples, could be processed by the computer in a few minutes. All of these simulations would likely produce different means from their data sets of 150 sample means, but these dozen values, like our mean of 7.0233, would be close to the population mean of 7 because ultimately, as we've seen, $\mu_{\bar{x}} = \mu$.

As the devil's advocate, we may say "So what?" to the fact that $\mu_{\bar{x}} = \mu$. No one in a realistic situation really takes all possible sample combinations and calculates the sample means. In practice only one sample is taken. What benefit is there in discussing the sampling distribution? Shouldn't we really be concerned with the proximity of a single sample mean to the population mean? In essence, the discussion of the sampling distribution *is* concerned with the proximity of a sample mean to the population mean.

You can see from the example in Figures 6-5 and 6-6 that the possible values of the sample means *tend toward* the population mean. Since these values have frequencies of occurrence, the sampling distribution is essentially a probability distribution. If the sample size is *sufficiently large (n more than 30)*, the sampling distribution approximates the *normal distribution whether or not the population is normally distributed*. And the sampling distribution is *normally distributed regardless of sample size if the population is normally distributed*. Figure 6-7 illustrates the sampling distribution as a normal distribution.

You'll recall that in a normal probability distribution the likelihood of an outcome is determined by the number of standard deviations from the mean of the distribution. (You do recall all those light bulb problems, don't you?) Therefore, as you can see in Figure 6-7, there's a 68.3 percent chance that a sample selected at random will have a mean that lies within *one* standard deviation $(\sigma_{\bar{x}})$ of the population mean. Also, there's a 95.4 percent chance that the sample mean will lie within *two* standard deviations of the population mean. Thus, a knowledge of the properties of the sampling distribution tells us the probable proximity of a sample mean outcome to the value of the population mean. With a knowledge of the sampling distribution, probability statements can be made about the range of possible values a sample mean may assume. This range of possible values can be calculated if a value for the standard deviation of the sampling distribution $(\sigma_{\bar{x}})$ is available. The computation of $\sigma_{\bar{x}}$ is shown in the next section.

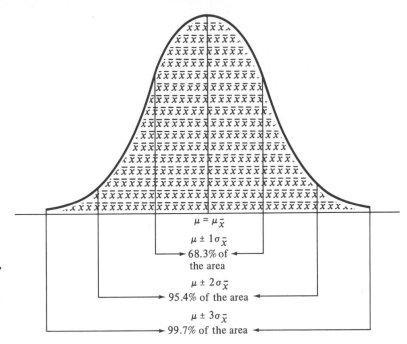

FIGURE 6-7
μ, $\mu_{\bar{x}}$, and $\sigma_{\bar{x}}$ for
the areas under the
sampling distribu-
tion of means.

$\mu = \mu_{\bar{x}}$

$\mu \pm 1\sigma_{\bar{x}}$

→ 68.3% of ←
the area

$\mu \pm 2\sigma_{\bar{x}}$

→ 95.4% of the area ←

$\mu \pm 3\sigma_{\bar{x}}$

→ 99.7% of the area ←

Standard Deviation of the Sampling Distribution of Means

In order to determine the extent to which a sample mean might differ from
the population mean, we need some *measure of dispersion.* In other
words, we must be able to compute the likely deviation of a sample mean
from the mean of the sampling distribution. The standard deviation of the
sampling distribution, in statistical jargon, is given the rather intimidating
name of **standard error of the mean** and, as we have seen, is represented by
the symbol $\sigma_{\bar{x}}$. For the data given in Figure 6-5, the calculation of this measure is similar to the calculation of any other standard deviation about a
mean:

$$\sigma_{\bar{x}} = \sqrt{\frac{\Sigma(\bar{x} - \mu_{\bar{x}})^2}{N}}$$ (formula 6-2)

$$= \sqrt{10.00/10}$$
$$= \sqrt{1.00}$$
$$= 1.00$$

where N = total number of possible samples.

However, as pointed out in an earlier paragraph, no one (other than an
eccentric statistician or a student with an assigned problem) ever deals

with all the possible sample combinations. Therefore, an alternative method to compute $\sigma_{\bar{x}}$ must exist.

Since we've seen the relationship that exists between $\mu_{\bar{x}}$ and μ, we might intuitively assume that there is a relationship between the standard error of the mean and the population standard deviation that will produce a shortcut method of computing the standard error. As a matter of fact, we are right, and the standard error may be computed for a *finite population* with the following formula:

$$\sigma_{\bar{x}} = \frac{\sigma}{\sqrt{n}} \sqrt{\frac{N - n}{N - 1}}$$
(formula 6-3)

where σ = the population standard deviation
N = population size
n = sample size
$\sqrt{\dfrac{N - n}{N - 1}}$ = finite population correction factor

From the data in Figure 6-4, the standard deviation of the population is computed as follows:

STUDENT	X	$(X - \mu)$	$(X - \mu)^2$
A	7	1	1
B	3	-3	9
C	6	0	0
D	10	4	16
E	4	-2	4
			30

$$\sigma = \sqrt{\frac{\Sigma(X - \mu)^2}{N}}$$
$$= \sqrt{\frac{30}{5}}$$
$$= 2.45$$

Therefore, the standard error for the data given in Figure 6-5 is computed as follows:

$$\sigma_{\bar{x}} = \frac{2.45}{\sqrt{3}} \sqrt{\frac{5 - 3}{5 - 1}}$$
$$= 1.4145 \, (.7071)$$
$$= 1.00$$

We can now see that the results of the computations based on formulas 6-2 and 6-3 are equal. Thus, $\sigma_{\bar{x}}$ may be determined with a knowledge of the population standard deviation, the sample size, and the population size.

If the population is *infinite in size,* as, for example, in the case of items from an assembly-line operation, the standard error does not require a finite correction factor and may be computed as follows:

$$\sigma_{\bar{x}} = \frac{\sigma}{\sqrt{n}}$$

(formula 6-4)

If the population is infinite, there's no need for the correction factor. However, a finite population doesn't necessarily mean that the correction has to be used. At this point the last statement may cause you to wince, since there's an apparent contradiction. Let's look at the following example and, we hope, relieve your pain.

Suppose we have a finite population of approximately 200 million, and we have taken a sample of 2,000. If we followed the rules strictly, we would have:

$$\sigma_{\bar{x}} = \frac{\sigma}{\sqrt{n}} \sqrt{\frac{N-n}{N-1}}$$

$$= \frac{\sigma}{\sqrt{n}} \sqrt{\frac{200,000,000 - 2,000}{200,000,000 - 1}}$$

$$= \frac{\sigma}{\sqrt{n}} (.99999)$$

However, the size of the population is so large that the finite correction factor for all practical purposes is 1. If the population size is extremely large compared to the sample size, formula 6-4 may be used to calculate the standard error of a finite population.

The Relationship between n and σ_x

The standard error of the mean is, of course, *a measure of the dispersion* of sample means about the population mean. If the degree of dispersion *decreases,* the range of probable values a sample mean may assume also *decreases,* meaning the value of any single *sample mean* will probably be closer to the value of the *population mean* as the standard error decreases. And with formula 6-3 or 6-4, the value of $\sigma_{\bar{x}}$ obviously must decrease as the size of *n* increases. That is,

$$\downarrow \sigma_{\bar{x}} = \frac{\sigma}{\sqrt{n} \uparrow}$$

To add meaning to this mathematical manipulation, let's look at the relationship between *n* and $\sigma_{\bar{x}}$ intuitively. Let's assume we wish to estimate some parameter of a population of 100, and we are initially entertaining the thought of taking a sample of 10 items. Ten items may provide adequate information, but it's clear that more information could be obtained from a larger sample such as 20. More information provides a more precise estimate of the population parameter. As a matter of fact, a sample of 50 or 60 items would provide even more information and thus a more precise estimate. The ultimate option is that we could sample the entire population and obtain complete information, and thus there would be no difference between the sample statistic and the population parameter. In our example, if we were to estimate the population mean, a sample of 100 would have the following standard error:

$$\sigma_{\bar{x}} = \frac{\sigma}{\sqrt{n}} \sqrt{\frac{N - n}{N - 1}}$$

$$= \frac{\sigma}{\sqrt{n}} \sqrt{\frac{100 - 100}{99}}$$

$$= 0$$

The general principle is that *as n increases, $\sigma_{\bar{x}}$ decreases.* As the sample size increases, we have more information on which to estimate the population mean, and thus the probable difference between the true value and any sample outcome decreases. Figure 6-8 summarizes the points made in this section.

The Central Limit Theorem

Up to this point we've explained the concept of the sampling distribution of means in a rather intuitive way. Now, though, we're ready to formalize

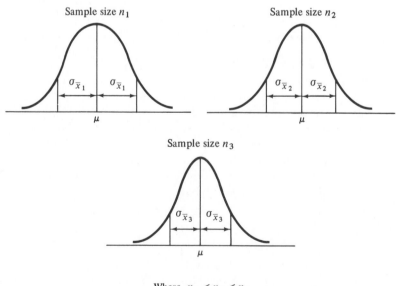

FIGURE 6-8
The relationship
between n and $\sigma_{\bar{x}}$.

Where $n_1 < n_2 < n_3$

$\sigma_{\bar{x}_1} > \sigma_{\bar{x}_2} > \sigma_{\bar{x}_3}$

the concepts developed in previous sections and attribute the properties of the sampling distribution to the Central Limit Theorem. The **Central Limit Theorem** basically states:

The mean of the sampling distribution of means is equal to the population mean. The standard deviation of the sampling distribution of means ($\sigma_{\bar{x}}$) is equal to $\sigma/\sqrt{n}$ for an infinite population, and it's equal to $\sigma/\sqrt{n}\ \sqrt{(N-n)/(N-1)}$ for a finite population. If the sample size (n) is sufficiently large, the sampling distribution approximates the normal probability distribution. If the population is normally distributed, the sampling distribution is normal regardless of sample size.

A considerable amount of important theoretical material has been presented in the last few pages. Let's look now at some example problems to illustrate certain basic concepts.

Example Problems

Example 6-1   The Bigg Truck Company has a fleet of five trucks which have monthly maintenance costs of $200, $175, $185, $210, and $190. An estimate of the average monthly cost for a truck is to be obtained from a simple random sample of 3. What are the mean and the standard deviation of this sampling distribution of size 3?

The population mean is $192 − ($200 + $175 + $185 + $210 + $190)/5. Since the population mean is equal to the mean of the sampling distribution, $\mu_{\bar{x}}$ = $192. The standard deviation of the population is computed as follows:

$$\sigma = \sqrt{\frac{\Sigma(X-\mu)^2}{N}}$$

$$= \sqrt{\frac{8^2 + 17^2 + 7^2 + 18^2 + 2^2}{5}}$$

$$= \sqrt{\frac{64 + 289 + 49 + 324 + 4}{5}}$$

$$= 12$$

Consequently,

$$\sigma_{\bar{x}} = \frac{\sigma}{\sqrt{n}} \sqrt{\frac{N-n}{N-1}}$$

$$= \frac{12}{\sqrt{3}} \sqrt{\frac{5-3}{5-1}}$$

$$= 4.9$$

Example 6-2   Sam and Janet Evening want to estimate the average dollar amount of each order filled by their catering company. They obtain their estimate by selecting a simple random sample of 49 orders. Sam and Janet

don't know it, but their orders are normally distributed. The population mean value of these orders is $120, and the population standard deviation is $21. What's the standard deviation of the sampling distribution—that is, what's the standard error? What's the likelihood that the sample mean will fall between $\mu - 1\sigma_{\bar{x}}$ and $\mu + 1\sigma_{\bar{x}}$? Within what range of values does $\bar{x}$ have a 95.4 percent chance of falling?

Well, $\sigma_{\bar{x}}$ is $\sigma/\sqrt{n}$ = $21/$7 = $3.00. The likelihood that $\bar{x}$ will fall between $120.00 ± $3.00, or between $117.00 and $123.00, is 68.3 percent. There's a 95.4 percent change that $\bar{x}$ will fall between $\mu - 2\sigma_{\bar{x}}$ and $\mu + 2\sigma_{\bar{x}}$, or between $114.00 and $126.00.

Let's carry out another computer simulation to see how reasonable Sam and Janet's estimate is likely to be. The first two lines of Figure 6-9 show that this simulation is based on 100 random samples of size 49 drawn from a normal population with a mean of $120.00 and a standard deviation of $21.00. The stem-and-leaf display shows the 100 sample means generated in this simulation, the histogram plots these 100 sample means generated in a bar chart, and the DESCRIBE command instructs the program to generate the descriptive values shown in Figure 6-9. As you can see, the mean of the 100 sample means—$120.16 in this simulation—is close to the population mean of $120.00, as it must be. The smallest sample mean produced in this simulation is $112.55, and the largest sample mean is $128.46. The middle 50 percent of the sample means lie between $118.38 and $122.33 (the Q_1 and Q_3 values), but what sample mean values bound the middle 68 percent of the data set?

Our previous analysis showed that about the middle two-thirds of the sample means should fall between $117.00 and $123.00. Do they in this simulation? Well, the middle 68 percent of the sample means fall between sample number 16 in the stem-and-leaf display and sample number 84 in that display. The value of sample mean number 16 is about $117.30, and the value of sample mean number 84 is about $122.80, so the simulation and the theory agree. And 94 percent of the sample means in our simulation fall between $113.70 and $125.40—very close to our previously calculated 95.4 percent chance that a sample mean would lie between $114.00 and $126.00.

Notice, too, in the descriptive measures computed for our data set of 100 sample means at the bottom of Figure 6-9 that the standard deviation (STDEV) value is $3.06. This measure of dispersion for our group of 100 sample means is close to the measure of dispersion we've computed for *all* possible sample means. That is, the $3.06 figure is close to the standard error value of $3.00, as it should be.

As noted in our earlier simulation example, if we were to conduct another simulation and instruct the statistical package to randomly select another 100 samples of size 49, we would get different sample items, different sample means, a different stem-and-leaf display, a different histogram, and different descriptive values. But the mean (and other values) of this next simulated data set will be close to the values we've just examined, and they'll be close to the population values of interest.

```
MTB > RANDOM 100 SAMPLES INTO C1-C49;
SUBC> NORMAL  MU = 120    SIGMA = 21.
MTB > RMEAN C1-C49 INTO C50
MTB > STEM-AND-LEAF C50

Stem-and-leaf of C50          N = 100
Leaf Unit = 0.10


     1   112 5
     6   113 16789
     6   114
    10   115 3356
    15   116 24789
    23   117 36677899
    31   118 03444689
    44   119 1222355577789
   (18)  120 000123344446777789
    38   121 00233455788
    27   122 12333566788
    16   123 13334457
     8   124 02
     6   125 044
     3   126 39
     1   127
     1   128 4




MTB > HISTOGRAM C50

Histogram of C50   N = 100

Midpoint    Count
    112        1   *
    114        5   *****
    116        9   ********
    118       16   ****************
    120       31   *******************************
    122       22   **********************
    124       10   **********
    126        5   *****
    128        1   *

MTB > DESCRIBE C50

               N      MEAN    MEDIAN    TRMEAN     STDEV
C50          100    120.16    120.35    120.18      3.06

             MIN       MAX        Q1        Q3
C50       112.55    128.46    118.38    122.33
```

FIGURE 6-9
A computer simulation of 100 samples of size 49 taken from a normal distribution with a μ of $120.00 and a σ of $21.00. The sample means of the 100 samples are shown in the stem-and-leaf display, and are plotted in the histogram.

Example 6-3   The Write-On Pen Company wants to estimate the average number of pens sold per month on the basis of the mean of a sample of 100 months. If the true population mean is 5,650 pens per month and the population standard deviation is 700 pens, what are the chances that the mean of a random sample will have a value within 200 pens of the true mean?

The problem basically asks: "What are the chances that the value of the sample mean will be between $\mu - 200$ and $\mu + 200$?" Remember that the width has the general form $\mu - Z\sigma_{\bar{x}}$ and $\mu + Z\sigma_{\bar{x}}$. (The interval is more commonly expressed as $\mu \pm Z\sigma_{\bar{x}}$.) Consequently, we want $Z\sigma_{\bar{x}}$ to be 200. The $\sigma_{\bar{x}}$ can be determined as follows:

$$\sigma_{\bar{x}} = \frac{\sigma}{\sqrt{n}} = \frac{700}{\sqrt{100}} = 70$$

Therefore, Z may be calculated in the following manner:

$$Z\sigma_{\bar{x}} = 200$$
$$Z(70) = 200$$
$$Z = 2.86$$

Since $Z = 2.86$, we can consult Appendix 2 and see that the probability of a value occurring within 2.86 standard deviations to one side of a true mean is 49.79 percent. Since we are concerned with 2.86 standard deviations to both sides of the mean, the total likelihood of a sample mean value between $5,650 \pm 200$ is 99.58 percent.

Self-Testing Review 6-5

1   A _____ distribution is a distribution of all the possible sample means, while a _____ distribution is the distribution of individual items of a single sample.

2   If Greek letters are to be used only with population parameters, why are they used to denote characteristics of a sampling distribution of sample means?

3   What method of sample selection is assumed when we discuss concepts concerning the sampling distribution of means?

4   How is the mean of the sampling distribution of means related to the population mean?

5   When may we omit the finite correction factor in calculating $\sigma_{\bar{x}}$?

6   When will the sampling distribution approximate the normal probability distribution?

7   Why is it possible for us to make probability statements concerning the value of a sample mean if we know the population mean?

SAMPLING DISTRIBUTION OF PERCENTAGES

We're often interested in estimating population percentages.[1] For example, a company might want to estimate the percentage of defective items produced by a machine, or it might be interested in the percentage of minority employees in its work force. And we might be interested in the percentage of students who hope to become statisticians. As in the case of estimating a population mean, we estimate a population percentage on the basis of sample results. Let's look now at the relationship between a population percentage and the possible values a sample percentage may assume. That is, let's look at the **sampling distribution of percentages**—a distribution of the percentages of all possible samples that could be taken in a given situation, where the samples are simple random samples of fixed size n.

Mean of the Sampling Distribution of Percentages

The Greek letter π (pi) is used to denote the *population percentage,* while the lowercase letter p denotes the *sample percentage.* The symbol μ_p refers to the mean of the sampling distribution of percentages. The sample percentage is defined as $p = x/n$, where x is the number of items in a sample possessing the characteristic of interest, and n is the sample size.

Suppose we have a population of 5 students and we wish to take a simple random sample of 3 students and estimate the true percentage of students who have made the dean's list. Figure 6-10 lists the population and each student's status concerning the dean's list. What would the sampling distribution look like? Well, in Figure 6-11 the possible percentages a sample might have are listed. And you can see that the mean of the sampling distribution in Figure 6-11 is equal to the population percentage calculated in Figure 6-10. Thus, *the mean of a sampling distribution of percentages with simple random samples of size n is equal to the population percentage*—that is, $\mu_p = \pi$.

Standard Deviation of the Sampling Distribution of Percentages

The standard deviation of the sampling distribution of percentages (more frequently called the **standard error of percentage**) may be computed with knowledge of the population percentage, population size, and sample size. The symbol for this standard deviation is σ_p. The computation of σ_p is as follows for a *finite population:*

---

[1] Many texts deal with the material in the discussion that follows in terms of *proportions* rather than percentages. We prefer to use percentages because many students seem to find the arithmetic easier, and because percentages are more frequently used in every day discussion. Of course, if you prefer to use proportions, you can simply move the decimal two places to the left in all computations. The ultimate results are identical.

FIGURE 6-10   Population of Students and Dean's List Status

STUDENT	DEAN'S LIST
A	Yes
B	No
C	Yes
D	No
E	No

$X = 2$ (the number of students on the dean's list)

$\pi = \dfrac{X}{N} = \dfrac{2}{5}$ or 40 percent

where π = population percentage
N = population size
X = number of students on the dean's list

FIGURE 6-11   Sampling Distribution of Percentages

SAMPLE COMBINATIONS	SAMPLE DATA	SAMPLE PERCENTAGE (p)
1. A, B, C	Yes, no, yes	66.7
2. A, B, D	Yes, no, no	33.3
3. A, B, E	Yes, no, no	33.3
4. A, C, D	Yes, yes, no	66.7
5. A, C, E	Yes, yes, no	66.7
6. A, D, E	Yes, no, no	33.3
7. B, C, D	No, yes, no	33.3
8. B, C, E	No, yes, no	33.3
9. B, D, E	No, no, no	00.0
10. C, D, E	Yes, no, no	33.3
		$\Sigma_p = 400.0$

$\mu_p = \dfrac{\Sigma_p}{N} = \dfrac{400.0}{10}$ or 40 percent

where μ_p = mean of the sampling distribution of percentages
N = number of sample combinations—i.e., $_NC_n$

$$\sigma_p = \sqrt{\frac{\pi(100 - \pi)}{n}} \sqrt{\frac{N - n}{N - 1}} \qquad \text{(formula 6-5)}$$

where π = population percentage possessing a particular characteristic
$100 - \pi$ = population percentage not possessing a particular characteristic
N = population size
n = sample size

$\sqrt{\dfrac{N - n}{N - 1}}$ = correction factor for a finite population

As you've probably anticipated, the finite correction factor may be omitted if the population is infinite, or if the size of the population relative to the sample size is extremely large. The computation of the standard deviation of the sampling distribution for *an infinite population* is:

$$\sigma_p = \sqrt{\frac{\pi(100 - \pi)}{n}}$$ (formula 6-6)

One last comment should be made about the sampling distribution of percentages. The Central Limit Theorem also has application for sample percentages. *If the sample size is sufficiently large, the sampling distribution will approximate a normal probability distribution.* The implication of such a distribution is that probability statements can be made about the possible value of a sample statistic based on the knowledge of the population percentage. For example, there's a 95.4 percent chance that a sample percentage will fall within $\pm 2\sigma_p$ of π. And there's approximately a 99.7 percent chance that a sample percentage will assume a value within $\pm 3\sigma_p$ of π.

A few example problems illustrating some basic concepts of the sampling distribution of means were presented earlier. Let's now do the same thing with some problems related to the material just presented.

Example Problems

Example 6-4   The Kane and Abel Fraternal Organization has a total of eight members whose ages are 27, 32, 33, 26, 43, 52, 28, and 25. The organization has a weird rule which requires a minimum age of 33 for a member to be president. Assume a simple random sample of size 4 is selected to provide an estimate of the population percentage eligible for the presidency. What would be the mean and the standard deviation of the sampling distribution?

The population percentage is computed as follows:

$$\pi = \frac{\text{the number possessing the age qualification}}{\text{the population size}} = \frac{3}{8} \quad \text{or} \quad 37.5 \text{ percent}$$

Since $\mu_p = \pi$, the mean of the sampling distribution is also 37.5 percent. And since we have a *finite* population, the standard error of percentage is computed as follows:

$$\sigma_p = \sqrt{\frac{\pi(100 - \pi)}{n}} \sqrt{\frac{N - n}{N - 1}} = \sqrt{\frac{37.5(100 - 37.5)}{4}} \sqrt{\frac{8 - 4}{8 - 1}}$$

$$= 18.3 \text{ percent}$$

Example 6-5   The Tack Nail Company has selected a random sample of 100 nails to estimate the percentage of nails in a production run which are acceptable. Assume that the population is 90 percent acceptable. What are

the chances that the sample percentage will be within 5 percent of the population percentage?

The basic question is: What are the chances that the value of p will be between $\pi - 5$ percent and $\pi + 5$ percent? In this case, $Z\sigma_p$ equals 5 percent because the basic format in determining the width of the interval is $\pi - Z\sigma_p$ and $\pi + Z\sigma_p$. (The interval is commonly expressed in the form $\pi \pm Z\sigma_p$.) We can solve for σ_p in the following manner:

$$\sigma_p = \sqrt{\frac{\pi(100 - \pi)}{n}}$$

$$= \sqrt{\frac{(90)\,(10)}{100}}$$

$$= 3 \text{ percent}$$

And since $\sigma_p = 3$ percent, we can solve for Z as follows:

$$\begin{aligned} Z\sigma_p &= 5 \text{ percent} \\ Z\,(3 \text{ percent}) &= 5 \text{ percent} \\ Z &= 1.67 \end{aligned}$$

With a Z value of 1.67, we see in Appendix 2 that the area under the normal curve corresponding to a Z of 1.67 is .4525. And with 1.67 standard errors to *each side* of π, the likelihood of a sample percentage within 5 percent of the population percentage is 90.5 percent.

Self-Testing Review 6-6

1 How is the mean of the sampling distribution of percentages related to the population percentage?
2 What population parameters need to be known in order to calculate the standard deviation of the sampling distribution of percentages?
3 When may we omit the finite correction factor in calculating σ_p?
4 When does the sampling distribution of percentages approximate the normal probability distribution?

LOOKING BACK

1 A population is the total of any kind of unit under consideration by an analyst. A finite population is one where the total is a limited, specific number, while an infinite population is unlimited in size. A sample is any portion of the population selected for study. You've seen in this chapter that if a sample statistic is representative of a population parameter, it's possible to make an inference about the population measure from the sample figure.
2 Although complete information, in the form of a census, may be desirable, sampling advantages such as reduced cost, faster response to questions, and acceptable accuracy often outweigh the disadvantage of sampling error. A sample should be representative of the population, but there are several ways that sam-

ples can be selected. For example, a judgment sample is based on someone's expertise about the population. Although a judgment sample is easy to select, there's no objective way to assess its results.

3   A random (or probability) sample is one in which the probability of selection of each element in the population is known prior to sample selection. Unlike the use of a judgment sample, the use of a random sample results in evaluations of population characteristics that can be objectively judged. A simple random sample is one in which each possible combination has an equal probability of occurrence, and each item in the population has an equal chance of being included in the sample. A computer may be programmed to carry out a random selection process, or a table of random numbers may be used for this purpose. Another example of a probability sample is the stratified sample. (If a population is divided into relatively homogeneous groups or strata and a sample is drawn from each group to produce an overall sample, this overall sample is known as a stratified sample.) And a third type of sample—the cluster sample—is one in which the individual units are groups or clusters of single items.

4   Although sampling variation exists so that a statistic will not provide an exact value of a parameter, it's adequate for decision-making purposes to know that sample values are governed by population characteristics. When a simple random sample is used, the mean of a sampling distribution of either sample means or sample percentages is equal to the population parameter being sought—that is, the population mean or the population percentage. The standard deviation of the sampling distribution of sample means—the standard error of the mean $(\sigma_{\bar{x}})$—is determined by the standard deviation of the population and the size of the sample. And the standard deviation of the sampling distribution of percentages (σ_p)—the standard error of percentage—may be computed with knowledge of the population percentage, population size, and sample size. Different samples that can be taken from a population will have different values, as we've seen in two computer simulations involving 150 and 100 samples. But these sample values tend toward the population value being sought.

5   The standard error of the mean is a measure of the dispersion of sample means about the population mean. If this value decreases, the range of probable values a sample mean may assume also decreases, meaning the value of any single sample will probably be closer to the value of the unknown population mean as the standard error decreases. How can this desirable result be achieved? Unfortunately, reducing the standard error requires that we increase the sample size (and the sample cost, and the time required to take the sample . . .).

6   The Central Limit Theorem summarizes several very important facts presented in this chapter: the mean of the sampling distribution of means is equal to the population mean; if the sample size is sufficiently large, the sampling distribution approximates the normal probability distribution; and, if the population is normally distributed, the sampling distribution is normal regardless of sample size. Given this normality property, it's possible to make probability statements concerning the possible values a statistic may assume. The value of a sample statistic will tend toward the parameter value, and a probability statement can be made about the proximity of the statistic to the parameter.

7   The characteristics of the two sampling distributions (means and percentages) presented in Chapter 6 are summarized in Figure 6-12.

FIGURE 6-12   Properties of Sampling Distributions

	SAMPLING DISTRIBUTION OF	
POPULATION	MEANS ($\bar{x}$)	PERCENTAGES (p)
Finite	$\mu_{\bar{x}} = \mu$	$\mu_p = \pi$
	$\sigma_{\bar{x}} = \dfrac{\sigma}{\sqrt{n}} \sqrt{\dfrac{N-n}{N-1}}$	$\sigma_p = \sqrt{\dfrac{\pi(100-\pi)}{n}} \sqrt{\dfrac{N-n}{N-1}}$
Infinite	$\mu_{\bar{x}} = \mu$	$\mu_p = \pi$
	$\sigma_{\bar{x}} = \dfrac{\sigma}{\sqrt{n}}$	$\sigma_p = \sqrt{\dfrac{\pi(100-\pi)}{n}}$

* Both sampling distributions approximate the normal probability distribution, if the simple random sample size is sufficiently large. As a general rule of thumb "sufficiently large" is over 30.

KEY TERMS AND CONCEPTS

Finite population   *214*
Infinite population   *214*
Judgment (purposive) sample   *219*
Random (probability) sample   *219*
Simple random sample   *221*
Table of random numbers   *221*
Stratified sample   *222*
Cluster sample   *222*
Sampling variation   *223*
Sampling distribution of means   *224*

$$\mu_{\bar{x}} = \frac{\Sigma(\bar{x}_1 + \bar{x}_2 + \bar{x}_3 + \cdots + \bar{x}_N C_n)}{{}_N C_n}$$
(formula 6-1) *226*

Standard error of the mean   *229*

$$\sigma_{\bar{x}} = \sqrt{\frac{\Sigma(\bar{x} - \mu_{\bar{x}})^2}{N}} \quad \text{(formula 6-2)} \quad 229$$

$$\sigma_{\bar{x}} = \frac{\sigma}{\sqrt{n}} \sqrt{\frac{N-n}{N-1}} \quad \text{(formula 6-3)} \quad 230$$

$$\sigma_{\bar{x}} = \frac{\sigma}{\sqrt{n}} \quad \text{(formula 6-4)} \quad 231$$

Central Limit Theorem   *233*
Sampling distribution of percentages   *237*
Standard error of percentage   *237*

$$\sigma_P = \sqrt{\frac{\pi(100-\pi)}{n}} \sqrt{\frac{N-n}{N-1}} \quad \text{(formula 6-5)} \quad 238$$

$$\sigma_P = \sqrt{\frac{\pi(100-\pi)}{n}} \quad \text{(formula 6-6)} \quad 239$$

PROBLEMS

1   Assume that a population consists of 10 items. What's the probability of selection for each possible sample if a simple random sample of 3 is taken?

2   If we have a population of 8, what's the probability of selection for each possible sample if a sample of 5 is taken? What's the probability of selection for each item of the population?

3   A population consists of 5 students. The number of hours they spend watching television is as follows:

STUDENT	HOURS
a	7
b	16
c	20
d	12
e	22

A simple random sample of 3 is to be taken to estimate the population mean, that is, the average number of hours spent watching television.

a Calculate the population mean and the population standard deviation.

b What is the mean of the sampling distribution?

c Calculate the standard deviation of the sampling distribution.

4 We have a population of 5 motorists in Saudi Arabia. The price they each pay for a gallon of gasoline is as follows:

MOTORIST	PRICE
a	$0.52
b	0.48
c	0.54
d	0.50
e	0.53

A simple random sample of 3 is to be taken, and an average price per gallon is to be estimated.

a Obtain the sampling distribution of $\bar{x}$.

b Calculate the mean and the standard deviation of the sampling distribution.

c Verify the values of $\mu_{\bar{x}}$ and $\sigma_{\bar{x}}$ with the use of the population parameters.

5 The Tite Wire Company manufactures wires for circus acts. It has taken a random sample of 100 pieces of wire and wants to see if the thickness of a batch of wire meets minimum specifications. Assume that $\mu = 0.45$ inches, with a standard deviation of 0.03 inches.

a Calculate the mean and standard deviation of the sampling distribution.

b What may be said about the shape of the sampling distribution?

c Within what range of values does the sample mean have a 68.3 percent chance of falling?

d Within what range of values does the sample mean have a 95.4 percent chance of falling?

6 Assume we have an infinite population with a mean of 200 and a standard deviation of 15.

a Within what range of values will a random sample mean have a 95.4 percent chance of falling if we have a sample of 45?

b Within what range of values will there be a likelihood of 95.4 percent occurrence for the sample mean if we have a sample of 36? What about a sample of 49? What about a sample of 64?

c What relationship can you observe between the sample size and the dispersion of the sampling distribution?

7 The Keep On Trucking Company wants to estimate the average tonnage of freight handled per month, and it has taken a random sample of 36 months. Assume the true average tonnage per month is 225 tons, with a standard deviation of 30 tons. What are the chances that the sample mean will have a value within 7 tons of the true mean?

8   Dr. D. Zees would like to estimate the average amount charged per patient each visit. He has a random sample of 40 patients. Assume that μ = $13 and σ = $4. What are the chances that the sample mean will have a value within $1 of the true mean?

9   Assume we have a population of 20 high school students and we take a random sample of 5 students to estimate the proportion of students who intend to enter college. Assume the true percentage is 60 percent.

   a   What will be the mean of the sampling distribution of percentages? Why?

   b   Is the finite correction factor needed to calculate σ_p? What's the value of the finite correction factor?

   c   Calculate σ_p.

10 The Vanity Press Company wants to estimate the percentage of books printed that are defective and cannot be sold. Assume we have a simple random sample of 100 and the true percentage is 8.5 percent. What are the chances that the sample percentage will be within 1 percent of the population percentage?

11 Let's assume that the average verbal SAT score of all students attending McGuire College is 540, and the population standard deviation is 30 SAT points. What's the probability that the mean verbal SAT score of a random sample of 36 students is greater than 535?

12 If the dollar values of the accounts receivable of the Rice Corporation are normally distributed with an arithmetic mean of $10,000 and a standard deviation of $2,000, and if a random sample of 400 accounts is selected, what's the probability that the sample mean will be between $10,000 and $10,200 in size?

13 Suppose that 64 percent of the population in Jamesburg favor a certain candidate for an elective office. What's the probability that a random sample of 100 voters would have a sample percentage favoring the candidate of between 60 and 68 percent?

14 If the true average gasoline consumption of families in Rocktown is 16.9 gallons per week with a population standard deviation of 3.2 gallons per week, what's the probability that the mean of a random sample of 50 families would exceed 17.5 gallons per week?

15 The mean SAT scores of all students attending a large university is 980 points with a population standard deviation of 50 points. What's the probability that the mean SAT score of a random sample of 100 students is greater than 990 points?

16 If the hourly wages of Meyers Corporation workers have a population mean value of $5.00 per hour and a population standard deviation of $0.60, what's the probability that the mean wage of a random sample of 50 Meyers workers will be between $5.10 and $5.20?

17 If 55 percent of a television viewing population watched a program called "Name that Poem" one evening, what's the probability that, in a random sample of 100 viewers, less than 50 percent of the sample watched the program?

18 The average purchase in Jim's furniture store over the past 24 months has been $236 with a standard deviation of $48. If a random sample of 36 sales made during this period were selected, what's the probability that the sample mean would be $220 or less?

19 All fourth-graders in Hamburgh take the Blocker Interval Test. The population

mean score is 200 with a population standard deviation of 20. What's the probability that a sample of 16 children chosen at random from fourth-grade students will have a mean score greater than 205?

20 Miller Flour Company fills sacks of flour using automated machinery. The population mean weight of the sacks is 50 pounds and the population standard deviation is 2 pounds. The sacks are shipped in pallets of 64 sacks. What proportion of the pallets will have an average weight per sack of more than 50.5 pounds?

21 The Western Wire Company produces rolls of barbed wire. If 10 percent of the rolls produced are defective, what's the probability of drawing a random sample of 400 rolls and having 14.5 percent or more of them turn out to be defective?

22 A population consists of four students. The exam scores made by these students on a college entrance test are given below:

STUDENT	EXAM SCORE
A	990
B	1,403
C	1,059
D	1,213

a What is the value of the population mean and population standard deviation?

b How many simple random samples of size 2 can be taken from this population?

c What is the mean of all the sample means that could be taken?

d What is the value of the standard deviation of the sampling distribution of sample means?

23 Let's assume that we have an infinite population with a mean of 50 and a standard deviation of 3.

a If samples of 100 are repeatedly taken from this population, what's the likely value of the mean of all the sample means that could be computed in a reasonable period of time?

b What's the value of the standard error of the sample means?

24 A population of 1,000 posts has a mean length of 50 inches and a standard deviation of 3 inches. If a simple random sample of 100 posts is selected and the mean of this sample is computed, what's the interval (centered on the mean of the sampling distribution) within which this sample mean has a 50 percent chance of lying?

25 A population of six politicians and their ages are listed below:

POLITICIAN	AGE
A	30
B	50
C	60
D	34
E	33
F	29

To serve as President of the United States, a politician must be 35 years of age or older.

   a   What percentage of the above population is eligible to serve as President?

   b   If simple random samples of size 2 are taken from the above population, what's the value of the standard error of percentage?

26   It's known that, in a large population of politicians, 50 percent are eligible to serve as President of the United States. If a random sample of 100 politicians is selected from this population, what's the probability that the sample percentage eligible to serve as President lies between 45 and 55 percent?

27   Suppose that 53 percent of a voting population favor candidate X in an election. Let's assume, though, that only 50 people turn out to vote. If we can consider the voters to be a random sample of the electorate, what's the probability that candidate X loses?

28   Seventy percent of a population of 50 medical students plan to become surgeons. A random sample of 5 students is taken to determine their future career plans.

   a   What's the mean of the sampling distribution of percentages?

   b   What's the standard error of percentage in this case?

29   A population of 1,000 high school students has an average IQ of 100 with a standard deviation of 20. If a random sample of 30 students is selected, what's the probability that the mean IQ of the group exceeds 110?

30   A population of 5 students and their grades on an exam are listed below:

STUDENT	GRADE
A	92
B	83
C	70
D	53
E	77

   a   How many possible samples of size 3 can be taken?

   b   What's the mean of the sampling distribution of sample means?

   c   What's the standard deviation of the sampling distribution of means (assuming that samples of size 3 are taken)?

31   A food packaging company fills sacks of cereal using automated machinery. The population mean weight of the sacks is 2 pounds with a standard deviation of 0.1 pound. A dozen sacks are placed in a box for shipment. What's the probability that the contents of a box equals or exceeds 25 pounds?

TOPICS FOR REVIEW AND DISCUSSION

1   "Statisticians maintain the distinction between parameters and statistics through the use of different symbols." Discuss this statement, identify three parameters and three statistics, and give the symbols for the parameters and statistics you've chosen.

2 a Why is sampling important?
  b What are some of the reasons why a decision maker might prefer to take a sample rather than a census?

3 Identify and discuss three types of samples.

4 How can a random sample be selected?

5 What's the difference between a sample distribution and a sampling distribution?

6 In what situations may your statistics class be a population, and under what conditions may your class be a sample?

7 "We should never be satisfied with a sample unless it's completely representative of the population." Discuss this comment.

8 "Since sampling errors will always exist, it's difficult to have any confidence in an estimation." Discuss this statement.

9 "In practice we take only one sample. It's therefore not necessary to be concerned about the sampling distribution, because no one ever takes all possible samples." Do you agree with this sentiment? Defend your answer.

10 "A finite correction factor must always be used in the calculation of the standard error of the mean if we have a finite population." Why is this statement incorrect?

11 What effect does sample size have on the dispersion of sample means about the population mean?

PROJECTS/ISSUES TO CONSIDER

1 Team up with several of your classmates and locate a population data set in your library or elsewhere that's of interest to the group. You might want to join with those you worked with to complete the projects following Chapters 3 and 4, and it might be possible for you to use the same data set that you analyzed in those projects. After you've settled on a population data set, carry out the following steps *as a group:*
  a Compute the population mean and population standard deviation. (If you use the earlier data set, this step has already been done.)
  b Select a simple random sample of *n* items from your data set. Be prepared to discuss how you picked your sample items from the population and why you selected the sample size you used.
  c Compute the mean for your sample, and the standard error of the mean.
  d With the data you now have, analyze the relationship between your sample mean and the population mean and present this analysis to the class.

2 A computer simulation of 150 samples is shown on the next page. You'll recognize the similarity between this simulation and the one presented in Figure 6-9. To analyze this simulation, first explain the meaning of the values that have been computed at the bottom of the simulation output (ignore TRMEAN). Then compare these values to those shown in Figure 6-9 and explain the differences. Finally, prepare a brief written summary of your analysis.

```
MTB > RANDOM 150 SAMPLES INTO C1-C49;
SUBC> NORMAL MU = 120 SIGMA = 21.
MTB > RMEAN C1-C49 INTO C50
MTB > HISTOGRAM C50

Histogram of C50   N = 150

Midpoint    Count
     112       1   *
     114       5   *****
     116      12   ************
     118      37   *************************************
     120      38   **************************************
     122      30   ******************************
     124      18   ******************
     126       6   ******
     128       3   ***

MTB > DESCRIBE C50

                 N      MEAN     MEDIAN    TRMEAN     STDEV
C50            150    120.15     119.89    120.12      2.93

               MIN       MAX         Q1        Q3
C50         112.78    128.34     117.98    122.24
```

ANSWERS TO SELF-TESTING REVIEW QUESTIONS

6-1
1 a   Population.
   b   Sample.
   c   Sample.
   d   Population.
2 Yes: the population may be the major leagues.
3 a   The parameter is the total membership, 300. It may also be the mean age,
       42.
   b   The sample size, 25, and the sample mean, 39, are statistics.
   c   Population mean.
6-2
1 The statement is incorrect. Sampling occurs frequently in the course of daily
   events.
2 The purpose of sampling is to provide sufficient information so that judgments
   may be made concerning the characteristics of the population.
3 The goal is maximum representation of population characteristics.
4 False. If the errors can be objectively assessed, we have some idea of the precision of the estimate.
5 It can be determined by objectively assessing the amount of sampling error.

6-3

1 This statement is true. Representation can be achieved if the right method of sample selection is employed.

2 There's justification for a sample if the sample costs are less than census costs, but costs should never be the sole reason for sampling.

6-4

1 There's no one best method for sample selection. The nature of the population and the skills of the researcher determine the appropriate method.

2 It's difficult to say whether a probability sample is more representative than a judgment sample because it's impossible to assess objectively the error in a judgment sample.

3 The probability sample is desirable because it can be objectively assessed.

4 There are 45 possible combinations. Each sample combination will have a 1 in 45 chance of selection. Each individual item will have a 1 in 5 chance of selection.

5 The basic assumption in cluster sampling is that the items within a cluster are representative of the population.

6 We must have some prior knowledge about the structure of the population.

6-5

1 Sampling; sample.

2 The sampling distribution is a population of all possible sample means of a fixed sample size.

3 In discussing any sampling distribution, we assume simple random samples.

4 The mean of the sampling distribution is always equal to the population mean.

5 The finite population correction factor may be omitted when there is an infinite population or when the population is extremely large relative to the sample size.

6 The sampling distribution will approximate the normal distribution if the sample size is sufficiently large.

7 The sampling distribution is a probability distribution, and in cases of a large sample size, the sampling distribution approximates a *normal* probability distribution. Since we know the shape and characteristics of a normal distribution, we can make probability statements.

6-6

1 The mean of the sampling distribution of percentages, μ_p, is equal to the population percentage.

2 We need to know the population percentage, and in the case of a finite population, we need to know the population size.

3 The finite correction factor may be omitted with an infinite population, or the finite correction factor may be omitted if the population is extremely large relative to the sample size.

4 The sampling distribution approximates the normal probability distribution when the sample size is large.

A CLOSER LOOK

CENSUS BUREAU GEARS UP FOR COUNTDOWN TO 1990

SUITLAND, Md.—When the first U.S. census was authorized in 1790, 17 U.S. marshals and 600 assistants were charged with traveling around the country to count all its inhabitants. They had to provide their own pens and paper, and did much of their work through door-to-door interviews.

In 1990, the 200th anniversary of the census, most of the census-taking will be done by mail. Still, 400,000 temporary government employees will help in the effort, including ones who will fan out across the nation—by car, airplane, horse and even sled dog team—to knock at doors, glance under bridges and peer into trees to make sure that every person is included.

"We say it's the largest peacetime activity we have," said Richard Bitzer, assistant division chief for census, field division. "It is like organizing the troops."

The countdown to Census Day, April 1, 1990, began in earnest this spring. A "dress rehearsal" was held in selected regions last month; 13 regional census centers, including one in Dallas, have been readied in recent weeks; and the April deadline for Congress to approve the 1990 questions has been met.

The work is being coordinated at the Suitland Federal Center, headquarters of the Census Bureau. The bureau will spend about $2.6 billion over 10 years on the 1990 census and plans to open 450 offices across the country to help conduct the count.

The census provides important lifestyle statistics as well as a head count. Results are used to determine state and national political representation—including the distribution of seats in the House of Representatives—and the allocation of about $35 billion annually in federal funding, affecting everything from roads to schools.

"It is important from the standpoint of political power and money—the two big ones," said Charles Jones, associate director for the decennial.

The census bureau is already compiling address lists for the approximately 100 million forms to be mailed. Five out of every six households will receive the short form, with about 13 questions, while the rest will receive the long form, with about 55 questions.

The contents of the questionnaire approved by Congress on April 1 were settled on after extensive research and hundreds of public meetings across the country since the 1980 census.

"There's a host of people out there who would like us to ask all sorts of probing personal questions, which we don't touch," said Arthur Norton, assistant chief of the population division.

Each question is worded so that it elicits the most specific information in the least offensive manner. The term "head of the household," for instance, has not been used since 1970 because "some people were offended," Norton said. Instead, respondents are asked to "give the name of the person in whose name the home is owned or rented."

Though the decennial census has been conducted primarily through the mail since 1970, that hasn't made the census taker or "enumerator" obsolete. In 1980, 83 percent of the people mailed in completed forms within several weeks of Census Day, but the 17 percent who didn't made the bureau's job difficult.

To make sure no one was missed in the first census 200 years ago, papers were tacked to the walls of taverns and other public places for voluntary signing. In 1990, those who fail to return the form, or don't respond in full, will be visited by a census taker. Four attempts will be made to contact residents of known addresses, and if all else fails, neighbors will be queried. (People can be fined up to $500 for not responding to the census.)

Those living in remote areas without regular mail service, from the Alaska wilderness to the Southwestern desert, also will be visited by a census taker. But it is actually urban locations that are most difficult to survey, Bitzer said.

New York City, for instance, "provides a lot of challenges," he said, noting that people in large cities are busy and mobile, and therefore difficult to find at home. Many urban inhabitants also have unusual living arrangements, he said, such as those who live on the street.

Not having a permanent address, however, is no excuse for not participating in the census. Whenever a census is taken, one night is designated for tracking down those who live under bridges, in missions or on the streets. Census takers distribute forms everywhere from hotels to all-night theaters.

"We don't look for homeless people," said C. Louis Kincannon, deputy director of the bureau. "We look for all the people, wherever they are."

That includes legal and illegal residents. Though it has become a controversial issue in recent years, the census does not distinguish between the two.

Five states, including Texas, have more than 80 percent of all illegal aliens, according to the 1980 census. Those residents do not vote, but their inclusion in the census results in greater congressional representation for those areas, which some politicians feel is unfair to states without large illegal alien populations.

Rep. Thomas J. Ridge, R-Pa., has introduced a bill in Congress to exclude illegal aliens in 1990 for the purposes of reapportionment, and has also filed a lawsuit challenging the practice of counting illegal aliens for reapportionment.

"He feels it's disenfranchisement as far as votes are concerned," said Ridge aide Kevin Feather.

Though the bill has 73 co-sponsors, its chances of passing are considered slim.

Another perennial criticism of the census is that it undercounts the population—particularly minorities. In 1980 the bureau estimated that it missed 1.4 percent of the population, including about 5 percent of the black population.

Rep. Mervyn M. Dymally, D-Calif., is spon-soring a bill that would force the bureau to adjust the 1990 census to compensate for the expected minority undercount. "It is important because the central cities pay a very high price financially for the undercount," Dymally said. "They must provide services for which they do not receive federal reimbursement. Second, if we are to expand the base of minority representation in America, we must have an accurate count."

Jones said that the bureau would like to compensate but does not have the statistical techniques to do so.

While Congress debates those issues, the bureau is concentrating on preparations for the census—including hiring the 400,000 temporary workers. In the past, most census takers have been homemakers looking for short-term work. With the growth of two-income families, such workers are increasingly hard to find, said assistant division chief Bitzer.

Approximately 1.6 million people will be tested for the jobs, which start at $5.50 an hour. An estimated 400,000 will be hired, including about 30,000 in the Dallas-based region of Texas, Louisiana and Mississippi, said Mickey Cole, assistant regional census manager.

The work can be difficult, Bitzer noted, because some respondents are uncooperative. The most common concern is confidentiality. Illegal aliens, for instance, may think that responding to the census could result in their deportation.

But bureau officials note that information-sharing is illegal. Personal census records are kept private for 72 years, the average life span.

Bureau employees take their oath of confidentiality to heart, officials said. In 1980, Jones recalled, the bureau received a frantic telephone call from a Colorado office. A bureau employee had locked herself in the building because FBI agents were knocking on the door, requesting information about an individual. The bureau contacted the FBI director, who called off his agents.

"This was a temporary employee," Jones said. "She was not one of our permanent cadre, but she'd sworn an obligation. The confidentiality is something everybody is quite proud of—and we go to the mat."

1990 Census

SHORT FORM

Questions are expected to include ones such as:
- What is the name of each person living here?
- How are these people related?
- What is their sex, race, age, year of birth and marital status?
- What type of dwelling is this?

LONG FORM

Questions are expected to include ones such as:
- Do you have complete plumbing facilities in this house or apartment; that is 1) hot and cold piped water, 2) a flush toilet and 3) a bathtub or shower?
- How many automobiles, vans, and trucks of one-ton capacity or less are kept at home?
- How much is your regular monthly mortgage payment?
- What time did various members of the household usually leave for work last week?
- Because of a health condition that has lasted for six or more months, does anyone in the household have any difficulty 1) going outside the home alone, for example, to shop or visit a doctor's office? 2) Taking care of his or her own personal needs, such as bathing, dressing, or getting around inside the home?
- Is a household member a citizen of the United States?
- How much school have household members completed?
- Where did household members live five years ago?

Source: *The Dallas Morning News,* Jan Brunson

After the census is completed, the information will be processed by about 600 computers across the country, and reports that will total several hundred thousand pages will be published over a two-year period. The final head count, expected to total about 250 million, must be done by Dec. 31, 1990, for congressional reapportionment. More detailed counts will be provided to governors three months later.

The census bureau, meanwhile, will continue with its other duties: The bureau—which has about 3,500 permanent employees at headquarters and several thousand more across the country—is constantly conducting surveys and produces about 2,000 reports a year, covering everything from unemployment to agriculture statistics.

For Jones and other census employees, April 1, 1990, is important, but not as crucial as the preparations over the next few months. "April 1 will be an easy day," he said. "It's the months before, and the months after."

—Diane Jennings, "Census Bureau Gears Up For Countdown to 1990," *The Dallas Morning News,* April 25, 1988, pp. 1A and 8A. Reprinted with permission.

7

ESTIMATING MEANS AND PERCENTAGES

If the Gallup Poll People Call, I Am Not above Lying to Them

Quite frankly, there is no scientific method of computing precisely how many people I have met in my life, but I have developed a formula suitable, perhaps, for the task at hand.

First, I figured out how many days I have spent on Earth and, not counting today, it came to 19,499. It is logical to assume that, during that time, I have averaged meeting three new acquaintances per day. Some days, of course, I don't meet any new people. But on other days, I meet six or seven or more, so three seems to be a reasonable average.

Multiplying the number of days I have been around by three, I calculate that I have met 58,497 people in my life. And not one of those people, to the best of my knowledge, has EVER been called by the George Gallup organization, not even on a wrong number.

Therefore, I hope I may be excused if I don't put too much stock in many of the results the Gallup Poll disseminates with annoying regularity. I mean, I run around with a pretty average crowd, yet few of the findings published by the Gallup people are applicable to any of us.

Take the most recent poll, for instance. The Gallupers based their findings on "in-person interviews with 1,509 adults, 18 and older, conducted in more than 300 sci-entifically selected localities across the nation during the period Nov. 9–12." Yet as far as I know, not one of the 58,497 people I have met in my lifetime was invited to participate.

Which is probably why I find the poll results quite far removed from what I perceive to be reality.

The topic of this most recent sampling was recreational activities, and the pollsters discovered that swimming, bicycling and fishing were Americans' top three leisure-time activities. With the exception of a brief period during last summer's visit to Hawaii when I waddled cautiously out into the Pacific to wash away sand kicked in my face by a passing bully, I have participated in none of those activities during the past year, nor do I anticipate doing so in the foreseeable future.

I have neither the time nor the resources to check with all 58,497 of my acquain-tances, but I doubt seriously that more than six of them have been astride a bicycle in the past three decades. Could it be that the other 58,491 are all out of step?

The remainder of the leisure-time list isn't all that realistic either. Following The Big Three come camping, jogging, aerobics or dancersize, weight lifting, bowling, bil-liards or pool, softball, calisthenics, motor-boating and volleyball. The list continued downhill from there, including such outlandish pastimes as roller skating, canoing and playing touch football.

It is an undisputed fact that at least 90 percent of my 58,000-plus acquaintances have, over the past few years, been transformed into gigantic fat globules, and they would probably be arrested if they showed up for a dancersize class in leo-tards, as well they should be.

The point of all this is quite simple: Either the Gallup folks are trying to pull a fast one on us or, as I suspect, the people being polled are lying through their teeth. People being what they are, they tend to give answers that (A) they think the ques-tioner wants to hear or (B) make them look good.

I suspect that, if the chips were down, I would follow the same devious course. If I were asked about my favorite leisure-time activities, knowing the results were going to be made public, I, too, would say shinnying up Mount Everest in midwinter, dog-paddling across the Hellespont and bench-pressing a Ford pickup instead of con-fessing that most of my spare time is spent reducing the nation's dangerous beer surplus, watching Godzilla movies and trying to remember where I parked the car.

So, go ahead and call, Mr. Gallup. I'm in the book, and I promise not to embarrass either of us.

—Bill Youngblood, "If the Gallup Poll People Call, I Am Not above Lying to Them," *Fort Worth Star-Telegram*, January 10, 1985, p. 10A. Reprinted with permission.

LOOKING AHEAD

This tongue-in-cheek vignette dealing with polls touches on one of the topics—confidence interval estimates of percentages—discussed in this chapter. In addition to the type of poll mentioned in the vignette, sample results are also used during an election campaign to estimate the percentage of voters that will vote for particular candidates, and they're used to estimate the percentage of television viewers watching a particular program.

Learning about procedures for estimating the population percentage is just one of the topics in this chapter. You'll also learn (1) some new terms, (2) some basic concepts about the estimation of the population mean, (3) the procedures for actually estimating the population mean, and (4) the approaches used to determine the size of a sample.

Thus, after studying this chapter, you should be able to:

- Explain the basic theories and concepts underlying the estimation of population means and population percentages.
- Compute estimates of the population mean at different levels of confidence when the population standard deviation is unknown as well as when it's available.
- Understand when and how to use the appropriate probability distributions needed for estimation purposes.
- Compute estimates of the population percentage at different levels of confidence.
- Determine the appropriate sample size to use to estimate the population mean or percentage at different levels of confidence.

CHAPTER OUTLINE

NEED FOR ACCURATE ESTIMATES

To guess is cheap,
To guess wrongly is expensive.
    —An old Chinese proverb

As the proverb implies, a guess or an estimate is easily made. Anyone can offer an opinion. The naive as well as the expert can produce an estimate if requested to do so. Martha Q. Public or the famous economist Monte Terry Drane can give an estimate of next year's gross national product. But the task in estimation *isn't* just to produce an estimate; the challenge is to produce one that has some degree of accuracy.

Let's assume that a sales manager who must make a sales forecast for the next period asks you, her trusted assistant, to estimate the average dollar purchase made by a typical customer. Perhaps you base your estimate on a method such as rolling a pair of dice, drawing from a deck of cards, or reading a cup of soggy tea leaves. Now maybe your luck holds and your hunch doesn't cause later embarrassment. But consider the alternative: on the basis of your wild guess, your boss makes a grossly inaccurate forecast which leads to the loss of tens of thousands of dollars. This result brings the wrath of the company president down on your boss, and she, in turn. . . . Well, you get the picture.

In this chapter we'll see how to estimate the population mean and the population percentage with some degree of confidence that the figures arrived at approximate the true values. Of course, our methods won't produce exact population values. Some error is inevitable in estimation (as the Roman poet Ovid once wrote, "The judgment of man is fallible"[1]), but the

---

[1] *Fasti*, chap. V, line 191. (There is nothing like a little Ovid to mollify the poets who are required to read this text.)

amount of error can be objectively assessed and controlled, as you'll see in later pages.

ESTIMATOR, ESTIMATE, ESTIMATION, ET CETERA

Although no formal definitions of the words "estimate" and "estimation" have appeared, you probably have an intuitive feeling for their meanings. In case you don't, though, this section clarifies some of the terms we'll be using.

Suppose the Adam and Eve Apple Orchards want to estimate the average dollar sales per day, and a sample of days has produced a sample mean of $300. In this case the statistic (the sample mean) may be used to estimate the parameter (the population mean). The sample value of $300 is an estimate of the population value. Any statistic used to estimate a parameter is an **estimator.** Thus the sample mean is an estimator of the population mean, and the sample percentage is an estimator of the population percentage. But remember that any *specific value* of a statistic is an **estimate.**

There are several reasons for selecting a particular statistic to be an estimator. A complete discussion of all the reasons is beyond the scope of this book, but we'll mention one important criterion for the selection of a statistic as an estimator. The sample mean is selected as an estimator of the population mean, and the sample percentage is an estimator of the population percentage because these statistics are *unbiased.* An **unbiased estimator** is one that produces a sampling distribution that has a mean that's equal to the population parameter to be estimated. We've seen in Chapter 6 that the mean of the sampling distribution of sample means equals the population mean, and the mean of the sampling distribution of sample percentages equals the population percentage. Thus, sample means and sample percentages are unbiased estimators. They are distributed about, and they tend toward, the population mean and the population percentage. This unbiased tendency of an estimator is, of course, highly desirable.

The entire process of using an estimator to produce an estimate of the parameter is known as **estimation.**

Point Estimation

A single number used to estimate a population parameter is called a **point estimate,** and the process of estimating with a single number is known as **point estimation.** Thus, the sample mean of $300 in our previous example is a point estimate because the value is only one point along a scale of possible values. But how likely is a single estimate to be correct?

Interval Estimation

A parameter is usually estimated to be within a *range* of values, rather than as a single number. It's unlikely that any particular sample mean will be exactly equal to the population mean, so allowances must be made for sampling error. A range of values used to estimate a population parameter is called an **interval estimate**, and the process of estimating with a range of values is known as **interval estimation.**

With two methods of estimation, which one should be used? To answer this, let's look at the precision of a point estimate. You've seen in the computer simulations in Chapter 6 (and in Figures 6-6 and 6-9) that the means of different random samples taken from a population have different values. A few sample means may equal the population mean, but most won't. Furthermore, we can't tell by looking at a point estimate whether it's equal to the population mean or falls some distance away from that value. A point estimate is not only highly subject to error in estimation, it also doesn't allow us to evaluate the precision of the estimate.

The precision of an estimate is determined by the degree of sampling error. And although a point estimate will probably be incorrect as a result of sampling error, this doesn't prevent us from placing considerable confidence in the estimate that the parameter will be within a given range of values. For example, we may say that the daily average sales for the orchards is likely to be between $285 and $315 instead of simply saying that the true mean may be around $300. Thus, the parameter is estimated to be within a certain range or interval. And since allowance is made for sampling error, the precision of the estimate can be objectively assessed. Of course, an interval estimate can be wrong, like any other estimate. But in contrast to the point estimate, the probability of error for the interval estimate can be objectively determined.

Don't get the impression, though, that a point estimate is of little value in the estimation process. As you'll see in the following sections, the interval estimate is actually based on the point estimate. In essence, the *point estimate is adjusted for sampling error to produce an interval estimate*. The remainder of this chapter discusses interval estimations of the population mean and population percentage, and the accuracy of these estimations can be determined with some degree of confidence.

Self-Testing Review 7-1

1 The sample mean is an _____ of the population mean.
2 A sample of 36 items has produced a sample percentage of 82 percent. Which is the estimator and which is the estimate?
3 What's the difference between a point estimate and an interval estimate?
4 What's the disadvantage of a point estimate?
5 "Since sampling error is considered in an interval estimate, the parameter value will always fall within the range." Comment on this statement.
6 Why is an unbiased estimator desirable?
7 "Since a point estimate is rarely used as an estimate, it's rarely calculated." Comment on this statement.

INTERVAL ESTIMATION OF THE POPULATION MEAN: SOME BASIC CONCEPTS

In practice, only one sample of a population is taken, the sample mean is calculated, and an estimate of the population mean is made. To estimate the population mean, we must know something about the relationship between the sample mean and that value.

The Sampling Distribution—Again

A quick review of the concepts of the sampling distribution of means shows the theoretical basis for the interval estimation of the population mean. Suppose we have a sample size that is sufficiently large so that the sampling distribution is approximately normal. Figure 7-1 shows that 95.4 percent of the possible outcomes of $\bar{x}$ are within $2\sigma_{\bar{x}}$ to each side of the mean of the sampling distribution. This means that if a mad statistician takes 1,000 samples of the same size from a population, about 954 of the sample means will fall within two standard errors to both sides of the population mean. (If this last sentence—and Figure 7-1—puzzles you, review Chapter 6 again.)

Interval Width Considerations

If 95.4 percent of the possible values of the sample mean fall within two standard errors of the population mean as shown in Figure 7-1, then obviously μ will not be farther than $2\sigma_{\bar{x}}$ from 95.4 percent of the possible values of $\bar{x}$. Now let's show in nonstatistical terms the logic of the preceding sentence. Let's assume we have 1,000 towns located various distances from

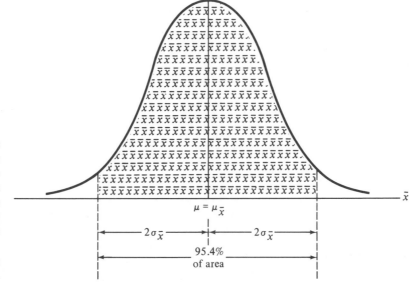

FIGURE 7-1
An educational schematic of the sampling distribution of the means when the sample size is large. There is a 95.4 percent chance that a $\bar{x}$ will have a value between $\mu \pm 2\sigma_{\bar{x}}$.

the city of Boston and it happens that 95.4 percent of these towns are within a 50-mile radius of Boston. If 954 towns are within a 50-mile radius of Boston, then logically Boston must fall within a 50-mile radius of each of these 954 towns. If Hingham is within 50 miles of Boston, then Boston will certainly be no farther than 50 miles from Hingham (see Figure 7-2). If we randomly select a large number of towns from the 1,000, Boston will be within a radius of 50 miles of 95.4 percent of all the towns selected. If all this appears simple and trite, then we've accomplished something.

In returning to the statistical world, substitute the population mean for Boston, let the possible sample means be the towns, and use $2\sigma_{\bar{x}}$ in place of the 50-mile radius. To repeat, then, if 95.4 percent of the sample means are within $2\sigma_{\bar{x}}$ of μ, then certainly μ must be within $2\sigma_{\bar{x}}$ of 95.4 percent of the sample means. Thus, if we use the method of $\bar{x} \pm 2\sigma_{\bar{x}}$ to estimate the population mean, and if we construct a large number of intervals, 95.4 percent of the interval estimates will include μ.

Now let's assume that we have 1,000 possible samples and thus 1,000 sample means, 3 of which are shown in Figure 7-3. The population mean will be located within 95.4 percent of the 1,000 possible intervals that could be constructed using $\bar{x} \pm 2\sigma_{\bar{x}}$. Any specific single interval may or may not contain μ (note that in Figure 7-3 the intervals produced using $\bar{x}_1$ and $\bar{x}_2$ do include μ, but the interval constructed using $\bar{x}_3$ fails to reach μ), but the method employed assures that if a *large number of intervals* are constructed, μ will be included in 95.4 percent of them.

We'll not be limited to using a 95.4 percent probability of estimating the population mean for the rest of this book. Thus, we need to generalize what has been discussed so far so that we can apply different interval estimates to a variety of situations. If the sampling distribution is normal, an interval estimate of μ may be constructed in the following manner:

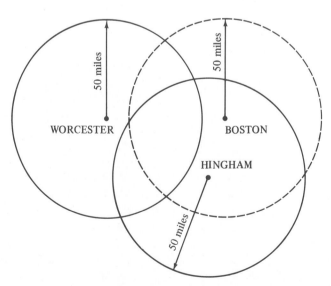

FIGURE 7-2
An illustration of distance relationships.

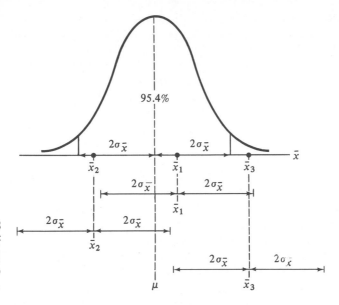

FIGURE 7-3
An illustration of
the interval
relationship
between μ and $\bar{x}$.

$$\bar{x} - Z\sigma_{\bar{x}} < \mu < \bar{x} + Z\sigma_{\bar{x}}$$
Lower limit   Upper limit   (formula 7-1)
of estimate   of estimate

where $\bar{x}$ = sample mean (and point estimator of μ)
       $\sigma_{\bar{x}}$ = standard error of the mean
       Z = value determined by the probability associated with the interval esti-
             mate—that is, the value associated with a certain likelihood that μ will
             be included in a large number of interval estimates

The Level of Confidence

The degree of probability associated with an interval estimate is known as
the **confidence level** (or **confidence coefficient**). The word "confidence" is
used because the probability is an indicator of the degree of certainty that
the particular method of interval estimation will produce an estimate
which includes μ. The higher the level of probability associated with an
interval estimate, the more certainty there is that the method of estimation
will produce an estimate containing the population mean.

In practice, the confidence level is generally specified before estimation, and the appropriate Z value is then used to construct the interval estimate. For example, 90 percent confidence means a *45 percent area on
each side of the normal distribution.* Let's now check Appendix 2 at the
back of the book. We look in the body of that table to find the figure nearest to .4500 (or 45 percent). The Z value corresponding to the table figure

of .4495 is 1.64, and the Z value for the table figure of .4505 is 1.65. The most accurate Z value is thus 1.645, but let's agree to settle on the 1.64 figure for our purpose. That is, we'll say that the Z value corresponding to an area of .45 (or 45 percent) is about 1.64. Thus, the interval estimate of μ using a 90 percent confidence level is:

$$\bar{x} - 1.64\sigma_{\bar{x}} < \mu < \bar{x} + 1.64\sigma_{\bar{x}}$$

The confidence levels generally used in interval estimation are 90, 95, and 99 percent. The Z values and the general forms of the interval estimates associated with these confidence levels are shown in Figure 7-4. To summarize, the interval estimates based on specified confidence levels are known as **confidence intervals,** and the upper and lower limits of the intervals are known as **confidence limits.**

At this point you may wonder why it's necessary to have various confidence levels when it seems logical that the highest level of confidence is desirable in estimating μ. Your thoughts may be characterized in the following words: "If I am required to provide an accurate estimate of the true mean, why shouldn't I always go with a 99 percent confidence level? It makes intuitive sense to have as much confidence as possible in my estimate!"

Unfortunately, intuitive sense is only partially correct in this matter. There's no question that it's highly desirable to have as much confidence as possible in the estimate. However, if more confidence is desired in an estimate, the allowance for sampling error must be increased. In a nutshell, *a higher confidence level produces a wider interval estimate, and thus the precision of the estimate decreases.* Consider the following exchange among three dormitory students anxiously waiting for the mail.

First student: "I have a *feeling* they'll pass out the mail around 2:30 like they usually do."

Second student: "I'm *almost sure* we'll get the mail some time between 2:15 and 2:45."

Third student: "I'm *absolutely sure* we'll get the mail between now and never."

If you want more confidence in your estimate, you must allow for more sampling error. The widths of the intervals will increase, and the estimate

FIGURE 7-4   Commonly Used Confidence Coefficients and Confidence Intervals for a Large Sample

CONFIDENCE COEFFICIENT	Z VALUE	GENERAL FORM OF THE INTERVAL ESTIMATE
90	1.64	$\bar{x} - 1.64\sigma_{\bar{x}} < \mu < \bar{x} + 1.64\sigma_{\bar{x}}$
95	1.96	$\bar{x} - 1.96\sigma_{\bar{x}} < \mu < \bar{x} + 1.96\sigma_{\bar{x}}$
99	2.58	$\bar{x} - 2.58\sigma_{\bar{x}} < \mu < \bar{x} + 2.58\sigma_{\bar{x}}$

will lose some precision. Figure 7-4 shows this relationship between the confidence coefficient and the interval width. If an interval estimate is too wide, the estimate will have no utility.

Let's assume you must submit a budget request to the finance department for advertising, and the advertising expenditures for the next period must be 10 percent of sales. A random sample has produced a mean sales figure of $\bar{x} = \$250,000$ with a $\sigma_{\bar{x}}$ of $2,000. Using the general forms of the interval estimates shown in Figure 7-4, you construct intervals for both the 90 and 99 percent levels of confidence. If you base your advertising budget on the estimate with a coefficient of 90 percent, you tell the finance department you'll need between $21,720 and $28,280 [$25,000 ± 1.64($2,000)]. But if your advertising budget is based on an estimate with a 99 percent confidence level, you allow more room for error and tell the finance department to provide between $19,840 and $30,160 [$25,000 ± 2.58($2,000)]. As the range in your budget request increases, the planning in the finance department becomes more uncertain. Financial people may then be forced to tie up more funds than necessary. In short, an increase in the confidence level might produce an estimate that isn't useful.

One caution should be mentioned here. The confidence coefficient should be stated *before* the interval estimation. Sometimes a novice researcher calculates a number of interval estimates on the basis of a single sample while varying the confidence level. After obtaining these estimates, he or she then selects the one that seems most suitable. Such an approach is really manipulating data so that the results of a sample are the way a researcher would like to see them. This approach introduces the researcher's bias into the study, and it should be avoided.

Self-Testing Review 7-2

1 What's the theoretical basis for the interval estimation of μ with $\bar{x} \pm Z\sigma_{\bar{x}}$?
2 What's the difference between a confidence coefficient and a confidence level?
3 What does a 95 percent confidence level mean?
4 What relationship exists between the confidence level and the interval width?

ESTIMATING THE POPULATION MEAN: σ KNOWN

Moving on now from the general form of (and the theoretical basis for) the interval estimate, the remainder of this chapter discusses interval estimation of parameters under specific conditions.

When the *population standard deviation* (σ) *is known*, we may directly compute the standard error of the mean. Thus, *the interval estimate may be constructed in the following manner:*

$$\bar{x} - Z\sigma_{\bar{x}} < \mu < \bar{x} + Z\sigma_{\bar{x}}$$

Lower confidence limit   Upper confidence limit

And you'll recall from Chapter 6 that $\sigma_{\bar{x}}$ may be found by:

$$\sigma_{\bar{x}} = \frac{\sigma}{\sqrt{n}} \qquad \text{for an infinite population}$$

or $\quad \sigma_{\bar{x}} = \frac{\sigma}{\sqrt{n}} \sqrt{\frac{N-n}{N-1}} \qquad \text{for a finite population}$

So let's now use this estimation procedure to consider some example problems.

Example 7-1 The Papyrus Paper Company wants to estimate the average time required for a new machine to produce a ream of paper. A random sample of 36 reams required an average machine time of 1.5 minutes for each ream. Assuming $\sigma = 0.30$ minute, construct an interval estimate with a confidence level of 95 percent.

We have the following data from the problem: $\bar{x} = 1.5$, $\sigma = .30$, $n = 36$, and confidence level = 95 percent. The standard deviation of the sampling distribution ($\sigma_{\bar{x}}$) is computed as follows:

$$\sigma_{\bar{x}} = \frac{\sigma}{\sqrt{n}}$$
$$= \frac{.30}{\sqrt{36}}$$
$$= .05$$

With a 95 percent confidence coefficient, the Z value equals 1.96. Thus, the interval estimate of the true average time (μ) is constructed as follows:

$\bar{x} - Z\sigma_{\bar{x}} < \mu < \bar{x} + Z\sigma_{\bar{x}}$
$1.5 - 1.96(.05) < \mu < 1.5 + 1.96(.05)$
$1.5 - .098 < \mu < 1.5 + .098$
$1.402 < \mu < 1.598$

We can use the data in this example, along with *Minitab* computer simulations, to verify some of the concepts that have now been presented. Let's assume in our Papyrus Paper Company example that the true population mean time for the machine to produce a ream is actually 1.48 minutes. The σ, we've seen, is .30 minute. If we were to take 15 random samples, with each sample containing the time data required to produce 36 reams, about how many of the 15 sample confidence intervals would include the μ of 1.48 minutes if we use the 95 percent confidence level? We could answer this question without much hesitation if we were taking 1,000 random samples (the answer would be about 950 of them), but with just 15 samples we can't always be sure.

Figure 7-5a shows the results obtained when the statistical package sim-

```
MTB > RANDOM 36 SAMPLE ITEMS INTO C1-C15;
SUBC> NORMAL MU = 1.48, SIGMA = .30.
MTB > ZINTERVAL 95 PERCENT, SIGMA = .30, DATA IN C1-C15

THE ASSUMED SIGMA =0.300

            N      MEAN    STDEV   SE MEAN   95.0 PERCENT C.I.
C1          36    1.4602   0.2641   0.0500   ( 1.3621,  1.5584)
C2          36    1.3522   0.2893   0.0500   ( 1.2541,  1.4503)
C3          36    1.4846   0.3192   0.0500   ( 1.3865,  1.5828)
C4          36    1.5185   0.3065   0.0500   ( 1.4203,  1.6166)
C5          36    1.4471   0.2573   0.0500   ( 1.3490,  1.5453)
C6          36    1.5056   0.3172   0.0500   ( 1.4075,  1.6037)
C7          36    1.5157   0.3037   0.0500   ( 1.4176,  1.6139)
C8          36    1.5327   0.3038   0.0500   ( 1.4346,  1.6309)
C9          36    1.4700   0.2783   0.0500   ( 1.3718,  1.5681)
C10         36    1.4726   0.3246   0.0500   ( 1.3745,  1.5708)
C11         36    1.4092   0.3084   0.0500   ( 1.3110,  1.5073)
C12         36    1.4958   0.2891   0.0500   ( 1.3976,  1.5939)
C13         36    1.4691   0.2818   0.0500   ( 1.3710,  1.5673)
C14         36    1.5007   0.3578   0.0500   ( 1.4026,  1.5988)
C15         36    1.4271   0.2573   0.0500   ( 1.3290,  1.5253)
```

(a)

```
MTB > RANDOM 15 SAMPLES INTO C1-C36;
SUBC> NORMAL  MU = 1.48  SIGMA = .30.
MTB > RMEAN C1-C36 INTO C37
MTB > LET K1 = 1.96 * .30/SQRT(36)
MTB > LET C40 = C37 - K1
MTB > LET C41 = C37 + K1
MTB > SET C42
MTB > END
MTB > MPLOT C40, C42  C41, C42


        -
        -    B
        -
  1.65+                        B                    B
        -
        -        B      B  B        B  B             B
        -           B                    B
        - B                     B            B
  1.50+     A  B
        -                          A
        -                                         A
        -        A       A        A  A             A
        -           A                 A
  1.35+        A            A
        - A    A                              A
        -
        -

        ------+---------+---------+---------+---------+---------+
            2.5       5.0       7.5      10.0      12.5      15.0

   A = C40 vs. C42          B = C41 vs. C42
```

(b)

FIGURE 7-5   (a) The 95 percent confidence intervals produced when 15 random samples of size 36 are simulated using the Papyrus Paper Company data introduced in Example 7-1. (b) A plot of the lower (A) and upper (B) values of 15 more simulated random samples of size 36.

ulates taking 15 random samples, with 36 sample items in each sample. The first line in Figure 7-5a instructs the program to randomly select the data items and to then put the data for each sample in a separate column. The first sample data goes in column 1 (C1), the second in C2, and so on. The second line in Figure 7-5a gives the package the values of μ and σ. A ZINTERVAL command is then used in line 3 to produce a 95 percent confidence interval for each of the 15 samples. The means of each of the simulated samples are listed in the program output. The SE MEAN gives the standard error figured earlier in this example, and the lower and upper confidence limits are given for each of the 15 intervals produced.

You'll see that 14 of the 15 samples have intervals that *include* the μ of 1.48 minutes. Only the sample in C2 has failed to include this parameter. Notice, too, that the sample in C14 has duplicated the results we noted earlier (this is just a coincidence.) All intervals, of course, have the same width, but they have different lower and upper limits, as you would expect.

To graphically demonstrate how these lower and upper limits vary, Figure 7-5b shows *another simulation* of 15 samples. You can ignore the program commands at the top of Figure 7-5b and concentrate on the plot. The vertical scale shows the average time required to produce a ream of paper. A solid line is drawn horizontally through the middle of the plot to show the position of the μ of 1.48 minutes. The baseline in this case is simply a count of the samples. Each letter A represents the location of the *lower limit* of the confidence interval for one sample, and if you draw a vertical line up to the corresponding letter B, you'll have located the position of the *upper limit* of the sample's confidence interval. With one exception, all A's fall below the population mean line of 1.48, and all B's are scattered above that line. In the one case (sample two again) where both A and B are above the μ line, the interval fails to include the μ. Thus, in this second simulation, our samples have different means, different lower limits, and different upper limits, but 14 of the 15 intervals include the population mean. Don't think that every simulation of 15 samples will always produce one sample that fails to include the μ in its intervals. The next simulation might easily have no "failures" or two failures.

Example 7-2   The Ledd Pipe Company has received a shipment of 100 lengths of pipe, and it wants to estimate the average diameter of the pipes to see if they meet minimum standards. A random sample of 50 pipes produced an average diameter of 2.55 inches. In the past, the population standard deviation of the diameter has been 0.07 inches. Construct an interval estimate with a 99 percent degree of confidence.

We have the following data from the problem situation: $\bar{x} = 2.55$, $\sigma = .07$, $n = 50$, $N = 100$, and confidence level = 99 percent. The standard error of the mean is computed as follows:

$$\sigma_{\bar{x}} = \frac{\sigma}{\sqrt{n}} \sqrt{\frac{N-n}{N-1}}$$

$$= \frac{.07}{\sqrt{50}} \sqrt{\frac{100 - 50}{100 - 1}}$$
$$= .007$$

With a 99 percent confidence coefficient, the Z value is 2.58. Therefore, the interval estimate of μ, the true average diameter of the shipment of pipes, is constructed as follows:

$\bar{x} - Z\sigma_{\bar{x}} < \mu < \bar{x} + Z\sigma_{\bar{x}}$
$2.55 - 2.58(.007) < \mu < 2.55 + 2.58(.007)$
$2.55 - .018 < \mu < 2.55 + .018$
$2.532 < \mu < 2.568$

Note that the preceding examples were based on situations in which σ *was known (or could be identified), the sample size exceeded 30, and the sampling distribution was thus normally distributed.* The general procedure for interval estimation under such conditions is illustrated in Figure 7-6.

Self-Testing Review 7-3

1 When does the sampling distribution approximate the normal distribution?
2 Determine the Z value for the following confidence levels:
  a 91 percent
  b 73 percent
  c 86 percent
3 Construct a confidence interval with the following data: $\bar{x} = 48$, $\sigma = 9$, $n = 36$, and confidence level = 90 percent.
4 Construct a confidence interval with the following data: $\bar{x} = 104$, $\sigma_{\bar{x}} = 13$, and confidence level = 80 percent

ESTIMATING THE POPULATION MEAN: σ UNKNOWN

In most situations, not only is the population mean unknown but the population standard deviation is also unknown. In fact, it's only in isolated cases that σ is known, so it usually must be estimated along with the population mean.

The Estimator of σ

You've seen in Chapter 4 that the formula needed to compute the standard deviation for a set of *population* values is:

$$\sigma = \sqrt{\frac{\Sigma(X - \mu)^2}{N}}$$

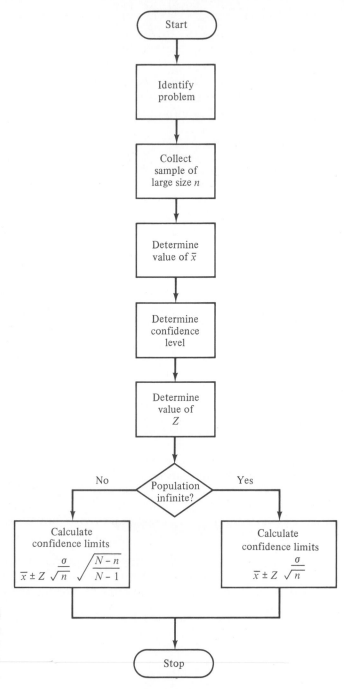

FIGURE 7-6
Procedure for the interval estimation of μ with σ known and $n > 30$.

Now if the data set comes from a *sample* rather than a population, how is the standard deviation computed? It seems reasonable to simply substitute sample symbols for population symbols and use the following formula:

$$s = \sqrt{\frac{\Sigma(x - \bar{x})^2}{n}}$$

This formula is fine if we're computing the sample standard deviation for its own sake. But we seldom do that. Rather, s is almost always computed to provide an estimate of an unknown population standard deviation.

But, alas, if we use this formula to estimate σ, we run into a bias problem. Why is that? Well, you'll recall that a sample mean is an unbiased estimator of μ because the mean of the sampling distribution of means from which x is taken equals the population mean—that is, because $\mu_{\bar{x}} = \mu$. But if we were to use the sample standard deviation formula above to compute all the s values needed to produce a sampling distribution of standard deviations, the mean of that sampling distribution wouldn't equal σ. Instead, the mean of the sampling distribution of standard deviations will be less than the population standard deviation, and so there's a tendency for a sample standard deviation computed with the above formula to also be less than σ. That is, the above formula is biased toward understating σ, and so this tendency must be removed.

It's easy to get bogged down in statistical details and assumptions at this point. But we'll simplify and note that we *could* reduce the inherent bias by modifying the above formula for s with a bias correction factor in the following way to get a useful estimator of σ:

$$\hat{\sigma} = s\sqrt{\frac{n}{n-1}}$$

There are two things to note about this correction formula. First, the inherent bias diminishes as the sample size increases, and second, the ^ mark over σ or any other symbol shows that we have an *estimated value*. Note further that the wonders of mathematical manipulation allow us to show that

$$\hat{\sigma} = \sqrt{\frac{\Sigma(x - \bar{x})^2}{n}}\sqrt{\frac{n}{n-1}} \quad \left(\text{or } \hat{\sigma} = s\sqrt{\frac{n}{n-1}} \right)$$

or

$$\hat{\sigma} = \sqrt{\frac{\Sigma(x - \bar{x})^2}{n}} \frac{n}{n-1}$$

or

$$\hat{\sigma} = \sqrt{\frac{\Sigma(x - \bar{x})^2}{n-1}}$$

In view of these perplexing considerations about the bias of an unadjusted sample standard deviation, how do statisticians typically deal with the problem? Very easily. They simply use the following formula when computing the sample standard deviation to arrive at a *useful estimator of* σ that needs no further manipulation:

$$s = \sqrt{\frac{\Sigma(x - \bar{x})^2}{n - 1}}$$

(formula 7-2)

Since this formula has wide applicability, it's the one commonly used by statistical software packages (and calculators with a built-in standard deviation function) to compute standard deviation values. Unless otherwise indicated, *all references to a sample standard deviation in future problems will assume that the value was computed using formula 7-2.*

With that detail out of the way, the estimated standard error needed to approximate the population mean when σ is unknown is

$$\hat{\sigma}_{\bar{x}} = \frac{s}{\sqrt{n}} \qquad \text{for an infinite population}$$

(formula 7-3)

or

$$\hat{\sigma}_{\bar{x}} = \frac{s}{\sqrt{n}} \sqrt{\frac{N - n}{N - 1}} \qquad \text{for a finite population}$$

(formula 7-4)

Notice that the calculation of $\hat{\sigma}_{\bar{x}}$ is the same as the computation of $\sigma_{\bar{x}}$ except that the population standard deviation is replaced by the sample standard deviation (s), and it's assumed that s is computed with formula 7-2 to be a useful estimator of σ.

If the population standard deviation is unknown, the sampling distribution of means can be assumed to be approximately normal *only when the sample size is relatively large (over 30).* With estimated values of σ and $\hat{\sigma}_{\bar{x}}$, the general form of the interval estimate *for large samples* is altered slightly so that it appears as follows:

$$\bar{x} - Z\hat{\sigma}_{\bar{x}} < \mu < \bar{x} + Z\hat{\sigma}_{\bar{x}}$$

(formula 7-5)

As you can see, we merely substituted the estimated value for the true value of the standard error of the mean. Again, let's look at some examples to illustrate the above points.

Example 7-3   The Wayside Tavern wants to estimate the average dollar purchase per customer. A sample of 100 customers spent an average of $3.50 each with a *sample* standard deviation of $0.75. Estimate the true average expenditure with a 90 percent confidence level.

We have the following data from the problem situation: $\bar{x} = \$3.50$, $s = .75$, $n = 100$, confidence level $= 90$ percent. Therefore the estimate of $\hat{\sigma}_{\bar{x}}$ is computed as follows:

$$\hat{\sigma}_{\bar{x}} = \frac{s}{\sqrt{n}}$$
$$= \frac{.75}{\sqrt{100}}$$
$$= .075$$

With a 90 percent confidence level the Z value is 1.64. The interval estimate is thus

$\bar{x} - Z\hat{\sigma}_{\bar{x}} < \mu < \bar{x} + Z\hat{\sigma}_{\bar{x}}$
$\$3.50 - 1.64(.075) < \mu < \$3.50 + 1.64(.075)$
$\$3.50 - .123 < \mu < \$3.50 + .123$
$\$3.38 < \mu < \$3.62$

Example 7-4   The Rogers Poultry Company has received a shipment of 100 hens, and the manager wants to estimate the true average weight of a hen in order to determine if the hens meet Rogers' standards. A sample of 36 hens has shown an average weight of 3.6 pounds with a *sample* standard deviation of .6. Construct an interval estimate of the true average weight per chicken with a 99 percent confidence level.

We have the following data: $\bar{x} = 3.6$, $s = .6$, $n = 36$, confidence level $= 99$ percent, $N = 100$. The estimate of the standard error is computed as follows:

$$\hat{\sigma}_{\bar{x}} = \frac{s}{\sqrt{n}}\sqrt{\frac{N-n}{N-1}}$$
$$= \frac{.6}{\sqrt{36}}\sqrt{\frac{100-36}{100-1}}$$
$$= .08$$

With a 99 percent confidence level, the Z value is 2.58. Therefore, the interval estimate is:

$\bar{x} - Z\hat{\sigma}_{\bar{x}} < \mu < \bar{x} + Z\hat{\sigma}_{\bar{x}}$
$3.6 - 2.58(.08) < \mu < 3.6 + 2.58(.08)$
$3.6 - .21 < \mu < 3.6 + .21$
$3.39 \text{ pounds} < \mu < 3.81 \text{ pounds}$

When σ is unknown, and when the *sample size is large*, the sampling distribution is approximately normally distributed. But if $\sigma_{\bar{x}}$ must be estimated, and if the sample size is *30 or less*, the sampling distribution will not be normally distributed, and therefore the interval estimate *cannot* be calculated with the use of the Z distribution. What distribution should then be used? The following section provides us with the answer.

Estimation Using the t Distribution

Instead of following a normal distribution curve, the sampling distribution of means with a small sample size follows a *t* distribution. A *t* **distribution** is similar to a Z distribution with a zero mean and a symmetrical shape. But unlike the shape of the Z distribution, the *t* distribution's shape depends on the sample size. (There's a different distribution for each sample size.) In general, the shape of a *t* distribution is flatter than that of a Z distribution. As the sample size increases and approaches 30, however, the shapes of the *t* distributions lose their flatness and approximate the shape of the Z distribution (see Figure 7-7). Figure 7-8 summarizes the conditions under which the *t* or Z distributions are used for estimation purposes. Note that it's possible to still use Z values when the sample size is 30 or less *if* the σ is known, and *if* it's also known that the items in the population are normally distributed. Knowing both these facts is unlikely, and we'll consider more plausible situations in the next few pages.

If σ is unknown, and if the sample size is small, the interval estimate of the population mean has the following form:

$$\bar{x} - t_{\alpha/2}\hat{\sigma}_{\bar{x}} < \mu < \bar{x} + t_{\alpha/2}\hat{\sigma}_{\bar{x}}$$
Lower confi-   Upper confi-   (formula 7-6)
dence limit   dence limit

Like the Z value, the value of *t* depends on the confidence level.

Appendix 4 at the back of the book is a table of *t* distribution values. Unfortunately, the *t* table is constructed in a different manner of presentation than the Z table. If, for example, we are interested in making an estimate at the 95 percent confidence level, the *t*-table format is not designed to em-

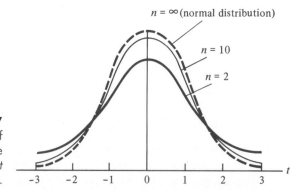

FIGURE 7-7
The effect of sample size on the shape of the *t* distributions.

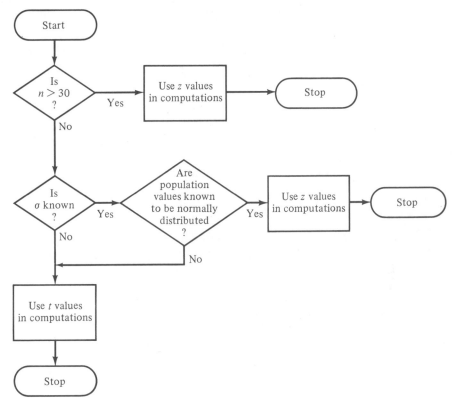

FIGURE 7-8
Which distribution to use?

phasize the 95 percent chance of including μ in the estimate; rather, the presentation focuses attention on the 5 percent chance of *not including* μ. *This chance of error is labeled α (alpha) and in decimal form equals 1.00 minus the confidence coefficient.* For example, if the confidence coefficient is .95 (or 95 percent), then α will be 1.00 − .95 or .05. Since α represents the *total* chance of error—that is, the chance of not including μ—and since the particular t distribution being used is symmetrical, the total error is divided evenly between the chance of overestimation and the chance of underestimation. As indicated by the shaded portion in the figure at the top of Appendix 4, however, the t table *only deals with areas to one side of the distribution.* Consequently, the subscript $\alpha/2$ follows t in the interval formula 7-6 shown above. With a 95 percent confidence level, we want only a 5 percent chance of error, and therefore we look under the *column* designated by t_{025} in Appendix 4.

The use of a t table is admittedly rather confusing to beginning students. Some find it convenient to locate the correct column by looking at the last row in the table (where df is ∞). This row presents Z values, since the last distribution in the family of t distributions is the normal distribution. Thus, when you know that a Z value of, say, 1.96 would have been used if the sample size had been large, you can locate the proper column by finding 1.96 in the last row.

Another factor that must be known before an appropriate t value can be determined is the **degrees of freedom** (df), a rather imposing term that in our present situation simply has a value of $n - 1$. Thus, for our purposes here, the df values identify the appropriate row in the table. (Each df row refers to a different t distribution in the family of curves.) Let's assume that we have a sample size of 17 and we want a 95 percent confidence level for an interval estimate. The α value is .05, and thus we refer to *column* t_{025} in Appendix 4. And since $df = 16$ ($17 - 1$), the necessary t value of 2.120 is found at the intersection of the 16 df row and the .025 column.

Let's now look at some examples of the use of the t distribution in estimating the population mean.

Example 7-5   The Dew Drop Inn wants to estimate the average number of gallons of a product sold per day. Twenty business days were monitored, and an average of 32 gallons was sold daily. The sample standard deviation was 12 gallons. Calculate the confidence limits at the 95 percent confidence level.

We have the following information: $\bar{x} = 32$, $s = 12$, $n = 20$, and confidence level = 95 percent. With a confidence level of 95 percent, and with a sample size of 20, α is 5 percent and the degrees of freedom are 19. Thus, from Appendix 4 under the t_{025} column, we see that the t value is 2.093. The estimate of $\sigma_{\bar{x}}$ is computed as follows:

$$\hat{\sigma}_{\bar{x}} = \frac{s}{\sqrt{n}}$$
$$= \frac{12.0}{\sqrt{20}}$$
$$= 2.68$$

The interval estimate is:

$$\bar{x} - t_{\alpha/2}\hat{\sigma}_{\bar{x}} < \mu < \bar{x} + t_{\alpha/2}\hat{\sigma}_{\bar{x}}$$
$$32 - t_{025}(2.68) < \mu < 32 + t_{025}(2.68)$$
$$32 - 2.093(2.68) < \mu < 32 + 2.093(2.68)$$
$$26.38 < \mu < 37.62$$

Finally, the confidence limits are 26.38 and 37.62 gallons.

Example 7-6   The Kelly Bread Company wants to estimate its average daily usage of flour. With a sample of 14 days, the point estimate of μ is 173 pounds, with $s = 45$ pounds. Construct a confidence interval with a 99 percent confidence coefficient.

The data from the problem are as follows: $\bar{x} = 173$, $s = 45$, $n = 14$, and confidence coefficient = 99 percent. The estimate of $\sigma_{\bar{x}}$ is computed as follows:

$$\hat{\sigma}_{\bar{x}} = \frac{s}{\sqrt{n}}$$
$$= \frac{45}{14}$$
$$= 12.03$$

With a 99 percent confidence coefficient, α is 1 percent. With a sample size of 14, there are 13 degrees of freedom. Therefore, consulting Appendix 4 under t_{005} we see that the t value is 3.012. The interval estimate is:

$$\bar{x} - t_{005}\hat{\sigma}_{\bar{x}} < \mu < \bar{x} + t_{005}\hat{\sigma}_{\bar{x}}$$
$$173 - 3.012(12.03) < \mu < 173 + 3.012(12.03)$$
$$173 - 36.23 < \mu < 173 + 36.23$$
$$136.77 < \mu < 209.23$$

Example 7-7   Let's assume that you've looked at the table in your newspaper that shows the high temperatures recorded for cities in the United States on the fifth day of August. You select a random sample of five cities located in the southern tier of states and record the following temperatures: 101, 88, 94, 96, and 103. You now decide to construct an interval estimate, at the 95 percent level of confidence, of the average temperature recorded in all southern communities on that date. Since you only have sample data items, you must first calculate the sample mean and sample standard deviation. The following effort gives you those values:

x	$x - \bar{x}$	$(x - \bar{x})^2$
101	4.6	21.16
88	−8.4	70.56
94	−2.4	5.76
96	−.4	.16
103	6.6	43.56
482		141.20

$$\bar{x} = \frac{482}{5} = 96.4$$

$$s = \sqrt{\frac{141.2}{(5 - 1)}} = \sqrt{35.3} = 5.94$$

Next, you compute your estimated standard error:

$$\hat{\sigma}_{\bar{x}} = \frac{s}{\sqrt{n}} = \frac{5.94}{\sqrt{5}} = \frac{5.94}{2.236}$$
$$= 2.66$$

Finally, you prepare your interval estimate:

$$\bar{x} - t_{025}\hat{\sigma}_{\bar{x}} < \mu < \bar{x} + t_{025}\hat{\sigma}_{\bar{x}}$$
$$96.4 - 2.776(2.66) < \mu < 96.4 + 2.776(2.66)$$
$$96.4 - 7.384 < \mu < 96.4 + 7.384$$
$$89.02 < \mu < 103.78$$

You show your interval estimate of 89.02 to 103.78 degrees—a range that has taken several minutes to prepare—to a friend with a statistical software package. The friend keys in your sample data in a program column as shown in Figure 7-9, gives a TINTERVAL command, and receives the same results in a few seconds.

The general approach to constructing an interval estimate with an unknown σ is summarized in Figure 7-10. Note that the procedures for large and small sample sizes are basically the same. The only difference is the use of a t or a Z distribution.

Self-Testing Review 7-4

1   Suppose we have a sample of 24 items and σ is unknown. Would you use a t or a Z distribution in your interval estimation?
2   What is the t value in interval estimation if the sample size is 27 and the desired confidence level is 95 percent?
3   Assume that a sample of 100 customer charge accounts showed an average balance of $42 with a sample standard deviation of $16. Construct a confidence interval with a coefficient of 95 percent.
4   If you have been given a sample of 20 candles from a shipment of 1,000 candles and are asked to provide an interval estimate of their average burning life, how would you proceed and what information would be needed?
5   How does the shape of a t distribution differ from the shape of a Z distribution?
6   Identify the factor that affects the shape of a t distribution.

FIGURE 7-9   A statistical software package can compute confidence intervals using t distributions in a few seconds.

```
MTB > SET IN C1
MTB > END
MTB > PRINT C1

C1
    101     88     94     96     103

MTB > TINTERVAL 95 PERCENT, C1

                N       MEAN    STDEV   SE MEAN    95.0 PERCENT C.I.
C1              5       96.40    5.94      2.66   (   89.02,  103.78)
```

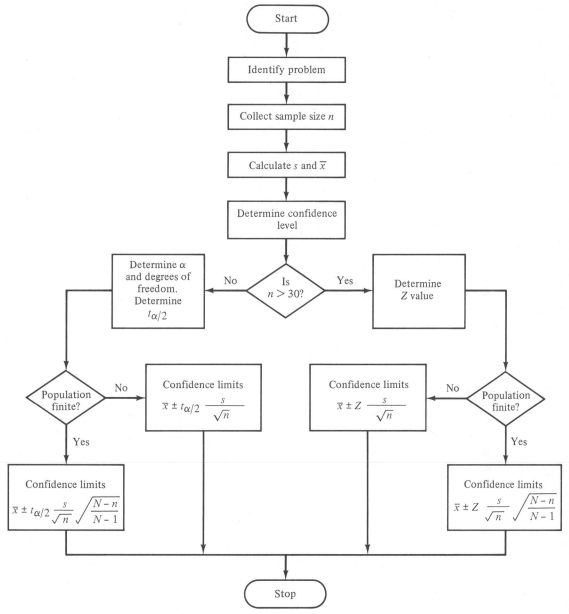

FIGURE 7-10   Procedure for the interval estimation of μ with σ unknown.

7   The Z table is constructed so that we are concerned with the probability of in-
    cluding the population mean in the interval estimate. How could the format of
    the t table in Appendix 4 be characterized?

8   How can the sample standard deviation be used as an estimator of the popula-
    tion standard deviation?

INTERVAL ESTIMATION OF THE POPULATION PERCENTAGE

Since the mean of the sampling distribution of percentages is equal to the population percentage, the *sample* percentage *(p)* is an unbiased estimator of the *population* percentage (π). If the sample size is sufficiently large, the sampling distribution approximates the normal distribution, and thus we are able to make probability statements about the interval estimates of π that are based on sample percentages. In this section we'll discuss only the large-sample case in the interval estimation of π, since the small-sample approach is beyond the scope of this book.

As indicated below, the *sample percentage* serves as the basis for constructing the interval estimate of the population percentage:

$$p - Z\hat{\sigma}_p < \pi < p + Z\hat{\sigma}_p$$

Lower confi-   Upper confi-   (formula 7-7)
dence limit   dence limit

A Z value is used here in exactly the same way it was used in estimating a population mean. The symbol $\hat{\sigma}_p$ is the *estimate* of the standard deviation of the sampling distribution of percentages—i.e., an estimate of the standard error of percentage. An unbiased estimate of the standard error of percentage may be computed in the following manner:

$$\hat{\sigma}_p = \sqrt{\frac{p(100 - p)}{n - 1}} \sqrt{\frac{N - n}{N - 1}} \quad \text{for a finite population} \qquad \text{(formula 7-8)}$$

or

$$\hat{\sigma}_p = \sqrt{\frac{p(100 - p)}{n - 1}} \quad \text{for an infinite population} \qquad \text{(formula 7-9)}$$

The *estimate* of the standard error is *always used* in the construction of an interval estimate. Why? Because the true standard error cannot be computed for an interval estimate of π. This fact is obvious from the following formula:

$$\sigma_p = \sqrt{\frac{\pi(100 - \pi)}{n}}$$

where, as you can see, the calculation of σ_p requires knowledge of π. Yet π is what we are trying to estimate! In order to resolve this dilemma, we must therefore use formulas (7-8) and (7-9).

We are now in a position to illustrate the similarity of the procedures used to estimate population means and percentages by considering the following example problems.

Example 7-8   The Highland Fling Scottish Boomerang Company wants to estimate the percentage of credit customers who have submitted checks in payment for boomerangs and whose checks have bounced. A random sample of 150 accounts showed that 15 customers have passed bad checks. Estimate at the 95 percent confidence level the true percentage of credit customers who have passed bad checks.

We have the following data: $p = 15/150 = 10$ percent, $n = 150$, and confidence coefficient $= 95$ percent. The estimate of σ_p is computed as follows:

$$\hat{\sigma}_p = \sqrt{\frac{p(100 - p)}{n - 1}}$$

$$= \sqrt{\frac{10(90)}{149}}$$

$$= 2.46 \text{ percent}$$

With a confidence coefficient of 95 percent, the Z value is 1.96. Therefore, the interval estimate of the true percentage of credit customers who pass bad checks is:

$p - Z\hat{\sigma}_p < \pi < p + Z\hat{\sigma}_p$
10 percent $- 1.96(2.46$ percent$) < \pi < 10$ percent $+ 1.96(2.46$ percent$)$
10 percent $- 4.82$ percent $< \pi < 10$ percent $+ 4.82$ percent
5.18 percent $< \pi < 14.82$ percent

Example 7-9   A high school student counselor was interested in the proportion of male students who would volunteer for military service. Out of 600 male students, she randomly sampled 50 and found that 15 of them expressed a favorable opinion toward enlistment. Use a 99 percent confidence coefficient to estimate the true percentage.

The data are: $p = 15/50 = 30$ percent, $n = 50$, confidence coefficient $= 99$ percent. The estimate of the standard error is computed as follows:

$$\hat{\sigma}_p = \sqrt{\frac{p(100 - p)}{n - 1}} \sqrt{\frac{N - n}{N - 1}}$$

$$= \sqrt{\frac{30(70)}{49}} \sqrt{\frac{600 - 50}{600 - 1}}$$

$$= 6.28 \text{ percent}$$

EXCUSE ME, WHAT'S THE POLSTERS' BIG PROBLEM?

You're cornered in the street by a pollster who hands you a photo of Ronald Reagan standing next to David A. Stockman. "You know who this man is, don't you?" he demands. If you are like most people, your answer is prompt: "The President of the United States." But what if he asked: "Do you have any idea at all who this man is?" When Stanford University psychologist Herbert H. Clark tried that on 15 volunteers, only one identified Reagan. Most identified Stockman—or asked for more information.

Polls and market surveys are a black art. Ask the questions one way, and the answers reflect what the public is thinking. But phrase them just slightly differently, and the results can be totally out of touch with reality. In 1984 a panel of experts from the National Research Council concluded that survey designers "simply do not know much about how respondents answer questions." Adds sociologist Judith Tanur of the State University of New York at Stony Brook: "Survey designers have taken a seat-of-the-pants approach to constructing questionnaires. It hasn't been guided by a framework or a theory."

BAD MEMORY
But just as the ancient Greeks consulted oracles, companies and government agencies are more frequently turning to professional pollsters to augur everything from election results to product introductions. "Business is definitely up," says Harry W. O'Neill, vice-chairman of Opinion Research Corp. in Princeton, N.J. Last year the top 44 U.S. polling companies pulled in more than $1.5 billion in revenues. While revenues at most polling companies have been growing by at least 8% annually since 1983, some pollsters, such as Gallup Organization Inc. in Princeton, report gains of 15% to 20%.

Policy decisions and billions of dollars in corporate funds ride on the accuracy of those polls. So the people who design them are teaming up with researchers in cognitive psychol-

ogy, anthropology, linguistics, and decision theory. At least a dozen research projects are under way at universities such as Yale, Stanford, and the University of Chicago.

Already, some of the major pollsters as well as several federal agencies, including the Bureau of Labor Statistics and the Census Bureau, are using the new techniques to design better surveys. Gallup President Andrew Kohut says his company has just made a "major investment" in a company that builds devices for measuring eye movement. Using it, he says, Gallup researchers hope to study the ways people respond to television commercials and magazine ads.

The researchers have discovered that it's a big mistake to rely on people's ability to remember events. Stanford psychologist Lee D. Ross discovered that students were only marginally better at recalling details, such as how many checks they'd written during the course of a month, than they were at predicting the future behavior of their roommates. In addition, the mind tends to "telescope" past events: They often seem to be more recent than they really are. "Most of us just don't have a good internal calendar for dating events," observes David J. Mingay, a cognitive psychologist at the National Center for Health Statistics in Hyattsville, Md.

Mingay is part of a team that is designing better questions for an annual health survey that polls 135,000 Americans. He is exploring a technique that measures the time lag between when a question is asked and when it is answered. Studies show that a too-slow response may mean the question is too difficult to understand, while an immediate response probably means that it was either misunderstood or hurriedly answered. Pollsters are also turning to a technique known as "benchmarking" to help subjects place events in time. For example, surveyors found that answers to questions such as "How many times have you been robbed since the Chernobyl disaster?" are more accurate than answers to "How many times have you been robbed within the past 12 months?"

JUST A GUESS

The survey designers are also trying to unravel the thought processes that lead to an answer. By taping test subjects' answers and then asking them to explain why they responded in a particular way, researchers can determine whether a question is eliciting the information that pollsters are trying to examine. Others ask subjects to rate answers according to how confidently they feel about them. A low score usually means the answer is just a guess.

Pollsters have an added incentive for honing their questions. The public is so barraged by surveyors that people refuse to be guinea pigs about 40% of the time. Improving accuracy, however, is still the most important goal of pollsters. This year the BLS will spend about $200 million on more than 50 surveys, which Commissioner Janet L. Norwood says will be "the most politically and economically sensitive surveys in the U.S. government." And with so much riding on the results, she adds, "we can't afford to be wrong."

—Sana Siwolop, "Excuse Me, What's the Pollsters' Big Problem." Reprinted from February 16, 1987 issue of *Business Week*, by special permission, copyright © 1987 by McGraw-Hill, Inc.

With a confidence coefficient of 99 percent, the Z value is 2.58. Therefore, the interval estimate is:

$p - Z\hat{\sigma}_p < \pi < p + Z\hat{\sigma}_p$

30 percent − 2.58(6.28 percent) < π < 30 percent + 2.58(6.28) percent

30 percent − 16.2 percent < μ < 30 percent + 16.2 percent

13.8 percent < π < 46.2 percent

(Notice the large width of this interval estimate as a result of the high level of confidence specified and the relatively small sample size.)

The general procedure for constructing an interval estimate of π in the large-sample case is summarized in Figure 7-11.

Example 7-10   Political polls represent one of the major uses of interval estimation of π. Let's assume that Governor Batson D. Attica faces a tough reelection campaign and orders a poll to learn how the voters view his candidacy. A random sample of 1,200 voters reveals that 532 are likely to vote for Governor Attica, while the others polled prefer his opponent or are undecided. At the 95 percent level of confidence, what's the population percentage of voters who express a preference for the Governor?

The data are: $p = 532/1{,}200 = 44.33$ percent, $n = 1{,}200$, confidence coefficient = 95 percent. The estimate of σ_p is:

$$\hat{\sigma}_p = \sqrt{\frac{p(100 - p)}{1{,}200 - 1}}$$

$$= \sqrt{\frac{44.33\,(55.67)}{1199}}$$

$$= \sqrt{2.058}$$

$$= 1.43 \text{ percent}$$

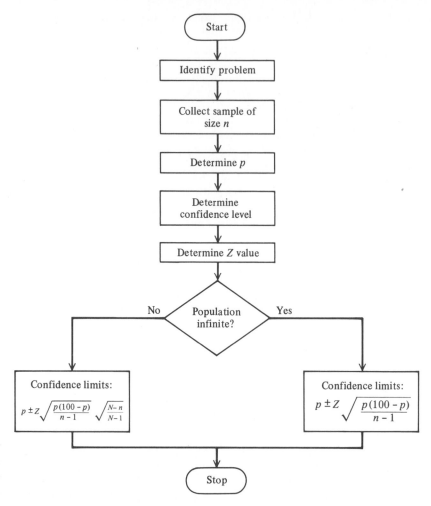

FIGURE 7-11
Procedure for
interval estimation
of π using large
samples.

With a confidence coefficient of 95 percent, the Z value is 1.96. Thus, the interval estimate is:

$$p - Z\hat{\sigma}_p < \pi < p + Z\hat{\sigma}_p$$
$$44.33 - 1.96\,(1.43) < \pi < 44.33 + 1.96\,(1.43)$$
$$41.53 \text{ percent} < \pi < 47.13 \text{ percent}$$

Governor Attica had better try to swing those undecided voters into his camp! For more information on the practical problems associated with polling, see the Closer Look reading at the end of this chapter.

Self-Testing Review 7-5

1   What assumption is made about the sample size when the Z distribution is used in calculating an interval estimate of π?

2 Compare the formulas for $\hat{\sigma}_p$ and σ_p. Why is there a difference in the denominators?

3 Why must we always use $\hat{\sigma}_p$ rather than σ_p in the interval estimation of π?

DETERMINATION OF SAMPLE SIZE

In this chapter, the basic problem situation in the estimation of population means (or percentages) can be summarized as follows: "A sample of size n has been collected. We have the calculated values for the sample mean and standard deviation (or for the sample percentage). Compute an interval estimate with a confidence level of _____ percent."

The sample was assumed to have been collected, and our task was to calculate an estimate based on the sample data given. We had to live with whatever confidence interval resulted from a specified confidence level, no matter how wide it might have been. Of course, one method of controlling the interval width is to change the confidence level, but it's improper to manipulate the confidence level so that the interval range comes out the way we want it to appear.

Reprinted with permission of Stephen B. Randolph.

General Considerations about Sample Size

It's very often the case that the precision of the estimate must be specified before a sample is ever taken. For example, you may be checking the average diameter of machined parts which should not have too much error or else they cannot be used in finely machined equipment. In such a case, you would take a sample of parts, but you would want an estimate with as little sampling error as possible in the interval estimate. You would want a precise estimate. Too much sampling error produces an interval width too large to be of any use.

We can control sampling error by selecting a sample of adequate size. Remember that sampling error arises because the entire population is not studied; someone or something is always left out of the investigation. Whenever sampling is performed, we always miss some bit of information about the population which would be helpful in our estimation. If we want a high level of precision, we must sample enough of the population to provide the necessary and sufficient information. The following sections will discuss the methods of determining the sample size necessary to achieve a specified level of precision.

Determining the Sample Size for the Estimation of μ

Consider the following situation: A hardware wholesaler receives a shipment of 10,000 widgets, and before he pays for the shipment he would like to know if the widgets were properly made and if they meet tolerance specifications. He'd like to estimate the average diameter of the widgets. He would also like the estimate to be within ± 0.01 inches of the true average diameter, and he'd like a 95 percent confidence level associated with his estimate. How does he determine the sample size?

First, look at what's actually desired by the hardware wholesaler. With a sample mean of $\bar{x}$, he wants the interval estimate to have *limits* which are no more than 0.01 inches *above* the point estimate and no more than 0.01 inches *below* the point estimate. He wants this interval estimate to have a 95 percent confidence level of containing the true average. Thus, the desired confidence limits have been specified to be

$\bar{x} \pm 0.01$ inches

Since the general form of the confidence limits is

$\bar{x} \pm Z\sigma_{\bar{x}}$

the hardware wholesaler is saying that he wants $Z\sigma_{\bar{x}}$ to equal .01.

We are now in a position to determine the necessary sample size by solving the equation $Z\sigma_{\bar{x}} = .01$. Since a confidence level of 95 percent is desired in our estimation of μ, the Z value is 1.96. Therefore,

$$Z\sigma_{\bar{x}} = .01$$
$$1.96\sigma_{\bar{x}} = .01$$
$$\sigma_{\bar{x}} = \frac{.01}{1.96}$$
$$\sigma_{\bar{x}} = .005$$

and the standard error should be .005. Assuming the finite correction factor is not applicable, the formula for $\sigma_{\bar{x}}$ is

$$\sigma_{\bar{x}} = \frac{\sigma}{\sqrt{n}}$$

With some mathematical manipulation which we shall omit, the sample size is then

$$n = \frac{\sigma^2}{\sigma_{\bar{x}}^2} \qquad \text{(formula 7-10)}$$

We know that $\sigma_{\bar{x}}$ should be .005, but what about the value of σ? At this point it's necessary that we make an assumption concerning the value of the population standard deviation. *In determining a sample size for an interval estimation, it's always necessary to make an assumption about the value of σ.*

On the basis of previous shipments, it might be possible to assume that the population standard deviation of the diameter of widgets is 0.05 inches. The necessary sample size for the wholesaler's desired level of precision is then computed as follows:

$$n = \frac{\sigma^2}{\sigma_{\bar{x}}^2}$$
$$= \frac{(.05)^2}{(.005)^2}$$
$$= 100$$

Figure 7-12 summarizes the general procedure for determining the sample size in the estimation of μ with a specified amount of precision.

Determining the Sample Size for the Estimation of π

The procedure for determining the sample size for the estimation of π is very similar to the procedure for determining the sample size for the estimation of μ. Consider the general form of the interval estimate, which is:

$$p - Z\sigma_p < \pi < p + Z\sigma_p$$

If it's specified that π must be estimated within a certain amount of *desired error,* it is essentially required that the confidence limits be:

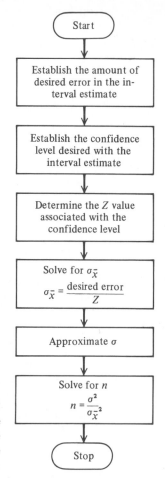

FIGURE 7-12
General procedure for determining the sample size for the estimation of μ.

$$p \pm Z\sigma_p = p \pm \text{desired error}$$

where $Z\sigma_p$ must, of course, equal *desired error*.

Following through with some mathematical manipulation which we will not illustrate, we see that σ_p is computed as follows:

$$\sigma_p = \frac{\text{desired error}}{Z}$$

And since the formula for σ_p is:

$$\sigma_p = \sqrt{\frac{\pi(100 - \pi)}{n}}$$

then

$$\sigma_p^2 = \frac{\pi(100 - \pi)}{n}$$

Therefore, you may now verify for yourself that

$$n = \frac{\pi(100 - \pi)}{\left(\dfrac{\text{desired error}}{Z}\right)^2} \qquad\qquad \text{(formula 7-11)}$$

where σ_p = desired error/Z

It's necessary at this point to approximate the value of π. The skeptics among you may wonder how it's possible to approximate π when π is what we want to estimate. Well, many times you have a rough idea of the true population percentage. For example, you may not know the true percentage of United States citizens who are black, but you know that the percentage is more than 10 percent but less than 20 percent. In many cases an experienced researcher has enough knowledge about the population to approximate the true percentage. Judgment provides an approximation of the parameter, but the sample results provide an estimate of the parameter which can be objectively assessed.

What if we are completely ignorant concerning characteristics of the population and are unable to make any approximation of the parameter? Since one picture is worth a thousand words, let's look at Figure 7-13 and assume that we've specified a desired amount of error and confidence level and our computations show that σ_p should be 2 percent. Figure 7-13 shows the necessary sample size under various assumptions about the population percentage. You can see *the symmetry in results.* The necessary sample size for an assumption of $\pi = 20$ percent is the same as that for an assumption of $\pi = 80$ percent. (A glance at column 2 of Figure 7-13 will show you why this is the case.)

As you can see in Figure 7-13, the largest sample size arises when the population percentage is assumed to be 50 percent. *When you have absolutely no idea about the true population percentage, you should assume that* $\pi = 50$ *percent and obtain the largest sample size possible which also*

FIGURE 7-13   Illustration of the Relationship between the Assumed Value of π and the Sample Size*

ASSUMED VALUE OF π (%) (1)	$\pi(100 - \pi)$ (2)	$n = \dfrac{\pi(100 - \pi)}{\sigma_p{}^2}$ (3)
20	(20)(80) = 1,600	400
40	(40)(60) = 2,400	600
50	(50)(50) = 2,500	625
60	(60)(40) = 2,400	600
80	(80)(20) = 1,600	400

*Given: $\sigma_p{}^2 = (2 \text{ percent})^2 = 4$ percent.

gives you as much information as possible to make an estimate of the population parameter.

Perhaps a problem example would be helpful. Suppose you wished to estimate the percentage of students at a university who would be willing to donate a pint of blood. Since the Red Cross is planning its schedule for the coming months, it would like you to provide an estimate that would be within ±5 percent of the true percentage. A confidence level of 95 percent is desired. How big should the random sample be? Assume you have no idea of the true percentage.

With the data available, you can say the $Z\sigma_p$ should equal 5 percent, since the confidence limits are computed as $p \pm Z\sigma_p$. With a confidence level of 95 percent you know that Z is 1.96, and therefore

$$Z\sigma_p = 5 \text{ percent}$$
$$1.96\sigma_p = 5 \text{ percent}$$
$$\sigma_p = \frac{5 \text{ percent}}{1.96}$$
$$\sigma_p = 2.55 \text{ percent}$$

Since we have no knowledge of the true percentage at all, we must obtain the largest sample size possible by assuming that $\pi = 50$ percent. Therefore the necessary sample size is computed as follows:

$$n = \frac{\pi(100 - \pi)}{\sigma_p{}^2}$$
$$= \frac{50(50)}{2.55^2}$$
$$= \frac{2,500}{6.50}$$
$$= 385$$

Figure 7-14 summarizes the general procedure for determining the sample size for the estimation of π.

Self-Testing Review 7-6

1 A Chamber of Commerce wishes to determine the mean price of new single-family residences built in the Chamber's city in the last 12 months. Data are available from local builders, realtors, and the building permit office of the city. A recent survey in a nearby city showed that the σ amount for single-family housing in that city was $5,000. The Chamber manager wants to be 90 percent confident that the results of a study will yield an estimate that is within $1,000 of the true mean price. Since you are assigned to conduct the study, what sample size would you use?

2 Executives of the Surface Transit Company are considering a new policy of reducing bus fares for senior citizens (over 65 years of age) during specified periods of the year. Before making a final decision, however, they would like to estimate what percentage of their passengers are senior citizens. The execu-

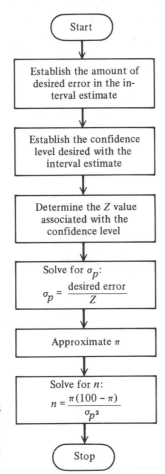

Start

Establish the amount of desired error in the interval estimate

Establish the confidence level desired with the interval estimate

Determine the Z value associated with the confidence level

Solve for σ_p:

$$\sigma_p = \frac{\text{desired error}}{Z}$$

Approximate π

Solve for n:

$$n = \frac{\pi(100 - \pi)}{\sigma_{p^2}}$$

Stop

FIGURE 7-14
General procedure for determining the sample size for the estimation of π.

tives want to be 95 percent confident that the estimate obtained is within 3 percent of the true figure. What size sample of passengers should be taken?

LOOKING BACK

1 Any specific value of a statistic is an estimate, and any statistic used to estimate a parameter is an estimator. An unbiased estimator is one that produces a sampling distribution that has a mean that's equal to the population parameter to be estimated. Thus, sample means and sample percentages are unbiased estimators of population means and percentages. The entire process of using an estimator to produce an estimate of the parameter is known as estimation.

2 A single number used to estimate a population parameter is called a point estimate, and the process of estimating with a single number is known as point estimation. Although an unbiased estimator tends toward the value of the population parameter, it's unlikely that the value of a point estimate will be exactly equal to the

parameter value. Thus, an interval estimate—one that uses a range of values to estimate a parameter—is desired over a point estimate because allowances are made for sampling error. As you've seen in this chapter, a point estimate is adjusted for sampling error to produce an interval estimate.

3 On the basis of the properties of their sampling distributions, it's possible to construct an interval estimate of μ or π with some degree of certainty. The width of the interval estimate increases as the level of confidence increases, since allowance must be made for more sampling error. The degree of probability associated with an interval estimate is known as the confidence level (or confidence coefficient). The confidence levels generally used in interval estimation are 90, 95, and 99 percent. The interval estimates based on specified confidence levels are known as confidence intervals, and the upper and lower limits of the intervals are known as confidence limits.

4 The formulas needed to estimate the population mean, when σ is known or unknown and when the population is finite or infinite, are summarized in Figure 7-15. You've seen that the sample size must be considered in estimation, since it affects the width of the confidence interval. If the sample size is sufficiently large (over 30), the sampling distribution of means approximates the normal distribution and Z values are used in the estimation process. On rare occasions, Z values can also be used if it's known that the population is normally distributed and the σ is also known. In most cases, though, if we are estimating μ and the sample size is 30 or less, the sampling distribution approximates a t distribution. The shape of a t distribution is determined by the sample size.

5 The formulas needed to estimate the population percentage are also summarized in Figure 7-15. Only samples of over 30 are considered here. The procedure for the interval estimation of π is similar to the procedure for estimating μ. Interval estimates of π values can be found almost every week in the polls published by newspapers and magazines.

6 Discussions of how to determine the appropriate sample size when estimating the population mean or population percentage have been covered in this chapter. Determining the sample size for an interval estimation of μ requires an assumption about the value of σ. Likewise, determining the sample size to estimate the value of

FIGURE 7-15   Summary of Interval Estimation under Various Conditions*

POPULATION	ESTIMATING μ				ESTIMATING π
	σ KNOWN	σ UNKNOWN			$n>30$
		$n \leq 30$	$n>30$		
Finite	$\bar{x} \pm Z \dfrac{\sigma}{\sqrt{n}} \sqrt{\dfrac{N-n}{N-n}}$	$\bar{x} \pm t_{\alpha/2} \dfrac{s}{\sqrt{n}} \sqrt{\dfrac{N-n}{N-1}}$	$\bar{x} \pm Z \dfrac{s}{\sqrt{n}} \sqrt{\dfrac{N-n}{N-1}}$		$p \pm Z \sqrt{\dfrac{p(100-p)}{n-1}} \sqrt{\dfrac{N-n}{N-1}}$
Infinite	$\bar{x} \pm Z \dfrac{\sigma}{\sqrt{n}}$	$\bar{x} \pm t_{\alpha/2} \dfrac{s}{\sqrt{n}}$	$\bar{x} \pm Z \dfrac{s}{\sqrt{n}}$		$p \pm Z \sqrt{\dfrac{p(100-p)}{n-1}}$

*Note: $s = \sqrt{\dfrac{\Sigma(x-\bar{x})^2}{n-1}}$

π requires an approximate value of π. If the researcher has absolutely no idea of the approximate value of π, then that value is assumed to be 50 percent.

KEY TERMS AND CONCEPTS

Estimator   257
Estimate   257
Unbiased estimator   257
Estimation   257
Point estimate   257
Point estimation   257
Interval estimate   258
Interval estimation   258

$\bar{x} - Z\sigma_{\bar{x}} < \mu < \bar{x} + Z\sigma_{\bar{x}}$
Lower limit   Upper limit
of estimate   of estimate   (formula 7-1)   *261*

Confidence level (confidence coefficient)   *261*
Confidence intervals   262
Confidence limits   262

$s = \sqrt{\dfrac{\Sigma(x - \bar{x})^2}{n - 1}}$ (formula 7-2)   *270*

$\hat{\sigma}_{\bar{x}} = \dfrac{s}{\sqrt{n}}$ (formula 7-3)   *270*

$\hat{\sigma}_{\bar{x}} = \dfrac{s}{\sqrt{n}}\sqrt{\dfrac{N - n}{N - 1}}$ (formula 7-4)   *270*

$\bar{x} - Z\hat{\sigma}_{\bar{x}} < \mu < \bar{x} + Z\hat{\sigma}_{\bar{x}}$ (formula 7-5)   *270*

t distribution   272

$\bar{x} - t_{\alpha/2}\hat{\sigma}_{\bar{x}} < \mu < \bar{x} + t_{\alpha/2}\hat{\sigma}_{\bar{x}}$
    (formula 7-6)   *272*

Degrees of freedom (df) *274*

$p - Z\hat{\sigma}_p < \pi < p + Z\hat{\sigma}_p$ (formula 7-7)   *278*

$\hat{\sigma}_p = \sqrt{\dfrac{p(100 - p)}{n - 1}}\sqrt{\dfrac{N - n}{N - 1}}$ (formula 7-8)   *278*

$\hat{\sigma}_p = \sqrt{\dfrac{p(100 - p)}{n - 1}}$ (formula 7-9)   *278*

$n = \dfrac{\sigma^2}{\sigma_{\bar{x}}^2}$ (formula 7-10)   *285*

$n = \dfrac{\pi(100 - \pi)}{\left(\dfrac{\text{desired error}}{Z}\right)^2}$ (formula 7-11)   *287*

PROBLEMS

1   The following set of data represents a simple random sample of temperatures from 24 cities in a southern state. These temperatures were observed at the same hour and on the same day.

105	97	101	88	96	100	87	110
99	92	99	93	93	87	101	101
103	107	95	92	89	95	102	100

a   What estimator would you use to estimate the true average temperature within the state?
b   What's the point estimate of the true average temperature?
c   Calculate the sample standard deviation.

2   The following set of data represents a simple random sample of IQ scores of 32 students at an eastern university:

137	141	128	132	129	122	140	119
126	133	121	138	111	124	121	116
120	127	129	122	113	125	126	118
117	132	124	116	135	123	126	131

a   Calculate a point estimate of the true average IQ of the students at the eastern university.

b   Calculate the sample standard deviation.

3   Using the data in problem 1, construct a confidence interval of μ with a confidence coefficient of 99 percent.

4   Using the data in problem 2, construct a confidence interval of μ with a confidence coefficient of 95 percent.

5   Assume you are constructing an interval estimate with a large sample size. Determine the Z value for each of the following confidence coefficients:

a   75 percent

b   93 percent

c   88 percent

d   95 percent

6   Determine the t value for the following:

a $n = 15$, 99 percent confidence level

b $n = 23$, 90 percent confidence level

c $n = 28$, 95 percent confidence level

d $n = 27$, 95 percent confidence level

e $n = 25$, 95 percent confidence level

f $n = 20$, 95 percent confidence level

7   Look at the results of parts c through f in problem 6. What relationship do you see between the interval width and the sample size?

8   Assume a student is interested in determining the average amount of money she spends per day in the month of September. A random sample of 10 days shows an average of $6.24 per day with a standard deviation of $1.20. Calculate an interval estimate of μ with a 90 percent confidence level.

9   The Howe and Dee Answering Service wants to estimate the average number of calls handled daily. A random sample of 50 days produced a mean of 326 calls per day with $s = 48$.

a   Compute an estimate of the standard deviation of the sampling distribution of means.

b   Construct a confidence interval with a confidence coefficient of 90 percent.

10   The Automated Automaton Assembly Plant has had a large number of employees quit their jobs soon after they start employment. The personnel manager, Mr. Rowe Botts, would like to estimate the average stay with the company of each employee who quits. A random sample of 15 former employees' records indicated that the average stay with the company was 54 days. The sample standard deviation was 16 days. Construct an interval estimate of the true average with a 99 percent confidence level.

11   Suppose the average height of 20 students in a peculiar high school is 5 feet 8 inches with a sample standard deviation of 6 inches. Construct a confidence interval of the true average with only a 5 percent chance of error.

12   The Bill Fold Wallet Company wants to know the average number of wallets sold each day. A random sample of 36 days produced an average of 114 wallets sold daily. Assume $\sigma = 17$. Construct an interval estimate of μ with a 95 percent confidence coefficient.

13   The Russell Caddle Beef Company wants to estimate the average tonnage of

beef processed daily. A simple random sample of 50 days showed that the average daily tonnage was 100 tons. The population standard deviation is 22 tons. Construct an interval estimate of the true average with a 90 percent confidence coefficient.

14 A sample of 50 farm workers contained 36 workers who indicated that they would like to unionize.

   a What's the estimator of π?

   b What's the point estimate?

   c What further information is needed for an interval estimate?

15 The Just-for-the-Halibut Fish Store received a shipment of fish, and the store would like to estimate the percentage of fish which meet its weight standards. Of a random sample of 37 fish, 6 were too small. Estimate the true percentage of acceptable fish, using a 99 percent confidence level.

16 The fish friar and chip monk of a monastery kitchen would like to know what percentage of the 200 monks in residence actually like fish and chips. A random sample of 40 friars contained 30 who like fish and chips. Estimate at the 95 percent confidence level the true percentage of monks who actually like fish and chips.

17 Bjorn Talooz, a student at a Norwegian university, wants to determine the feasibility of campaigning for the presidency of the students' association. A random sample of 50 students showed that 22 percent of the students would vote for him. Estimate the true percentage with a confidence level of 99 percent.

18 In order to respond to a federal government survey, the director of a state office of education needs to estimate the percentage of women teachers in the state public school system. The director wants to be 99 percent confident that the estimate of the percentage of women teachers is within 3 percent of the true population percentage. What size sample should be selected from the state files?

19 In problem 10 above, what size sample would be needed if Rowe Botts wanted to be 95 percent confident that an estimate of the average employment time would be within 3 days of the true mean time? (Use the sample standard deviation of 16 in problem 10 as an approximation of the population standard deviation.)

20 At the Marshall Department Stores a random sample of 400 accounts receivable from a population of 300,000 accounts receivable produced the following sample results: Mean balance due = $25.00; standard deviation balance due = $2. At the 95 percent level of confidence, what's the range within which the population mean balance should be found?

21 A random sample of 101 components is selected to determine the life expectancy of a certain type of electronic part. The sample mean life expectancy was 495 hours and the sample standard deviation was 64 hours. At the 95 percent level, estimate the true life expectancy of the components.

22 In a random sample of 1,000 television viewers, 340 are watching "Name that Poem." At the 95 percent level, what's the estimate of the percent of all viewers who are watching the program?

23 Over a random 10-day period, Marsha Winn, a salesperson, kept records of her expenses for meals and lodging. The sample mean for expenses per day was

$49.11 and the sample standard deviation was $9.33. At the 90 percent level, what's the estimate of the population mean daily expenses?

24   A random sample of 50 stocks traded on the New York Stock Exchange had a sample mean of $67.42, and the value of the sample standard deviation was $6.28. What's the 90 percent interval estimate of the mean price of all stocks traded on the NYSE?

25   The city manager of a city located near a large factory wants to know the mean daily amount of sulfur oxides emitted by the plant. A random sample of 16 days is used and the sample mean is 5.3 tons per day. The sample standard deviation is 3 tons per day. At the 90 percent level, what's the estimate of the population mean emissions?

26   In a randomly selected public opinion poll, 320 out of 400 persons sampled supported Senator Adams' position on public works programs. At the 95 percent level, estimate the population percentage of persons favoring Adams' position.

27   The mean nicotine content of a random sample of five Wheezer brand cigarettes is 31 milligrams, and the sample standard deviation is 2 milligrams. At the 99 percent level, what's the estimate of the average nicotine content of all Wheezer cigarettes?

28   The Badd-Bolt Manufacturing Company selected a random sample of 300 bolts from production to estimate the percentage of defectives. There were 30 defective bolts in the sample. What's the 99 percent interval estimate of the percentage of defective parts in the population?

29   Assume that you want to estimate the average golf score of your 200 fellow club members. You take a random sample of size 100 and find that the sample mean is 98 strokes with a sample standard deviation of seven strokes. At the 95 percent level, what's the estimate of the true mean score for all members?

30   In a community of 10,000 families, a random sample of 400 families showed that 120 of the families surveyed have an annual income exceeding $30,000. What's the 95 percent confidence interval estimate of the percentage of all families in the community that earn over $30,000?

31   The following sample temperatures are recorded in a northern state at noon on New Year's day:

10   8   12   11   9   12   15   13   4   0   19   17   19   19
22   11   7   9   7   15   7

a   Compute a point estimate of the average temperature in the state at the time these values were recorded—that is, compute the sample mean temperature.

b   Compute the sample standard deviation temperature.

c   Using the sample mean and standard deviation temperature values found in parts a and b above, compute the 95 percent confidence interval for the mean statewide temperature.

d   Suppose it was known ahead of time that the population standard deviation temperature was actually 5 degrees. Compute the 95 percent confidence interval for the mean statewide temperature.

32 Let's assume that you're asked to conduct a poll to determine the percentage of voters who plan to vote for a referendum. The poll must be conducted so that there's a 95 percent probability that the results are accurate to within ±3 percent.

a What's the minimum number of voters who must be polled to achieve this goal?

b Suppose that your poll had to be accurate to ±1.5 percent at the same 95 percent confidence level. What must the sample size be in this case?

33 The following set of data represents entrance exam scores taken by a group of applicants at a western university:

```
137  141  128  132  129  122  140  119  126  133  121
138  111  124  121  116  120  127  129  122  113  125
126  118  117  132  124  116  135  123  126  131  221
150   98  166  129  124  117  132
```

a If a school official wants to estimate the population mean exam score by taking a random sample from the above group, how many scores must be sampled if the official wants to be 90 percent sure that the true mean lies within 5 points of the sample mean score? (Assume the population is infinite and has a standard deviation value of 10 points.)

b Using the results of part a above, select a sample of the right size from the group of scores listed above and estimate the population mean score.

34 A shipment of 1,000 hammers is delivered and a sample of 100 is examined. It's found that the sample contains 5 defective hammers. What is the 95 percent confidence interval for the percentage of defective hammers in the shipment?

35 A random sample of 600 accounts receivable taken from a population of 150,000 such accounts at a department store chain produced a sample mean balance of $20.00 and a sample standard deviation of $3.00. What is the range within which the population mean balance lies at the 99 percent confidence level?

36 A random sample of 200 hospital patients is taken at Wiley Memorial Hospital, and 8 are found to be carrying a newly discovered bacterium. There are a total of 5,000 patients in the hospital. At the 95 percent level, what is the interval estimate of the percentage of all Wiley hospital patients carrying the bacterium?

37 A sample of 5 resistors is chosen at random from a population of 100 resistors. The measured resistances (in ohms) are listed below:

```
100.03    99.80    102.71    101.60    100.36
```

a What is the point estimate of the population mean resistance?

b What is the value of the sample standard deviation?

c What is the estimated value of the standard error?

d At the 95% level, what's the range of values that is likely to include the population mean resistance?

38 A minimum score of 50 is required on an entrance exam at a private military academy. A random sample of 30 entrance exam scores taken from a population of 500 scores is listed below:

70	30	60	80	77	99	26	97	42	33	55	65
85	63	73	98	11	83	88	99	65	80	74	69
91	80	44	12	63	61						

a What's the point estimate of the percentage of applicants who fail the exam?

b What is the estimate of the standard error of percentages?

c At the 95 percent level of confidence, what's the interval estimate of the percentage of failing applicants?

39 In a random sample, 1,000 people were asked to taste several brands of cola. In this test, 667 people preferred the taste of Longneck Cola. At the 90 percent level of significance, what percentage of all cola drinkers are likely to prefer Longneck?

40 A sample of 100 iron rods is subjected to destructive testing. The sample mean breaking load is 3,050 pounds, and the sample standard deviation is 500 pounds. What's the 95 percent interval estimate of the breaking load of all rods in the population?

TOPICS FOR REVIEW AND DISCUSSION

1 Why is an unbiased estimator desirable?

2 "If a point estimate is rarely used in estimation, it is rarely computed." Discuss the fallacy of this statement.

3 "A 95 percent confidence level means that there is a 95 percent chance that μ will fall within the computed estimate." Comment on this statement.

4 What effect does an increase in the confidence level have on the width of the confidence interval?

5 "We should always use the highest confidence coefficient in interval estimation." Discuss this statement.

6 When does the sampling distribution approximate the normal distribution?

7 What can be said about the sample size when σ is unknown and when the Z distribution applies to the interval estimation?

8 If a t distribution is applicable, what must be known in order to determine the t value for an interval estimation?

9 How can the sample standard deviation be used as an estimator of σ?

10 If the parameter $\sigma_{\bar{x}}$ may sometimes be used in an interval estimation of μ, why isn't it possible to sometimes use the parameter σ_p in an interval estimation of π?

11 "If an interval estimate is too wide, we should lower the specified confidence level so that the interval range narrows." Why is this statement incorrect?

PROJECTS/ISSUES TO CONSIDER

1   In project 1 in Chapter 6, your team located a population data set, selected a random sample of size n from that population, and then computed the sample mean and standard error of the mean. Now, using the sample data you produced in that project, your team should:

 a   Construct a 95 percent confidence interval estimate of the population mean. (You know the population σ, but your sample size and knowledge about the shape of the population distribution will determine if you use the Z or t distributions.)

 b   Compare your interval estimate with the population mean you've computed for the Chapter 6 project. Did your population mean fall within your confidence interval? If not, why not?

 c   Present your results to the class.

2   Now your team should identify some research question or topic that interests team members. (One of many possibilities is "What percent of the student body would vote for _____ for the Student House of Representatives?") Calculate the sample size needed to give an interval estimate at the level of precision the team wants. Prepare a brief written report summarizing your research question and the sample-size requirements.

3   On the next page are some of the findings of a *Business Week/Harris Poll.* You'll notice that the sample data were obtained by polling 1,250 adults in October, 1987, and April, 1988. The sample percentages shown in the poll are point estimates; the confidence interval estimates (at the 95 percent level) of the appropriate population percentages are found by adding about 3 percentage points to the point estimate to get the upper limit, and by subtracting about 3 percentage points from the point estimate to get the lower limit. These interval estimates aren't shown. Now your job is to locate the results of another poll in a newspaper or magazine. Study the poll data, the methodology used (that is, the sample size and other factors), the point and interval estimates produced, and the conclusions reached. Write a brief summary of your findings. You may want to refer to the Closer Look reading found at the end of this chapter in preparing your summary.

ANSWERS TO SELF-TESTING REVIEW QUESTIONS

7-1

1   Unbiased estimator.

2   p is an unbiased estimator of π, while 82 percent is a point estimate of π.

3   An interval estimate is actually a point estimate with an allowance for sampling error. An interval estimate estimates the parameter to be within a range of values, while a point estimate is a single value.

4   A point estimate by itself gives no indication of the precision of the estimate.

5   The statement is false. Although allowance is made for sampling error, an interval estimate, like any other estimate, can be wrong.

Business Week/Harris Poll
Who's Afraid of a Little Ol' Crash?

The Crash of '87 doesn't seem to have diminished public confidence in the economy, though Americans' anxieties about inflation, interest rates, employment, and recession remain high. The public has become more bearish about the stock market. Why the relative calm? Perhaps it's because 85% of the public said the crash had not affected them financially.

SHORT-TERM PROSPECTS

	OCT. 1987	APR. 1988
◗ Six months from now, do you think the general business conditions in your area will be better, the same, or worse?	Better...................26%	27%
	Same 51%	56%
	Worse.................17%	13%
	Not sure.............. 6%	4%

A BIT FURTHER DOWN THE ROAD

◗ A year from now, do you feel or not that . . . ?	YES	NO	NOT SURE
Prices will be going up much faster	54%	41%	5%
Unemployment will start going up	55%	40%	5%
Interest rates will begin to climb again	73%	22%	5%
You will be more cautious as a consumer	76%	22%	2%
The country will be going into a recession	42%	49%	9%

WHERE IS THE MARKET GOING?

	OCT. 1987	APR. 1988
◗ Over the next year, do you think stocks will go up, stay about the same, or go down?	Will go up.......................50%	29%
	Stay about the same.......31%	42%
	Will go down..................12%	20%
	Not sure 7%	9%

Polls of 1,250 adults conducted Oct. 23–25, 1987, and Apr. 1–5, 1988. Surveys were conducted by Louis Harris & Associates Inc. for BUSINESS WEEK. Overall results should be accurate to within three percentage points.
Source: *Business Week/Harris Poll,* edited by Stuart Jackson. Reprinted from April 18, 1988 issue of *Business Week* (p. 59) by special permission, copyright © 1988 by McGraw-Hill, Inc.

6 The mean of the sampling distribution of an unbiased estimator is equal to the population parameter to be estimated. This means that the possible outcomes of samples tend toward the value of the parameter.

7 The statement is false. The point estimate serves as a basis for construction of the interval estimate.

7-2

1 The theoretical basis for estimating μ with the interval lies in the properties of the sampling distribution of means.

2 There is no difference between a confidence coefficient and a confidence level.

3 A 95 percent confidence level means that the particular method being used for interval estimation can produce a large number of intervals of which approximately 95 percent will contain the parameter.

4 A positive relationship exists. A higher confidence level will produce a wider confidence interval.

7-3

1 The sampling distribution approximates the normal distribution when the population is normally distributed or when the sample size is more than 30.

2 a $Z = 1.70$
   b $Z = 1.10$
   c $Z = 1.48$
3 With $\sigma = 9$ and $n = 36$,

$$\sigma_{\bar{x}} = \frac{\sigma}{\sqrt{n}} = \frac{9}{\sqrt{36}} = 1.5$$

   With the confidence coefficient equal to 90 percent, $Z = 1.64$. The confidence interval is $\bar{x} \pm Z\sigma_{\bar{x}} = 48 \pm 1.64(1.5)$. The confidence limits are 45.54 and 50.46.
4 With the confidence coefficient equal to 80 percent, $Z = 1.28$. The confidence interval is $\bar{x} \pm Z\sigma_{\bar{x}} = 104 \pm 1.28(13)$. The confidence limits are 87.36 and 120.64.

7-4
1 A t distribution should apply because the sample size is less than 30.
2 With 26 degrees of freedom, $t_{025} = 2.056$.
3 $\$38.86 < \mu < \$45.14$.
4 You should calculate the sample mean and the sample standard deviation. You should then use the sample standard deviation to calculate $\hat{\sigma}_{\bar{x}}$. In order to construct the interval estimate you must specify a confidence level.
5 A t distribution is generally flatter than a Z distribution.
6 The sample size affects the shape of a t distribution.
7 The format of the t table is based on the total chance of error in estimation.
8 It can be computed using formula 7-2.

7-5
1 When we use the Z value in interval estimation, we assume a large sample size—i.e., $n > 30$.
2 The denominator $n - 1$ in $\hat{\sigma}_p$ makes $\hat{\sigma}_p$ an unbiased estimator of σ_p.
3 The formula for σ_p requires knowledge of π. However, π is what is to be estimated.

7-6
1 Since $Z\sigma_{\bar{x}} = \$1,000$, and since $Z = 1.64$, $\sigma_{\bar{x}} = \$1,000/1.64$ or $\$609.76$. Therefore, since σ may also be estimated to be $\$5,000$ in the Chamber's city, the necessary sample size is computed as follows:

$$n = \frac{\$5,000^2}{\$609.76^2} = 67.24 \text{ or } 68$$

2 Since $Z\sigma_p = 3$ percent, and since $Z = 1.96$, or $\sigma_p = 3$ percent/1.96 or 1.53 percent. Therefore, since we have no idea what the value of the population percentage is, the necessary sample size is computed as follows:

$$n = \frac{(50)(50)}{1.53^2} = \frac{2,500}{2.34} = 1,068$$

A CLOSER LOOK

POLLS APART: A PRIMER ON WAYWARD SURVEYS

In a national poll of July 13–15, the Gallup Organization found Ronald Reagan seemingly well ahead of Walter Mondale, 55% to 36% (the remaining 9% being undecided or for some other candidate). But just one week later another Gallup Poll put Mr. Mondale and Geraldine Ferraro ahead of Mr. Reagan and George Bush 48% to 46%. The Democratic Convention had been held, but had it really made that much difference? The big bounce raised anew questions about the accuracy and reliability of the election picture drawn by polls.

All during the 1984 campaign we have seen striking disparities in the findings of different polls. Some of these are persisting differences between survey organizations. For example, in eight CBS News/New York Times polls taken nationally from January through August, Mr. Reagan maintained a consistently large edge over Mr. Mondale in trial heats—from 12 points on the low side to 20 on the high. But according to four national polls by ABC News and the Washington Post during the same span, the Reagan margin has been consistently smaller, two to seven points.

CAVEAT EMPTOR

Sometimes polls conducted at about the same time agree in their overall distributions but differ greatly on the presidential preferences of key groups or sections. The accompanying table shows, for example, massive differences in three national surveys on Mr. Reagan's and Mr. Mondale's regional strength. All three were taken at about the same time—before the Democratic Convention and the Ferraro announcement. One put Mr. Reagan 29% ahead of Mr. Mondale in the South, another 13% ahead and the third had Messrs. Reagan and Mondale dead even there.

The fact is that such disparities are neither happenstance nor products of special occurrences this election year. They are unavoidable, permanent features of the polling business—which means that poll users need to practice the hallowed maxim: Caveat emptor. Here is a checklist of guidelines and cautions:

1　Above all else, remember that public-opinion polls do not chart a firmly fixed reality. An animal's red-blood-cell count can be measured precisely not just because the measurement tools are adequate but because there is something precise to be measured. Not so with Reagan-Mondale preferences—because many voters aren't paying a lot of attention and haven't arrived at firm judgments. No amount of technical sophistication in polling can change this, and consumers should stop beating pollsters over the head when the latter prove unable to supply precise readings of still-unformed opinions. But some pollsters need to be a lot more sensible, too. In a July statement, Louis Harris proclaimed that "the selection of Rep. Geraldine Ferraro from New York as the vice presidential choice of Walter Mondale increases the Democratic chances of winning November's election. When paired with Mondale on the ticket, Ferraro narrows a 52-44% Reagan lead to 51-45%." One doesn't know whether to laugh or cry. Polls can never, repeat *never*, achieve a measure of precision to such an extent that one percentage point means anything at all.

2　There are signs that the electorate may be harder to read now than it was 40 or 50 years ago, because of the weakening of voter loyalties to political parties. The impact on polling of this well-documented development is greatest among those voters who don't pay much attention to national politics. In times past when these voters were confronted with the choice for president, they in effect answered "my party." But many of them now no longer have a party. When a pollster asks them about their likely vote, they will often answer—and answer truthfully—for one candidate or another. That professed choice may, however, hang by a thread so thin that a day later for no special reason it has broken altogether.

3 It is more, though, than the squishiness of public sentiment that makes for discrepancies in poll findings. The techniques of polling produce countless forms of variability. One way pollsters do this is through question wording and question placement. Though it isn't nearly as hard to ask about presidential preference as it is to tap opinion on a complex issue such as the U.S. role in Central America, question wording can influence responses even in the former area. Some of my polling colleagues think, for instance, that it may make a difference whether the question refers to "President Reagan," just "Ronald Reagan" or "Ronald Reagan, the Republican." They also think that placing a trial-heat question after a string of issue questions may skew responses, if the issues covered are framed in a way that has an implicit bias toward Mr. Reagan or toward Mr. Mondale. The best advice seems to be: Put the trial-heat question(s) right up front in the survey, so that there will be no preceding questions to contaminate it.

4 Interviewer effects cannot be discounted. Do different polls report contrasting pictures of the presidential race because those operating them have different electoral preferences? *No.* Polling has plenty of problems, but "cooking" data to suit the taste of the polling chefs isn't one of them. We can be less confident, however, that individual interviewers do not themselves slant results by voice, tone or inflection, either knowingly or quite unintentionally. Every reputable polling organization is aware of this potential problem and takes steps to minimize its actual occurrence—from addressing it when training interviewers to monitoring telephone interviews. But it is unlikely that all of these efforts at quality control are equally strenuous. Interviewers are enormously important and not wholly controllable actors in polling, and it may be expected that sometimes interviewers have strong preferences on an election or policy question. Some skewing may result—and no one knows for sure whether the amount is statistically trivial or significant.

We do know that other kinds of consequential interviewer effects are very hard to eradicate. For example, a field experiment conducted by Eleanor Singer (editor of the Public Opinion Quarterly), Martin Frankel and Marc Glassman found that "interviewers' . . . expectations of response rates . . . influence the response rates they achieved." I personally have seen dramatic evidence that even polite, well-spoken, fully trained interviewers who for some reason are pessimistic about getting people to respond to certain questions, or in a particular context, get markedly higher refusal rates than those who are confident about the interviews. Other factors such as the age and sex of interviewers may affect responses.

Reagan's Margin over (+) or under (−) Mondale

	PENN AND SCHOEN POLL OF JUNE 12–17	CBS NEWS/ NEW YORK TIMES POLL OF JUNE 23–28	ABC NEWS/ WASHINGTON POST POLL OF JULY 5–8
Northeast	+4	+26	−4
South	+29	+13	0
Midwest	+8	+10	+9
West	+6	+14	+29
National	+13	+15	+7

5   Factors related to sampling may also contribute to contrasting poll results. If there is any bias, however unintentional, in the way an organization completes its final sample, it is likely to persist over a number of polls because basic methodologies are not frequently changed. The question of sample bias is complex, but here is one way a problem can develop. After conducting their interviews, some survey firms compare the distributions of the achieved sample with known national distributions—on such variables as sex, region and socio-economic position—and then "weight" the sample to make it correspond to the nation in these characteristics. If, for instance, an achieved sample is 60% female while the known distribution of the population age 18 and over is 52% female, women's answers are weighted by a fraction that reduces their proportion from 60% to 52%. This weighting assumes that those initially interviewed from a particular group are a random sample of the entire group. If so, fine. But if they are not for any reason, weighting may significantly *exaggerate* the initial error.

Also, whether a survey is conducted through face-to-face interviews in the home, or by telephone, has great implications for how the sample is drawn and who actually gets included. Sampling theory is straightforwardly valid—but applying the theory in the rushed, cost-conscious everyday world of polling allows great variety.

LOOK FOR A PATTERN

Together the above elements can produce a great deal of variation from one survey to another—and some of this may be patterned or persistent from one organization to another. One reason we may now be more conscious of the problem is simply the dramatically expanded volume of polling.

In 1948 there were really only two national political pollsters—Gallup and Roper—and one of them, Roper, didn't poll during the last two months before the election. Of course there were bound to be fewer apparent discrepancies among the polls in that situation than today, when we have a multiplicity of surveys by the media, the traditional independent polling organizations, academic centers and candidates. But the present-day numbers are in fact a great boon. The pluralism of the American polling industry permits the careful reader to discount individual survey findings and look for a pattern among the many contending polls. In such patterns we find the closest thing to truth that opinion research can deliver.

—Everett Carll Ladd, "Polls Apart: A Primer on Wayward Surveys," *The Wall Street Journal*, August 16, 1984, p. 22. Reprinted with permission of The Wall Street Journal, © 1989 Dow Jones & Company, Inc. All rights reserved.

TESTING HYPOTHESES AND MAKING DECISIONS: ONE-SAMPLE PROCEDURES

Statistically Significant?

A common misuse of statistical words lies in the transplantation of reasonably precise technical terms into inappropriate contexts. The term "statistically significant" is a frequent victim. In a recent letter to the *New York Times*, for example, a letter about our postal system, the writer says:

> The vastness of this country, the high mobility rate of many of its inhabitants and its statistically significant immigrant population all contribute to the need for an efficient postal service.

That "statistically significant" gives the sentence a lot of rhetorical weight, although all it appears to mean is "large." A similar case appears in a *New York Times* article by Harold M. Schmeck: " . . . the numbers involved in this comparison were considered too small to be statistically significant."

This difficulty is not limited to the ephemeral columns of newspapers; one may find it in highly regarded scientific journals. For example, in a 1975 issue of *Science,* the official journal of the American Association for the Advancement of Science, an article on temperature trends says that

> It is difficult to test our results . . . against observations because no statistically significant *global* record of temperature back to 1600 has been constructed.

Here the significant phrase apparently means "accurate," but I am not certain.

A 1973 example from the same journal says that

... if we take n pictures of N_0 particles, ... (we have) $6nN_0$ coordinates, a number which can easily be made sufficiently large to have statistical significance.

Here again the term is used in a vague, almost boastful sense.

In fact the phrase "statistical significance" has a widely understood, rather precise meaning that is taught in introductory statistics textbooks and courses. It refers to a so-called tail probability that is surprisingly small under a relevant, if sometimes tacit, hypothesis.

—From William Kruskal, "Formulas, Numbers, Words: Statistics in Prose." Reprinted from THE AMERICAN SCHOLAR, Volume 47, Number 2, Spring, 1978. Copyright © 1978 by the author. By permission of the publisher.

LOOKING AHEAD

You guessed it. The concept of "statistical significance" mentioned in the opening vignette is one that's developed in Chapter 8. You'll learn procedures in this chapter for objectively determining, under various conditions, whether sample results support a hypothesis about a parameter value or whether the results indicate that the hypothesis should be rejected. That is, you'll learn that if the sample results differ from the hypothetical population value being tested by an amount that *exceeds* what might be expected because of sampling variation, the difference is called a statistically significant difference, and this difference is the basis for rejecting the hypothesis being tested.

But we're getting ahead of ourselves. Before you are asked to test hypotheses and make decisions, Chapter 8 will outline a general hypothesis-testing procedure for you to use. Then you'll be in a better position to conduct one-sample hypothesis tests of means and percentages. (In the next three chapters, we'll look at hypothesis-testing procedures that involve two or more samples.)

Thus, after studying this chapter you should be able to:

- Explain the necessary steps in the general hypothesis-testing procedure.
- Compute one-sample hypothesis tests of means (both one- and two-tailed versions) when the population standard deviation is unknown as well as when it's available.
- Compute one-sample hypothesis tests of percentages for both one- and two-tailed testing situations.

CHAPTER OUTLINE

LOOKING AHEAD
THE HYPOTHESIS-TESTING PROCEDURE IN GENERAL

THE HYPOTHESIS-TESTING PROCEDURE IN GENERAL

In the previous chapter on estimation, the value of the population parameter was unknown, and the results of a sample were manipulated to provide some insight into the true value. In this chapter the sample results are used for a different purpose. Although the exact value of a parameter may be unknown, there's often some hunch or hypothesis about the true value. Sample results may bolster the hypothesis, or the sample results may indicate that the assumption is untenable. For example, Dean I. V. Leeg may state that the average IQ of the students at her university is 130. This statement is an assumption on her part, and there should be some way of testing the dean's claim. One possible method of validation (assuming that the dean really knows) is to strap her in a chair and administer a lie detector test.[1] Another method, which is more feasible and attractive to the dean and the researcher, is sampling. If a random sample of these students had

---

[1] This method further assumes that you can locate the dean. Father Damian Fandal, formerly academic dean at the University of Dallas, has *facetiously* formulated two rules for deans: Rule 1—Hide!!!; Rule 2—If they find you, lie!!! See Thomas L. Martin, Jr., *Malice in Blunderland*, McGraw-Hill Book Company, New York, 1973, p. 90.

an average IQ of 104, it would be easy to reject the assumption that the true average is 130 because of the large discrepancy between the sample mean and the assumed value of the population mean. Similarly, if a sample mean were 131, it would be reasonable to accept the dean's statement. Unfortunately, life's decisions are not always as easy as this. Many times decisions fall under the general category of ulcer inducers. In the case of testing a hypothesis, the difference between the values of the sample statistic and the assumed parameter is usually neither too large nor too small, and thus obvious and clear-cut decisions are often rare. Suppose, for example, the average IQ of a sample was 134; or suppose it was 127. Does either value warrant rejection of the statement that $\mu = 130$? Obviously such a decision cannot be eyeballed. There must be some sort of criterion on which the decision process can be focused.

Before we present a formal enumeration of the steps in the hypothesis-testing procedure, let's consider another example. Suppose the mayor of a rural town has stated that the average per-capita income of the town's citizens is $10,000, and you have an emerging statistician friend—Stan Strate—who has been assigned by the town council to verify or discredit the major's claim. Obviously, Stan's knowledge about sampling variation tells him that even if the true mean were $10,000 as stated, a sample mean would *most likely not equal* the parameter value. As an educated person versed in the rudiments of statistics, Stan realizes there will probably be a difference between the sample mean and the assumed value. The immediate problem confronting him is how large or *significant* should the difference between $\bar{x}$ and the assumed value be to provide sufficient reason to dismiss the mayor's claims? Is a difference in values of $100 significant? Is a difference of $1,000 significant? Well, the significant differences can be determined through statistical techniques.

Still Another Look at the Sampling Distribution of Means

Let's look at Figure 8-1 and assume that we have a sampling distribution of means where (1) the true mean (μ) is actually equal to the hypothesized value (μ_{Ho}) of $10,000 and (2) the standard error is equal to $200. *In other words, we are assuming that the mayor was actually correct and μ is indeed $10,000.* (Of course, Stan and the town council are not aware of this fact.) Suppose further that Stan takes a sample of townspeople, with the result that the sample mean per-capita income is equal to $10,200. Is it reasonable for Stan to expect this result with a μ_{Ho} of $10,000 and a $\sigma_{\bar{x}}$ of $200? How likely is it that a $\bar{x}$ of $10,200 will occur in this situation? As a more general question, what are the chances of Stan's getting a sample mean that differs from the μ_{Ho} of $10,000 by $200?

Since the sampling distribution in Figure 8-1 is approximately normal, Stan can check the likelihood that a sample mean will equal $10,200 or $9,800 by determining how many standard errors from the μ of $10,000 a difference of $200 represents. How many Z values does $10,200 or $9,800 lie from the true and assumed population mean of $10,000—that is, what's

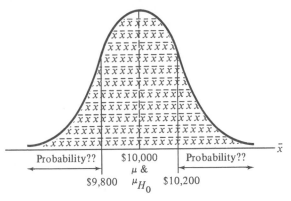

Where: $\sigma_{\bar{x}}$ = $200

the standardized difference or the number of *standard units?* Stan can calculate the standard units in this way:

$$Z = \frac{\bar{x} - \mu_{H0}}{\sigma_{\bar{x}}}$$

$$Z = \frac{\$10,200 - \$10,000}{\$200} \quad \text{and} \quad Z = \frac{\$9,800 - \$10,000}{\$200}$$

$$= 1.00 \qquad\qquad\qquad = -1.00$$

Thus, we can see that if a sample mean in our example differs from the assumed value by $200, it differs by one standard unit or one standard error. Consulting the Z table in Appendix 2, we see that the area under one side of the distribution up to $Z = 1$ is .3413, and the total area between $Z = \pm 1$ is .6826. This means that there's a .1587 chance that a sample mean could be *larger* than the true and assumed population mean by *one or more standard errors*, and there is also a .1587 chance that $\bar{x}$ may be *less than* the population mean by *one or more standard errors*. All this is demonstrated in Figure 8-2 where it's shown that there's a total chance of 31.74 percent that $\bar{x}$ will differ from μ by one standard unit or more. Consequently, Stan could report to the town council that a sample mean of $10,200 is likely to occur in this example and that the $200 difference is not sufficiently significant for him to reject the mayor's claim.

Suppose Stan's sample mean had been $10,400 instead of $10,200. Would he reject the major's claim with this sample result? (Remember, he doesn't know the true value of the population mean.) Converting this $400 difference between $\bar{x}$ and μ_{H0} into standard units, we get:

$$Z = \frac{\bar{x} - \mu_{H0}}{\sigma_{\bar{x}}}$$

$$= \frac{\$10,400 - \$10,000}{\$200}$$

$$= 2.00$$

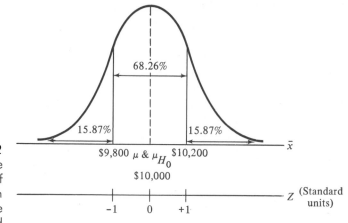

FIGURE 8-2
Illustration of the
likelihood of
obtaining a $\bar{x}$ which
differs from the true
mean by 1 standard
error or more.

Where: $\sigma_{\bar{x}} = \$200$

Thus, the total chance that a $\bar{x}$ will differ from our true mean of \$10,000 by two or more standard errors is only approximately 4.6 percent, as shown in Figure 8-3. Given such a low chance of occurrence, Stan would probably be justified in *rejecting* the mayor's claim. There's sufficient statistical evidence for him to conclude that the mayor's claim is incorrect.

As you can now see, the difference between the value of an obtained sample mean and an assumed value of a hypothetical population mean is considered significantly large to warrant rejection of the hypothesis if the likelihood of the value of a sample mean is too low. The criterion of "too low" will vary with the standards of researchers. At this point it's sufficient to state that all hypothesis tests must have some established rule which re-

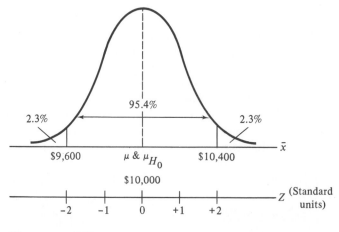

FIGURE 8-3
Illustration of the
likelihood of
obtaining a $\bar{x}$ which
differs from the true
mean by 2 or more
standard errors.

Where: $\sigma_{\bar{x}} = \$200$

jects a hypothesis if the likelihood of a value of $\bar{x}$ falls below a minimum acceptable probability level.

Unfortunately, if Stan were unaware in the above example that the true population mean was indeed $10,000, he might have justifiably but erroneously rejected the mayor's claim if he had obtained a sample mean of $10,400. As a matter of fact, if he had established a rule that any sample mean value which differed from the assumed mean of $10,000 by two or more standard errors in either direction of the sampling distribution would cause rejection of the hypothesis, and *if* indeed the true mean were $10,000, he would erroneously reject the mayor's claim 4.6 percent of the time in a large number of tests. In other words a particular sample mean may be a part of a sampling distribution in which the value of the true mean happens to be equal to the assumed value, but the likelihood of that particular sample mean occurring may be so low that there is sufficient reason not to accept the hypothesized value as the true value. In short, the minimum acceptable likelihood of a sample mean is also the *risk of rejecting a statement which is actually true.*

With this basic example in mind, we're now ready to study the formal steps in a hypothesis-testing procedure.

Steps in the Hypothesis-Testing Procedure

Stating the Null and Alternative Hypotheses The *first* step in hypothesis testing is to state specifically the assumed value of the parameter *prior* to sampling. This assumption to be tested is known as the **null hypothesis.** Suppose we want to test the hypothesis that the population mean is equal to 100. The format of this hypothesis would be:

$H_0: \mu = 100$

As we've noted earlier, the hypothesized value of the population mean when used in calculations is identified by the symbol μ_{H0}.

If the sample results do not support the null hypothesis, we must obviously conclude something else. The conclusion that is accepted contingent on the rejection of the null hypothesis is known as the **alternative hypothesis.** There are three possible alternative hypotheses to the null hypothesis stated above:

$H_1: \mu \neq 100$
$H_1: \mu > 100$
$H_1: \mu < 100$

The selection of an alternative hypothesis depends on the nature of the problem at hand, and later sections of this chapter will discuss these alternative hypotheses. As with the null hypothesis, the alternative hypothesis should be stated *prior* to actual sampling.

Selecting the Level of Significance Having stated the null and alternative hypotheses, the *second* step is to establish a criterion for rejection or acceptance of the null hypothesis. If the true mean is actually the assumed value, we know that the probability of the differences between sample means and μ_{H0} diminishes as the size of the difference increases. That is, extremely large differences are unlikely. We must state, *prior* to sampling, the minimum acceptable probability of occurrence for a difference between $\bar{x}$ and μ_{H0}. In our previous example involving the mayor's claim, a difference between $\bar{x}$ and μ_{H0} with a likelihood of only 4.6 percent or less was considered unlikely, and thus Stan felt there was sufficient reason to reject the hypothesis. In such a case, a 4.6 percent chance of occurrence would have been the minimum acceptable probability level.

As noted earlier, if the true mean is indeed the assumed value, the minimum acceptable probability level is also the risk of *erroneously* rejecting the null hypothesis when that hypothesis is *true*. Therefore, this next step in the hypothesis-testing procedure is to state the level of risk of rejecting a true null hypothesis. This risk of erroneous rejection is known as the **level of significance,** which is denoted by the Greek letter α (alpha). Of course, the costlier it is to mistakenly reject a true hypothesis—maybe because you might be sued and lose a lot of money—the smaller α should be. Technically, α is known as the risk of a **Type I error**—that is, the risk that a true hypothesis will be rejected. When a *false* hypothesis is erroneously *accepted* as true, it's known as a **Type II error.** (Some students were unkind enough to suggest to the author a few years ago that registering for his course was known on campus as a Type III error.)

Determining the Test Distribution to Use   Once the level of significance is selected, it's then necessary to determine the appropriate probability distribution to use for the particular test. In this chapter and the next one, we'll be concerned only with the normal (Z) and t distributions; in Chapters 10 and 11, other test distributions are introduced and used. For our purposes in this chapter, then, if $n > 30$, or if the σ is known in a smaller sample and it's also known in that case that the population values are normally distributed, then the Z table is used. If these conditions aren't met, then the t distributions table is needed. (See Figure 7-8, page 273, for a graphic presentation of these rules.)

Defining the Rejection or Critical Regions   Once the appropriate test distribution is known, it's then possible to move to the next step. Suppose in a test using the Z distribution that it's known that the acceptable risk of erroneous rejection of the null hypothesis is $\alpha = .05$. This means that the hypothesis will not be acceptable if the difference expressed in standard units between $\bar{x}$ and μ_{H0} has only a 5 percent or less chance of occurring. Since the hypothesis can be rejected if $\bar{x}$ is too high or if $\bar{x}$ is too low, we may want a .025 chance of erroneous rejection in each tail of the sampling distribution if the true mean is equal to the assumed value. In this case, an α value of .05 represents the *total* risk of error. Figure 8-4 indicates how the normal curve is partitioned. With .025 in each tail, the remaining area *in each half* of the sampling distribution is .4750 (.500 − .025). Appendix 2 indicates that the appropriate Z value corresponding to an area figure of .4750 is 1.96.

What does the partitioning of the normal curve in Figure 8-4 mean? Figure 8-5 shows that if a sample mean differs from the hypothetical mean by 1.96 or more standard errors in either direction, there's sufficient reason to reject the null hypothesis at the .05 level of significance. Thus, a Z value of 1.96 represents the level in standard units at which the difference between $\bar{x}$ and μ_{H0} becomes significant enough to raise doubt that $\mu_{H0} = \mu$. The **significant difference** is the degree of difference between $\bar{x}$ and μ_{H0} that leads to the rejection of the null hypothesis. The **rejection region** (or **critical region**) is thus that part of the sampling distribution—equal in total area to the level of significance—that's specified as being unlikely to contain a sample statistic if the H_0 is true. (There may be more than one rejection region.) The **acceptance region,** of course, is the remainder of the sampling distribution under consideration.

After the level of significance has been stated and the proper test distribution has been selected, the *fourth step* in our procedure is to determine the rejection region (or regions) of the sampling distribution, which is represented in standard units. If the difference between an obtained $\bar{x}$ and the assumed μ_{H0} mean has a value which falls into a rejection region, the null hypothesis is rejected. (If the difference doesn't fall into a critical region, of course, there's no statistical reason to doubt the hypothesis.) But a word of caution concerning conclusions about the validity of a null hypothesis is needed here. Theoretically, a test *never proves* that a hypothesis is true. Rather, a test merely provides statistical evidence for not rejecting a hypothesis. The only standard of truth is the population parameter, and since the true value of that parameter is unknown, the assumption can never be proved. Thus, when it's said that a hypothesis is accepted, it merely means that there's no statistically valid reason to reject the assumed parameter value.

Stating the Decision Rule After we've stated the hypotheses, selected the level of significance, determined the test distribution to use, and defined

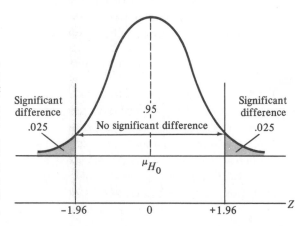

FIGURE 8-4
With a total desired risk of erroneous rejection of a true null hypothesis of .05, the standardized difference between $\bar{x}$ and μ_{H0} becomes significant at +1.96 or −1.96.

Significant difference
.025

Significant difference
.025

.95
No significant difference

μ_{H0}

−1.96   0   +1.96

Z

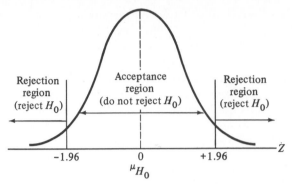

FIGURE 8-5
Construction of
acceptance and
rejection regions
with a significant
level of .05.

the rejection region(s), the *fifth step* is to prepare a **decision rule**—a formal statement that clearly states the appropriate conclusion to be reached about the H_0 based on sample results. The general format of a decision rule (only one of the following is needed) is:

Accept H_0 if the standardized difference between $\bar{x}$ and μ_{H0} falls into the acceptance region.

or

Reject H_0 if the standardized difference between $\bar{x}$ and μ_{H0} falls into a rejection region.

Making the Necessary Computations After all the ground rules have been laid out for the test, the *next step* is the actual data analysis. A sample of items must be collected, and an estimate of the parameter must be calculated. Assuming that we're testing a hypothesis about the value of the population mean, *we first* calculate the value of a sample mean. To convert the difference between $\bar{x}$ and μ_{H0} into a standardized value, it's then necessary to compute the standard error of the mean. The standardized difference between the statistic and the assumed parameter is called the **critical ratio** (**CR**), because this value is critical in determining the acceptance or rejection of the null hypothesis. The critical ratio for a hypothesis test of a population mean might be determined as follows:

$$CR = \frac{\bar{x} - \mu_{H0}}{\sigma_{\bar{x}}}$$ (formula 8-1)

Making a Statistical Decision *If the value of the critical ratio falls into a rejection region, the null hypothesis is rejected.* For example, Figure 8-6 shows the rejection regions of a normal curve with $\alpha = .01$. Referring to the Z table, a *total* risk of 1 percent corresponds to Z values of -2.58 and $+2.58$. Suppose a sample produced a critical ratio of 2.60. Since the

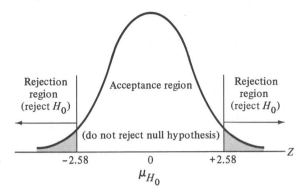

FIGURE 8-6
Acceptance and
rejection regions
with $\alpha = .01$.

CR falls into a rejection region, there's sufficient reason to reject the null hypothesis, and the risk of erroneous rejection is only 1 percent.

At this point, your head may be dizzy with definitions and procedural steps. In order to help you sort out your head, Figure 8-7 summarizes the general procedure for a hypothesis test.

Managerial Decisions and Statistical Decisions: A Caution

Let's conclude this section on a nonstatistical note. Although statistical laws may provide convenient and objective methods for assessing hypotheses, a statistical conclusion by no means represents a final decision in decision making. Consumers of statistical reports use quantitative results merely as one form of input in a complex network of factors that affect an ultimate decision. Undoubtedly decision making is full of uncertainty, and statistical results serve to reduce and control some of the uncertainty, but they do not completely eliminate uncertainty. Problems may be quantified and a result obtained, but the solution is only as good as the input that has gone into structuring the problem. Statistical results, although objectively determined, should not be accepted with blind faith. Other situational factors must be considered. For example, a statistical test may tell a production manager that a machine is not producing as much as the manager had assumed. This result, however, does not tell the manager the appropriate course of action to be taken. The manager may replace the machine, fix the machine, or leave the machine in its present condition. The ultimate decision is made by consideration of the available money for replacement, the repair record of the machine, the availability of new machines, and so forth. Thus, *the statistical conclusion is not necessarily the managerial conclusion;* it is simply one factor which must be considered in the context of the whole problem.

Self-Testing Review 8-1

1 What is a null hypothesis?
2 What is an alternative hypothesis?

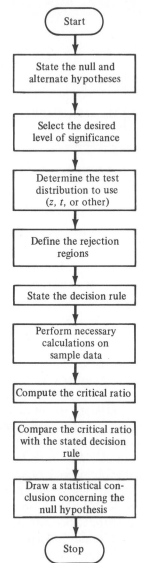

FIGURE 8-7
The general
hypothesis-testing
procedure.

3 What is a significance level?

4 When is a Z distribution applicable in hypothesis testing?

5 What is meant by a significant difference in hypothesis testing?

6 What is a rejection region?

7 How is the difference between $\bar{x}$ and μ_{H0} standardized?

8 "If a critical ratio is not within a rejection region, there is proof that the hypothesis is true." Comment on this statement.

9 "Since a hypothesis test is based on statistical laws, and since the conclusions have been objectively determined, decision making becomes easier." Discuss this statement.

ONE-SAMPLE HYPOTHESIS TESTS OF MEANS

This section considers the one-sample testing procedure for means under various circumstances. You'll recall that there are three possible alternative hypotheses, and the selection of an appropriate alternative conclusion if the null hypothesis is rejected depends on the nature of the problem. This section discusses circumstances under which the possible alternative hypotheses are appropriate.

Two-Tailed Tests When σ Is Known

When the null and alternative hypotheses are in this format:

H_0: μ = assumed value
H_1: $\mu \neq$ assumed value

then, if the null hypothesis cannot be accepted, it's merely concluded that the population mean does not equal the assumed value. It *doesn't matter* if the true value might be more or less than the assumed value. The only conclusion is that the true value and the hypothesized value aren't the same.

The nature of the above hypotheses requires a two-tailed test. A **two-tailed test** is one that rejects the null hypothesis if the sample statistic is significantly *higher or lower* than the assumed value of the population parameter. With a two-tailed test, therefore, there are *two* rejection regions, as shown in Figure 8-8. Since the hypothesis may be rejected with a sample that's too high or too low, the total risk of error in rejecting μ_{H_0} is evenly distributed for each tail. That is, the area in *each* rejection region is $\alpha/2$.

If $n > 30$, or if α is known and the population is normally distributed, the boundaries of the rejection regions are found through the use of the Z table. These boundaries are determined by the Z value corresponding to the probability $.5000 - \alpha/2$. For example, with a two-tailed test and $\alpha = .05$, the area in each tail is .025 and the boundaries of the rejection region are $Z = -1.96$ and $Z = +1.96$ (see Figure 8-8).

The appropriate *decision rule* in this example for a two-tailed test using the Z distribution is:

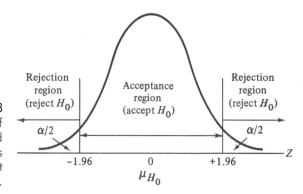

FIGURE 8-8
Illustration of acceptance and rejection regions for a two-tailed test with α = .05.

Accept H_0 if CR falls between[2] ± 1.96.

or

Reject H_0 and accept H_1 if CR < -1.96 or CR $> +1.96$.

The following examples show the use of two-tailed tests when σ is known.

Example 8-1   The owner of the Kate and Edith Cake Company stated that the average number of buns sold daily was 1,500. A worker in the store wants to test the accuracy of the boss's statement. A random sample of 36 days showed that the average daily sales were 1,450 buns. Using a level of significance of $\alpha = .01$ and assuming $\sigma = 120$, what should the worker conclude?

The *hypotheses are:*

H_0: $\mu = 1,500$ buns
H_1: $\mu \neq 1,500$ buns

and this is a two-tailed test because a sample mean which is significantly too high or too low is sufficient to reject the null hypothesis. The interest of this test is only whether or not $\mu = 1,500$; no other conclusion is to be drawn.

The Z distribution is applicable here because n is > 30 and σ is known. And with $\alpha = .01$, the risk of erroneous rejection is .005 in each tail. This means that the chance of correct acceptance of H_0 on one side of the normal curve is .4950. Consulting the Z table, the Z value corresponding to an area of .4950 is approximately 2.58. (The rejection regions for this problem are illustrated in Figure 8-9.)

---

[2] What if the CR is exactly 1.96? In this and other similar situations throughout the book, we will interpret the decision rule to include 1.96 in the acceptance region. A value of 1.97, of course, would be in the rejection region.

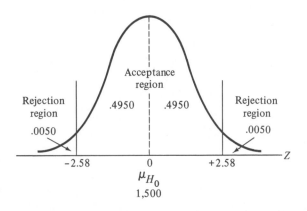

FIGURE 8-9

Under these circumstances, the *decision rule* is stated as follows:

Accept H_0 if CR falls between ±2.58.

or

Reject H_0 and accept H_1 if CR < -2.58 or CR $> +2.58$.

The computation of the critical ratio (CR) *is:*

$$
\begin{aligned}
\text{CR} &= \frac{\bar{x} - \mu_{H0}}{\sigma_{\bar{x}}} \\
&= \frac{1{,}450 - 1{,}500}{\sigma/\sqrt{n}} \\
&= \frac{-50}{120/\sqrt{36}} \\
&= -2.5
\end{aligned}
$$

Conclusion: Since CR $= -2.5$, which is between ±2.58, the null hypothesis should not be rejected at the .01 level of significance.

Example 8-2   A local insurance agent claimed that the average amount paid for personal injury in an automobile accident is \$8,500. A policyholder wants to check the accuracy of the claim and is allowed to sample randomly 36 cases involving personal injury. The sample mean is \$9,315. Assuming that $\sigma = \$2,600$, test the claim of the agent with $\sigma = .05$.
    The *hypotheses are:*

H_0: $\mu = \$8,500$
H_1: $\mu \neq \$8,500$

This is a two-tailed test because the policyholder is only interested in concluding nonequality between the assumed value and the true value if the null hypothesis is rejected.
    Since $n > 30$ and σ is given, the Z table is used. With $\alpha = .05$, there's a risk of .025 in each tail. The corresponding Z value for an area of $.5000 - .0250$ or $.4750$ is 1.96. Therefore, the *decision rule should be:*

Accept H_0 if CR falls between ± 1.96.

or

Reject H_0 and accept H_1 if CR < -1.96 or CR $> +1.96$.

The *critical ratio* is calculated as follows:

$$
\text{CR} = \frac{\bar{x} - \mu_{H0}}{\sigma_{\bar{x}}}
$$

$$= \frac{\$9,315 - \$8,500}{\$2,600/\sqrt{36}} = \frac{\$815}{433.33}$$
$$= 1.88$$

Conclusion: Since CR falls between ± 1.96, there's not enough evidence to reject the null hypothesis at the .05 level of significance. The sample mean of $9,315 lies 1.88 standard errors to the right of the assumed mean, and sampling variation could account for this fact.

Example 8-3   **Statistical process control** (**SPC**) is the name given to the sampling techniques that are used to monitor a controlled production process, and to signal when that process fails to behave in the desired way. Let's assume that your summer job includes checking the output of an automatic machine that produces thousands of bolts each hour. This machine, when properly adjusted, makes bolts with a diameter of 14.00 millimeters (mm). The μ diameter, then, should be 14.00 mm. Bolts that vary too much in either direction from this μ size aren't suitable for their intended use. It's known from past experience that σ is 0.15 mm, and it's also known that the machine makes bolts with diameters that are normally distributed about the population mean. You take a random sample of 6 bolts from the machine's output each hour, and your latest sample has bolts with the following diameters (in millimeters): 14.15, 13.85, 13.95, 14.20, 14.30, and 14.35. At the .01 level, is the machine properly adjusted?

Your *hypotheses are:*

H_0: μ = 14.00 mm
H_1: $\mu \neq$ 14.00 mm

This is a two-tailed test because bolts that vary significantly in either direction can't be used for their primary purpose.

You can use the Z *distribution* in this case, even though n is small, because the σ is known and because you also know that the population is normally distributed. With α = .01, the corresponding Z value is 2.58. Thus, you establish the following *decision rule:*

Accept H_0 if CR falls between $\pm$ 2.58.

or

Reject H_0 and accept H_1 if CR < -2.58 or CR > $+2.58$.

To calculate the *critical ratio*, you must first find the value of the mean of your sample of 6 bolts. You find that this mean is as follows: (14.15 + 13.85 + 13.95 + 14.20 + 14.30 + 14.35)/6 = 14.1333 mm. Now, you compute the critical ratio:

$$\text{CR} = \frac{x - \mu_{H0}}{\sigma_{\bar{x}}} = \frac{14.1333 - 14.00}{.15/\sqrt{6}} = \frac{.1333}{.0612}$$
$$= 2.18$$

Conclusion: Since your CR falls between ±2.58, you accept the null hypothesis that the machine is operating properly at the .01 level of significance.

Only after you've done this work by hand do you learn that the company has a computer statistical package that would have done the job for you. As you can see in Figure 8-10, a package user merely keys in the sample data, uses a ZTEST command, supplies the values of μ_{H0} and σ, and receives a printout giving the same values you've computed. The CR value of 2.18 is shown under the column labeled Z in the printout. And the figure under the P VALUE heading gives the significance level or probability figure for the sample mean if the hypothetical population mean is true. Thus, in this example, the figure of .03 tells us that a sample with a mean of 14.1333 or larger could be expected to happen 3 percent of the time when the μ is 14.00 mm. Since the P VALUE of .03 doesn't fall beyond our α value of .01, we know we should accept the null hypothesis.

A brief follow-up to this example is in order. After you've taken your sample of bolts and made your computations, your employer might expect you to plot your results on a **statistical process control chart** like the one shown in Figure 8-11. As you can see, this chart has horizontal upper and lower control limit (UCL and LCL) lines, and another horizontal line drawn in the center of the chart. There are numerous ways to lay out these charts using sample means, percentages, ranges, and other values, but let's assume that in your company's chart the middle line represents the value of μ_{H0}, or, in this case, 14.00 mm. The UCL line shows the highest acceptable sample mean value, and in this situation that is:

$$\mu_{H0} + Z\sigma_{\bar{x}} = 14.00 + 2.58(.0612) = 14.16 \text{ mm}$$

And the LCL line shows the lowest sample mean value that's acceptable—in this case:

$$14.00 - 2.58(.0612) \text{ or } 13.84 \text{ mm}$$

FIGURE 8-10   A *Minitab* computer printout of the two-tailed Z test for Example 8-3.

```
MTB > SET C2
MTB > END
MTB > PRINT C2

C2
   14.15    13.85    13.95    14.20    14.30    14.35

MTB > ZTEST   MU = 14,   SIGMA = .15, DATA IN C2

TEST OF MU = 14.0000 VS MU N.E. 14.0000
THE ASSUMED SIGMA = 0.150
```

	N	MEAN	STDEV	SE MEAN	Z	P VALUE
C2	6	14.1333	0.1966	0.0612	2.18	0.030

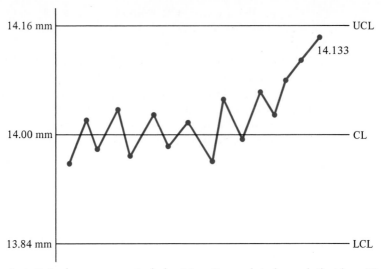

FIGURE 8-11   A statistical process control chart is a time-related graph that has (1) an upper control limit (UCL) that shows the maximum acceptable random variation obtained from sample results, (2) a lower control limit (LCL) that shows the minimum acceptable variation, and (3) a center line (CL) that represents the average quality characteristic being measured. In this example, the UCL and LCL values are set at $+2.58$ ($\sigma_{\bar{x}}$) from the μ_{H_0} value of 14.00 mm that's obtained when the machine is operating properly. In most factory situations, though, the UCL and LCL lines are located at a distance of ± 3 measures of dispersion from the CL.

Each sample mean is plotted on the process control chart as shown in Figure 8-11. A process that's in control should show sample mean points scattered randomly about the center line. If a trend develops, as may be the case in Figure 8-11, then production should be stopped and the problem corrected before products are produced that fail to meet quality standards. For more information on SPC—a vital tool for companies that hope to compete in global markets—see the Closer Look reading at the end of this chapter.

Self-Testing Review 8-2

1   What is a two-tailed test?
2   Determine the appropriate Z values for a two-tailed test for the following:
    a $\alpha = .01$
    b $\alpha = .08$
    c $\alpha = .05$
    d $\alpha = .03$
3   The population mean is assumed to be 500, with $\sigma = 50$. A sample of 36 had $\bar{x} = 475$. Conduct a two-tailed test with $\alpha = .01$.

One-Tailed Tests When σ Is Known

Many times it's unsatisfactory simply to conclude that the true value is *not equal to* the assumed value. If the null hypothesis is not tenable, we often want to know if the rejection occurs because the true value is *probably higher* or *probably lower* than the assumed value. In other words, was the hypothesis rejected because the true value is likely to be greater than or less than the assumed value? In such situations, the null hypothesis is still of the form:

H_0: μ = assumed value

But the *alternative hypothesis* may be one of the following:

H_1: μ > assumed value

or

H_1: μ < assumed value

The nature of either hypothesis indicates a one-tailed test. In a **one-tailed test,** there's only one rejection region and the null hypothesis is rejected only if the value of a sample mean falls into this single rejection region.

Right-Tailed Tests   If the alternative hypothesis is:

H_1: μ > assumed value

the rejection region is in the right tail of the sampling distribution, and this is known as a **right-tailed test.** The null hypothesis will be rejected in this case *only* if the value of a sample mean is *significantly high.* If the value of a sample mean is extremely low compared to the assumed value, the null hypothesis will not be rejected. The *major pitfall with a right-tailed test* is that the true value may be less than the assumed value, but because of the structure of the right-tailed test the null hypothesis will not be rejected. In such a test, the attention is focused on rejecting H_0 *solely* on the basis that the true value might be greater than the assumed value.

If you are confused by the above paragraph, consider this analogy. Suppose you and a friend are guessing a third person's age. Your friend hypothesizes that the third person is 23 years old, while you believe that he is older. Finally, you ask the third person, "Are you more than 23 years old?" He says, "No." As a result, you cannot reject your friend's assertion, but, of course, the assertion may be incorrect because you didn't ask the third person if he was less than 23 years of age. The nonrejection of H_0 in a right-tailed test is similar to this analogy.

Left-Tailed Tests   If the alternative hypothesis is:

H_1: μ < assumed value

we are interested in determining if the true value is *less* than the assumed value, and this is indicative of a **left-tailed test.** In a left-tailed test the single rejection region is on the left side of the sampling distribution. In this case, the null hypothesis will be rejected only if the value of a sample mean is *significantly low.* In a left-tailed test, the null hypothesis will not be rejected if the true value is likely to be more than the assumed value.

The distinctions between a left-tailed test and a right-tailed test are illustrated in Figure 8-12.

Level of Significance Considerations   The level of significance (α) is the *total risk* of erroneously rejecting H_0 when it's actually true. In a two-tailed test, the total risk is evenly divided between each tail. However, in a one-tailed test (since there's only one rejection region) *an area in the single tail is assigned the total risk or α.* If the Z distribution is applicable, the appropriate Z value is thus determined by the one-tailed probability of .5000 − α. For example, if the level of significance were .05 for a left-tailed test, the boundary of the rejection region would be a Z value of − 1.64 (see Figure 8-13).

Decision Rule Statements   If it's assumed that the Z distribution is applicable, the decision rule for a *left-tailed test* is of the following form (remember the elementary rules of first-year algebra here):

Accept H_0 if CR $\geq$ −Z value.

or

Reject H_0 and accept H_1 if CR $<$ −Z value.

The decision rule for a *right-tailed test* is:

Accept H_0 if CR $\leq$ Z value.

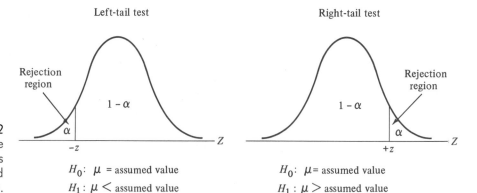

FIGURE 8-12
Illustration of the rejection regions for left-tailed and right-tailed tests.

Left-tail test

Rejection region

$1 - \alpha$

α

$-z$

Z

H_0: μ = assumed value

H_1 : $\mu <$ assumed value

Right-tail test

Rejection region

$1 - \alpha$

α

$+z$

Z

H_0: μ = assumed value

H_1 : $\mu >$ assumed value

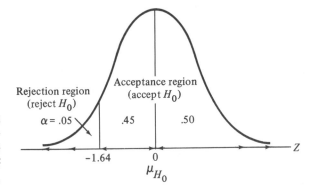

FIGURE 8-13
Rejection region for
a left-tailed test at
the .05 level of
significance.

or

Reject H_0 and accept H_1 if CR > Z value.

The computation of the critical ratio is the same for a one-tailed test as for a two-tailed test.

Let's now look at the following examples, in which a one-tailed test is applicable and in which σ is known.

Example 8-4   Mr. Tyrone Hops, the supervisor of the local brewery, wants to make sure that the average volume of the Super-Duper can is 16 ounces. If the average volume is significantly less than 16 ounces, customers (and various regulatory agencies) would likely complain, prompting undesirable publicity. The physical size of the can does not allow an average volume significantly above 16 ounces. A random sample of 36 cans showed a sample mean of 15.7 ounces. Assuming σ is 0.2 ounces, conduct a hypothesis test with α = .01.

The *hypotheses are:*

H_0: μ = 16 ounces
H_1: μ < 16 ounces

The nature of the problem indicates that if the null hypothesis is rejected, Mr. Hops wants to conclude that the hypothesis was rejected because the sample mean was significantly low.

With $n > 30$ and σ known, the Z distribution is applicable. Thus, with α = .01, and with a left-tailed test, the rejection region begins at a Z tail value beyond −2.33. Therefore, *the decision rule is:*

Accept H_0 if CR ≥ −2.33. (Remember, −2.32 > −2.33.)

or

Reject H_0 and accept H_1 if CR < −2.33. (Also, remember that −2.34 < −2.33.)

The *critical ratio* is computed as follows:

$$CR = \frac{\bar{x} - \mu_{H0}}{\sigma_{\bar{x}}}$$
$$= \frac{15.7 - 16}{.2/\sqrt{36}}$$
$$= -9.00$$

Conclusion: Since CR < −2.33, the brewery must reject H_0 and improve its filling process. It's very unlikely that a sample selected from a sampling distribution that had a true mean of 16 ounces would have a mean located 9.00 standard errors to the left of the true mean!

Example 8-5   Mr. Hal I. Tosis, a mouthwash distributor, has stated that the average cost to process a sales order is $13.25. Ms. Minnie Mize, cost controller, fears that the average cost of processing is more than $13.25. She is interested in taking action if costs are high, but she does not care if the actual average is below the assumed value. A random sample of 100 orders had a sample mean of $13.35. Assuming the σ is $0.50, conduct a test at the .01 level of significance.

The *hypotheses are:*

H_0: μ = $13.25 cost
H_1: μ > $13.25 cost

This is a right-tailed test because only a significantly high sample mean value will reject the null hypothesis. In the problem, there has been no concern expressed for a sample mean value which might be too low. With $n > 30$, the Z distribution is applicable. With α = .01, the appropriate Z value is 2.33. The appropriate *decision rule is:*

Accept H_0 if CR ≤ 2.33.

or

Reject H_0 and accept H_1 if CR > 2.33.

The *critical ratio* is computed as follows:

$$CR = \frac{\bar{x} - \mu_{H0}}{\sigma_{\bar{x}}} = \frac{\$13.35 - \$13.25}{\$0.50/\sqrt{100}} = 2.00$$

Conclusion: Since CR < 2.33, Ms. Mize has no evidence to reject Mr. Tosis' statement at the .01 level of significance.

Self-Testing Review 8-3

1   What is a one-tailed test?
2   How do the alternative hypotheses of a left-tailed and a right-tailed test differ?

3 Determine the appropriate Z values for the following:
   a Two-tailed test, $\alpha = .01$
   b Two-tailed test, $\alpha = .05$
   c Left-tailed test, $\alpha = .05$
   d Right-tailed test, $\alpha = .01$
4 Assume that you have the following null hypothesis: H_0: $\mu = 100$. Conduct a left-tailed test with $\alpha = .05$, $\sigma = 15$, $n = 36$, and $\bar{x} = 88$.
5 Assume that you have the following data: H_0: $\mu = 24$, $\sigma = 3$, $n = 16$, $\bar{x} = 26$, and $\alpha = .01$. Conduct a right-tailed test. (Also assume that the population values are normally distributed.)
6 Mr. X stated that the true mean was 500. Mr. Y disagreed, saying that Mr. X was overstating the value. What are the null and alternative hypotheses?
7 A widget machine must produce widgets with a width of 2.5 inches. A mean width of a batch of widgets that's less than 2.5 inches requires the batch to be destroyed. If the mean width is larger than 2.5 inches, the widgets can still be sold for the same price but for different uses. What are the hypotheses?
8 "If H_0 is not rejected in a right-tailed test, we can say that μ equals the assumed value (μ_{H0})." Discuss this statement.

Two-Tailed Tests When σ Is Unknown

Up to now, we've been working hypothesis tests with σ known. However, as you saw in the previous chapter, knowledge of σ is rare. More often than not, the sample standard deviation (s) must be used in the testing procedure.

With σ unknown, *the following* aspects of the hypothesis-testing procedure *are affected:*

1 The appropriate sampling distribution can no longer be assumed to be approximately normally shaped if n is 30 or less.
2 In the computation of the critical ratio (CR), $\hat{\sigma}_{\bar{x}}$ must be used instead of $\sigma_{\bar{x}}$.

In other words, when σ is unknown, the Z distribution (and Appendix 2) can only be used to determine the rejection regions when the sample size is more than 30. When the sample size is 30 or less in this situation, the sampling distribution takes the shape of a t distribution. The t value used to determine the boundary of a rejection region is based on the level of significance and the degrees of freedom (which are $n - 1$). For example, suppose you are making a *two-tailed test* at the .05 level of significance with a sample size of 16. In the t table in Appendix 4, the t value with 15 degrees of freedom is 2.131. That is, $t_{\alpha/2}$ or $t_{.025} = 2.131$. The t table is set up to show the rejection region in one tail.

As we shall see in the following examples, the testing procedure is the same with an unknown σ as with a given σ, with the exceptions (1) that the appropriate test distribution (Z or t) must be determined and (2) that the correct method of calculating the critical ratio must be used.

STATISTICS: A NEW CO-CARCINOGEN

On December 21, 1979, the National Cancer Institute and the Food and Drug Administration released to the media some of the findings of a $1.5 million epidemiological study on the relationship between the use of artificial sweeteners (AS) and the incidence of human bladder cancer. This Congressionally instigated study is the latest in a series of surveys, most of which have failed to incriminate AS as carcinogens. Animal and metabolic studies do not support the notion that either of the popular AS is a traditional carcinogen, although some scientists believe that at high doses, and under special conditions (such as exposure *in utero*), saccharin may be either a weak carcinogen or an abettor of cancer-producing factors.

As expected, media coverage of the release was broad but not deep. Reliance on the brief FDA-NCI press handout was bound to generate distorted stories, because it highlighted subsidiary calculations of questionable validity and de-emphasized the negative overall findings. A conscientious reporter would have had to obtain and critically analyze the full study to write a fair story, but what TV or newspaper reporter has the time (or background) for that?

What, in fact, does this extensive study of over 3,000 bladder cancer patients and almost 6,000 controls show? One important conclusion reached by the authors (and unquestionably supported by the data) is that *for the total population studied the use of AS was not associated with any increased risk*. This is the case for males as well as females, and for exposure to AS by any route (diet drink, "table top" use or diet food). Reassuring? One would think so.

But the study's statisticians decided to go a bit further, and it is these statistical side trips that prompted the news release to cite "some evidence that sweeteners may be hazardous"— specifically for heavy AS users, heavy smokers and women who would ordinarily be at low risk for bladder cancer.

A reading of the actual report leaves one underwhelmed by this "evidence." The press release's "60 percent increased risk of bladder cancer" in heavy AS users is based on seven cases. (Yes, Virginia, seven.) Furthermore, people who drank less than two cans of diet soda and used three to six servings of table top sweeteners per day had a 24 percent *decreased* risk of bladder cancer! Curiouser and curiouser, as Alice said. (The full report acknowledges that "none of the individual risk ratios taken by itself would be statistically significant.")

The handling of tobacco usage by the report is fascinating. *The clearest positive relationship in the whole study is that between lifetime cigarette consumption and bladder cancer.* Every additional increment of "pack-years" from less than 1 to 60 + shows an added increase of risk, the ratios running from 1.27 in the least exposed smokers to 3.31 for the most. In a logical world, one would expect FDA-NCI outrage over *cigarette smoking*, an undisputed source of carcinogens, rather than a will o' the wisp like saccharin, but who believes that the AS controversy is governed by the rules of logic?

The two tables accusing cigarettes and AS of a sinister joint villainy are unconvincing. For men there is a case to be made only at the two packs or more per day level, and for women the case depends on one's willingness to buy the use of one-tailed tests of significance. Such statistical usage has the effect of doubling the statistical impact of any given P value (i.e., a one-tailed "significant" 0.05 is an "insignificant" two-tailed 0.10). The one-tailed approach is used throughout the report as "appropriate for regulatory purposes since increased risks are the principal concern." (One can anticipate a new generation of texts on "Political Statistics: How to Make a Poor Scientific Case Look a Bit Better.")

In fact, a knowledgeable scientist could readily hypothesize that AS usage might be linked with dietary habits that could in turn lead to *decreased* cancer risk, since nutrition seems related to cancer. Indeed, the Canadian data that triggered the current U.S. concerns about AS suggested that *female* users of AS were actually *protected* from bladder cancer.

Furthermore, logic (and arithmetic) require one to consider that if the total population of AS users is *not* at increased risk of bladder cancer, but a subpopulation truly *is*, then *another* subpopulation might be at *decreased* risk.

FDA Commissioner Goyan, in the press release, appropriately expresses concern about cancers that might surface decades from now, but the FDA-NCI study in fact shows *no* relation between duration of AS use and bladder cancer, and a very tenuous relationship between amount of AS intake and the risk of cancer. There is, therefore, an absence of the sort of dose-response relationship that strengthens suspicions of causality.

The press release does refute the Canadian estimate of a 60 percent increased risk of bladder cancer in AS-consuming males, and states that the increased risk probably would not exceed 18 percent. But what is *not* said is that the data also permit one to assert that AS usage could easily be associated with a 16 percent *reduction* in risk. These are the confidence limits of the *best* estimate, which is that there is neither more nor less risk.

The report justifies its use of many, many statistical comparisons by asserting that all of the spotted associations in subgroups were specified as possible prior to the study. It also adds, to the credit of the authors, that "the positive associations seen herein [may] represent merely chance variations in subgroups of a study which, overall, fails to demonstrate an association between AS use and bladder cancer." I suspect that many neutral observers will opt

for the latter possibility. For instance, of the 42 P values given for AS associated risks (and there is no way to tell how many others were computed but not described) only 11 are statistically significant with one-tailed tests, and five of these fall by the wayside if one prefers two-tailed tests. Place your bets, friends.

In fairness to the report, it must be said that these (and some other) sources of error and bias are at least mentioned. But the press release and the media coverage ignore these caveats, and thus represent journalism that is at best sloppy and at worst dishonest.

The statisticians involved have thus become allies of those waging a holy war against AS. Saccharin and cyclamates seem to be such weak carcinogens that it requires a horde of statistician-epidemiologists, plus $1.5 million of public monies, to keep the carcinogenic pot boiling. Meanwhile, the Canadians have recanted, muttered "mea culpa" (not very loudly), and exonerated cyclamates because "there is not a shred of evidence" of the carcinogenicity which once seemed to demand regulatory action.

Miguel Cervantes, let thy shade rest in peace. Don Quixote is alive and well.

—From Louis Lasagna, "Statistics: A New Co-Carcinogen." This article is reprinted by permission of *THE SCIENCES* and is from the May/June 1980 issue. Individual subscriptions are $13.50 per year. Write to The Sciences, 2 East 63rd Street, New York, NY 10020 or call 1-800-THE-NYAS.

Example 8-6   Mr. Drinkwater of the D-T Liquor Shoppe thinks that his business sells an average 17 pints of Border Ale daily. His partner, Mr. Taylor, thinks this estimate is wrong. A random sample of 36 days showed a mean of 15 pints and a sample standard deviation (s) of 4 pints. Test the accuracy of Mr. Drinkwater's statement at the .10 level of significance.

The *hypotheses are:*

H_0: μ = 17 pints
H_1: $\mu \neq$ 17 pints

This is a two-tailed test because we only wish to determine the validity of

Mr. Drinkwater's statement. An extremely high value of $\bar{x}$ or an extremely low value of $\bar{x}$ will reject the null hypothesis. With $n = 36$, the sampling distribution approximates the normal distribution, and thus the Z *distribution applies*.

Since this is a two-tailed test and $\alpha = .10$, the risk of error in each tail is .05. The Z value corresponding to $.5000 - .05 = .45$ is approximately 1.64. The appropriate *decision rule is*:

Accept H_0 if CR falls between ± 1.64.

or

Reject H_0 and accept H_1 if CR < -1.64 or CR $> +1.64$.

With $s = 4$ and $n = 36$, $\hat{\sigma}_{\bar{x}}$ is estimated in the following manner:

$$\hat{\sigma}_{\bar{x}} = \frac{s}{\sqrt{n}} = \frac{4}{\sqrt{36}} = .667$$

Therefore, the *critical ratio* is computed as follows:

$$CR = \frac{\bar{x} - \mu_{H0}}{\hat{\sigma}_{\bar{x}}} \qquad \text{(formula 8-2)}$$

$$= \frac{15 - 17}{.667}$$
$$= -3.00$$

Conclusion: Since CR < -1.64, it's necessary to reject Mr. Drinkwater's claim at the .10 level of significance.

Example 8-7 A journal article claims that the average height of female adults in Biglandia is 64 inches. To test this claim, a sociologist took a random sample of 16 Biglandian women and found that the mean was 62.9 inches and the standard deviation was 2.5 inches. Can it be concluded that the article is correct at the .05 level of significance?

The *hypotheses are*:

H_0: $\mu = 64$ inches
H_1: $\mu \neq 64$ inches

The nature of the problem indicates a *two-tailed test*, with a risk of erroneous rejection of .025 in each tail. The *t distribution* is applicable in this case because the sample size is only 16. With 15 degrees of freedom and a .025 risk in each tail, $t_{025} = 2.131$.

The *decision rule for this problem is*:

Accept H_0 if CR falls between ±2.131.

or

Reject H_0 and accept H_1 if CR < −2.131 or CR > +2.131.

With $s = 2.5$ and $n = 16$,

$$\hat{\sigma}_{\bar{x}} = \frac{s}{\sqrt{n}} = \frac{2.5}{\sqrt{16}} = .625$$

And the *critical ratio* is computed as follows:

$$CR = \frac{\bar{x} - \mu_{H0}}{\hat{\sigma}_{\bar{x}}} = \frac{62.9 - 64.0}{.625} = -1.76$$

Conclusion: Since CR falls between ±2.131, there is no reason to reject the article's statement at the .05 level of significance.

Example 8-8   Let's go back to the problem in Example 8-3 and change some assumptions. Let's suppose now that the bolts made by the automatic machine should still have a μ diameter of 14.00 mm, and that bolts that vary significantly in either direction from this standard aren't suitable for their intended use. But let's assume now that the σ isn't known, and we must instead calculate a sample standard deviation. As before, the latest sample has 6 bolts with the following diameters (in millimeters): 14.15, 13.85, 13.95, 14.20, 14.30, and 14.35. Our previous decision was that the bolt machine was operating properly. Is that decision still the correct one at the .01 level?
   The *hypotheses* are still:

H_0: $\mu = 14.00$ mm
H_1: $\mu \neq 14.00$ mm

This is still a two-tailed test, but we must now use the *t distribution* because the sample size is only 6 and we don't know σ. With five degrees of freedom and a .005 risk in each tail, $t_{005} = 4.032$. The *decision rule* for this problem is:

Accept H_0 if CR falls between ±4.032.

or

Reject H_0 and accept H_1 if CR < −4.032 or CR > +4.032.

The value of s must be computed as follows (remember from Example 8-3 that $\bar{x} = 14.1333$):

x	$(x - \bar{x})$	$(x - \bar{x})^2$
14.15	.0167	.00028
13.85	−.2833	.08026
13.95	−.1833	.03360
14.20	.0667	.00445
14.30	.1667	.02779
14.35	.2167	.04696
		.19334

$$s = \sqrt{\frac{\Sigma(x-\bar{x})^2}{n-1}} = \sqrt{\frac{.19334}{5}} = .1966$$

And now that we have s, we can find:

$$\hat{\sigma}_{\bar{x}} = \frac{s}{\sqrt{n}} = \frac{.1966}{\sqrt{6}} = \frac{.1966}{2.449} = .0803$$

The *critical ratio* is then:

$$CR = \frac{\bar{x} - \mu_{HO}}{\hat{\sigma}_{\bar{x}}} = \frac{14.1333 - 14.00}{.0803} = 1.66$$

Conclusion: Since CR falls between ±4.032, we'll conclude, as in Example 8-3, that the machine is in adjustment. The sample results were such that we came close to rejecting H_0 in Example 8-3, but not this time because the t distribution is flatter than the z distribution and has tails that stick out farther from its mean.

The completion of this test depended on our carrying out the chore of computing the sample standard deviation. But, as in Example 8-3, our tedious computations can be turned over to a statistical software package and the same results can be achieved in seconds (see Figure 8-14). After the sample data are keyed into program storage (C2 here), the TTEST command used with this software is given and the μ_{HO} is supplied (MU = 14). The program responds with a printout giving all the values we've computed.

Self-Testing Review 8-4

1 How do the hypothesis-testing procedures with σ known and σ unknown differ?
2 Assume that you have the following data: H_0: $\mu = 612$, $\bar{x} = 608$, $s = 5$, $n = 13$, and $\alpha = .05$. Conduct a two-tailed test.
3 Assume that you have the following data: H_0: $\mu = 243$, $\bar{x} = 269$, $s = 15$, $n = 36$, and $\alpha = .01$. Conduct a two-tailed test.

```
MTB > PRINT C2

C2
   14.15    13.85    13.95    14.20    14.30    14.35

MTB > TTEST  MU = 14, DATA IN C2

TEST OF MU = 14.0000 VS MU N.E. 14.0000

                N      MEAN     STDEV    SE MEAN         T    P VALUE
C2              6   14.1333    0.1966    0.0803      1.66       0.16
```

FIGURE 8-14   A *Minitab* computer printout of the two-tailed *t* test for Example 8-8.

One-Tailed Tests When σ Is Unknown

The two preceding examples were two-tailed tests. The following examples are one-tailed tests made when σ is unknown. You'll notice that the testing procedure is essentially unchanged.

Example 8-9   The manager of the Granite Rock Company is under the impression that the average truckload delivered is 4,500 pounds. A stockholder, Mr. Chip Stone, contends that this is an overinflated figure to lure new investors. Mr. Stone randomly sampled the records of 25 trucks and found the mean load to be 4,460 pounds with a standard deviation of 250 pounds. Can Mr. Stone reject the manager's claim using a significance level of .05?
   The *hypotheses are:*

H_0: μ = 4,500 pounds
H_1: μ < 4,500 pounds

This is a *left-tailed test* because only a sample mean that is significantly low will cause rejection of the null hypothesis. And since this is a one-tailed test, the area in the single rejection region is equal to the significance level of .05.
   With n = 25, the t distribution applies, there are 24 degrees of freedom, and t_{05} = 1.711. The *decision rule then is:*

Accept H_0 if CR $\geq$ −1.711.

or

Reject H_0 and accept H_1 if CR < −1.711.

With s = 250 and n = 25,

$$\hat{\sigma}_{\bar{x}} = \frac{s}{\sqrt{n}} = \frac{250}{\sqrt{25}} = 50.00$$

The *critical ratio* is computed as follows:

$$CR = \frac{\bar{x} - \mu_{H0}}{\hat{\sigma}_{\bar{x}}} = \frac{4{,}460 - 4{,}500}{50.00} = -.80$$

Conclusion: Since $CR > -1.711$, there's no significant reason for Mr. Stone to doubt the manager's claim. It is quite possible that a sample could be selected with a mean located only -0.80 standard error from a true mean.

Example 8-10   Mr. Hiram N. Fyrem of the H & F Employment Agency believes that the agency receives an average of 16 complaints per week from companies which hire the agency's clients. Mr. B. S. DeGree, an interviewer, is concerned that the true mean is higher than Mr. Fyrem believes. If Fyrem's hypothesis is an understatement, something must be done about the agency's screening procedures. A sample of 10 weeks yielded an average of 18 complaints with a standard deviation of 3 complaints. Conduct a test at the .01 level.
    The *hypotheses are:*

H_0: $\mu = 16$ complaints
H_1: $\mu > 16$ complaints

If the statistical evidence cannot support the null hypothesis, Mr. DeGree would want to conclude that the parameter value is more than assumed. Thus, this is a *right-tailed* test.
    With $n = 10$ and $\alpha = .01$, the *t value* with 9 degrees of freedom is $t_{01} = 2.821$. The *decision rule* therefore is:

Accept H_0 if $CR \leq 2.821$.

or

Reject H_0 and accept H_1 if $CR > 2.821$.

With $s = 3$ and $n = 10$,

$$\hat{\sigma}_{\bar{x}} = \frac{s}{\sqrt{n}} = \frac{3}{\sqrt{10}} = .95$$

and the *critical ratio* is computed as follows:

$$CR = \frac{\bar{x} - \mu_{H0}}{\hat{\sigma}_{\bar{x}}} = \frac{18 - 16}{.95} = 2.11$$

Conclusion: Since $CR < 2.821$, there's no sufficient reason at the .01 level of significance to doubt Mr. Fyrem's hypothesis.

Example 8-11   We've seen how the *Minitab* statistical program processes two-tailed Z and *t* tests in earlier examples. But one-tailed tests are also handled with ease. Let's suppose that Mr. DeGree isn't satisfied with the results obtained in the preceding example and decides to sample another 10-week period. The sample gave the following complaint data: 20, 14, 12, 24, 17, 22, 13, 16, 15, and 19.

The *hypotheses* are still:

H_0: μ = 16 complaints
H_1: μ > 16 complaints

And at the .01 level, a *t* value of 2.821 is still needed. The *decision rule* remains:

Accept H_0 if CR $\leq$ 2.821.

or

Reject H_0 and accept H_1 if CR > 2.821.

The sample mean and sample standard deviation are found as follows:

x	$(x - \bar{x})$	$(x - \bar{x})^2$
20	2.8	7.84
14	−3.2	10.24
12	−5.2	27.04
24	6.8	46.24
17	−.2	.04
22	4.8	23.04
13	−4.2	17.64
16	−1.2	1.44
15	−2.2	4.84
19	1.8	3.24
172		141.60

$$\bar{x} = \frac{172}{10} = 17.2$$

$$s = \sqrt{\frac{\Sigma(x-\bar{x})^2}{n-1}} = \sqrt{\frac{141.60}{9}} = 3.967$$

With s = 3.967, we can now compute $\hat{\sigma}_{\bar{x}}$:

$$\hat{\sigma}_{\bar{x}} = \frac{3.967}{\sqrt{10}} = \frac{3.967}{3.162} = 1.254$$

And the *critical ratio* is:

$$CR = \frac{\bar{x} - \mu_{H0}}{\hat{\sigma}_{\bar{x}}} = \frac{17.2 - 16}{1.254} = .96$$

Conclusion: Since CR is < 2.821, there's still no reason to doubt Mr. Fyrem's hypothesis.

All of these results are duplicated in the *Minitab* computer output shown in Figure 8-15. The sample data are entered, a TTEST command is used, the μ_{H_0} of 16 complaints is recorded, and an ALTERNATIVE subcommand is specified. The ALTERNATIVE = +1 subcommand tells the program that it's to perform a right-tail test. (If we had wanted a left-tail test, the proper subcommand would have been ALTERNATIVE = −1.)

So far in this chapter, we've discussed one-sample hypothesis tests of means under various conditions. The general testing procedure is essentially the same under all conditions. The testing differences that exist are reflected in the differences that appear in the decision rules. These differences are summarized in Figure 8-16.

Self-Testing Review 8-5

1 If the null hypothesis is H_0: μ = assumed value, determine the appropriate t value for the following:

   a $n = 23$, $\alpha = .01$, H_1: $\mu <$ assumed value
   b $n = 16$, $\alpha = .05$, H_1: $\mu \neq$ assumed value
   c $n = 26$, $\alpha = .01$, H_1: $\mu >$ assumed value
   d $n = 27$, $\alpha = .05$, H_1: $\mu >$ assumed value

2 Assume that you have the following data: H_0: $\mu = 400$, $\bar{x} = 389$, $\hat{\sigma}_{\bar{x}} = 8$, $n = 23$, and $\alpha = .01$. Conduct a left-tailed test.

3 Assume that you have the following data: H_0: $\mu = \$6,425$, $\bar{x} = \$6,535$, $\hat{\sigma}_{\bar{x}} = \$55$, $n = 27$, and $\alpha = .05$. Conduct a right-tailed test.

ONE-SAMPLE HYPOTHESIS TESTS OF PERCENTAGES

The hypothesis-testing procedure for percentages in the large-sample case is essentially the same procedure employed for testing means with a large

FIGURE 8-15   A *Minitab* computer printout of the one-tailed t test for Example 8-11.

```
MTB > SET C1
MTB > END
MTB > PRINT C1

C1
    20     14     12     24     17     22     13     16     15     19

MTB > TTEST  MU = 16,  DATA IN C1;
SUBC> ALTERNATIVE = +1.

TEST OF MU = 16.000 VS MU G.T. 16.000

                N      MEAN     STDEV    SE MEAN        T    P VALUE
C1             10    17.200     3.967      1.254     0.96       0.18
```

FIGURE 8-16   Decision Rules under Various Conditions of Hypothesis Testing of Means

	$n > 30$, OR σ KNOWN AND POPULATION VALUES KNOWN TO BE NORMALLY DISTRIBUTED	$n \leq 30$, AND σ UNKNOWN
Two-tailed test	Accept H_0 if CR is between $\pm Z$ value Reject H_0 and accept H_1 if CR $> +Z$ value or CR $< -Z$ value	Accept H_0 if CR is between $\pm t_{\alpha/2}$ value Reject H_0 and accept H_1 if CR $> +t_{\alpha/2}$ value or CR $< -t_{\alpha/2}$ value
Left-tailed test	Accept H_0 if CR $\geq -Z$ value Reject H_0 and accept H_1 if CR $< -Z$ value	Accept H_0 if CR $\geq -t_\alpha$ value Reject H_0 and accept H_1 if CR $< -t_\alpha$ value
Right-tailed test	Accept H_0 if CR $\leq +Z$ value Reject H_0 and accept H_1 if CR $> +Z$ value	Accept H_0 if CR $\leq +t_\alpha$ value Reject H_0 and accept H_1 if CR $> +t_\alpha$ value

sample size. (This section will discuss only the large-sample case for percentage testing because the complexity of the small-sample case is beyond the scope of this book.)

The only significant change in conducting a test of percentages rather than a test of means is in the computation of the critical ratio. The critical ratio for percentages is computed as follows:

$$CR = \frac{p - \pi_{H0}}{\sigma_p} \qquad \text{(formula 8-3)}$$

where π_{H0} = hypothesized value of the population percentage

You may recall from Chapter 6 that the correct value of the standard error of percentage is found with this formula:

$$\sigma_p = \sqrt{\frac{\pi(100 - \pi)}{n}}$$

Of course, we don't know π now; if we did we wouldn't be making a test! But we'll use the hypothesized value of π in computing the standard error. That way, *if* our hypothesis is true, we'll calculate the correct value of σ_p. Thus, σ_p is found as follows:

$$\sigma_p = \sqrt{\frac{\pi_{H0}(100 - \pi_{H0})}{n}} \qquad \text{(formula 8-4)}$$

To demonstrate the testing procedure for percentages, let's first examine a two-tailed test situation and then look at one-tailed test examples.

Two-Tailed Testing

Example 8-12   A local newspaper, *The Weekly Daily,* has stated that only 25 percent of the college students in its circulation area read newspapers daily. A random sample of 200 of these college students showed that 45 of them were daily readers of newspapers. Test the accuracy of the newspaper's statement, and use a significance level of .05.

The *hypotheses are:*

H_0: π = 25 percent readership
H_1: $\pi \neq$ 25 percent readership

This is a two-tailed test because H_0 may be rejected if the sample percentage is too high or too low. With a large sample size, the Z *distribution* is applicable for determining the rejection regions. With α = .05, there is an area of .025 in each rejection region, and the appropriate Z value is 1.96. Thus, the *decision rule* is:

Accept H_0 if CR falls between ± 1.96.

or

Reject H_0 and accept H_1 if CR < -1.96 or CR > $+1.96$.

With an assumed value of 25 percent, the standard error of percentage is computed as follows:

$$\sigma_p = \sqrt{\frac{\pi_{H0}(100 - \pi_{H0})}{n}} = \sqrt{\frac{(25)(75)}{200}} = 3.1 \text{ percent}$$

And the *critical ratio* is computed as follows:

$$CR = \frac{p - \pi_{H0}}{\sigma_p} = \frac{22.5 \text{ percent} - 25 \text{ percent}}{3.1 \text{ percent}} = -.806$$

Conclusion: Since the CR falls between ± 1.96, there's not enough evidence to reject the newspaper's assertion.

One-Tailed Testing

Example 8-13   The manager of the Big-Wig Executive Hair Stylists, Hugo Bald, has advertised that 90 percent of the firm's customers are satisfied with the company's services. Polly Tician, a consumer activist, feels that this is an exaggerated statement that might require legal action. In a random sample of 150 of the company's clients, 132 said they were satisfied. What should be concluded if a test were conducted at the .05 level of significance?

The *hypotheses are:*

H_0: π = 90 percent satisfied
H_1: π < 90 percent satisfied

This is a *left-tailed test* because Bald's claim will be discredited only if the value of the sample percentage is significantly low. With the .05 level of significance, the rejection region is bounded by $Z = -1.64$. Thus, the *decision rule* is:

Accept H_0 if CR ≥ -1.64.

or

Reject H_0 and accept H_1 if CR < -1.64.

With $\pi_{H_0} = 90$ percent and $n = 150$, the standard error of percentage is computed as follows:

$$\sigma_p = \sqrt{\frac{\pi_{H0}(100 - \pi_{H0})}{n}} = \sqrt{\frac{(90)(10)}{150}} = 2.4 \text{ percent}$$

The *critical ratio* is computed as follows:

$$CR = \frac{p - \pi_{H0}}{\sigma_p} = \frac{88 \text{ percent} - 90 \text{ percent}}{2.4 \text{ percent}} = -.833$$

Conclusion: Since the CR is greater than -1.64, there's no sufficient reason to doubt Hugo's claim. Polly should look for another cause.

Example 8-14   The Howard Hurtz Patent Medicine Company assumes that the bottling machine is operating properly if only 5 percent of the processed bottles are not full. A random sample of 100 bottles had 7 bottles which were not full. Using a significance level of .01, conduct a statistical test to determine if the machine is operating properly.
   The *hypotheses are:*

H_0: $\pi = 5$ percent not full
H_1: $\pi > 5$ percent not full

This is a right-tailed test because the company is concerned that the true percentage might be more than anticipated. With a .01 level, the Z *value* is 2.33, and therefore the *decision rule* is:

Accept H_0 if CR ≤ 2.33.

or

Reject H_0 and accept H_1 if CR > 2.33.

With $\pi_{H0} = 5$ percent and $n = 100$, the standard error of percentage is computed as follows:

$$\sigma_p = \sqrt{\frac{\pi_{H0}(100 - \pi_{H0})}{n}} = \sqrt{\frac{(5)(95)}{100}} = 2.18 \text{ percent}$$

The *critical ratio* is computed as follows:

$$CR = \frac{p - \pi_{H0}}{\sigma_p} = \frac{7 \text{ percent} - 5 \text{ percent}}{2.18 \text{ percent}} = .917$$

Conclusion: Since the CR is less than 2.33, the machine appears to be operating properly.

Self-Testing Review 8-6

1 What is assumed in this section concerning the sample size in hypothesis tests of percentages?
2 What values are used in the calculation of σ_p?
3 Other than the different standard errors, is there any significant difference between the test procedure for means and the test procedure for percentages?

LOOKING BACK

1 Rather than simply estimate a parameter value, statisticians often take a sample to confirm or reject some hypothesis made about the value of a parameter. The assumption about the value of a parameter to be tested is called the null hypothesis. The conclusion that's accepted if sample results fail to support the null hypothesis is called the alternative hypothesis. Stating these hypotheses prior to sampling is the first step in the hypothesis-testing procedure.

2 Having stated the null and alternative hypotheses, the second step in the testing procedure is to establish a criterion for acceptance or rejection of the null hypothesis. This criterion is called the level of significance—the risk of rejecting a true null hypothesis. If a true hypothesis is rejected, a Type I error is made, and if a false hypothesis is accepted, a Type II error is made.

3 The third step in the testing procedure is to determine the correct probability distribution to use for the particular test. In this chapter, if $n > 30$, or if the σ is known in a smaller sample and it's also known in that case that the population values are normally distributed, then the Z table is used. If these conditions aren't met, then the t distribution table is needed. Other distributions are introduced in later chapters. Once the appropriate test distribution is known, it's then possible to move to the fourth step in the testing procedure. This step is to establish the rejection region(s)—that part of the sampling distribution (equal in total area to the level of significance) that's specified as being unlikely to contain a sample statistic if the H_0 is true. The acceptance region, of course, is the remainder of the sampling distribution under consideration. If the difference between an obtained sample statistic and the assumed parameter has a value that falls into a rejection region, the null hypothesis is rejected.

4 The fifth step in the testing procedure is to prepare a decision rule—a formal statement that clearly states the appropriate conclusion to be reached about the H_0 on

the basis of sample results. After all these steps have been completed, the sixth step is the actual data analysis. Sample values and the appropriate standard error must be calculated. The critical ratio—the standardized difference between the statistic and the assumed parameter—is then computed. Finally, the last step is to make the statistical decision. If the critical ratio falls into a rejection region, the null hypothesis is rejected; otherwise, it's accepted.

5   One-sample hypothesis tests of means or percentages are two-tailed or one-tailed. A two-tailed test is one that rejects the null hypothesis if the sample statistic is significantly higher or lower than the assumed value of the population parameter. With a two-tailed test, there are two rejection regions. The only concern in a right-tailed test is that the true value may be greater than the assumed parameter value, while the interest in a left-tailed test is to see if the true value is less than the assumed value. Both one- and two-tailed examples are found in this chapter in one-sample tests of hypotheses about population means and percentages. Several examples are computed by hand to show the procedures used, and are then repeated using a statistical software package. Statistical process control concepts are demonstrated in an example of a two-tailed test of a population mean, and further information on statistical quality control is given in the Closer Look reading at the end of the chapter.

KEY TERMS AND CONCEPTS

Null hypothesis   *309*
Alternative hypothesis   *309*
Level of significance   *310*
Type I error   *310*
Type II error   *310*
Significant difference   *311*
Rejection region (critical region)   *311*
Acceptance region   *311*
Decision rule   *312*
Critical ratio (CR)   *312*

$$CR = \frac{\bar{x} - \mu_{H0}}{\sigma_{\bar{x}}} \quad \text{(formula 8-1)} \quad 312$$

Two-tailed test   *315*
Statistical process control (SPC)   *318*
Statistical process control chart   *319*
One-tailed test   *321*
Right-tailed test   *321*
Left-tailed test   *322*

$$CR = \frac{\bar{x} - \mu_{H0}}{\hat{\sigma}_{\bar{x}}} \quad \text{(formula 8-2)} \quad 328$$

$$CR = \frac{p - \pi_{H0}}{\sigma_p} \quad \text{(formula 8-3)} \quad 335$$

$$\sigma_p = \sqrt{\frac{\pi_{H0}(100 - \pi_{H0})}{n}} \quad \text{(formula 8-4)} \quad 335$$

PROBLEMS

1   A Chamber of Commerce has stated that the average price per acre of land in Suburbanville is $3,125. A real estate salesperson, Mr. Sel N. Aker, would like to determine the veracity of the statement. A random sample of 36 acres on sale was priced at $\bar{x} = \$3,250$. Assume $\sigma = \$310$.
   a   What are the null and alternative hypotheses?
   b   Is this a two-tailed or a one-tailed test?
   c   Is it possible to conduct a hypothesis test with the data given? Why or why not?

2 Refer to problem 1. Conduct a hypothesis test with $\alpha = .05$.

3 Determine the Z value for the following:
  a Left-tailed test, $\alpha = .05$.
  b Two-tailed test, $\alpha = .10$.
  c Right-tailed test, $\alpha = .01$.
  d H_1: $\mu <$ assumed value, $\alpha = .01$
  e H_1: $\pi >$ assumed value, $\alpha = .05$

4 Establish the decision rules for hypothesis tests with the following data:
  a $n = 36$, $\alpha = .05$, H_1: $\pi <$ assumed value
  b $n = 14$, $\alpha = .01$, H_1: $\mu \neq$ assumed value
  c $n = 23$, $\alpha = .05$, H_1: $\mu >$ assumed value
  d $n = 46$, $\alpha = .05$, H_1: $\pi >$ assumed value

5 The population mean has been assumed to be 600. Given the following data, conduct a left-tailed test: $\bar{x} = 592$, $n = 36$, $s = 10$, and $\alpha = .05$.

6 It has been stated that $\mu = 69$. Conduct a right-tailed test with the following data: $\bar{x} = 75$, $n = 19$, $s = 6$, and $\alpha = .10$.

7 The population percentage has been assumed to be 82 percent. Conduct a right-tailed test with the following data: $p = 85$ percent, $n = 81$, and $\alpha = .05$.

8 Know-It-All Consultants, Ltd., has stated in its promotional brochure that the average cost for its advice is \$5,600 per client. Assume a random sample of 36 clients had $\bar{x} = \$5,750$ with $s = \$175$. Conduct a test of the consultants' statement with $\alpha = .05$.

9 Professor O.D. Statt declared that only 33 percent of a college's students have a job while attending school. A student, Fuller Doutt, thinks the professor has underestimated the zeal of his peers. A random sample of 49 students showed that 17 of them worked after school. Using $\alpha = .01$, determine who is correct.

10 A physical education instructor, Mr. Wate Lifter, claims that his method of exercising permits an individual to do an average of 60 consecutive sit-ups after 1 week of training. Ms. Mussel, a fellow instructor, does not think Lifter is correct. A random sample of 25 individuals who have undergone the course of training had $\bar{x} = 69$ with $s = 7$. Conduct a test with $\alpha = .01$.

11 The Big Cluck Chicken Farm manager states that the average weight of his chickens is 3.6 pounds. A wholesaler believes the stated value is too high. A sample of 24 chickens had $\bar{x} = 3.3$ pounds with $s = .25$. Conduct a test with $\alpha = .05$.

12 On the average, a well-functioning machine should produce items which are 90 percent acceptable. If the percentage is less than 90 percent, the machine must be repaired. A sample of 100 items has $p = 87$ percent. Determine if this machine should be repaired, using a significance level of .05.

13 Wild Guess Surveys, Inc., reported in a preelection poll that 68 percent of the voters favor Mr. M. T. Hedd for public office. A sample of 36 voters showed that 26 of them would vote for Mr. Hedd. What can be concluded about the accuracy of the report, using a significance level of .05?

14 A hospital receives a large shipment of vials of serum. These vials are supposed to contain 50 milligrams (mg) of serum each, and it's undesirable for the contents to be either above or below that value. A random sample of 64 vials shows a mean content of 49.25 mg. It's known that the population standard de-

viation will be about 2 mg. At the .01 level, should the hospital accept the shipment?

15 A television station manager claims that 70 percent of the TV sets in a city were tuned to Name That Poem on a Wednesday evening. A competitor doubted this claim and thus took a random sample of 200 families and found that the sample percentage of viewers was only 65 percent. At the .05 level, can the competitor conclude that the station manager was stretching the truth?

16 A tire manufacturer claims that the average life of a certain grade of tire is 50,000 miles when used under normal driving conditions. A random sample of 17 tires is selected and tested, and the sample mean and standard deviation are 49,000 miles and 5,000 miles respectively. At the .05 level, would you accept the manufacturer's claim?

17 Let's assume that you make hammocks and buy ropes for them from the Hemphill Rope Company. A random sample of 26 ropes from a new shipment has a sample mean breaking strength of 427 pounds with a sample standard deviation of 5 pounds. The company specifications require that the breaking strength be at least 430 pounds. At the .05 level, would you accept the shipment?

18 A light bulb manufacturer (is there no end to light bulb problems?) claims her bulbs have a population mean life of 1,000 hours. Is this claim justified at the .05 level if a random sample of 25 bulbs has a sample mean of 994 hours with a standard deviation of 30 hours?

19 A fertilizer distributor has been selling fertilizer in 50-pound bags for several months. The population weight is normally distributed with a standard deviation weight of .60 pound. A customer, Ms. Grassco, believes she has been sold underweight bags and is thinking of reporting the distributor to the Federal Trade Commission. To support her hunch, Ms. Grassco buys 4 bags of fertilizer at random and finds that the average weight is 49.65 pounds. At the .05 level, does Grassco have a case against the distributor?

20 The Clark Department Store's owner, Mr. Joe Clark, claims that the company's accounts receivable average is $290 per account. An auditor has just completed taking a sample of 50 accounts receivable. The sample mean is $280 and the sample standard deviation is $35. At the .05 level, is it likely that Clark's accounts receivable average is as high as Joe claims?

21 The sponsor of the Blackberry Ben TV program believes the program should be canceled if the program's share of the viewing audience is less than 25 percent. In a random sample of 1250 viewers, 260 watch the program. At the .05 level, should the program be canceled?

22 When properly adjusted, an automatic machine should produce parts that have a mean diameter of 50 millimeters (mm). It's undesirable for parts to vary significantly in either direction from this mean value. A random sample of 10 parts is used to check on machine operation. The sample mean is 50.02 mm and the sample standard deviation is .024 mm. At the .05 level, is the machine in adjustment?

23 The Speak-Easy Company produces speaker magnets for stereo systems. The magnets should weigh 2.6 ounces. Weight deviations in either direction are undesirable. In a recent quality control check, a random sample of 12 magnets taken from a large lot had a sample mean of 2.58 ounces, and the sample stan-

dard deviation was .035 ounces. At the .10 level, do the magnets in the lot meet quality standards?

24 Melvin claims that his average golf score is 75. Melvin lies a lot, so you observe his game for a random sample of nine rounds and find that the sample mean is 80 strokes with a sample standard deviation of four strokes. At the .01 level, should you buy Melvin's claim?

25 Senator Wilson claims that 60 percent of the population is in favor of passing a strict gun control law, but a hunting club member feels that the percentage is much less than that. You take a random sample of 200 people and find that 116 favor a strict law. At the .01 level, which person—Wilson or the club member—is more likely to be right?

26 To meet design specifications, steel rods produced in a factory must have a mean diameter of 1.5 centimeters (cm). (It's undesirable for the rods to vary in either direction from this standard.) The population standard deviation is known to be .01 cm. Samples of 50 rods are taken each hour and the sample mean diameters are computed. In the latest sample, the mean diameter is 1.5005 cm. At the .05 level, is the production process in control?

27 Light bulbs with a stated mean lifetime of 750 hours have been sitting in a warehouse for years. A sample of 10 bulbs is selected at random and destructively tested. The sample mean and standard deviation values are found to be 710 and 40 hours respectively. At the .10 level, has the prolonged storage significantly reduced the life expectancies of these bulbs?

28 Suppose you want to decide if a coin used in a coin toss game is loaded. (Any significant deviation of the number of heads from 50 percent indicates a loaded coin.) You flip the coin 300 times and note that the coin comes up heads 51 percent of the time. At the .05 level, is the coin fair?

29 An attorney is considering a class action lawsuit against a company for racial discrimination in its hiring practices. It's known that racial minorities make up 15 percent of the qualified workforce and that 12 of 200 workers hired by the company are members of racial minorities. Is there sufficient statistical evidence to suggest that a lawsuit is called for at the .05 level of significance?

30 A drug company produces vials of serum. These vials should contain 50 mg each, and it's undesirable for the contents to be above or below this value. To assure proper quality control, samples of 100 vials are periodically selected at random and the sample mean weight is computed. The population standard deviation is known to be 2 mg. What center line (CL) value and upper and lower control limit values (UCL and LCL) should be used on a statistical process control chart in this case? Assume that the UCL and LCL lines are located at a distance of ± 3 measures of dispersion from the CL.

31 Suppose that steel rods are used in the manufacture of bench presses. A random sample of 13 rods taken from a new shipment has a sample mean breaking strength of 3,950 pounds with a sample standard deviation of 100 pounds. The manufacturing specifications require that the breaking strength be at least 4,000 pounds. At the .05 level, should the shipment be accepted?

32 Suppose that in problem 31 the sample mean breaking strength had been greater than 4,000 pounds. Should the shipment be accepted in that case?

33 A contractor claims that its consulting services cost the government an average

of $10,000 per consultation. A random sample of 15 consultation fees is examined, and it's found that the sample mean is $10,575. The standard deviation of this sample is $600. At the .05 level, is there sufficient evidence to suggest that the contractor averages more than $10,000 per consultation?

34 A company manager believes that average business travel expenses should not exceed $1,700 per trip. A study of 10 randomly selected travel account records yielded the following travel expenses (in dollars):

1,750	1,693	1,710	1,730	1,650
1,720	1,688	1,703	1,680	1,760

At the .05 level of significance, is it likely that the mean of all travel expenses is too high?

35 A medical researcher claims that 21 percent of all French skiers catch winter colds. Seeking to verify this claim, Yves Jacquard interviews 50 French skiers and learns that 13 of them caught colds last winter. At the .05 level of significance, what should Yves conclude?

TOPICS FOR REVIEW AND DISCUSSION

1 What is meant by a significant difference in hypothesis testing?
2 What does a significance level of .05 actually mean?
3 Why should the term "accept" be taken lightly in hypothesis testing?
4 What is a rejection region?
5 When should a one-tailed test be employed?
6 Indicate whether the Z or the t distribution applies in each of the following:
   a $n = 16$, $s = 24$
   b $n = 19$, $\sigma = 15$
   c $n = 43$, $\sigma = 98$
   d $n = 102$, $s = 48$
7 How can a hypothesis be proved?
8 "The hypothesis-testing procedure consists of a series of steps." Discuss this statement.
9 What is statistical process control, and why is it important?
10 Explain how a statistical software package can be used in one-sample hypothesis tests of means.

PROJECTS/ISSUES TO CONSIDER

1 In an article entitled "How Often Are Our Statistics Wrong? A Statistics Class Exercise," that appeared in the April 1987 issue of *Teaching of Psychology,* Professor Joseph Rossi of the University of Rhode Island described an interesting class project. Students in a graduate class and an undergraduate class were assigned the task of searching journal articles to find published values for statistical tests. Let's team up and carry out a similar project. More specifically, each team should:

a   Locate five or six articles that include published statistical test results in journals that interest team members.

b   If possible, use the journal author's data, recompute the value of the statistical test, and summarize your findings.

c   If insufficient data are supplied in the article to allow you to check the author's results, note this finding.

d   Prepare a brief team presentation for the class.

e   Keep these project steps in mind; we'll return to them in later chapters.

You might be interested to know that the students in Professor Rossi's undergraduate class located 21 articles with test results that could be checked, and they found (and Rossi verified) that 5 of the 21 test values appeared to be inaccurate.

ANSWERS TO SELF-TESTING REVIEW QUESTIONS

8-1

1   A null hypothesis is the assumption of a parameter value which is subject to a statistical test.

2   An alternative hypothesis is the conclusion to be drawn contingent on the rejection of the null hypothesis.

3   A significance level is the level at which the null hypothesis would be rejected when it's actually true.

4   The Z distribution is applicable when $n > 30$ or in smaller samples when σ is known and population values are normally distributed.

5   A significant difference is the size of the difference between an obtained statistic and the assumed value which warrants rejection of the null hypothesis.

6   A rejection region is the area of the sampling distribution which leads to rejection of a null hypothesis.

7   The difference is standardized through the use of the critical ratio.

8   The statement is false because a hypothesis can never be proved. Since knowledge of the parameter value is the means of proving the hypothesis, a statistical test merely provides evidence not to reject a hypothesis.

9   This statement is only partially correct. Statistical results merely serve as one form of information input in decision making.

8-2

1   A two-tailed test rejects the null hypothesis if $\bar{x}$ is significantly too low or significantly too high compared to the assumed value. The conclusion on rejection of the null hypothesis is simply that the assumed value is not the true value.

2   a $Z = 2.58$
    b $Z = 1.75$
    c $Z = 1.96$
    d $Z = 2.17$

3   The hypotheses are:

H_0: $\mu = 500$
H_1: $\mu \neq 500$

and with $\alpha = .01$, $Z = 2.58$.
The decision rule is:

Accept H_0 if CR falls between ± 2.58.

or

Reject H_0 and accept H_1 if CR > 2.58 or CR < -2.58.

The critical ratio is computed as follows:

$$CR = \frac{\bar{x} - \mu_{H0}}{\sigma/\sqrt{n}}$$
$$= \frac{475 - 500}{50/\sqrt{36}}$$
$$= -3.00$$

Conclusion: Reject H_0.

8-3

1 In a one-tailed test, there's only one rejection region. The alternative hypothesis states that the true value is likely to be higher or lower than the assumed value.

2 For a left-tailed test, H_1: $\mu <$ assumed value. For a right-tailed test, H_1: $\mu >$ assumed value.

3 a 2.58
   b 1.96
   c -1.64
   d 2.33

4 The hypotheses are:

H_0: $\mu = 100$
H_1: $\mu < 100$

and with $\alpha = .05$, $Z = -1.64$.
The decision rule for this left-tailed test is:

Accept H_0 if CR ≥ -1.64.

or

Reject H_0 and accept H_1 if CR < -1.64.

The critical ratio is computed as follows:

$$CR = \frac{88 - 100}{15/\sqrt{36}} = -4.8$$

Conclusion: Reject H_0.

5 The hypotheses are:

H_0: μ = 24
H_1: μ > 24

and with α = .01, Z = +2.33.
The decision rule for this right-tailed test is:

Accept H_0 if CR $\leq$ 2.33.

or

Reject H_0 and accept H_1 if CR > 2.33.

The critical ratio is computed as follows:

$$CR = \frac{26 - 24}{3/\sqrt{16}} = -2.67$$

Conclusion: Reject H_0.

6 H_0: μ = 500
   H_1: μ < 500
7 H_0: μ = 2.5
   H_1: μ < 2.5
8 This statement is false. With the structure of a right-tailed test, the true value may be less than the assumed value, but the null hypothesis will not be rejected.

8-4

1 They differ in the following manner:
   a The normal (Z) distribution is used when n > 30. If σ is known and a smaller sample is used, the Z distribution may be appropriate if the population is normally distributed. Otherwise, the t distributions are used.
   b The critical ratio must be computed with $\hat{\sigma}_{\bar{x}}$ instead of $\sigma_{\bar{x}}$.
2 The hypotheses are:

H_0: μ = 612
H_1: $\mu \neq$ 612

and with α = .05 and n = 13, the t value with 12 degrees of freedom is t_{025} = 2.179.
The appropriate decision rule is:

Accept H_0 if CR falls between ± 2.179.

or

Reject H_0 and accept H_1 if CR > 2.179 or CR < -2.179.

The estimate of $\sigma_{\bar{x}}$ is computed as follows:

$$\hat{\sigma}_{\bar{x}} = \frac{s}{\sqrt{n}} = \frac{5}{\sqrt{13}} = 1.39$$

The critical ratio is computed as follows:

$$CR = \frac{\bar{x} - \mu_{H0}}{\hat{\sigma}_{\bar{x}}} = \frac{608 - 612}{1.39} = -2.88$$

Conclusion: Reject H_0.

3 The hypotheses are:

H_0: $\mu = 243$
H_1: $\mu \neq 243$

and with $\alpha = .01$ and $n = 36$, the Z value is 2.58.
The decision rule is:

Accept H_0 if CR falls between ± 2.58.

or

Reject H_0 and accept H_1 if CR > 2.58 or CR < -2.58.

The estimate of $\sigma_{\bar{x}}$ is computed as follows:

$$\hat{\sigma}_{\bar{x}} = \frac{s}{\sqrt{n}} = \frac{15}{\sqrt{36}} = 2.50$$

The critical ratio is computed as follows:

$$CR = \frac{\bar{x} - \mu_{H0}}{\hat{\sigma}_{\bar{x}}} = \frac{269 - 243}{2.50} = 10.40$$

Conclusion: Reject H_0.

8-5
1 a -2.508
  b 2.131
  c 2.485
  d 1.706
2 The hypotheses are

H_0: $\mu = 400$
H_1: $\mu < 400$

and with $\alpha = .01$, $n = 23$, and a left-tailed test, the t value with 22 degrees of freedom is $t_{01} = -2.508$.
The decision rule is:

Accept H_0 if CR ≥ -2.508.

or

Reject H_0 and accept H_1 if CR < -2.508.

The critical ratio is computed as follows:

$$CR = \frac{\bar{x} - \mu_{H0}}{\hat{\sigma}_{\bar{x}}} = \frac{389 - 400}{8} = -1.38$$

Conclusion: Accept H_0.

3 The hypotheses are:

H_0: $\mu = \$6,425$
H_1: $\mu > \$6,425$

and with $\alpha = .05$, $n = 27$, and a right-tailed test, the t value with 26 degrees of freedom is $t_{05} = 1.706$.
The decision rule is:

Accept H_0 if CR ≤ 1.706.

or

Reject H_0 and accept H_1 if CR > 1.706.

The critical ratio is computed as follows:

$$CR = \frac{\bar{x} - \mu_{H0}}{\hat{\sigma}_{\bar{x}}} = \frac{\$6,535 - \$6,425}{\$55} = 2.00$$

Conclusion: Reject H_0.

8-6

1 In this book we deal with only the large-sample case in the testing of percentages.
2 The values of π_{H0} and n must be used in the calculation of σ_p.
3 Essentially there's no difference.

A CLOSER LOOK

QUALITY TESTING: A PRIMER FOR BUYERS

Here's a quick test of your quality-consciousness. The hypothetical situation: Your company is planning to produce a new product.

Your assignment: Find a supplier you can trust to supply a certain part, say a bearing race, that unfailingly falls within a 6.0 cm diameter design spec. Engineering says variation of plus or minus 0.12 cm is OK; anything outside that variation, however, is a no-no—the product will self-destruct the moment it's started up.

You go to a supplier you've dealt with successfully in the past and ask for a sample run of 50 bearing races, plus copies of his test data on that sample run. His test results are shown in Lot Plot *a* in Figure CL8-1.

At the same time, you ask for samples from another supplier who has bid the job significantly lower—say 20% less—than the supplier you know and trust. His test results are shown in Lot Plot *b* in the figure.

In both cases, all the sample products fall within the acceptable range set by engineering. Which supplier do you go with?

If you chose the more expensive supplier, you'd be right—and you'd probably save your company a lot of money and grief in the bargain.

The test results show that supplier A's production process can consistently produce the bearing race well within tolerances, with only the statistically-expected random variation in size.

But the data on supplier B indicates he is probably using several low-quality machines in the production process, only one of which can even begin to approach your specs; the other machines are set to produce either too large a bearing race or too small a race. The supplier is making up for his low quality output with 100% inspection, and is rejecting all the output that falls outside your specs.

Result: Unless you yourself implement 100% inspection of incoming goods, you will always take the risk that some of those off-size items from supplier B will make their way into your products; whereas with supplier A, you can be certain that under normal circumstances his production controls won't even permit him to produce a single bearing race outside specification.

How can you convince management to back you in the decision to spend 20% more? It's all in knowing how to read the test results.

TOOLS OF QUALITY

That's the true test of a firm's quality commitment, in fact; management is still tempted to believe what is really too good to be true—high quality goods for a low-quality price. But armed with the tools of statistics, purchasing managers can help avoid costly mistakes by separating out the suppliers who can actually manufacture quality goods from those who only fabricate quality test results.

As more and more firms require suppliers to implement their own quality control programs, the use of standard statistical methods to measure conformance to quality has become increasingly widespread. Many suppliers now provide the results of ongoing production tests on a regular basis, for example. But unless you know how to use these tools for quality—and how they can be abused, whether knowingly or not—all the statistical process control reports in the world won't help improve the quality of incoming goods at your company. What follows is a brief refresher course on the most popular tools used to measure and report on product quality.

Lot plots, also known as histograms, are used to record the size or other critical characteristics of a particular production run. The plot shows the specific measurement of each item produced in a particular production run. Lot plots are also often used in the product design stage before production gets underway in order to help adjust production machinery to spec during sample runs.

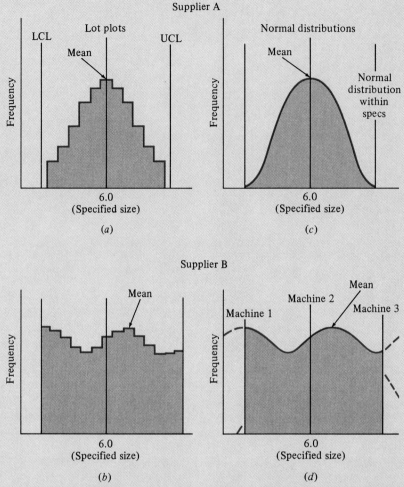

FIGURE CL8-1 **Reading Supplier A's test results:** Graph c is the normal distribution from lot plot a. It shows variation in the size of the bearing race, but all the variation is within spec. The mean of the curve is equal to the size specified by engineering. **Reading Supplier B's test results:** Graph d is an approximation of a normal distribution from lot plot b. It actually contains three normal distributions, one from each machine being used. Two of the machines cannot produce within spec. The mean does not equal the size specified by engineering. Dotted lines show units removed by supplier's quality control. (Illustrations originally drawn by Gayle Levee.)

What is a lot plot? Consider a lot plot as a simple *x-y* graph, where the size of each item produced is graphed along the *x* or horizontal axis, while the number of items of each size produced—the frequency of a particular size being made—is graphed along the *y* or vertical axis.

In a perfect world, lot plots always would be a straight line—size would never vary, so every item produced would plot at the same point; in the real world, however, variation is a given. But if a machine operator is shooting for a particular size item, his ability to get close to the specification more often than not is the true measure of quality.

Engineering specifications usually state the variation allowed in the size of a particular item; a bearing race, for example, could be specified to measure 6.0 cm in diameter "plus or minus 0.12 cm." The resulting lot plot from a supplier is shown in the original example sited earlier and shown in Figure CL8-1.

NORMAL CURVES

Also shown in Figure CL8-1 is an example of one of the most important laws of statistics: all random events in life exhibit a predictable pattern, the bell shaped curve known as a "normal distribution." If you repeat any operation an infinite number of times, the results will eventually form a continuous lot plot that looks like the normal curve in graph c in the illustration.

The science of statistics also shows that all normal distributions behave the same, mathematically, regardless of what it is they happen to be measuring. As a result, statisticians using mathematical techniques can measure the "standard deviation" of any normal distribution. You're probably familiar with the "mean" or average of a series of events. The average age of the U.S. population, for example, is 31.8 years—even though many are a lot younger or older than that.

In the case of the normal distribution, the "mean" is the highest point in the bell-shaped curve, and is a measure of the central tendency of whatever it is you're measuring. In a good-quality production line, if you were to measure every item produced and plot the results, the mean results should also coincide with the desired results—most bearing races produced should actually average 6.0 cm in diameter.

The "standard deviation" is a measure of how much variation exists around that central tendency—how close to the required spec the machine operator can produce the bearing race, and how far off he is when he misses. A large standard deviation means his misses are big ones; a small standard deviation means his misses are well within spec.

PREDICTING QUALITY

But finding the mean or average result isn't the only thing you can do with the normal distribution. You can also use it to predict how far off the mean results will be. Because it so happens (and has been proved many times over by

doubting mathematicians) that nearly all the results of a particular series of repeated events—99.7% of them, in fact—take place within 3 standard deviations, plus or minus, of the mean or central tendency in a normal distribution.

Once you have a sample production run, therefore, you can pretty much figure out exactly what the results will be on a regular basis (assuming nothing important changes—no new operator, no change in hours of operation, standard maintenance procedures followed, etc.). And you can select your suppliers accordingly.

In simplistic terms the primary goal of statistical process control [SPC] is to help you and your supplier to "minimize standard deviation." But there's more to lot plots and other SPC tools than that; you've got to learn to read the results of SPC to get the most from it. For a sampling of the power of SPC in helping you to determine the quality of materials your suppliers are shipping—and why they're on- or off-spec—see the six lot plots in Figure CL8-2. With the correct analysis of SPC charts, most buyers can easily avoid unqualified or untrustworthy suppliers.

CONTROL CHARTS

But what if your supplier tested out fine in the initial go-through, but something changes at his factory? Meanwhile, you're still relying on old test results that only prove he has the capability of producing on-spec—not whether he is doing so on a regular basis. How can you cover yourself? You need yet another statistical tool to track actual performance over time. Process control charts are used to do just that.

Think of a control chart as a series of lot plots; a sample of a production run is taken every hour or so, measured, the results averaged, and plotted on the control chart. Figure CL8-3 shows examples of process control charts. The UCL (upper control limit) and LCL (lower control limit) are the predetermined "six standard deviations" within which the particular machine being tested normally operates; with 99.7% certainty, everything it makes is within 3 standard deviations above the norm or 3 standard deviations below the norm.

Any output which exceeds these limits is the

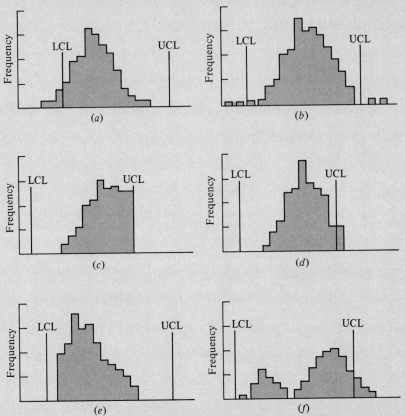

FIGURE CL8-2 Sample lot plots for analysis. (a) This plot shows the machine could produce within the specified range if it were adjusted properly. (b) Although this appears to be a normal distribution, too much random variation exists outside the acceptable limits; machine could need servicing. (c) Machine is producing beyond acceptable limits, but results appear to have been sorted out of the batch under test. (d) Same as c, except sorter appears to have read measuring device incorrectly and some defectives have gotten through. (e) Overzealous sorting of partially defective lot; some good items have been rejected. (f) Parts from two different lots have been mixed together; one lot is not within specifications. (Illustrations originally drawn by Gayle Levee.)

result of something abnormal, something other than random variation—a broken part in the machine, operator error, improper machine setup or even poor quality materials—any one or combination of which can result in production outside the control limits despite the established capabilities of a particular machine.

Properly used, control charts actually do control the production process; a good supplier will shut down any machine that produces outside its control limits, and won't start it back up again until the reason for variation is deter-

mined and corrections are made. Any supplier can produce reams of process control charts; the critical test is whether he'll shut down the process when the limits are exceeded.

And just as it is important for buyers to know how to read lot plots, buyers need to be able to know what to look for when reading process control charts. Figure CL8-4 shows just a few of the many control chart results possible when output is tested over time.

But the use of process control charts isn't just limited to post-production quality control; con-

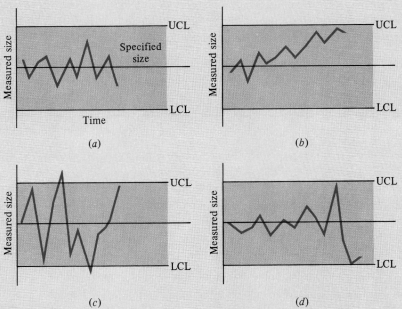

FIGURE CL8-3 Every chart tells a story. (a) In this chart, the process is under control; production continues to take place well within the specified limits for the item. (b) Although control limits have not been exceeded, the chart clearly indicates that limits will soon be breached. Recommendation: Stop production now, and correct the problem, before materials and time are wasted in making unacceptable products. The gradual deterioration of the product indicates either a tool-wear or an overheating problem. (c) Both upper and lower limits have been exceeded. Either the tool being used is incapable of meeting specs, or it is seriously out of adjustment. The process should have been stopped long before this amount of information could have been gathered in order to avoid production of costly rejects. (d) Extremely abnormal variation after a long period of within-spec production indicates a serious problem on the production line, perhaps operator-related. Stop production and investigate. (Illustrations originally drawn by Gayle Levee.)

trol charts are also becoming very useful in the development of production processes and in the early stages of product design in efforts to boost quality before production even begins.

In fact, two recent movements in the design of quality production methods—the Taguchi method developed by Dr. Genichi Taguchi of Japan and a new, simpler method developed by Dorian Shainin, also of Japan—use a combination of process control charts, statistical analysis of results, and controlled variation of production techniques to reduce costs and minimize variation at the earliest stages of the design of the product and the production process.

One other statistical tool—the scatter diagram—can be of use in tracking supplier conformance to quality. A process control chart, in fact, is itself a type of scatter diagram (see Figure CL8-4 for examples of scatter diagrams).

Scatter diagrams are useful in determining the relationship between two events or outcomes. A process control chart measures the relationship between the time of day and the size of the item produced, for example. Other scatter diagrams might show the relationship between the steel content of a powder metal part and its hardness, or the relationship between the change in the operating temperature of a machine tool and the variation in the size of its

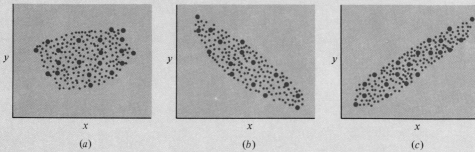

FIGURE CL8-4 What can a scatter diagram tell you? (a) No correlation between x and y. Events are taking place independently of each other. (b) Negative correlation between two events. As x increases, y decreases. Example: Operator experience and error-rate. (c) Positive correlation between two events. As x increases, so does y. Example: Size and weight. (Illustrations originally drawn by Gayle Levee.)

output. As the machine tool heats up, variation could increase sharply; as it reaches normal operating temperature, variation might fall back to within spec.

Although suppliers may not provide scatter diagrams in the normal course of reporting on the quality of their goods, they are a handy tool to help pinpoint unexpected variation in an apparently up-to-standards production process.

DECISION FACTORS

All these sophisticated statistical tools can't help buyers answer the most important quality questions, though. Who does the testing, and how extensive is it? Should suppliers self-test? Should 100% of incoming items be tested? Is a sample test adequate? Quality guru W. Edwards Deming developed a basic decision-formula on whether to test or not some years ago. The rule of thumb: If, in a preliminary sample, the fraction of defective goods coming in is smaller than the ratio of inspection costs to removal/repair costs of defective items, it pays to inspect none of the incoming goods; if the fraction of defective goods is greater than the ratio of inspection costs to replacement/repair costs, it pays to inspect 100% of the incoming items.

Deming states the rule mathematically as:

If $P < K_1/K_2$, perform no incoming tests

If $P > K_1/K_2$, perform 100% incoming tests

where P = the fraction of defective incoming goods

K_1 = the cost to inspect or test one item

K_2 = the cost to later remove and replace or repair a defective item

The formula would work fine except for a certain unquantifiable cost glossed over in the definition of terms: That is, what is the actual cost of replacing or repairing the defective product resulting from defective materials? In the early days of quality programs in the U.S., this cost was more often than not grossly underestimated; even today, when the true costs of quality are being documented in detail at plants across the nation, another previously undocumented cost—the cost of lost customers and a reputation for poor quality—is still underestimated in the Deming formula.

As a result, more and more firms are opting for the 100% inspection route; but to minimize their own costs and free up production personnel, they're more often than not forcing suppliers to do the inspecting. New manufacturing technologies that include continuous self-inspection have helped make this a reality; in addition, computer-aided inspection systems (CAI) are automating many of the more tedious inspection activities and, by removing error-prone human judgement from the loop, increasing the quality of final products even further.

SAMPLING THE ALTERNATIVE

But there is an alternative to 100% inspection: Sample testing. Sample testing—testing a random sample of items as they come through the receiving room door—is particularly useful when a supplier has shown he is capable of consistently supplying on-spec materials and is providing his own process control charts with each lot. The sample test then functions as a "test of the test," a further control element for quality-conscious manufacturers.

The most widely-used method of sample testing is "acceptance testing." Using statistical theory, incoming quality inspectors determine the number of rejects that would result from a normal distribution with a 0.3% probability (the other side of the 99.7% quality standard), or whatever standard engineering has predetermined; if more than that number is present in a lot, the whole lot is rejected; if less, it is accepted.

Example: a large lot of several thousand items is received; in a sample of 500, only 3 defects should be present using the 99.7% standard. Six are found—the lot is rejected.

Although some companies use sample acceptance as their only quality control system, it has certain disadvantages which make it more useful—and economical—as a back-up test rather than a primary one. These include problems in choosing a sample—if it is a truckload, taking a random sample could prove difficult;

more often the sample is selected from the top of the shipment, making it easy for suppliers to salt the test with the best in the lot; statistical risks still exist of accepting a bad lot or rejecting a good lot; the method used to sample and test is particularly sensitive to error in small samples.

But an even bigger problem with acceptance sampling exists in the real everyday setting of a factory operation. In most factories, especially those with just-in-time inventory programs in place, there is great pressure to keep materials moving through the system. As a result, if a shipment tests very close to the acceptance standard—say only 4 rejects instead of the acceptable 3—production management may force quality inspectors either to perform 100% inspection on that particular lot or to accept the lot out of hand.

The solution to this problem is to make sure everyone involved in the production process—suppliers, inspectors, and production personnel—has agreed to honor the established standards for quality set by engineering and management early in the process. As insurance, a mechanism should be established in advance to assure that adequate, pretested supplies are on hand or available on short notice from the supplier as backup.

—J. William Semich, "Quality Testing: A Primer for Buyers," *Purchasing*, January 28, 1988, pp. 98–107. Reprinted with permission.

.003 × 500 = 1.5 not 3

TESTING HYPOTHESES AND MAKING DECISIONS: TWO-SAMPLE PROCEDURES

Two-Sample Procedures and Other Statistical Techniques

Statistical techniques serve two main purposes, description and inference. Descriptive statistical techniques define the characteristics of a group of data items either mathematically or graphically. Commonly used descriptive statistics include the mean, mode, median, range, standard deviation, variance, and skewness. Examples of descriptive graphical techniques are Cluster Analysis and leaf and stem analysis. At times a known pattern of distribution can also be used to describe a group of data items. For example, the Poisson distribution may be used to describe the occurrence of rare events. Inferential statistical techniques can then be used to predict the characteristics of the total group of items based on a sample of the items. Inferential techniques are used, for instance, in conducting a survey. The characteristics of a small group, collected at random, can be used to determine the characteristics of the entire population.

These techniques can also be used to determine the probability that two or more samples came from groups having similar characteristics. Examples of this situation are found when a group receiving treatment is compared to a control or untreated group before and after treatment. Commonly used inferential techniques include the standard error of a difference, t test, Mann-Whitney U test, analysis of variance, regression, and analysis of covariance.

—From Richard T. Dué, "Predicting Results with Statistics," *Datamation*, May 1980, p. 227. Reprinted with permission of Datamation ® Magazine, © copyright by Technical Publishing Company, A Dun & Bradstreet Company, 1980. All rights reserved.

LOOKING AHEAD

You're already acquainted with many of the terms mentioned in the opening vignette. And you'll learn about most of the unfamiliar procedures in following chapters. In Chapter 9, however, our focus is on "the probability that two . . . samples came from groups having similar characteristics." You'll learn here how to use the hypothesis-testing procedures introduced in Chapter 8 along with the data obtained from two *independent samples* to make relative comparisons between (1) *two population means* and (2) *two population percentages*. In order to maintain simplicity of presentation, however, we'll only consider situations where the samples are relatively large (over 30) and where it's thus possible (by the use of the Central Limit Theorem) to assume that the appropriate sampling distributions are approximated by normal curves.

Thus, after studying this chapter you should be able to:

- Explain (a) the purpose of two-sample hypothesis tests of means and percentages and (b) the procedures to be followed in conducting these tests.
- Understand the concepts associated with both the sampling distribution of the differences between sample means and the sampling distribution of the differences between sample percentages. You should also be able to explain how such sampling distributions may be created.
- Perform the necessary computations and make the appropriate statistical decisions in two-sample hypothesis-testing situations.

CHAPTER OUTLINE

SOME PRELIMINARY THOUGHTS

Decision makers often want to see if two populations are similar or different with respect to some characteristic. For example, an instructor may be curious as to whether male professors receive higher salaries than female professors for the same teaching load. Or a psychologist may want to see if one group responds differently than another to an experimental stimulus. Or the purchasing agent of a firm that manufactures cooling towers may want to know if the cooling fan motors of one supplier are more durable than those of another vendor. In short, there are many situations that require groups to be compared on the basis of a given trait.

Chapter 8 discussed ways to test the validity of an assumed value of a parameter. The assumed value was a single quantity that was subjected to statistical testing. In this chapter, though, we're concerned with the parameters of two populations, but we're not primarily interested in estimating the *absolute values* of the parameters. Rather, the topic of interest is the *relative values* of the parameters. That is, does one population appear to possess more or less of a trait than the other?

Our purpose in this chapter, then, is to use the data obtained from two samples to see if there's likely to be a statistically significant difference between the parameters of two populations. Before we begin our two-sample analyses, though, let's be sure we understand that there are two prerequisites to using the techniques we'll employ in the chapter. The *first* requirement is that both samples be *large*—that is, greater than 30. (Note, though, that a procedure for testing two *small* samples is discussed in the Closer Look reading at the end of the chapter.) And the *second* requirement is that **independent samples** be used—that is, that the samples be taken from different groups and that the sample selected from the first group not be related to the sample selected from the second. Thus, a study designed to measure the effects of an intensive statistics review course by first testing a sample group of students before they take the review course and by then retesting the same students after the course is completed could not use the procedures presented in this chapter. [You'll be delighted (?) to know, though, that such a "before and after" study can be made if nonparametric test procedures discussed in Chapter 15 are used.]

TWO-SAMPLE HYPOTHESIS TESTS OF MEANS

A two-sample hypothesis test of means begins just like the one-sample tests described in the last chapter: a null hypothesis must be prepared. This hypothesis takes the following form:

$$H_0: \mu_1 = \mu_2$$

The null hypothesis states that the true mean of the first population is equal to the true mean of the second population. *In essence, H_0 states that the populations are the same on the basis of a given characteristic.*

If the null hypothesis cannot be supported, there are three possible alternative hypotheses that may be accepted:

H_1: $\mu_1 \neq \mu_2$ two-tailed alternative
H_1: $\mu_1 > \mu_2$ right-tailed alternative
H_1: $\mu_1 < \mu_2$ left-tailed alternative

If the nature of the problem simply indicates that it's sufficient to reject the null hypothesis without further inferences about the differences between μ_1 and μ_2, a two-tailed test is used. If the nature of the problem shows that H_0 should be rejected only on the basis that μ_1 is significantly higher than μ_2, a right-tailed test is employed. And if there are indications that the first population might possibly be less than the second population on a given characteristic, a left-tailed test is applied.

The Sampling Distribution of the Differences between Sample Means

Now, let's discuss the underlying concepts in testing for equality between groups. When the null hypothesis states that the the true mean of group 1 is equal to the true mean of group 2 ($\mu_1 = \mu_2$), it's essentially asserting that the *difference between the parameters of the two groups is zero*—that is, $\mu_1 - \mu_2 = 0$. This idea makes it possible for us to visualize yet another type of sampling distribution.

A conceptual schematic of this new sampling distribution is shown in Figure 9-1. Distribution A in Figure 9-1 is the sampling distribution of the means for population 1, and distribution B is the corresponding sampling distribution for population 2. As we've seen, each of these theoretical distributions is developed from the means of all the possible samples of a given size that could be drawn from a population. Now if we were to select a single sample mean from distribution A and another sample mean from distribution B, we could subtract the value of the second mean from the value of the first mean and get a difference—that is, $\bar{x}_1 - \bar{x}_2$ = difference. This difference could be either a negative or positive value, as shown in the examples between distributions A and B in Figure 9-1. We could theoretically continue to select sample means from each population and continue to compute differences until we reached an advanced stage of senility. If we then constructed a frequency distribution of *all the sample differences*, we would have distribution C in Figure 9-1, which is the **sampling distribution of the differences between sample means**. And, as noted in Figure 9-1, *if H_0 is true*, and *if μ_1 is equal to μ_2, then* the value of the mean of the sampling distribution of the differences (μ_d) will be equal to $\mu_1 - \mu_2$. That is, μ_d will have a value of zero. In short, the negative differences and the positive differences will cancel, and the mean will be zero.

If it's assumed that the mean of the sampling distribution of differences is zero, the characteristics of the sampling distribution of differences allow for variation in the value of $\bar{x}_1 - \bar{x}_2$ from zero. If the parameters are truly equal, and if random samples are taken from the two populations, it's un-

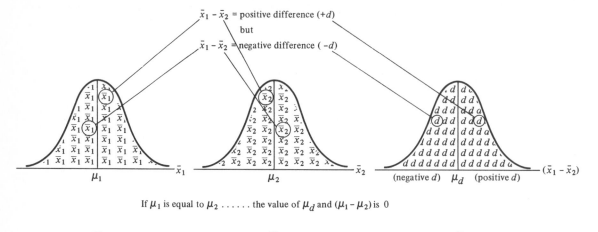

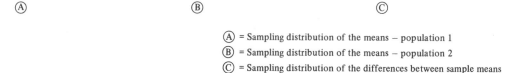

Ⓐ = Sampling distribution of the means – population 1
Ⓑ = Sampling distribution of the means – population 2
Ⓒ = Sampling distribution of the differences between sample means

FIGURE 9-1   Conceptual schematic of the sampling distribution of the differences between sample means.

likely that the difference of $\bar{x}_1 - \bar{x}_2$ will equal zero; there will usually be some sampling variation. However, if the true means are equal, the likelihood of an extremely large difference between $\bar{x}_1$ and $\bar{x}_2$ will be small, especially in two large samples. Thus, if an extremely large difference occurs between $\bar{x}_1$ and $\bar{x}_2$, it's justifiable to conclude that the true means aren't equal. The immediate problem, of course, is to figure out when the difference between samples becomes significant so that the null hypothesis of equality of μ_1 and μ_2 can be rejected.

If you've absorbed the discussion in Chapter 8, you'll notice the similarity between what's presented here and what was presented earlier. In a test for differences between means, we look for a significant difference between $\bar{x}_1$ and $\bar{x}_2$, which leads to rejection of the null hypothesis.

If we take a large sample from both populations, the shape of the sampling distribution of the differences between means is approximately normal. Thus, the middle 68.26 percent of the differences in that sampling distribution are found within 1 standard deviation of the mean, and 2 standard deviations to either side of the mean accounts for 95.4 percent of the differences. The standard deviation of the sampling distribution of differences goes by the name **standard error of the difference between means** and is identified by the symbol $\sigma_{\bar{x}_1 - \bar{x}_2}$ (see Figure 9-2).

With the use of the Z table, it's possible to establish the significant difference level once the risk of erroneously rejecting the null hypothesis has been chosen. And once a level of significance (α) has been picked, the

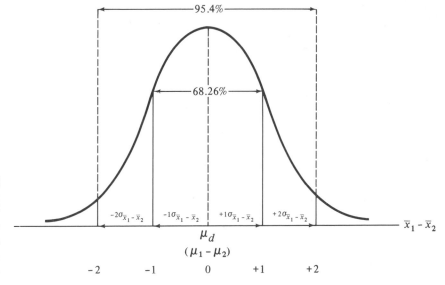

FIGURE 9-2
The sampling
distribution of
differences between
means with large
samples from each
group.

boundaries of the rejection region may be set. The process of establishing rejection regions and stating decision rules is exactly the same here as in Chapter 8, when the Z distribution was applicable. For example, with a two-tailed test and a .05 level of significance, the Z values for the boundaries of the rejection regions are ±1.96. For a left-tailed test and a .05 level, the Z value is −1.64.

Just as in other hypothesis tests, a *critical ratio* (CR) must also be calculated for two-sample test of means. You'll recall from Chapter 8 that one CR was found with this formula:

$$CR = \frac{\bar{x} - \mu_{H0}}{\sigma_{\bar{x}}}$$

That is, the ratio was the difference between actual and hypothetical values expressed in number of standard errors. Now, our critical ratio uses the standardized difference between $\bar{x}_1$ and $\bar{x}_2$ and is computed as follows:

$$CR = \frac{(\bar{x}_1 - \bar{x}_2) - (\mu_1 - \mu_2)}{\sigma_{\bar{x}_1 - \bar{x}_2}} \quad \text{or} \quad CR = \frac{\bar{x}_1 - \bar{x}_2}{\sigma_{\bar{x}_1 - \bar{x}_2}} \qquad \text{(formula 9-1)}$$

since $\mu_1 - \mu_2$ is assumed to be zero if H_0 is true. And (if it's assumed that we have two large independently drawn random samples) the standard error of the difference is computed as follows:

$$\sigma_{\bar{x}_1 - \bar{x}_2} = \sqrt{\frac{\sigma_1^2}{n_1} + \frac{\sigma_2^2}{n_2}} \qquad \text{(formula 9-2)}$$

Let's now look at the following examples, which illustrate two-tailed and one-tailed tests.

Two-Tailed Testing When σ_1 and σ_2 Are Known

As noted above, the steps that are followed when making two-sample hypothesis tests are the same as the ones introduced for the general hypothesis-testing procedure in Chapter 8. Let's illustrate this fact with the following two-tailed test.

The Russ Trate Traffic Signal Company has decided to install microcomputers in its traffic light assemblies to more efficiently monitor and control traffic flows. Microcomputers from two suppliers are judged to be suitable for the application. To have more than one source of supply, the Trate Company prefers to buy microcomputers from both suppliers, provided there's no significant difference in durability. A random sample of 35 computer assemblies of brand A and a sample of 32 computers of brand B are tested. The mean time between failure (MTBF) for the brand A computers is found to be 2,800 hours, and the MTBF for the brand B units is found to be 2,750 hours. Information from industry sources indicates that the population standard deviation is 200 hours for brand A and 180 hours for brand B. At the .05 level of significance, is there a difference in durability?

The *hypotheses are:*

H_0: $\mu_1 = \mu_2$
H_1: $\mu_1 \neq \mu_2$

Since the Trate Company is only interested in testing for a significant difference, this is a two-tailed test. Also, the level of significance is specified at the .05 level, and the large sample sizes enable us to use the Z distribution. Thus, the rejection regions are bounded by $Z = \pm 1.96$, and the *decision rule is:*

Accept H_0 if CR falls between ± 1.96.

or

Reject H_0 and accept H_1 if CR < -1.96 or CR $> +1.96$.

With $\sigma_1 = 200$ hours, $n_1 = 35$, $\sigma_2 = 180$ hours, and $n_2 = 32$,

$$\sigma_{\bar{x}_1 - \bar{x}_2} = \sqrt{\frac{\sigma_1^{2}}{n_1} + \frac{\sigma_2^{2}}{n_2}}$$

$$= \sqrt{\frac{200^2}{35} + \frac{180^2}{32}}$$

$$= 46.43 \text{ hours}$$

Therefore, the *critical ratio* is computed as follows:

$$\text{CR} = \frac{\bar{x}_1 - \bar{x}_2}{\sigma_{\bar{x}_1 - \bar{x}_2}} = \frac{2,800 - 2,750}{46.43} = 1.08$$

Conclusion: Since the critical ratio falls within the acceptance region of ± 1.96, we can conclude that there's no significant difference in the durability of the two microcomputer brands.

One-Tailed Testing When σ_1 and σ_2 Are Known

Discount Stores Corporation owns outlet A and outlet B. For the past year, outlet A has spent more dollars on advertising widgets than outlet B. The corporation wants to see if the advertising has resulted in more sales for outlet A. A random sample of 36 days at outlet A had a mean of 170 widgets sold daily. A random sample of 36 days at outlet B had a mean of 165. Assuming $\sigma_1{}^2 = 36$ and $\sigma_2{}^2 = 25$, what can be concluded if a test is conducted at the .05 level of significance?

The *hypotheses are:*

H_0: $\mu_1 = \mu_2$
H_1: $\mu_1 > \mu_2$

This is a *right-tailed test* because the corporation wants to see if the performance of group 1 (outlet A) is better than the performance of group 2. With a .05 level, the Z value that bounds the rejection region is 1.64. Thus, the *decision rule is:*

Accept H_0 if CR ≤ 1.64.

or

Reject H_0 and accept H_1 if CR > 1.64.

The *critical ratio* is computed as follows:

$$\text{CR} = \frac{\bar{x}_1 - \bar{x}_2}{\sigma_{\bar{x}_1 - \bar{x}_2}} = \frac{170 - 165}{\sqrt{36/36 + 25/36}} = \frac{5}{1.3017} = 3.84$$

Conclusion: Since CR is more than 1.64, there's sufficient reason to believe that outlet A has sold more widgets than outlet B.

Self-Testing Review 9-1

1   "The purpose of the two-sample hypothesis-testing procedures presented in this chapter is to determine the absolute values of the parameters." Comment on this statement.
2   What conditions must be met by the samples if the procedures described in this chapter are to remain valid?
3   Two brands of truck tires are being compared by a trucking firm. A random sample of tires of brand X (group 1) had an average life of 45,000 miles, while the average life of a random sample of 40 brand Y tires (group 2) was 46,500 miles.

Assuming that σ_1 was 2,000 miles and σ_2 was 1,500 miles, is there any significant difference in quality at the .01 level?

4　Two brands of batteries are being compared by the same trucking company. A random sample of 36 Die Dead batteries had an average life of 42 months, while the average life of a random sample of 35 Crank-Up batteries was 45 months. Assuming that the σ of Die Dead batteries is 6 months and the σ of Crank-Up batteries is 3 months, are the Die Dead batteries likely to be inferior to the Crank-Up brand at the .05 level?

Two-Tailed Testing When σ_1 and σ_2 Are Unknown

When the population standard deviations are unknown—the usual situation—the sample standard deviations are used to estimate the population values. Remember, it's assumed that s is computed with the following formula so that it's a useful estimator of σ:

$$s = \sqrt{\frac{\Sigma(x - \bar{x})^2}{n - 1}}$$

Thus, $s_1 = \hat{\sigma}_1$, and $s_2 = \hat{\sigma}_2$.

An estimated standard error of the difference between means is then computed as follows:

$$\hat{\sigma}_{\bar{x}_1 - \bar{x}_2} = \sqrt{\frac{s_1{}^2}{n_1} + \frac{s_2{}^2}{n_2}} \qquad \text{(formula 9-3)}$$

Let's now consider the following example.

Dr. I. M. Sain, a psychologist, administered IQ tests to determine if female college students were as smart as male students. The random sample of 40 females had a mean score of 131 with a standard deviation of 15. The random sample of 36 males had a mean of 126 and a standard deviation of 17. At the .01 level of significance, is there a difference?

The *hypotheses are:*

$H_0: \mu_1 = \mu_2$

$H_1: \mu_1 \neq \mu_2$

Since Dr. Sain is only interested in concluding equality or nonequality between groups, this is a two-tailed test. And since our sample sizes are large, the Z distribution may be used. Thus, the rejection regions are bounded by $Z = -2.58$ and $+2.58$, and the *decision rule is:*

MARK TWAIN AND STATISTICAL INFERENCE

The man in the ticket-office said:

"Have an accident insurance ticket, also?"

"No," I said, after studying the matter over a little. "No, I believe not; I am going to be traveling by rail all day to-day. However, tomorrow I don't travel. Give me one for tomorrow."

The man looked puzzled. He said: "But it is for accident insurance, and if you are going to travel by rail—"

"If I am going to travel by rail I sha'nt need it. Lying at home in bed is the thing I am afraid of."

I hunted up statistics, and was amazed to find that after all the glaring newspaper headings concerning railroad disasters, less than *three hundred* people had really lost their lives by those disasters in the preceding twelve months.

By further figuring, it appeared that between New York and Rochester the Erie ran eight passenger-trains each way every day—16 altogether; and carried a daily average of 6,000 persons. That is about a six million in six months—the population of New York City. Well, the Erie kills from 13 to 23 persons out of its million in six months; and in the same time 13,000 of New York's million die in their bed! My flesh crept, my hair stood on end. "This is appalling!" I said. "The danger isn't in traveling by rail, but in trusting to those deadly beds. I will never sleep in a bed again."

You will excuse me from taking any more chances on those beds. The railroads are good enough for me.

And my advice to all people is, Don't stay at home any more than you can help; but when you have *got* to stay at home a while, buy a package of those insurance tickets and sit up nights. You cannot be too cautious.

Accept H_0 if CR falls between ± 2.58.

or

Reject H_0 and accept H_1 if CR < -2.58 or CR $> +2.58$.

With $s_1 = 15$, $n_1 = 40$, $s_2 = 17$, and $n_2 = 36$, the *critical ratio* is computed as follows:

$$\text{CR} = \frac{\bar{x}_1 - \bar{x}_2}{\sqrt{s_1^2/n_1 + s_2^2/n_2}} = \frac{131 - 126}{\sqrt{15^2/40 + 17^2/36}} = \frac{5}{3.695} = 1.35$$

Conclusion: Since the critical ratio falls between ± 2.58, we can conclude that there is no significant difference. (In IQ, that is.)

One-Tailed Testing When σ_1 and σ_2 Are Unknown

A Chamber of Commerce is seeking to attract new industry to its area. One argument it has been using is that average wages paid for a particular type of job are lower than in other parts of the nation. A rather skeptical company president assigns his brother-in-law the task of testing this claim. A random sample of 60 workers (group 1) performing the particular job in the Chamber's area is taken, and the sample mean is found to be $7.75 per hour with a sample standard deviation of $2.00 per hour. Another random

sample of 50 workers (group 2) taken in region NE produced a sample mean of $8.25 per hour with a sample standard deviation of $1.25 per hour. At the .01 level, what report should the brother-in-law give to the president?
The *hypotheses are:*

$H_0: \mu_1 = \mu_2$
$H_1: \mu_1 < \mu_2$

This is a one-tailed test because the validity of the Chamber's claim is being evaluated—i.e., that wages paid are less than in other areas. At the .01 level, the Z value that bounds the rejection region is −2.33. Thus, the *decision rule is:*

Accept H_0 if CR ≥ −2.33.

or

Reject H_0 and accept H_1 if CR < −2.33.

With $s_1 = \$2.00$, $n_1 = 60$, $s_2 = \$1.25$, and $n_2 = 50$, the *critical ratio* is computed as follows:

$$CR = \frac{\bar{x}_1 - \bar{x}_2}{\sqrt{s_1^2/n_1 + s_2^2/n_2}} = \frac{\$7.75 - \$8.25}{\sqrt{\$2.00^2/60 + \$1.25^2/50}} = \frac{-\$0.50}{\$0.313} = -1.60$$

Conclusion: Since CR falls into the area of acceptance, the Chamber's claim is not supported by the sample results at the .01 level. The brother-in-law is relieved to report that the test results support the president's doubts.

Self-Testing Review 9-2

1 Assume the following data are available:

	GROUP	
	1	2
$\bar{x}$	$5,600	$5,300
s	$1,120	$ 725
n	38	36

Conduct a two-tailed test at the .10 level.
2 What would be the hypotheses and the decision rule if a right-tailed test had been made at the .10 level using the data in problem 1 above?

TWO-SAMPLE HYPOTHESIS TESTS OF PERCENTAGES

The purpose of conducting two-sample hypothesis tests of percentages is to see, through the use of sample data, if there's likely to be a statistically significant difference between the percentages of two populations. The null hypothesis in such tests of the differences between percentages is:

H_0: $\pi_1 = \pi_2$

If the null hypothesis cannot be supported, there are three possible alternative hypotheses which may be accepted:

H_1: $\pi_1 \neq \pi_2$ two-tailed alternative

or

H_1: $\pi_1 > \pi_2$ right-tailed alternative

or

H_1: $\pi_1 < \pi_2$ left-tailed alternative

The Sampling Distribution of the Differences between Sample Percentages

The **sampling distribution of the differences between sample percentages** is theoretically analogous to the sampling distribution of the differences between sample means. The mean of the sampling distribution of the differences between percentages is zero *if* the null hypothesis is true—that is, if $\pi_1 = \pi_2$. If the sample size of each group is large, the shape of the sampling distribution is approximately normal. The standard deviation of the sampling distribution of the differences between percentages, which is called the **standard error of the difference between percentages**, is calculated as follows:

$$\sigma_{p1-p2} = \sqrt{\frac{\pi_1(100 - \pi_1)}{n_1} + \frac{\pi_2(100 - \pi_2)}{n_2}}$$

Unfortunately, the computation of σ_{p1-p2} requires a knowledge of the parameters. If these values were known in the first place, there would be no need to conduct a test! Therefore, *in a test of differences between percentages the estimate $\hat{\sigma}_{p1-p2}$ must always be used:*

$$\hat{\sigma}_{p1-p2} = \sqrt{\frac{p_1(100 - p_1)}{n_1 - 1} + \frac{p_2(100 - p_2)}{n_2 - 1}} \qquad \text{(formula 9-4)}$$

And the *critical ratio* is computed as follows:

$$\text{CR} = \frac{(p_1 - p_2) - (\pi_1 - \pi_2)}{\hat{\sigma}_{p1-p2}} \quad \text{or} \quad \text{CR} = \frac{p_1 - p_2}{\hat{\sigma}_{p1-p2}} \qquad \text{(formula 9-5)}$$

Since $\pi_1 - \pi_2$ will equal zero if the null hypothesis is true.

Testing the Differences between Percentages

The procedure for testing the differences between percentages is not any different than the test for differences between means. Consequently, there's little need here for further discussion of this matter. The following is an example of a two-tailed test. (You should by now be able to determine for yourself the procedure for a one-tailed test.)

Ken Kharisma, candidate for public office, feels that male voters as well as female voters have the same opinion of him. A random sample of 36 male voters had 12 persons favoring his election. A random sample of 50 female voters had a p_2 of 36 percent. Test the validity of Ken's assumption, using a significance level of .05.

The *hypotheses are:*

H_0: $\pi_1 = \pi_2$
H_1: $\pi_1 \neq \pi_2$

This is a two-tailed test because Ken is only interested in the equality or nonequality of opinions between the two groups. The Z distribution is applicable, and the rejection regions are bounded by $Z = \pm 1.96$.

The *decision rule is:*

Accept H_0 if CR falls between ± 1.96.

or

Reject H_0 and accept H_1 if CR < -1.96 or CR $> +1.96$.

And the *critical ratio* is calculated as follows:

$$\text{CR} = \frac{p_1 - p_2}{\sqrt{\dfrac{p_1(100 - p_1)}{n_1 - 1} + \dfrac{p_2(100 - p_2)}{n_2 - 1}}}$$

$$= \frac{33 \text{ percent} - 36 \text{ percent}}{\sqrt{\dfrac{(33)(67)}{35} + \dfrac{(36)(64)}{49}}}$$

$$= \frac{-3.00 \text{ percent}}{10.50}$$

$$= -.29$$

Conclusion: Since the critical ratio is between ± 1.96, there is no reason to reject Ken Kharisma's claim. Apparently, both sexes have about the same low opinion of Ken!

Self-Testing Review 9-3

1 Voters in two cities are polled to see if they are in favor of a proposal to limit the amount of property tax revenue that the state can collect. A random sample of 200 is taken in each city, and 120 voters in city A favor the proposal, while 109 voters in city B favor it. At the .05 level, is there a significant difference in voter opinion between the two cities?

2 A television producer believes her new program is more popular with urban viewers than with rural viewers. To test this claim, a network shows the program to 300 urban viewers and 100 rural viewers. It's noted that 65 of the urban viewers and 18 of the rural viewers enjoy the program. At the .05 level, should the producer's claim be accepted?

LOOKING BACK

1 Our purpose in this chapter is to see if two populations are similar or different with respect to some characteristic. Our interest isn't in estimating the absolute values of parameters; rather, our concern is to see if one population appears to possess more or less of a trait than the other. The prerequisites to using the techniques explained in this chapter are that the two samples be large (over 30), and that they be independently selected from different groups.

2 We've used testing procedures to make relative comparisons between two population means or two population percentages. The null hypothesis to be tested in both situations is that the two parameters are equal. That is, there's no significant difference between the two parameters. The alternative hypothesis may be either one-tailed or two-tailed depending on the logic of the situation.

3 The concepts associated with the sampling distributions of the differences between sample means and the differences between sample percentages are used in analyzing the data taken from two samples. The mean of each of these sampling distributions is zero *if* the two population means or two population percentages being compared are equal in size. If the null hypothesis is true, then there should only be a small difference between the values of the sample statistics. The purpose of the tests used in this chapter is to see if the standardized difference between sample statistics is acceptable when compared with some Z value that corresponds to a stated level of significance.

4 The computation of the critical ratio in the two-sample examples presented in this chapter requires the use of either the standard error of the difference between means or the standard error of the difference between percentages. Although these are new measures to consider, the other steps in the testing procedure are similar to those presented in Chapter 8.

KEY TERMS AND CONCEPTS

Independent samples   358

Sampling distribution of the differences between sample means   359

Standard error of the difference between means   360

$$CR = \frac{(\bar{x}_1 - \bar{x}_2) - (\mu_1 - \mu_2)}{\sigma_{\bar{x}_1 - \bar{x}_2}}$$

or

$$CR = \frac{\bar{x}_1 - \bar{x}_2}{\sigma_{\bar{x}_1 - \bar{x}_2}} \quad \text{(formula 9-1)} \quad 361$$

$$\sigma_{\bar{x}_1 - \bar{x}_2} = \sqrt{\frac{\sigma_1^2}{n_1} + \frac{\sigma_2^2}{n_2}} \quad \text{(formula 9-2)} \quad 361$$

$$\hat{\sigma}_{\bar{x}_1 - \bar{x}_2} = \sqrt{\frac{s_1^2}{n_1} + \frac{s_2^2}{n_2}} \quad \text{(formula 9-3)} \quad 364$$

Sampling distribution of the differences between sample percentages   367

Standard error of the difference between percentages   367

$$\hat{\sigma}_{p_1 - p_2} = \sqrt{\frac{p_1(100 - p_1)}{n_1 - 1} + \frac{p_2(100 - p_2)}{n_2 - 1}}$$
$$\text{(formula 9-4)} \quad 367$$

$$CR = \frac{(p_1 - p_2) - (\pi_1 - \pi_2)}{\hat{\sigma}_{p_1 - p_2}}$$

or

$$CR = \frac{p_1 - p_2}{\hat{\sigma}_{p_1 - p_2}} \quad \text{(formula 9-5)} \quad 368$$

PROBLEMS

1 The following random sample data have been gathered by a researcher employed by a department store chain:

	DOWNTOWN STORE	RITZY MALL STORE	DISCOUNT CITY STORE	RURAL RETREAT STORE	SKI TEXAS STORE	GUNN MALL STORE
Average purchase amount ($\bar{x}$)	$36.00	$40.00	$33.50	$28.25	$22.80	$26.00
Population standard deviations	$ 6.00	$ 8.20	$ 9.50	—	—	—
Sample standard deviations	—	—	—	$10.15	$10.50	$ 8.75
Sample size	40	38	32	42	50	58

a At the .05 level, is there a significant difference in average purchase amounts between the downtown store and the Ritzy Mall store?

b The manager of the downtown store is convinced that the average amount of purchases made at her store is higher than at the Discount City store. Is this belief supported at the .01 level of significance?

c At the .01 level, is there a significant difference in average purchase amounts between the Ski Texas store and the Gunn Mall store?

d Is the average purchase amount significantly greater at the Rural Retreat store than at the Gunn Mall store? Use the .10 level of significance.

2 The public relations manager of Tailspin Airlines (TA) is concerned about a recent increase in the number of customer claims of damage to luggage. A random sampling of the records at three TA terminals yields the following data:

	BAYBURG TERMINAL	PITT CITY TERMINAL	BEANTOWN TERMINAL
Sample count of luggage handled	760	610	830
Number of items damaged	44	53	60

a At the .05 level, is there a significant difference in damage claims between Bayburg and Pitt City?

b The terminal manager at Bayburg believes that the baggage handling at Beantown is sloppy and that his terminal experiences lower damage claims. At the .01 level, would you agree?

3 A manufacturer wants to increase the life of her product through a change in the machining process. Prior to the change in the process, a random sample of 100 items showed a mean life of 600 hours with a standard deviation (σ) of 40 hours. After the change, another random sample of 100 revealed a mean life of 612 hours with a standard deviation (σ) of 30 hours. At the .05 level, has there been a significant increase in product life?

4 Two machines are used to cut steel bars. Each machine can be adjusted to control the average bar length, but there will always be some slight variation in the length of individual bars. The variation in the production of each machine has been monitored in the past with the following results:

Machine 1 population standard deviation = 0.10 inches
Machine 2 population standard deviation = 0.12 inches

After the machines had been moved and adjusted, a machinist selected a random sample of 50 bars cut on each machine. The samples yield a mean of 55.60 inches for machine 1 and a mean of 55.50 inches for machine 2. At the .05 level, are the machines properly adjusted?

5 Two types of throat sprays are available for treating a throat ailment. Type 1 has no apparent side effects, but type 2 sometimes produces mild stomach disorders. A team of doctors has decided that it would prefer to use type 2 in spite of the side effect if the percentage of cures produced is higher with 2 than with 1. However, if the percentage of cures produced by type 1 is equal to or greater than the cures due to type 2, then type 1 will be the drug of choice. The results of two random samples are as follows: 26 of 40 persons treated with the type 1 drug were cured, and 35 of 50 persons treated with the type 2 drug were cured. At the .05 level, what should the doctors conclude?

6 A national brokerage company wants to see if the percentage of new accounts of over $30,000 has changed. A random sample of 900 accounts opened last year showed that 36 were over $30,000 in size. And a random sample of 1,000 accounts for the year before showed 44 of over $30,000 in size. At the .05 level, what should the brokerage company conclude?

7 A medical researcher claims she has discovered a cure for the common cold. When the cure is given to 50 sufferers, 13 of them recover within two days. A control group of 60 cold victims is given a placebo and within two days only 12 have recovered. At the .05 level, do these results support the researcher's claims?

8 A study is conducted to see if people living in high-altitude regions (population 1) have a greater life expectancy than those living at sea level (population 2). A random sample of 50 death certificates for individuals from population 1 yields a mean lifetime of 70 years, and a standard deviation of 11.2 years. A similar sample of death certificates for 130 persons from population 2 shows a mean lifetime of 65 years with a standard deviation of 12 years. At the .05 level, does population 1 have a greater life expectancy?

9   Does vitamin X help prevent the common cold? A random sample of 130 French skiers are given vitamin X, and another random sample of 128 French skiers are given a placebo. During the winter season 30 skiers who received vitamin X catch cold as do 39 of those who took the placebo.
   a   At the .05 level, does vitamin X significantly reduce the chances of catching cold?
   b   Can you reach the same conclusion at the .10 level? Explain.

10 A random sample of 150 men are polled, and 80 of them prefer brand X over competitive products. A random sample of 230 women shows that 130 of them also prefer brand X. At the .01 level, is there a significant difference in the preferences of men and women?

11 The Chamber of Commerce of Pitt City believes that the average cost of living in that city is lower than the average cost of living in Beantown. The living expenses of 250 Pitt City residents are sampled and the sample mean and standard deviations are found to be $33,000 and $12,000 respectively. A similar sample of 100 Beantown residents yields a sample mean and standard deviation of $39,000 and $18,000 respectively. At the .05 level is there evidence to support the belief of the Pitt City Chamber?

12 A telephone company wants to launch a communications satellite using either the Space Skuttle rocket plane or the Thor-Epsilon booster. The Thor-Epsilon is cheaper, but the Space Skuttle will be used if it's more reliable. Public information from the aerospace industry shows that 60 of 600 launches of the Thor-Epsilon were failures while 2 of 35 Space Skuttle attempts were thwarted. At the .05 level, which launch approach should the telephone company use?

13 A mountain climbing team wants to determine which of two brands of dehydrated food provides the most energy. Brand X is cheaper per pound, but brand Y will be used if it has more nutritional value (based on calories). A sample of 100 packages of brand X is tested in a calorimeter and it's found that the mean number of calories per four-ounce serving is 300 with a sample standard deviation of 30. The test is repeated on 110 packages of brand Y, and the sample mean and standard deviation for four-ounce servings are 309 and 40 calories respectively. At the .05 level, what should the team decide?

14 The president of a small college claims that 3 years after graduation, alumni of his school (group 1) are better able to find jobs with salaries greater than $40,000 than are graduates of a rival school (group 2). To test this claim, a survey of the two groups is made. A random sample of 200 alumni from group 1 shows that 55 earn more than $40,000 per year, and a similar sample from group 2 shows that 40 of the 150 alumni have salaries that exceed $40,000. At the .01 level, is the president's claim viable?

TOPICS FOR REVIEW AND DISCUSSION

1   "The purpose of this chapter is not to estimate the absolute values of parameters." Discuss this statement.

2   What assumptions must be made about the samples if the procedures discussed in this chapter are to be properly employed?

3 a What's a sampling distribution of the differences between sample means?
   b Explain how such a sampling distribution is created.
4 When will the mean of the sampling distribution of the differences between sample means equal zero?
5 a What's a sampling distribution of the differences between sample percentages?
   b Explain how such a distribution would be created.
   c If the null hypothesis is true, what would be the value of the mean of this distribution?
6 Why must an estimated standard error of the difference between percentages be used in hypothesis testing?

PROJECTS/ISSUES TO CONSIDER

1 There are usually several issues on campus that receive the attention of the student body at any given time. These issues may deal with student elections, food in the cafeteria, or dozens of other topics. Are there differences of opinion on these issues between freshman and sophomore students on the one hand and junior and senior students on the other? Do resident students agree with commuter students? Do men agree with women? Do athletes share the views of others? Let's carry out the following steps to get some answers:
   a We'll first need to divide the class into teams. Each team (or perhaps two teams) should identify a campus issue of interest to team members.
   b Team members should determine the nature of the study and the null hypothesis to be tested. For example, if the team wants to see if freshman and sophomore students take a different view of an issue than junior and senior students, then the null hypothesis can be that there's no difference between the population percentage of each group that favors the issue. The alternative hypothesis, of course, is that there *is* a significant difference.
   c Prepare the question(s) that will be used to obtain information about the issue from the respondents. Is the question clear or can it be misinterpreted?
   d Several members of a team (or one of the two teams) can sample one population (say freshman and sophomore students), while the other members (or team) poll the other population. Perhaps a telephone survey can be made and the students called can be randomly selected from the student directory. In any event, remember the material discussed in Chapter 6 and try to pick a random sample from each population. Be prepared to defend your sample selection method. Be sure to obtain large samples (over 30) from each population.
   e After the data have been gathered, the remaining step is to conduct a test of the difference between sample percentages that leads to the acceptance or rejection of the null hypothesis. Team members will pick the level of significance to use in this test.
   f Prepare a class presentation of the study methodology and the test results obtained.

ANSWERS TO SELF-TESTING REVIEW QUESTIONS

9-1

1 This statement is false. The purpose here is to determine the *relative* values of the parameters.

2 The samples must be independent samples selected from different groups, and they must be large—i.e., over 30.

3 The hypotheses are:

H_0: $\mu_1 = \mu_2$
H_1: $\mu_1 \neq \mu_2$

And at the .01 level, $Z = 2.58$.
The decision rule is:

Accept H_0 if CR falls between ± 2.58.

or

Reject H_0 and accept H_1 if CR $> +2.58$ or CR < -2.58.

The critical ratio is computed as follows:

$$CR = \frac{\bar{x}_1 - \bar{x}_2}{\sqrt{\sigma_1^2/n_1 + \sigma_2^2/n_2}} = \frac{45,000 - 46,500}{\sqrt{2,000^2/50 + 1,500^2/40}} = \frac{-1,500}{369.12} = -4.06$$

Conclusion: Reject H_0; there's a significant difference in tire quality at the .01 level.

4 The hypotheses are:

H_0: $\mu_1 = \mu_2$
H_1: $\mu_1 < \mu_2$

And at the .05 level, $Z = 1.64$.
The decision rule is:

Accept H_0 if CR ≥ -1.64.

or

Reject H_0 and accept H_1 if CR < -1.64.

The critical ratio is computed as follows:

$$CR = \frac{\bar{x}_1 - \bar{x}_2}{\sqrt{\sigma_1^2/n_1 + \sigma_2^2/n_2}} = \frac{42 - 45}{\sqrt{6^2/36 + 3^2/35}} = \frac{-3.00}{1.12} = -2.68$$

Conclusion: Reject H_0; it's likely that the Die Dead batteries have a shorter life expectancy than the Crank-Up brand.

9-2

1 The hypotheses are:

H_0: $\mu_1 = \mu_2$
H_1: $\mu_1 \neq \mu_2$

At the .10 level, $Z = 1.64$.
The decision rule is:

Accept H_0 if CR falls between ± 1.64.

or

Reject H_0 and accept H_1 if CR $> +1.64$ or CR < -1.64.

To compute CR:

$$CR = \frac{\$5{,}600 - \$5{,}300}{\sqrt{\$1{,}120^2/38 + \$725^2/36}} = \frac{\$300}{\$218} = 1.38$$

Conclusion: Since $1.38 < 1.64$, the null hypothesis would be accepted.

2

H_0: $\mu_1 = \mu_2$
H_1: $\mu_1 > \mu_2$

The decision rule is:

Accept H_0 if CR ≤ 1.28.

or

Reject H_0 and accept H_1 if CR > 1.28.

9-3

1 The hypotheses are:

H_0: $\pi_1 = \pi_2$
H_1: $\pi_1 \neq \pi_2$

This is a two-tailed test because the only concern is whether or not there is a significant difference. The Z distribution is used, and $Z = \pm 1.96$. The decision rule is:

Accept H_0 if CR falls between ± 1.96.

or

Reject H_0 and accept H_1 if CR < -1.96 or CR $> +1.96$.

The critical ratio is computed as follows:

$$CR = \frac{p_1 - p_2}{\sqrt{\dfrac{p_1(100 - p_1)}{n_1 - 1} + \dfrac{p_2(100 - p_2)}{n_2 - 1}}}$$

$$= \frac{60 - 54.5}{\sqrt{\dfrac{(60)(40)}{199} + \dfrac{(54.5)(45.5)}{199}}}$$

$$= \frac{5.50}{4.95}$$

$$= 1.13$$

Conclusion: Since the CR value falls into the area of acceptance, we can conclude that there's no significant difference of opinion between the two cities.

2 The hypotheses are:

H_0: $\pi_1 = \pi_2$
H_1: $\pi_1 > \pi_2$

This is a right-tailed test to see if the urban percentage is significantly greater than the rural percentage. The rejection region is bounded by $Z = 1.64$. The decision rule is:

Accept H_0 if CR ≤ 1.64.

or

Reject H_0 and accept H_1 if CR > 1.64.

The critical ratio is computed as follows:

$$CR = \frac{p_1 - p_2}{\sqrt{\dfrac{p_1(100 - p_1)}{n_1 - 1} + \dfrac{p_2(100 - p_2)}{n_2 - 1}}}$$

$$= \frac{21.67 - 18.00}{\sqrt{\dfrac{(21.67)(78.33)}{299} + \dfrac{(18)(82)}{99}}}$$

$$= \frac{3.67}{4.54}$$

$$= .81$$

Conclusion: The CR falls into the acceptance region—i.e., we must reject the producer's claim and accept the H_0 that there's no significant difference in preference between urban and rural viewers. (Neither group is very enthusiastic about the producer's program!)

A CLOSER LOOK

TWO-SAMPLE TESTS OF MEANS USING SMALL SAMPLES

The Z distribution was used in the examples presented in this chapter because the two samples were always large (over 30). Procedures are available, though, to apply similar tests to two *small* samples (30 or less) if it can be assumed (1) that the samples come from normally distributed populations, and (2) the standard deviations of both populations, although unknown, are equal.

When these assumptions hold, we can state our hypotheses just as we did in the chapter. That is, our null hypothesis is:

$$H_0: \mu_1 = \mu_2$$

And the alternative is one of the following:

$$H_1: \mu_1 \neq \mu_2$$
$$H_1: \mu_1 > \mu_2$$
$$H_1: \mu_1 < \mu_2$$

Values from the t table are used rather than Z values now when we define the rejection region(s). The degrees of freedom needed to use the t table when two small samples are tested is found with this formula:

$$df = n_1 + n_2 - 2$$

Thus, if there are 14 items in sample 1 and 16 items in sample 2, our df value is $(14 + 16 - 2)$ or 28. If we are making a *two-tailed test* at the .05 level of significance with these samples, then our t value is 2.048, and our decision rule is to accept H_0 if the critical ratio falls between ±2.048.

Before we can compute the critical ratio, we must first calculate the *estimated* standard error of the difference between means. Since it's assumed that the unknown population stan-

dard deviations are equal, an estimate of the population standard deviation is found by *pooling* or combining information from the two samples. The formula for the pooled standard deviation is:

$$s_{p_0} = \sqrt{\frac{s_1{}^2(n_1 - 1) + s_2{}^2(n_2 - 1)}{n_1 + n_2 - 2}}$$

And the formula for the estimated standard error is then:

$$\hat{\sigma}_{\bar{x}_1 - \bar{x}_2} = s_{p_0}\sqrt{\frac{1}{n_1} + \frac{1}{n_2}}$$

Once you've computed the standard error, the critical ratio is found in the following way:

$$CR = \frac{\bar{x}_1 - \bar{x}_2}{\hat{\sigma}_{\bar{x}_1 - \bar{x}_2}}$$

To summarize, the procedure to conduct a two-sample hypothesis test is quite similar to the procedure discussed in the chapter, with the only differences being (1) the use of the t rather than Z table, and (2) the different—or pooled—way to calculate the standard error. Now let's look at an example.

AN EXAMPLE USING SMALL SAMPLES

Let's suppose that two experimental diets designed to add weight to malnourished third-world children are being tested. It's assumed that each diet produces weight gains that are normally distributed, and it's further assumed that the population standard deviations are equal. The first diet (A) is given to 8 children, and the second diet (B) is supplied to 9 suffering from hunger. The weight gains for diet A (in pounds) after a six-week period are:

4.1   4.3   6.0   5.6   8.5   7.9   5.1   4.9

And the gains made by those fed diet B during the same time are:

7.3   6.7   8.3   7.0   6.6   6.8   9.2   7.6   5.9

At the .05 level, is there a significant difference in the weight gained by the two groups?
    The hypotheses are:

$H_0: \mu_1 = \mu_2$

$H_1: \mu_1 \neq \mu_2$

We're only interested in testing for a significant difference in weight gain. The t distribution must be used, and the rejection regions are bounded by a t value with $df = 15$ and $\alpha = .05$. This t value is 2.131. The *decision rule* is:

Accept H_0 if CR falls between ± 2.131.

or

Reject H_0 and accept H_1 if CR < -2.131 or CR $> +2.131$.

The values of the sample statistics are shown in the table below.

The pooled estimate of the population standard deviation is found next:

$$s_{p_0} = \sqrt{\frac{s_1^2(n_1 - 1) + s_2^2(n_2 - 1)}{n_1 + n_2 - 2}}$$

$$= \sqrt{\frac{2.6029(7) + .9800(8)}{15}}$$

$$= \sqrt{\frac{18.2203 + 7.84}{15}}$$

$$= 1.32$$

And the estimated standard error of the difference is then:

$$\hat{\sigma}_{\bar{x}_1 - \bar{x}_2} = s_{p_0} \sqrt{\frac{1}{n_1} + \frac{1}{n_2}}$$

$$= 1.32 \sqrt{\frac{1}{8} + \frac{1}{9}}$$

$$= .6413$$

Finally, the CR is:

$$\text{CR} = \frac{\bar{x}_1 - \bar{x}_2}{\hat{\sigma}_{\bar{x}_1 - \bar{x}_2}} = \frac{5.8 - 7.27}{.6413} = \frac{-1.47}{.6413} = -2.29$$

	SAMPLE 1			SAMPLE 2		
x	$(x - \bar{x})$	$(x - \bar{x})^2$		x	$(x - \bar{x})$	$(x - \bar{x})^2$
4.1	−1.7	2.89		7.3	.03	.0009
4.3	−1.5	2.25		6.7	−.57	.3249
6.0	.2	.04		8.3	1.03	1.0609
5.6	−.2	.04		7.0	−.27	.0729
8.5	2.7	7.29		6.6	−.67	.4489
7.9	2.1	4.41		6.8	−.47	.2209
5.1	−.7	.49		7.6	.33	.1089
4.9	−.9	.81		5.9	−1.37	1.8769
46.4		18.22		65.4		7.8401

$\bar{x}_1 = 5.8$ $S_1^2 = \dfrac{18.22}{8-1}$ $\bar{x}_2 = 7.27$ $S_2^2 = \dfrac{7.8401}{9-1}$

$\qquad\qquad\qquad\qquad = 2.6029$ $\qquad\qquad\qquad\qquad = .9800$

Conclusion: Reject H_0; CR < -2.131. The population mean weight gain of diet A isn't equal to the gain of diet B at the .05 level.

A COMPUTER MAKES IT EASY

As you've observed in Chapter 8, generating statistics from the raw sample data taken from single samples and then computing standard errors and critical ratios for those samples can be a tedious process. And in this example, the work is doubled because we must deal with two small samples. But as you also saw in Chapter 8, statistical software packages can quickly do these tiresome calculations.

Figure CL9-1 shows how the *Minitab* program processes the data for our two samples. As you can see in the first six program lines, weight-gain figures for the children fed diet A are entered into a program storage area (C1), and the gains of the children fed diet B are entered into C2. The program user then gives a TWOSAMPLE T command and specifies that the data to be used in the test are located in C1 and C2 (see line 7 of Figure CL9-1). A POOLED subcommand in line 8 tells the program how to calculate the estimated population standard deviation. The remaining lines in Figure CL9-1 are then generated by the program.

You'll notice that the package produces a 95 percent *confidence interval* for the *difference* between μ_1 and μ_2 that we don't need. (The interval -2.83 to -0.10 tells us, though, that we can be 95 percent sure that the population mean weight gain of diet A is from 2.83 pounds to 0.10 pounds *less than* the weight gain produced by diet B.) The last two lines of computer output show the same results that we've painstakingly calculated. The MU C1 = MU C2 is the null hypothesis, and the (VS NE) is the "not equal" alternative hypothesis. The pooled estimate of the population standard deviation is shown to be 1.32, and the CR (or T in this case) is -2.29. The P = 0.037 entry is the level of significance. Since our test is at the .05 level, we reject the H_0 that the population means are equal. But if a .01 test were called for, the H_0 would have been accepted.

FIGURE CL9-1 The *Minitab* output duplicating the results computed in our example problem.

```
MTB > SET C1
DATA> 4.1, 4.3, 6.0, 5.6, 8.5, 7.9, 5.1, 4.9
DATA> END
MTB > SET C2
DATA> 7.3, 6.7, 8.3, 7.0, 6.6, 6.8, 9.2, 7.6, 5.9
DATA> END
MTB > TWOSAMPLE T C1 VS C2;
SUBC> POOLED.

TWOSAMPLE T FOR C1 VS C2
       N       MEAN      STDEV    SE MEAN
C1     8       5.80      1.61      0.57
C2     9       7.267     0.990     0.33

95 PCT CI FOR MU C1 - MU C2: (-2.83, -0.10)

TTEST MU C1 = MU C2 (VS NE): T= -2.29  P=0.037  DF=  15

POOLED STDEV =          1.32
```

COMPARISON OF THREE OR MORE SAMPLE MEANS: ANALYSIS OF VARIANCE

Microcomputers, Middle Schools, and the Study of Statistics

The application of microcomputers in education is diverse but in the areas of science, mathematics and especially statistics, the power is awesome. The integration of microcomputer systems into school systems is moving at a rapid pace and special progress is being made at the middle school levels where the presence of microcomputers offers an air of excitement and interest on the part of students that has not been seen very often in any area of education. . . .

At the Miller Creek Middle School (Dixie School System) outside of San Rafael in Northern California, the integration of educational goals related to computer literacy has brought about curricula centered upon computers—an essential component of which has learning objectives requiring competency in statistics. To maximize learning transfer, the computer course was cross-correlated to learning competencies in science and mathematics courses. Students are encouraged to explore and flex their reasoning and deductive talents using microcomputers and statistical inference for projects and assignments in other courses of study. . . .

Students first utilize the graphics capabilities of the microcomputers to generate polygons, histograms, and tabular displays of data. Using simple BASIC programs written by students, there are further computations of measures of central tendency and measures of dispersion. During this time students also have contact with probability concepts and ideas related to the drawing of random samples from populations.

Students are further familiarized with statistics by detailed discussions regarding terminology and symbols—Greek letters and their statistical meanings. Students are introduced to scattergrams, correlation coefficients and regression equations. Examples of these statistical activities are taken mostly from ecology, medicine and psychology. Students are given assignments in forming hypotheses and testing hypotheses about means, proportions, and variability. Attention is given to theoretical distributions—the normal distribution, the student-*t* distribution, the binomial, the chi-square, and the *F* distribution. Using appropriate statistical programs, the students conduct statistical analyses using *z*-tests, *t*-tests, chi-square tests, and univariate and multivariate analysis of variance. Other statistical techniques are discussed and utilized as time and student progress dictate.

—From Ronald Saltinski, "Statistics Is Important: A Time to Stop Being Timid," *School Science and Mathematics*, March 1982, pp. 249–253. Reprinted with the permission of *School Science and Mathematics*.

LOOKING AHEAD

Once again you're acquainted with most of the terms used in our opening vignette. But once again there are terms such as "*F* distribution," "analysis of variance," and "chi-square" that haven't yet appeared in this book. Well, the precocious middle school students mentioned in the vignette won't have the advantage for long, because this chapter deals with analysis of variance and *F* distributions, and the next chapter is concerned with chi-square analysis. (The "scattergrams, correlation coefficients and regression equations" mentioned in the vignette are discussed in Chapter 14.)

In Chapter 9, you learned a technique for determining if there was a significant difference between the means of *two* independent samples. In this chapter, a technique which determines if there are significant differences between *three or more* sample means is discussed. This technique is called *analysis of variance* (ANOVA).

An analysis of variance permits a decision maker to conclude whether or not all means of the populations under study are equal based upon the degree of variability in the sample data. For example, if a plant supervisor wants to know if four stamping machines are producing parts within the same level of tolerance, the ANOVA technique can be used to analyze random samples drawn from the output of each machine.

In Chapter 10, we'll first consider the *hypotheses and assumptions* associated with the ANOVA technique. We'll then present an *overview of the reasoning behind analysis of variance*. Finally, we'll use the *ANOVA testing procedure* along with *F distributions tables* to make statistical decisions.

Thus after studying this chapter you should be able to:

◆ Explain the purpose of analysis of variance and identify the assumptions that underlie the ANOVA technique.

◆ Discuss the procedures for estimating the population variance by two independent approaches and the rationale for these procedures.
◆ Describe the ANOVA hypothesis-testing procedure.
◆ Use the ANOVA testing procedures and the F distributions tables to arrive at statistical decisions concerning the means of three or more populations.

CHAPTER OUTLINE

FIRST THINGS FIRST

Analysis of variance (ANOVA) is the name given to the approach that allows us to use sample data to test whether the values of two or more unknown population means are likely to be equal. If exactly *two means* are compared, the ANOVA procedure discussed in this chapter gives the same results obtained with the two-sample procedure for testing small samples that was outlined in the Closer Look reading in Chapter 9. You can verify this in the Closer Look section that concludes this chapter. Our purpose in this chapter, though, isn't to generate more two-sample tests. Rather, our purpose here is to look at situations where the need is to compare three or more unknown population means.

We'll limit our attention to cases in which the data taken from the populations deal with a *single variable or factor* such as customer traffic data handled at three sampled locations, or the speed of relief obtained by samples of people using four brands of pain suppressors. Our focus, then, is on a **one-way** ANOVA test of means, where only one classification factor or variable is considered. But techniques beyond the scope of this book also allow two or more factors to be studied. Your reaction about now might be to wonder how analyzing variances is going to help when our real interest is in comparing population means. Let's see in the following pages why the emphasis on variances isn't misplaced.

Hypotheses in ANOVA

The *null hypothesis* in analysis of variance is that the independent samples are drawn from different populations with the same mean. In other words, the null hypothesis is always:

$$H_0: \mu_1 = \mu_2 = \mu_3 = \cdots = \mu_k$$

where k = number of populations under study. And the *alternative hypothesis* in any analysis of variance is:

H_1: *Not all* population means are equal.

A careful reading of the alternative hypothesis will show you that if the alternative hypothesis is accepted, you may conclude that *at least one* population mean differs from the other population means. But the analysis of variance technique *cannot* tell you exactly *how many* population means differ; nor will it give you exact information about *which* means differ. For example, six populations could be under study, and if only one population mean differs from the other five means, which are equal, the null hypothesis may be rejected and the alternative hypothesis accepted.

Assumptions Which Must Be Met

Statistical techniques generally involve assumptions that must be valid if the techniques are to be correctly applied. In the case of analysis of variance, the *following assumptions must be true:*

1   The samples are drawn randomly, and each sample is independent of the other samples.

2 The populations under study have distributions which approximate the normal curve.

3 The populations from which the sample values are obtained all have the same unknown population variance (σ^2). That is, this third assumption is:

$$\sigma_1^{\,2} = \sigma_2^{\,2} = \sigma_3^{\,2} = \cdots = \sigma_k^{\,2}$$

where k = number of populations

Thus, *if* the null hypothesis is *true*, and *if* these three assumptions are valid, the net effect is conceptually equal to the case where all the samples are picked from the one population shown in Figure 10-1*a*. But *if* the null hypothesis turns out to be *false*, and *if* the three assumptions still remain valid, the population means will not be equal. In this event, the samples in

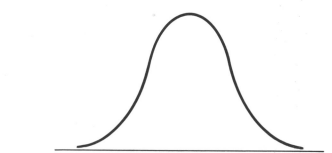

k number of normal populations with:

$$\mu_1 = \mu_2 = \mu_3 = \ldots \ldots \mu_k$$
$$\sigma_1^{\,2} = \sigma_2^{\,2} = \sigma_3^{\,2} = \ldots \ldots \sigma_k^{\,2}$$

(*a*)

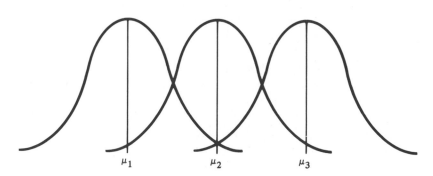

Three normal populations with:

$$\mu_1 \neq \mu_2 \neq \mu_3$$
$$\sigma_1^{\,2} = \sigma_2^{\,2} = \sigma_3^{\,2}$$

FIGURE 10-1   (*b*) an application might be taken from populations such as those shown in Figure 10-1b. Of course, these populations are still normally distributed, and they still have the same variance value. If there are any serious violations of the three assumptions, the use of the ANOVA procedure is ruled out.

THE REASONING BEHIND ANALYSIS OF VARIANCE

Before we describe the step-by-step procedure in ANOVA, you should understand the underlying conceptual framework of the technique. In an analysis of variance, *two estimates* of the population variance (σ^2) are computed on the basis of two independent computational approaches. A *first*[1] *approach* is to compute an estimator of σ^2 that remains appropriate regardless of any differences between population means. In other words, the means of the several populations may differ, but this estimator of σ^2 *isn't* affected by the possible fact that the H_0 is false. Because of this fact, this computed value cannot, by itself, be used to test the validity of the H_0. A second element is needed.

The *second approach* results in the computation of an appropriate estimate of σ^2 *if, and only if,* the population means are equal. This approach produces an estimate that does contain the effects of any differences between the population means. If there are *no differences* between means, this computed value of σ^2 doesn't differ too much from the first value (which is now used as a standard against which the second estimate is evaluated).

In short, the preceding two paragraphs may be summarized with the following *decision rules:*

If the two computed estimates are approximately the same, we may conclude that there are probably no differences between the population means. Thus, the null hypothesis should be accepted.

But

If the estimate computed by the second approach is significantly different from the estimate of the first approach, we may conclude that the second estimate contains effects of population mean differences. Thus, the H_0 should be rejected.

Let's now briefly examine these two estimates of σ^2 in a little more detail.

$\hat{\sigma}^2_{within}$: An Estimator of σ^2 Regardless of Population Mean Values

The first computed estimate of σ^2 is found by *pooling* and *averaging* the variances found *within* each of the samples. (Early in Chapter 4—and in

---

[1] The word "first" is used here in a conceptual sense. As you'll see later, it's not necessary for this value to be computed first in the ANOVA procedure.

the discussion of Figure 4-2—a variance was computed, and you might wish to review that material now before going on.) You learned in Chapter 7 that the formula for s—the sample standard deviation used as an estimator of σ—is:

$$s = \sqrt{\frac{\Sigma(x - \bar{x})^2}{n - 1}}$$

Thus, the formula for s^2 is $\Sigma(x - \bar{x})^2/(n - 1)$. Although any one of the sample variances (s^2) could be used as an estimator of σ^2, the variances of all the samples are usually pooled and averaged to estimate σ^2 because of the greater amount of data thus considered. Therefore, in the ANOVA procedure, a variance from each sample is computed, and the sample variances are then pooled to produce $\hat{\sigma}^2_{within}$—an estimator of the population variance that remains appropriate regardless of the equality or inequality of the population mean. Any deviation of this estimate from the population variance is due to random error. (Keep that word "error" in mind—we'll refer to it later.)

$\hat{\sigma}^2_{between}$: An Estimator of σ^2 if (and Only if) the H_0 Is True

This second approach to estimating σ^2 is based upon the variation *between* the sample means and is founded on the Central Limit Theorem. (Ha! And you had hoped that you had heard the last of that concept!) If the H_0 is true, then, as we saw in Figure 10-1a, it is as though all the samples are selected from the same normal population distribution with the same μ. And as we saw in Chapter 6, and as the Central Limit Theorem tells us, if the population is normally distributed, the sampling distribution of the sample means will also be normal. Furthermore, you will remember that the standard deviation of this sampling distribution—the standard error of the sample means—is found by this basic formula:

$$\sigma_{\bar{x}} = \frac{\sigma}{\sqrt{n}}$$

Now, if we square both sides of this basic equation, we get

$$\sigma_{\bar{x}}^2 = \frac{\sigma^2}{n}$$

which can be manipulated further to yield

$$n\sigma_{\bar{x}}^2 = \sigma^2$$

Thus, if we knew the square of the standard error ($\sigma_{\bar{x}}^2$), we could compute the precise value of σ^2 merely by multiplying $\sigma_{\bar{x}}^2$ by the sample size.

"Well, so what?" you may be thinking, since you don't expect to have any idea of the size of $\sigma_{\bar{x}}^2$. Fortunately for us all, you will be able (by a formula to be presented in just a few pages) to effectively (1) compute an estimate of the square of the standard error $(\hat{\sigma}_{\bar{x}}^2)$ and (2) multiply this estimate by the sample size to effect an estimate of σ^2. This second approach to estimating σ^2 produces $\hat{\sigma}_{between}^2$—an estimator of σ^2 that has merit if (and only if) the H_0 is true. Any deviation of this estimator from the σ^2 is due to factor differences among the samples. (Keep that word "factor" in mind too because we'll also get back to it later.)

In summary, then, if the H_0 is true, the $\hat{\sigma}_{between}^2$ value should be a good estimate of the population variance, and it should be approximately the same as the $\hat{\sigma}_{within}^2$ value. Should there be a significant difference between $\hat{\sigma}_{within}^2$ and $\hat{\sigma}_{between}^2$, however, it may be concluded that this difference is the result of differences between population means.

The F Ratio and the F Distributions Tables

When is the difference between the two σ^2 estimates statistically significant? When is the difference between $\hat{\sigma}_{within}^2$ and $\hat{\sigma}_{between}^2$ due to inequality between population means, and when is the difference simply due to random sampling error? For analytical purposes, a clever statistician has determined that the difference between $\hat{\sigma}_{within}^2$ and $\hat{\sigma}_{between}^2$ may be expressed as a **computed F ratio** in this way:

$$F = \frac{\hat{\sigma}_{between}^2}{\hat{\sigma}_{within}^2} \qquad \text{(formula 10-1)}$$

Ideally, from the standpoint of verifying the H_0, the computed F ratio should have a value of 1. However, some disparity between the two σ^2 estimates may be expected because of sampling variation even when the H_0 is true. How much disparity, as reflected by the computed F ratio, may be tolerated before the H_0 is rejected? The answer to this question is found in the **F distributions tables** in Appendix 6. p. 661

The *maximum* value that the computed F ratio may attain (at a chosen level of significance) before the H_0 must be rejected is specified in the F distributions tables. Thus, conclusions concerning the H_0 in analysis of variance tests are based on comparisons of computed F ratios with values from the F distributions tables. If the computed F ratio value $\leq$ the table value, the H_0 is accepted; if the computed F ratio $>$ the table value, the H_0 is rejected. Further comment on the use of Appendix 6 is given in a few pages.

Self-Testing Review 10-1

1 What's the purpose of an analysis of variance?
2 There are three major assumptions in an analysis of variance. What are they?
3 What is the null hypothesis in an analysis of variance test?

4   When the H_0 is rejected, it may be concluded that all the population means are unequal. True or false?

5   The $\hat{\sigma}^2_{within}$ value is an estimator of σ^2 if, and only if, the population means are equal. True or false?

6   If $\hat{\sigma}^2_{within}$ is 6 and $\hat{\sigma}^2_{between}$ is 8, the F ratio is .75. True or false?

PROCEDURE FOR ANALYSIS OF VARIANCE

The manager of the Nickel and Dime Savings Bank is reviewing employee performance for possible salary increases and position promotions. In evaluating tellers, the manager decides that the most important performance criterion is the number of customers served each day. The manager expects that each teller should handle approximately the same number of customers daily. Otherwise, each teller should be rewarded or penalized accordingly.

Six business days are sampled randomly, and customer traffic for each of the three tellers is recorded. The factor or variable of interest, then, is customer traffic. The sample (or teller) data are shown below:

Customer Traffic Data

DAY	TELLER 1 MS. MUNNY	TELLER 2 MR. COYNE	TELLER 3 MR. SENTZ
1	45	55	54
2	56	50	61
3	47	53	54
4	51	59	58
5	50	58	52
6	45	49	51

Stating the Null and Alternative Hypotheses

As you know by now, the first step in any hypothesis-testing situation is to state the null and alternative hypotheses. For the problem at hand, the *null hypothesis* is that the three tellers each serve the same average number of customers per day. That is, Ms. Munny, Mr. Coyne, and Mr. Sentz are assumed to have the same work load. Symbolically, we have

H_0: $\mu_1 = \mu_2 = \mu_3$

The *alternative hypothesis* is that not all the tellers are handling the same average number of customers per day. That is, at least one of the tellers is performing much better than the others, or, perhaps, at least one of the tellers is performing much worse than the others. Thus,

H_1: Not all the population means are equal.

Specifying the Level of Significance

You'll undoubtedly recall from preceding chapters that a criterion for rejection of the H_0 is necessary. In our case, if the bank manager is to make a judgment about the tellers' performance, and if the consequences of this judgment will be reflected directly in the tellers' paychecks, the manager should have some idea of the degree of error possible in the decision. Let's assume that the manager has stated that there should be at most only a 5 percent chance of erroneously rejecting a true H_0. Thus we will specify that $\alpha = .05$.

Computing Estimates of σ^2: $\hat{\sigma}^2_{between}$ and $\hat{\sigma}^2_{within}$

We are now into an essential phase of the ANOVA technique. Roll up your sleeves and concentrate.

Computing $\hat{\sigma}^2_{between}$ The computed $\hat{\sigma}^2_{between}$ value is an estimate of σ^2 based upon the variation between sample means. Furthermore, the $\hat{\sigma}^2_{between}$ value is likely to yield a reliable estimate of the population variance if, and only if, the H_0 is true. *The following steps should be completed.*

1   For the sample data on each teller, a mean must be computed. The means are shown in Figure 10-2 and have been designated $\bar{x}_1$, $\bar{x}_2$, and $\bar{x}_3$ for tellers 1, 2, and 3 respectively.

2   After computing the individual sample means, you should calculate the "total" or "grand" mean, $\bar{\bar{X}}$. This **grand mean** is simply the mean of all the sample values. It's computed using the following formula:

$$\bar{\bar{X}} = \frac{\text{total of all sample items}}{\text{number of sample items}} \qquad \text{(formula 10-2)}$$

In our example, and as shown in Figure 10-2, the grand mean is computed as follows:

$$\bar{\bar{X}} = \frac{\text{total of all sample items}}{\text{number of sample items}} = \frac{294 + 324 + 330}{18} = \frac{948}{18} = 52.67$$

3   After these easy steps, we are now ready to determine the value of $\hat{\sigma}^2_{between}$. It is computed by the following formula:

$$\hat{\sigma}^2_{between} = \frac{n_1(\bar{x}_1 - \bar{\bar{X}})^2 + n_2(\bar{x}_2 - \bar{\bar{X}})^2 + \cdots + n_k(\bar{x}_k - \bar{\bar{X}})^2}{k - 1} \qquad \text{(formula 10-3)}$$

FIGURE 10-2   Data Used in Computing $\hat{\sigma}^2_{between}$

DAY	TELLER 1 MS. MUNNY (1)	TELLER 2 MR. COYNE (2)	TELLER 3 MR. SENTZ (3)
1	45	55	54
2	56	50	61
3	47	53	54
4	51	59	58
5	50	58	52
6	45	49	51
Totals	294	324	330
	$\bar{x}_1 = 49$	$\bar{x}_2 = 54$	$\bar{x}_3 = 55$

$$\bar{\bar{X}} = \frac{\text{total of all sample items}}{\text{number of sample items}} = \frac{294 + 324 + 330}{18} = 52.67$$

$$\hat{\sigma}^2_{between} = \frac{n_1(\bar{x}_1 - \bar{\bar{X}})^2 + n_2(\bar{x}_2 - \bar{\bar{X}})^2 + n_3(\bar{x}_3 - \bar{\bar{X}})^2}{k - 1}$$

$$= \frac{6(49 - 52.67)^2 + 6(54 - 52.67)^2 + 6(55 - 52.67)^2}{3 - 1}$$

$$= \frac{6(13.469) + 6(1.769) + 6(5.429)}{2} = \frac{124.00}{2}$$

$$= 62.0$$

where n_1 = number of items or observations in sample 1
     n_2 = number of items or observations in sample 2
     n_k = number of items or observations in sample k
     k = number of samples under study
     $\bar{x}_1$ = mean of sample 1
     $\bar{x}_2$ = mean of sample 2
     $\bar{x}_k$ = mean of sample k
     $\bar{\bar{X}}$ = total mean or the average of all the items in the samples

If you closely examine formula 10-3, you'll notice that $\hat{\sigma}^2_{between}$ is basically $n\hat{\sigma}^2_{\bar{x}}$, which, as we've seen, is equal to $\hat{\sigma}^2$.

From the use of this formula in Figure 10-2, we can see that the total computed by using the squared differences of the sample means about the $\bar{\bar{X}}$ is 124.00. This total of 124 is often called a "sum of squares" value, as we'll see later. The computed value for $\hat{\sigma}^2_{between}$ given in Figure 10-2 is 62.0, and this figure is often called a "mean of squares" that's based on the factor or variable being considered. This 62.0 is an appropriate estimate of σ^2 *if, and only if*, the null hypothesis is true.

The procedure for computing $\hat{\sigma}^2_{between}$ by focusing on the variance between sample means is summarized in Figure 10-3.

Computing $\hat{\sigma}^2_{within}$ You'll recall that $\hat{\sigma}^2_{within}$ is an estimate of σ^2 based upon an average that's computed after the variance data within each sample are

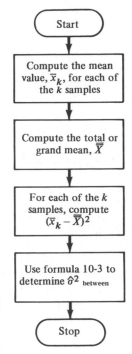

FIGURE 10-3
Procedure for the
computation of
$\hat{\sigma}^2_{between}$ (the "MS
factor").

pooled. More specifically, $\hat{\sigma}^2_{within}$ is the mean of the estimates of σ^2 that have been obtained from each sample.

The formula to compute $\hat{\sigma}^2_{within}$ (regardless of whether the samples are of equal size or not) is:[2]

$$\hat{\sigma}^2_{within} = \frac{\Sigma d_1{}^2 + \Sigma d_2{}^2 + \Sigma d_3{}^2 + \cdots + \Sigma d_k{}^2}{T - k} \qquad \text{(formula 10-4)}$$

where $\Sigma d_1{}^2$ = the sum of the squared differences—that is, $\Sigma(x_1 - \bar{x}_1)^2$—for the first sample

$\Sigma d_2{}^2$ = the sum of the squared differences—that is, $\Sigma(x_2 - \bar{x}_2)^2$—for the second sample, etc.

T = total number of all items in all samples $(n_1 + n_2 + n_3 + \cdots + n_k)$

k = number of samples

[2] If the samples are *equal in size,* the σ^2_{within} value can also be found with this formula:

$$\sigma^2_{within} = \frac{\sigma_1{}^2 + \sigma_2{}^2 + \cdots \sigma_n{}^2}{k}$$

where $\hat{\sigma}^2$ for each sample = $\Sigma(x - \bar{x})^2/(n - 1)$ and k = the number of samples.

Let's now work through the computational steps for $\hat{\sigma}^2_{within}$ for our bank teller example problem.

1   For *each sample,* we compute the *deviation* between each value within the sample and the mean of that sample—that is, compute $x - \bar{x}$ for each sample. Figure 10-4 shows the deviation computations for each teller. For teller 1, Ms. Munny, the average customer traffic ($\bar{x}_1$) is 49. Thus, the deviation for day 1 is $45 - 49$, or -4; the deviation for day 2 is $56 - 49 = 7$; and so forth.

2   After the deviation of each observation from its sample mean is computed, each deviation is *squared*—that is, $(x - \bar{x})^2$. These squared deviations are summed— $\Sigma(x - \bar{x})^2$—and the sums are labeled Σd^2. That is, $\Sigma(x - \bar{x})^2 = \Sigma d^2$. For tellers 1, 2, and 3, the sums of the squared deviations are 90, 84, and 72 respectively. This total of 246 is also often called a "sum of squares" value, as we'll see later.

3   The sum of squares figure of 246 is then divided by the quantity $T - k$, where T is the total of all sample items, or 18 in this example, and k is the number of samples (3). Thus, $T - k = 15$.

Finally, as you can see in Figure 10-4, $\hat{\sigma}^2_{within}$ is conveniently computed as follows, using formula 10-4:

$$\hat{\sigma}^2_{within} = \frac{\Sigma d_1^2 + \Sigma d_2^2 + \Sigma d_3^2}{T - k}$$

$$= \frac{90 + 84 + 72}{18 - 3} = \frac{246}{15}$$

$$= 16.4$$

This estimate of σ^2 is also referred to as a "mean of squares," and any deviation of this value from the true σ^2 is due to random error.

FIGURE 10-4   Data Used in Computing $\hat{\sigma}^2_{within}$

DAY	TELLER 1 MS. MUNNY x_1	$x_1 - \bar{x}_1$	$(x_1 - \bar{x}_1)^2$	TELLER 2 MR. COYNE x_2	$x_2 - \bar{x}_2$	$(x_2 - \bar{x}_2)^2$	TELLER 3 MR. SENTZ x_3	$x_3 - \bar{x}_3$	$(x_3 - \bar{x}_3)^2$
1	45	-4	16	55	1	1	54	-1	1
2	56	7	49	50	-4	16	61	6	36
3	47	-2	4	53	-1	1	54	-1	1
4	51	2	4	59	5	25	58	3	9
5	50	1	1	58	4	16	52	-3	9
6	45	-4	16	49	-5	25	51	-4	16
	294		$\Sigma d_1^2 = 90$	324		$\Sigma d_2^2 = 84$	330		$\Sigma d_3^2 = 72$
	$\bar{x}_1 = 49$			$\bar{x}_2 = 54$			$\bar{x}_3 = 55$		

$$\hat{\sigma}^2_{within} \frac{90 + 84 + 72}{18 - 3} = \frac{246}{15} = 16.4$$

Figure 10-5 summarizes the steps performed in this section. You'll be delighted to know that most of the tedious computations are now behind us. And you'll also be pleased to see in a few pages that programs are readily available that will allow you to process ANOVA problems on micro- or larger computers. All that remains now is for us to compute the F ratio and then compare this value with an appropriate F table figure. We'll then be in a position to make a decision about the H_0.

Computing the Critical F Ratio: CRF

Now that we've computed $\hat{\sigma}^2_{between}$ and $\hat{\sigma}^2_{within}$, we're ready to compare these two $\hat{\sigma}^2$ estimates and calculate the magnitude of the difference between the two. In short, we're now ready to calculate the computed F ratio, or CR_F. As we saw earlier in formula 10-1,

$$CR_F = \frac{\hat{\sigma}^2_{between}}{\hat{\sigma}^2_{within}}$$

On the basis of our problem data and our recent calculations, we may compute the F ratio as follows:

$$CR_F = \frac{62.0}{16.4} = 3.78$$

Thus, it's obvious that the estimate of $\hat{\sigma}^2_{between}$ is almost 4 times as large as the estimate of $\hat{\sigma}^2_{within}$. But is this calculated CR_F of such magnitude that we may conclude that the H_0 is not true? To answer this question, we must know the *critical value* that separates the area of acceptance from the area of rejection. That is, we must know the critical value shown in Figure 10-6.

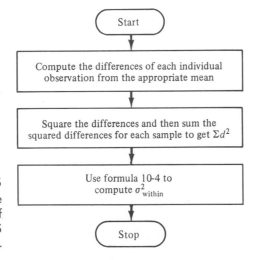

FIGURE 10-5
Procedure for the computation of $\hat{\sigma}^2_{within}$ (the "MS error").

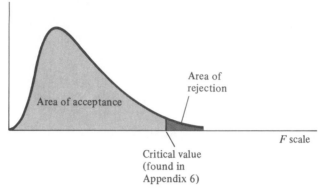

If the computed F ratio $\leq$ the critical value shown in Figure 10-6, it falls into the area of acceptance and the H_0 is accepted. And if the computed F value $>$ this critical value, it falls into the area of rejection and the H_0 is rejected. The critical value against which the computed F ratio is compared is found in the F distributions tables in Appendix 6. If a hypothesis test is being made at the .05 level, the critical value is simply the point on the F scale beyond which an F ratio value is expected to fall only 5 times in 100 if the H_0 is true.

As you can see in Figure 10-6, an F distribution is skewed to the right. Each F distribution is determined by the number of samples and the number of observations in the samples. This means that there's a *different F distribution* for every possible combination of sample number and sample size. And since there are different table critical values for each possible combination and for each selected level of significance, the size of an F distributions tables can quickly become unmanageable. Therefore, in Appendix 6, only a selected number of possible combinations are presented, and only the .01 and .05 levels of significance are available. Let's now see how to use Appendix 6.

Using the *F* Distributions Tables

In using the F distributions tables in Appendix 6, you first determine the degrees of freedom (df) for both the numerator and the denominator of the computed F ratio. The *df for the numerator is computed as follows:*

$$df_{\text{num}} = k - 1 \qquad \text{(formula 10-5)}$$

where k = the number of samples.

And the *df for the denominator is computed as follows:*

ANOVA IN QUALITY CONTROL

Although quality control is primarily a statistical discipline, it also covers methods of analysis and generation of data from which statistical probabilities, and statistical inferences in terms of predictions and deviations, may be derived. Thus, statistical manipulation of data obtained in a total quality control system is little more than the final analysis which assists in the decision as to whether action is to be taken, particularly in cases where results are not obvious. Stated another way, all that application of statistics does within a quality control system is provide the opportunity to make decisions on the basis of known risks instead of on the basis of intuition alone. . . .

Statistical quality control techniques were developed before the time of mini- and microcomputers. Therefore, the usefulness of a statistical control technique was based not only on its mathematical validity, but also on the opportunity for rapid computation, so that results could be reported and posted promptly and accurate... tion taken as soon as possible. Thus, it became of major importance to develop not only rapid analytical methods, but also rapid computational methods. For this reason, a whole array of nonparametric procedures were developed.

With the recent revolution in miniaturization of computers, it is no longer necessary to consider the length or tediousness of the mathematical calculations, or the frequency of recalculation of limits, or even the posting of the statistically analyzed data. Thus, more elegant and fewer "quick-and-dirty" methods need to be used. Where data follow, even approximately, a normal distribution, the analysis of variance (ANOVA) method can be used to great advantage. . . .

—From Amihud Kramer, "Application of Statistics to Quality Control," *Food Technology*, vol. 35, no. 4, April 1981, pp. 56ff. Copyright © by Institute of Food Technologists. Reprinted with permission.

$$df_{\text{den}} = T - k \qquad \qquad \text{(formula 10-6)}$$

where T = the total number of items in all samples, or $n_1 + n_2 + n_3 + \cdots + n_k$
  k = the number of samples

The degrees of freedom for our bank teller example, then, are determined as follows:

$df_{\text{num}} = 3 - 1 = 2$
$df_{\text{den}} = 18 - 3 = 15$

(As a check, it might be worth noting here that the total of the df in the numerator and denominator should equal $T - 1$—that is, $df_{\text{num}} + df_{\text{den}} = T - 1$. Since $15 + 2 = 18 - 1$, we can conclude that our math is correct.)

We're now ready to find the critical F value from the appropriate table for the specific combination of df_{num}, df_{den}, and α that is present in our teller problem. The critical F value in our case is based upon (1) 2 degrees of freedom in the numerator, (2) 15 degrees of freedom in the denominator, and (3) a level of significance of .05. In using the F tables in Appendix 6, we must first locate the table with the relevant α. (In this problem, it's the first table in that appendix.) Next, we must locate the critical value of F where the degrees of freedom for the numerator (found at the top of the *columns*) and the degrees of freedom for the denominator (shown to the left of the *rows*) intersect. For our example, the critical F value is 3.68:

$$F_{(2,15,\alpha=.05)} = 3.68$$

The final step in the ANOVA procedure is to draw a statistical conclusion concerning the validity of the null hypothesis.

Making the Statistical Decision

As indicated earlier, the following *decision rule* applies in the analysis of variance test:

If $\mathrm{CR}_F > F$ value found in the table, reject H_0 and accept the alternative hypothesis.

If the computed F ratio is greater than the table F value, it may be concluded that the difference between the two estimates of the population variance, $\hat{\sigma}^2_{\text{between}}$ and $\hat{\sigma}^2_{\text{within}}$, is so large that such a magnitude of difference has an extremely unlikely chance of occurring. Therefore, such a magnitude of difference implies that it's likely that the H_0 is not valid.

Since in our example problem we have $\mathrm{CR}_F = 3.78$, and since the critical table value = 3.68, we must conclude that the H_0 is unlikely and that the alternative hypothesis should be accepted. At least one of the tellers among Munny, Coyne, and Sentz is likely to be handling more or less work than the others. Additional research is needed to precisely define the nature of the difference in work performance.

THE ONE-WAY ANOVA TABLE AND COMPUTERS TO THE RESCUE

The One-Way ANOVA Table

It's often desirable to place the computations of the last several pages in a **one-way ANOVA table**—a summary listing of the values needed to produce an ANOVA test. The general format of such a table is shown in Figure 10-7. The first column lists the *source* or type of variation. As we've seen, this variation is measured *between* samples—the **factor variation**—and *within* samples—the **error variation.** Column 2 shows the *degrees of freedom* associated with each source of variation (the *df* values we've just computed). Column 3 records the *sum of squares* (SS) figures computed during our $\hat{\sigma}^2_{\text{between}}$ and $\hat{\sigma}^2_{\text{within}}$ calculations. (You were warned then that we would use these terms again.) The fourth column is labeled **mean of squares (MS)**—the designation in an ANOVA table for our computed values of $\hat{\sigma}^2_{\text{between}}$ and $\hat{\sigma}^2_{\text{within}}$. The $\hat{\sigma}^2_{\text{between}}$ value is often called **MS factor,** and the $\hat{\sigma}^2_{\text{within}}$ value is

FIGURE 10-7   The Format of a General One-Way ANOVA Table

SOURCE OF VARIATION	DEGREES OF FREEDOM (df)	SUM OF SQUARES (SS)	MEAN OF SQUARES (MS)	COMPUTED F VALUE (CR$_F$)
Between samples (factor variation)	$k - 1$	SS between (SSB) or SS factor	$\hat{\sigma}^2_{between}$ (MS factor)	$\dfrac{\hat{\sigma}^2_{between}}{\hat{\sigma}^2_{within}}$
Within samples (error variation)	$T - k$	SS within (SSW) or SS error	$\hat{\sigma}^2_{within}$ (MS error)	
Total	$T - 1$	SSB + SSW		

often referred to as **MS error.** Finally, column 5 presents the computed F value (CR$_F$), which, we've seen, is $\hat{\sigma}^2_{between}/\hat{\sigma}^2_{within}$ (or MS factor/MS error).

Computers to the Rescue

The *Minitab* statistical package makes short work of the procedures we've taken a chapter to explain. Figure 10-8 shows that the customer traffic data used in our example problem have been stored in three columns. Column 1 (C1) holds the number of customers served by the first teller, and columns 2 and 3 store the sample data for tellers 2 and 3 (see the top lines in Figure 10-8). Once the sample data are entered, the entire ANOVA procedure is carried out by the package in response to the single AOVONEWAY C1-C3 command. As you can see in Figure 10-8, a one-way ANOVA table is produced. The same values we've painstakingly calculated are reproduced in the computer output, along with some additional useful information. The p figure of .047 at the top right corner of the ANOVA table is the level of significance for the sample data. Since .047 is less than the specified α of .05, we know to reject the H_0. If p had been .05 or larger, the H_0 would have been accepted.

The 95 percent confidence intervals produced below the ANOVA table give us an idea of the intervals that are likely to include the three population means. Each interval is calculated with the formula:

$$\bar{x} \pm t\left(\frac{s_{p_o}}{\sqrt{n}}\right)$$

where s_{p_o} is the estimated population standard deviation found by pooling and averaging the sample variances. In this case the value of s_{p_o} is 4.05, which is the square root of the $\hat{\sigma}^2_{within}$ value of 16.4. Thus, the 95 percent confidence interval that's likely to include μ_1 is found as follows:

$$49.00 \pm t\left(\frac{4.05}{\sqrt{6}}\right) = 49.00 \pm 2.131\,(1.653) = 45.48 \text{ to } 52.52$$

(The t value from the t table in Appendix 4 is at the 95 percent level, and

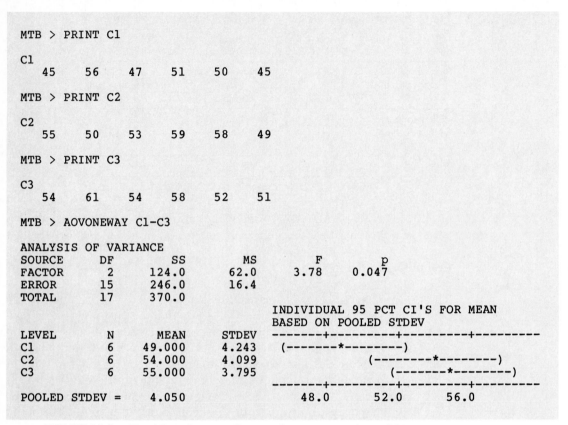

```
MTB > PRINT C1

C1
   45      56      47      51      50      45

MTB > PRINT C2

C2
   55      50      53      59      58      49

MTB > PRINT C3

C3
   54      61      54      58      52      51

MTB > AOVONEWAY C1-C3

ANALYSIS OF VARIANCE
SOURCE      DF          SS         MS         F         p
FACTOR       2       124.0       62.0      3.78     0.047
ERROR       15       246.0       16.4
TOTAL       17       370.0

                                   INDIVIDUAL 95 PCT CI'S FOR MEAN
                                   BASED ON POOLED STDEV
LEVEL       N        MEAN       STDEV    -------+---------+---------+---------
C1          6      49.000       4.243    (-------*---------)
C2          6      54.000       4.099              (---------*---------)
C3          6      55.000       3.795                  (-------*---------)
                                         -------+---------+---------+---------
POOLED STDEV =    4.050                     48.0      52.0      56.0
```

FIGURE 10-8   The *Minitab* output for our chapter example problem.

the *df* is that of the MS error value, or 15 in this case, so *t* is 2.131 in this example.) The other intervals likely to include μ_2 and μ_3 are found in the same way. Since there's a slight overlap of all three intervals, there's a slim chance that the H_0 is true, but, as we've seen, that chance is less than .05. The greater the interval overlap, of course, the more likely it is that H_0 is true.

Congratulations! You've made it through a rather detailed chapter. Figure 10-9 provides an outline of the ANOVA procedure that you've now successfully digested.

Self-Testing Review 10-2

1   The Jingle and Cliché Ad Agency has been commissioned to design a package for a new powdered soft-drink mix. According to Mr. Jones of Jingle and Cliché, a good package should lead to good sales. The creative staff, after 3 months' work, has produced three attractive packages. One is gold, another is red, and the third is orange. Prior to making a final package decision, however, the ad agency managers decided that each package should be placed on the shelves of supermarkets in a city for 14 business days. This was done, and the package sales for each day were recorded. The sales data are shown on the next page:

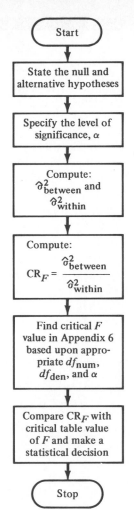

FIGURE 10-9
Procedure
for analysis of
variance.

Start

State the null and
alternative hypotheses

Specify the level of
significance, α

Compute:
$\hat{\sigma}^2_{\text{between}}$ and
$\hat{\sigma}^2_{\text{within}}$

Compute:

$$CR_F = \frac{\hat{\sigma}^2_{\text{between}}}{\hat{\sigma}^2_{\text{within}}}$$

Find critical F
value in Appendix 6
based upon appro-
priate df_{num},
df_{den}, and α

Compare CR_F with
critical table value
of F and make a
statistical decision

Stop

Package Sales

DAY	GOLD PACKAGE	RED PACKAGE	ORANGE PACKAGE
1	10	12	9
2	19	13	18
3	17	25	36
4	22	39	18
5	25	44	25
6	29	37	21
7	32	36	38
8	31	38	31
9	29	35	33
10	33	27	28
11	31	42	29
12	32	22	32
13	28	36	30
14	27	25	31

Using these data, conduct an analysis of variance at the .05 level of significance.

2   The *Minitab* program, when supplied with the package sales data given above, produces the output shown in Figure 10-10. Examine this output, explain the values given in the ANOVA table, and analyze the confidence intervals that are shown.

3   Professor Krusher overheard one of his students saying, "Krusher's exams are so tough, it makes no difference whether you take them after a good night's rest, at the end of a tiring day, or after 36 hours without sleep! You get the same grade." Professor Krusher decided to conduct an experiment to see whether the student's statement had any validity. He initially recruited a random sample of 36 student volunteers and divided the 36 students into three groups of 12. Each group was administered a typical 1-hour Krusher exam. One group was fresh and alert when they took the exam; another group took the exam at the end of a long day; and the third group took the test in a groggy state. During the testing, one student in the tired second group could not complete the exam, and two students in the groggy group fell asleep. The scores on the exam were as shown on the next page:

FIGURE 10-10   The *Minitab* output for the package sales data given in problem 1.

```
MTB > PRINT C1

C1
   10      19      17      22      25      29      32      31      29      33      31      32      28
   27

MTB > PRINT C2

C2
   12      13      25      39      44      37      36      38      35      27      42      22      36
   25

MTB > PRINT C3

C3
    9      18      36      18      25      21      38      31      33      28      29      32      30
   31

MTB > AOVONEWAY C1-C3

ANALYSIS OF VARIANCE
SOURCE      DF          SS          MS          F          p
FACTOR       2       172.8        86.4       1.21      0.310
ERROR       39      2790.2        71.5
TOTAL       41      2963.0
                                         INDIVIDUAL 95 PCT CI'S FOR MEAN
                                         BASED ON POOLED STDEV
LEVEL        N        MEAN       STDEV    -------+---------+---------+---------
C1          14      26.071       6.776    (----------*-----------)
C2          14      30.786      10.222                   (-----------*----------)
C3          14      27.071       8.014      (------------*----------)
                                         -------+---------+---------+---------
POOLED STDEV =      8.458                   24.0       28.0       32.0
```

	GROUP		
	ALERT	TIRED	GROGGY
Exam scores	43	56	84
	67	67	78
	76	69	69
	55	62	64
	71	61	67
	62	57	65
	39	59	66
	65	68	63
	61	72	58
	53	66	70
	58	64	
	62		

Using these data, conduct an analysis of variance at the .01 level of significance.

LOOKING BACK

1   The purpose of the one-way ANOVA technique discussed in this chapter is to enable a decision maker to compare three or more independent sample means to determine if there are statistically significant differences between the means of the populations from which the samples were selected.

2   The following three assumptions must be met before the ANOVA technique can be applied to a decision-making situation: (1) the samples are random and independent of each other, (2) the population distributions approximate the normal distribution, and (3) the variances of all populations are equal.

3   As is the case with other hypothesis-testing procedures, the ANOVA technique begins with a statement of the null and alternative hypotheses; it then requires that the test be made at a suitable level of confidence. Following these steps, two estimates of the population variance are computed by two independent computational approaches. In one approach, the $\hat{\sigma}^2_{within}$ approach, σ^2 is estimated by computing the variances found in each sample. These sample variances are pooled and averaged to get an estimate of the population variance. Any deviation of this estimate from the true population variance is due to random error. In the other approach, $\hat{\sigma}^2_{between}$, an estimate of σ^2 is computed by measuring the variation between the sample means. This estimate of σ^2 has merit if, and only if, the H_0 is true. Any deviation of this estimate from the σ^2 is due to factor differences among the samples. The two estimates of σ^2 are then used to compute an F ratio. Ideally, if the H_0 is true, this ratio will have a value of 1.00 since the two estimates of σ^2 would yield the same results.

4   Realistically, however, sampling variation will normally cause the F ratio to exceed 1.00 even when the H_0 is true. The amount by which the computed CR_F value may be permitted to exceed 1.00 before the H_0 is rejected is determined in the F distributions tables in Appendix 6.

5   If a computed CR_F value is found to exceed the appropriate F table value at a specified level of confidence, the ANOVA test procedure concludes with the rejection of the H_0; if the CR_F value $\leq$ the table value, the H_0 is accepted.

6   It's often desirable to prepare a one-way ANOVA table—a summary listing of the values needed to produce an ANOVA test. The first column of this table lists the sources of variation, and subsequent columns show degrees of freedom, sum of squares (SS) figures, mean of squares (MS) values (the $\hat{\sigma}^2_{between}$ and $\hat{\sigma}^2_{within}$ calculations), and the computed F ratio. Statistical software packages typically display ANOVA output in this table format.

KEY TERMS AND CONCEPTS

Analysis of variance (ANOVA)   382
One-way ANOVA   383
Computed F ratio   387

$$F = \frac{\hat{\sigma}^2_{between}}{\hat{\sigma}^2_{within}} \text{ (formula 10-1)} \quad 387$$

F distributions tables   387
Grand mean   389

$$\overline{\overline{X}} = \frac{\text{total of all sample items}}{\text{number of sample items}} \text{ (formula 10-2)} \quad 389$$

$$\hat{\sigma}^2_{between} = \frac{n_1(\overline{x}_1 - \overline{\overline{X}})^2 + n_2(\overline{x}_2 - \overline{\overline{X}})^2 + \cdots + n_k(\overline{x}_k - \overline{\overline{X}})^2}{k - 1}$$
(formula 10-3)   389

$$\hat{\sigma}^2_{within} = \frac{\Sigma d_1^2 + \Sigma d_2^2 + \Sigma d_3^2 + \cdots + \Sigma d_k^2}{T - k}$$
(formula 10-4)   391

$df_{num} = k - 1$ (formula 10-5)   394
$df_{den} = T - k$ (formula 10-6)   395
One-way ANOVA table   396
Factor variation   396
Error variation   396
Mean of squares (MS)   396
MS factor   396
MS error   397

PROBLEMS

1   A high school track coach has learned of two new training techniques which are designed to reduce the time needed to run a mile. Three random samples of novice runners have been selected for an experiment. Group A has trained under the old approach, group B has trained under one of the new techniques, and group C has trained under the other new technique. After a month of training, each runner was timed in a mile run, and the following times were recorded (in minutes).

GROUP A	GROUP B	GROUP C
4.81	4.43	4.38
4.62	4.50	4.29
5.02	4.32	4.33
4.65	4.37	4.36
4.58	4.41	4.74
4.52	4.39	4.42
4.73	4.64	4.40

Conduct an analysis of variance at the .05 level to determine if $\mu_A = \mu_B = \mu_C$.

2   A consumer advocate group wants to determine if the three leading brands of aspirin are really different from each other in terms of speed of relief. Consumers have been randomly selected and assigned to use either brand A, brand B, or brand C. Each subject was instructed to take the recommended

dosage when a headache began, and each subject was told to record the number of minutes that elapsed before relief occurred. The following data (in minutes) have been gathered:

BRAND A	BRAND B	BRAND C
7.3	6.7	7.4
8.5	7.1	7.8
6.4	9.0	6.9
7.9	8.4	8.5
6.7	7.8	7.4
9.1	6.9	8.2
7.4	8.7	7.4

Conduct an analysis of variance at $\alpha = .05$. Is there a significant difference in speed of relief?

3   A manufacturer of canned beans wonders if there is truly any difference in sales based upon the height of the shelves on which the product is displayed. With the cooperation of three retail stores, an experiment has been conducted. One store placed the cans at eye level on the shelves; another store placed the cans at waist level; and the third store placed the cans at knee level. Sales data over 8 days were recorded:

Sales of Cans of Beans

EYE LEVEL	WAIST LEVEL	KNEE LEVEL
98	106	103
106	105	95
111	98	87
85	93	94
108	96	92
86	98	82
83	97	87
109	104	83

Conduct an analysis of variance at $\alpha = .05$. Is there a significant difference in sales based on the shelf location of the product?

4   A large accounting firm wants to determine if the accuracy of its employees is related to the school from which the employees graduated. Accountants representing four schools were randomly selected, and the number of errors committed by each accountant over a 2-week period was recorded as shown below:

SCHOOL A	SCHOOL B	SCHOOL C	SCHOOL D
14	17	19	23
16	16	20	12
17	18	22	21
13	15	21	10
22	16	18	9
9	12	19	15
10	14	15	16

Conduct an analysis of variance at $\alpha = .01$. Is there a significant difference in accuracy?

5 A manufacturer wants to know if the four machines in operation are performing with equal efficiency. Random samples have been drawn from the machines, and the deviations of the samples from specifications have been recorded in millimeters as shown below:

MACHINE A	MACHINE B	MACHINE C	MACHINE D
50	66	50	70
50	61	75	75
55	57	65	73
45	72	60	80
61	68	55	72
56	55	52	78

What statistical conclusion can be reached at the .01 level?

6 A large retailer must make a choice between three sales locations within a shopping mall. The retailer is wondering if the daily traffic count is the same for all locations. The following data are traffic counts for a random 10-day period:

LOCATION X	LOCATION Y	LOCATION Z
643	249	458
542	404	513
569	378	475
552	337	482
607	426	539
514	298	491
576	345	468
585	362	487
581	425	464
600	376	476

At the .05 level, is there a significant difference in traffic count at the three locations?

7 What can be concluded about a three-sample experiment from the following information?

$n_1 = 13, n_2 = 12, n_3 = 8$, and $\alpha = .05$

$\hat{\sigma}^2_{between} = 164$ and $\hat{\sigma}^2_{within} = 43$

8 Draw a statistical conclusion based upon the following data:

$n_1 = 20, n_2 = 16, n_3 = 19, n_4 = 21$, and $\alpha = .05$

$\hat{\sigma}^2_{between} = 158$ and $\hat{\sigma}^2_{within} = 54$

9 At $\alpha = .01$, what decision can be reached if the following facts are known?

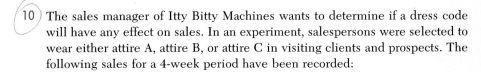

$n_1 = 22$, $n_2 = 22$, $n_3 = 22$, $n_4 = 23$, and $n_5 = 26$

$\hat{\sigma}^2_{\text{between}} = 374$ and $\hat{\sigma}^2_{\text{within}} = 93$

10 The sales manager of Itty Bitty Machines wants to determine if a dress code will have any effect on sales. In an experiment, salespersons were selected to wear either attire A, attire B, or attire C in visiting clients and prospects. The following sales for a 4-week period have been recorded:

ATTIRE A	ATTIRE B	ATTIRE C
26	19	22
37	24	33
41	31	34
35	28	19
29	23	25
33	25	29
40	24	31
	29	
	32	

What statistical decision can be made at the .05 level?

11 As a part of his research on smoking addiction, Nick O. Teen is interested in the abilities of cigarette, cigar, and pipe smokers to refrain from lighting up. Subjects were randomly selected in each category, and each subject was asked to wait as long as possible between smokes. The time interval was recorded in minutes for each smoker. The following data have been obtained:

CIGARETTE SMOKERS	CIGAR SMOKERS	PIPE SMOKERS
6	13	8
13	22	17
7	12	14
19	14	23
8	17	18
9	19	12
12	20	11
23	11	15
16		12
10		27
25		31
8		

What may be concluded about the ability of different smokers to refrain from smoking? Use $\alpha = .05$.

12 A major distributor of cameras suspects that consumers would be insensitive to price changes for the highest-quality camera. To test this suspicion, four retail outlets have been selected, and each outlet has sold the highest-quality camera at one of four predetermined prices. After some time, the following weekly unit sales at each store were reported:

PRICE I	PRICE II	PRICE III	PRICE IV
3	5	10	8
5	4	9	4
7	6	4	5
9	5	7	7
4	8	2	6
2	7	6	9
10	6	8	6
8	5	8	
		11	

What conclusion may be drawn at $\alpha = .05$?

13 A college professor believes that the season of the year affects the amount of time students spend studying. In a year-long experiment, students were selected randomly during the different seasons and were asked to estimate the average number of hours they spent per week studying. The hourly estimates of the students were:

SUMMER	FALL	WINTER	SPRING
4	6	7	7
3	8	11	5
6	7	12	6
7	9	8	4
5	6	13	4
3	8	6	3
4	5	5	4
	4	4	7
		7	

What may the professor conclude at the .05 level?

14 Due to clerical error on the part of a chicken farmer, four truckloads of chickens have arrived simultaneously at a processing plant. The receiving supervisor is thus confronted with the problem of selecting one of the truckloads of chickens. If the weights of the chickens in all four trucks can be assumed to be equal, any truck may be selected. Samples have been drawn randomly from each truck. (Equal samples were not possible because the crates on each truck were loaded differently.) The weights, in pounds, are as follows:

TRUCK 1	TRUCK 2	TRUCK 3	TRUCK 4
4.3	3.7	4.1	3.4
3.7	3.6	3.9	4.1
3.8	4.0	3.4	4.2
4.2	3.8	4.2	3.9
3.9	3.7	3.8	4.0
3.5			3.7

At the .05 level, what decision should the supervisor make?

15 The Tackey Toy Company plans to install special battery packs in its new line of Tackey robots. Three suppliers can produce packs that meet Tackey's needs. Although Tackey managers have price quotations from these three suppliers, they also want data on the life expectancy of each brand of battery pack before selecting a vendor. Thus, a sample of six packs produced by each vendor was selected at random and used to power Tackey robots. The useful life of each battery pack (in hours) was:

VENDOR 1	VENDOR 2	VENDOR 3
144	168	184
136	150	172
146	142	168
134	166	180
150	136	176

At the .05 level, is there a significant difference in the life expectancy of the different brands of battery packs?

TOPICS FOR REVIEW AND DISCUSSION

1 What is analysis of variance?
2 "The null hypothesis is the same in every one-way ANOVA test." Discuss this statement.
3 What are the three assumptions in an analysis of variance?
4 What does acceptance of the alternative hypothesis in an analysis of variance mean?
5 "In the ANOVA procedure, two estimates of the population variance are computed using two independent computational approaches." Discuss this statement, and explain the reasoning behind each of the approaches used.
6 a Which estimate of the population variance remains valid regardless of the population mean values?
   b Which estimate is true only if the population means are equal?
7 "A sum of squares value is the numerator in both the $\hat{\sigma}^2_{between}$ formula and the $\hat{\sigma}^2_{within}$ formula." Explain this sentence.
8 a What's another name for MS factor?
   b What's another name for MS error?
9 "In conducting an analysis of variance, as soon as we see that CR_F is less than 1, we may accept the null hypothesis and assume that there's no need for further computations." Discuss this statement.
10 Find $F_{(3,14,\alpha=.05)}$, $F_{(3,16,\alpha=.05)}$, $F_{(3,18,\alpha=.05)}$, and $F_{(3,20,\alpha=.05)}$.
11 In comparing the four F values found in the preceding question, what may be concluded about the relationship between the total sample size and the F value for the hypothesis test?
12 Identify the columns and rows in a one-way ANOVA table, and discuss the meaning of the values that are held in each cell of that table.

PROJECTS/ISSUES TO CONSIDER

1   In the project described in Chapter 8, you were asked to team up with other class members and then locate journal articles that contained published values for statistical tests. Review the steps taken to complete that project and follow the same procedure again. That is, locate a few journal articles of interest to team members that contain ANOVA results. If possible, recompute the F test figures presented in the articles, note any discrepancies you might find, and prepare a brief team presentation for the class.

ANSWERS TO SELF-TESTING REVIEW QUESTIONS

10-1

1   The purpose of an analysis of variance is to determine whether or not all means of the populations under study are equal.
2   a   The samples are random and independent of each other.
   b   The population distributions approximate the normal distribution.
   c   The variances of all populations are equal.
3   The null hypothesis states that all population means are equal.
4   False. Acceptance of the alternative hypothesis simply means that at least one mean differs from the other means.
5   False. The $\hat{\sigma}^2_{within}$ value is an unbiased estimator of σ^2 regardless of the truth of the null hypothesis.
6   False. $F = 1.33$

10-2

1   a   The hypotheses are: H_0: The packages are equally attractive and will result in equal average sales—that is, $\mu_1 = \mu_2 = \mu_3$. H_1: Not all population means are equal.
   b α is specified at .05.
   c   Computation of $\hat{\sigma}^2_{between}$:
       $\bar{x}_1$ = mean of gold package sales = 26.071
       $\bar{x}_2$ = mean of red package sales = 30.786
       $\bar{x}_3$ = mean of orange package sales = 27.071
       $\bar{X}$ = 27.976

$$\hat{\sigma}^2_{between} = \frac{14(-1.905)^2 + 14(2.81)^2 + 14(-.905)^2}{2}$$

$$= \frac{172.8}{2}$$

$$= 86.40$$

   d   Computation of $\hat{\sigma}^2_{within}$:
       Gold package: $\Sigma d_1^2 = 596.93$
       Red package: $\Sigma d_2^2 = 1,358.36$
       Orange package: $\Sigma d_3^2 = 834.93$

$$\hat{\sigma}^2_{within} = \frac{596.93 + 1,358.36 + 834.93}{42 - 3}$$

$$= \frac{2,790.22}{39}$$

$$= 71.54$$

e   Computation of CR_F:

$$CR_F = \frac{86.40}{71.54} = 1.21$$

f   Using the F table:

$$F_{(2,39, \alpha = .05)} \approx 3.24$$

g   Since the computed F ratio is less than 3.24, we cannot reject the null hypothesis. We may conclude that all the packages have the same effect on sales.

2   The computed value of $\hat{\sigma}^2_{between}$ is 86.4 and the value of $\hat{\sigma}^2_{within}$ is 71.5. The computed F ratio is thus 1.21 as we've seen in the answer to problem 1 above. The level of significance is high (.31) in this case—much higher than the specified α of .05—and so we accept the H_0. The 95 percent confidence intervals for the population means that are produced with the data taken from these three samples support this conclusion. There's considerable overlap in these intervals, and thus plenty of room for $\mu_1 = \mu_2 = \mu_3$. Each confidence interval is produced using the formula $\bar{x} \pm t(s_{p_o}/\sqrt{n})$. The sample mean $(\bar{x})$ for sample 1 is used to prepare the interval for μ_1, and so on. The s_{p_o} value is the pooled estimate of the σ that's assumed to be equal in all three populations. The value of s_{p_o} (8.458) used in each of the interval calculations is the square root of the MS error value of 71.54 that's computed above. And the t value used to figure each interval is found in the t table at the intersection of the 95 percent level column and the 39 df row. Since there isn't a $df = 39$ row in the table, we'll use the $df = 40$ value of 2.021. Thus, the interval for $\mu_1 = 26.071 \pm 2.021 (8.458/\sqrt{14})$ or 21.5 to 30.64 gold packages sold (ignoring the fact, of course, that fractional parts of packages can't be sold).

3   a   The hypotheses are: H_0: The students will perform equally on the exam—i.e., $\mu_1 = \mu_2 = \mu_3$. H_1: Not all population means are equal.

b α is specified at .01.

c   Computing $\hat{\sigma}^2_{between}$, we have:

$\bar{x}_1$ = mean score of alert group = 59.333
$\bar{x}_2$ = mean score of tired group = 63.727
$\bar{x}_3$ = mean score of groggy group = 68.400

$$\bar{\bar{X}} = \frac{2,097}{33} = 63.545$$

Finally,

$$\hat{\sigma}^2_{between} = \frac{12(-4.212)^2 + 11(.182)^2 + 10(4.855)^2}{2} = \frac{448.96}{2}$$

$$= 224.5$$

d   Computing $\hat{\sigma}^2_{within}$, we have:

$\Sigma d_1{}^2 = 1{,}262.67 = $ sum of squared deviations for alert group
$\Sigma d_2{}^2 = 268.18 = $ sum of squared deviations for tired group
$\Sigma d_3{}^2 = 514.40 = $ sum of squared deviations for groggy group

Thus,

$$\hat{\sigma}^2_{within} = \frac{1{,}262.67 + 268.18 + 514.40}{33 - 3}$$

$$= \frac{2{,}045.25}{30}$$

$$= 68.18$$

e   Computing CR_F:

$$CR_F = \frac{224.5}{68.18} = 3.29$$

f   Using the F table:

$$F_{(2,30,\alpha=.01)} = 5.39$$

g   Since the CR_F of $3.29 < 5.39$, we conclude that there's no real difference between the group means at the .01 level. Students may get the same grade on a Krusher exam whether they're alert or groggy! You might want to compare this answer to the *Minitab* output for this problem, shown below.

```
MTB > PRINT C1

C1
   43    67    76    55    71    62    39    65    61    53    58    62

MTB > PRINT C2

C2
   56    67    69    62    61    57    59    68    72    66    64

MTB > PRINT C3

C3
   84    78    69    64    67    65    66    63    58    70

MTB > AOVONEWAY C1-C3

ANALYSIS OF VARIANCE
SOURCE      DF        SS        MS        F         p
FACTOR       2      448.9     224.5     3.29     0.051
ERROR       30     2045.2      68.2
TOTAL       32     2494.2

                                INDIVIDUAL 95 PCT CI'S FOR MEAN
                                BASED ON POOLED STDEV
LEVEL        N       MEAN      STDEV   ----------+---------+---------+------
C1          12     59.333     10.714   (-------*-------)
C2          11     63.727      5.179        (-------*--------)
C3          10     68.400      7.560              (--------*--------)
                                       ----------+---------+---------+------
POOLED STDEV =      8.257                      60.0      66.0      72.0
```

A CLOSER LOOK

COMPARING PROCEDURES FOR TESTS OF TWO SMALL SAMPLES

The Z distribution was used in the examples presented in Chapter 9 to compare the population means of two samples because the two samples were always large (over 30). But in the Closer Look reading at the end of that chapter, a procedure was presented to apply similar tests to two *small* samples (30 or less). The assumptions required for this small-sample procedure are that (1) the two independent random samples come from normally distributed populations and (2) the standard deviations of both populations, although unknown, are equal. You'll recognize now that these are also the assumptions that apply when a one-way ANOVA test is made. In fact, if exactly two means are compared, the ANOVA procedure

discussed in this chapter gives the same results produced by the two-sample t test procedure outlined in the Chapter 9 Closer Look reading. To verify this fact, follow these steps:

1 Review the example of the two experimental diets shown in the Closer Look in Chapter 9.
2 Study the computer output shown in Figure CL9-1 that duplicates the hand calculations carried out in the reading.
3 Now that you've refreshed your memory about the problem situation and the two-sample t-test procedure, compare the results produced in Figure CL9-1 with those shown in Figure CL10-1 that are generated by the one-way ANOVA procedure.

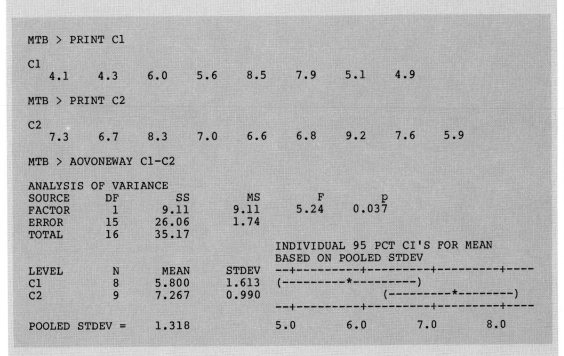

```
MTB > PRINT C1

C1
   4.1    4.3    6.0    5.6    8.5    7.9    5.1    4.9

MTB > PRINT C2

C2
   7.3    6.7    8.3    7.0    6.6    6.8    9.2    7.6    5.9

MTB > AOVONEWAY C1-C2

ANALYSIS OF VARIANCE
SOURCE      DF         SS         MS         F          p
FACTOR       1       9.11       9.11       5.24      0.037
ERROR       15      26.06       1.74
TOTAL       16      35.17
                                          INDIVIDUAL 95 PCT CI'S FOR MEAN
                                          BASED ON POOLED STDEV
LEVEL        N       MEAN      STDEV     --+---------+---------+---------+----
C1           8      5.800      1.613     (---------*---------)
C2           9      7.267      0.990                    (---------*--------)
                                         --+---------+---------+---------+----
POOLED STDEV =       1.318               5.0       6.0       7.0       8.0
```

FIGURE CL10-1 The results of using the one-way ANOVA procedure to process the sample data given in Figure CL9-1.

4 Except for some rounding differences, you'll notice that the relevant values are the same—that is, the pooled estimate of the population standard deviation is about 1.32, and the p value, or level of significance, is .037 in both tests. The conclusions, of course, are also the same in both examples: The null hypothesis that the population mean weight gain of diet A is equal to the mean weight gain of diet B is rejected at the .05 level. In the case of the t-test procedure, the computed value of -2.29 *didn't* fall into the acceptance region bounded by t values of ± 2.131. And in the ANOVA procedure, the computed F ratio of 5.24 falls *beyond* the table value of $F_{(1,15,\alpha=.05)} = 4.54$.

11

COMPARISON OF SEVERAL SAMPLE PERCENTAGES: CHI-SQUARE ANALYSIS

Trial by Number

Our world is inundated with statistics. Every medical fear or triumph is charted by a complex analysis of chances. Think of cancer, heart disease, AIDS: The less we know, the more we hear of probabilities. This daily barrage is not a matter of mere counting but of inference and decision in the face of uncertainty. No committee changes our schools or our prisons without studies on the effects of busing or early parole. Money markets, drunken driving, family life, high energy physics, and deviant human cells are all subject to tests of significance and data analysis.

This all began in 1900 when Karl Pearson published his chi-square test of goodness of fit, a formula for measuring how well a theoretical hypothesis fits some observations. The basic idea is simple enough. Suppose that you think a die will fall equally often on each of its six faces. You roll it 600 times. It seems to come up six all too frequently. Could this simply be chance? How well does the hypothesis— that the die is fair—fit the data? The result that would best fit your theory would be that each face came up just 100 times in 600. In practice the ratios are almost always different, even with a fair die, because even for many throws there will always be the factor of chance. How different should they be to make you suspect a poor fit between your theory and your 600 observations? Pearson's chi-square test gives one measure of how well theory and data correspond.

The chi-square test can be used for hypotheses and data where observations naturally fall into discrete categories that statisticians call cells. If, for example, you are testing to find whether a certain treatment for cholera is worthless, then the patients divide among four cells: treated and recovered, treated and died, untreated and recovered, and untreated and died. If the treatment is worthless, you expect no difference in recovery rate between treated and untreated patients. But chance and uncontrollable variables dictate that there will almost always be some difference. Pearson's test takes this into account, telling you how well your hypothesis—that the treatment is worthless—fits your observations.

There were earlier measures of the same sort of thing. Pearson's contribution was to provide a *standard* test, applicable to any discipline, which he institutionalized through the work of his laboratory, his students, and his journal, *Biometrika*. Pearson's test was based on well-known mathematics. And it was practical. In an era when one had to calculate by hand, it was essential that the arithmetic for applying the test be simple. It needed only one standard table, computed by Pearson's students and published in his journal.

The chi-square test was a tiny event in itself, but it was the signal for a sweeping transformation in the ways we interpret our numerical world. Now there could be a standard way in which ideas were presented to policy makers and to the general public.

—From Ian Hacking, "Trial By Number," *Science 84*, November 1984, pp. 69–70. Copyright 1984 by the AAAS. Reprinted with permission.

LOOKING AHEAD

In the passage that opens this chapter, you learned that Karl Pearson ushered in a new type of statistical test in 1900—a chi-square test that changed the way people reason, carry out experiments, and form opinions. (More information about Karl Pearson and other prominent statisticians is found in the Closer Look reading at the end of this chapter.)

In the pages that follow, we'll first present an overview of chi-square distributions and chi-square testing. This orientation is then followed by a discussion of the procedures involved in conducting a k-sample hypothesis test of percentages. Such a test involves the use of a contingency table. Finally, the steps in another procedure—the chi-square test for goodness of fit mentioned in the opening vignette—are presented. Thus, after studying this chapter you should be able to:

- Explain the purpose of (a) the k-sample hypothesis test of percentages and (b) the goodness of fit test.
- Describe the steps in the k-sample testing procedure, and use them to arrive at statistical decisions concerning the percentages of three or more populations.
- Describe the steps in the goodness of fit test, and use them to arrive at statistical decisions about whether or not a population under study follows a uniform distribution.

CHAPTER OUTLINE

CHI-SQUARE DISTRIBUTIONS AND CHI-SQUARE TESTING: AN OVERVIEW

In Chapter 9, we first considered two-sample hypothesis tests of means, and then in Chapter 10 we used analysis of variance to test for the existence of significant differences between three or more sample means. In Chapter 9, we also studied two-sample hypothesis tests of percentages. Now in this chapter, we'll use a technique to test the hypothesis that three or more *independent samples* have all come from populations that have the same *percentage* of a given characteristic.

When three or more sample percentages are compared, the techniques described in Chapter 9 aren't adequate for the job. Although the procedure discussed in this chapter can be used in two-sample situations to give the same results obtained in Chapter 9, our focus now is on problems involving the comparison of three or more samples.

The symbol for the Greek letter chi (which is pronounced as the first two letters in the word "kind") is χ. Therefore, the symbol that we'll use in this discussion of chi-square distributions and chi-square testing is χ^2.

In any hypothesis-testing situation, we're always concerned with the computation of the value of some critical ratio or statistic for which we know the appropriate distribution. For example, in the tests that have been considered in the previous three chapters, the appropriate distributions at one time or another were the normal distribution, the *t* distribution, and the *F* distribution. But none of these distributions is appropriate if we want to test the significance of the differences that may exist between three or more sample percentages. When such a test is made, a new distribution is needed. This new distribution is the **chi-square distribution**—a continuous probability distribution with a shape that's dependent on the number of degrees of freedom in a problem. Let's examine this distribution a little closer, and clarify this definition.

The χ^2 Distributions

You'll soon see that a χ^2 value is found by adding squared numbers, and so it always has a positive sign. As a matter of fact, the scale of possible χ^2 values extends from zero indefinitely to the right in a positive direction. Like the binomial, t, and F probability distributions that we've already studied, the χ^2 probability distribution is *not* a single probability curve. Rather, there's an entire family of χ^2 curves, and the *shapes* of these curves vary according to the number of *degrees of freedom (df)* that exist in a given problem. The *df* value, in fact, is the only parameter in a χ^2 distribution, and the mean of any χ^2 distribution is equal to the *df* value. For χ^2 distributions with over 2 *df*, the mode or peak of the curve is at $df - 2$. Thus, the mode for a χ^2 curve with 7 *df* is a χ^2 value of 5.

Figure 11-1 shows the shapes of several χ^2 probability distributions with different degrees of freedom. You'll notice that the curves with small degrees of freedom—say from 3 to 10 *df*—are skewed to the right. As the *df* number continues to increase, however, the χ^2 distributions begin to take on the appearance of a normal curve. Although there are different distributions, there's only a single χ^2 table to contend with. This table is found in Appendix 7 at the back of the book, and we'll see how to use it in a few pages.

An Introduction to Chi-Square Testing

In Chapter 8, the one-sample hypothesis test of percentages compared an observed sample percentage with an expected or hypothetical population percentage. Well, chi-square distributions are also used in a procedure that involves comparing the differences between the actual sample frequencies of occurrence or percentages, and the hypothetical population frequencies of occurrence or percentages that are expected if the hypothesis is true. The *steps in the general* χ^2 *testing procedure are as follows:*

1   *Formulate the null and alternative hypotheses* The null hypothesis in a particular application, for example, might be that there's *no* significant difference between the frequencies of occurrence or percentages that are of interest in the several populations under study. The alternative hypothesis, on the other hand,[1] might then be that *not all* the population percentages are equal.

---

[1] The following is the supplication of an executive who, after listening to his staff statistician, had once again published an embarrassingly inaccurate forecast: "Lord, please find me a one-armed statistician . . . so I won't always hear 'on the other hand. . . .'"

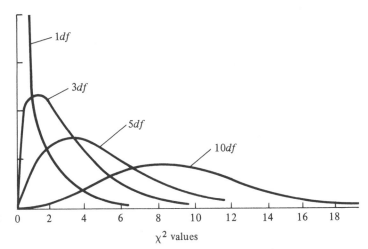

FIGURE 11-1
χ^2 distributions for
different degrees of
freedom.

2 *Select the level of significance* to be used in the particular testing situation.
3 Take random samples from the populations, and *record the observed fre-quencies* that are actually obtained.
4 *Compute the frequencies or percentages that would be expected if the H_0 is true.*
5 *Use the observed* (sample) *and expected* (hypothetical population) *frequencies to compute a χ^2 value with the following formula:*

$$\chi^2 = \sum \frac{(f_o - f_e)^2}{f_e} \qquad \text{(formula 11-1)}$$

where f_o = an observed (sample) frequency
  f_e = an expected (hypothetical) frequency if the H_0 is true

If the observed frequencies and the expected frequencies are *identical,* formula 11-1 shows us that the computed χ^2 value is *zero.* From the standpoint of verifying the null hypothesis, a computed χ^2 value of zero is ideal, since our sample data would be exactly what we had expected. But you're sophisticated enough by now in matters statistical to know that it's very unlikely that the f_o and f_e values in an application will be identical even when the H_0 is actually true. Sampling variation will, after all, usually cause some discrepancy between the f_o and f_e values.

6 *Compare the value of χ^2 computed in step 5 with a χ^2 table value* (found for the specified level of significance from the appropriate χ^2 distribution) to determine if the computed χ^2 value is *significantly* above zero. The maximum value that the computed χ^2 statistic may attain at a chosen level of significance before the H_0 is rejected is specified in Appendix 7. (Be patient; we'll look at the table soon.) This table value is the critical χ^2 value that separates the

area of acceptance from the area of rejection. That is, this table value is the critical value shown in Figure 11-2. If the computed χ^2 value $\leq$ the critical value shown in Figure 11-2, it falls into the area of acceptance and the H_0 is accepted. But if the computed χ^2 value $>$ the table value, this is cause for rejecting the H_0. If a hypothesis test is made at the .05 level, the critical table value is simply the point on the χ^2 scale beyond which a computed χ^2 value would be expected to fall only 5 times in 100 if the H_0 is true. (Since there's a separate χ^2 distribution for each df value, the size of a detailed χ^2 distributions table very quickly becomes too large for our purposes. Therefore, in Appendix 7, only a selected number of df and α values are available.)

Now that you have some idea of the general χ^2 hypothesis-testing procedure, let's see how this testing concept is applied to situations in which three or more sample percentages are evaluated.

k-SAMPLE HYPOTHESIS TEST OF PERCENTAGES

Our purpose now is to use the testing procedure to analyze the significance of the percentage differences that may exist among three or more—that is, k—independent samples. This type of test is called a **k-sample hypothesis test of percentages.** Let's look at an example of this test.

Three candidates are running for sheriff of Lawless County. These aspirants to public office are Larson E. Bound, Graff D. Lux, and Emma Nocruk. Bound's campaign manager has conducted candidate preference polls in the county's three towns. The results of these random samples of county voters are shown in the table in Figure 11-3. This table is called a **contingency table** because it shows all the cross-classifications of the variables being studied—that is, it accounts for all contingencies. As you can see, the samples of voters are classified by town of residence and by candidate preference. In planning future campaign strategies, Bound's manager would like to know if there's a significant difference in voter preference between the three towns.

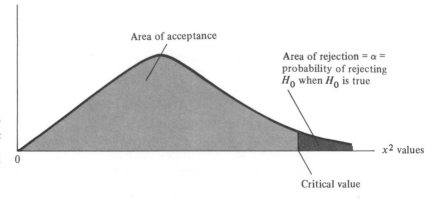

FIGURE 11-2
Areas of acceptance and rejection in a χ^2 distribution.

FIGURE 11-3   Survey of Voters Classified by Town of Residence and Candidate Preference

	TOWNS				
	WHITE LIGHTNING	CASINO CITY	SMUGGLERSVILLE		
Bound	50	40	35	125	row
Lux	30	45	25	100	totals
Nocruk	20	45	20	85	
	100	130	80	310	
		column totals		grand total	

Steps in the Hypothesis-Testing Procedure

Formulate the Null and Alternative Hypotheses   The *null hypothesis* is that the population percentages favoring each of the three candidates is unchanged from town to town. This does not mean that our H_0 is that each candidate has an equal population percentage of 33.33. Rather, our H_0 is that the population percentage of voters favoring Bound is the same in all three towns, the π favoring Lux is the same regardless of location of residence, and the π favoring Nocruk is equal in the three locations. Of course, the population percentage favoring Lux can be different from the π favoring Bound in the context of our null hypothesis. Thus, our H_0 may be stated as follows:

H_0: The population percentage favoring each candidate is the same from town to town.

The *alternative hypothesis* is:

H_1: The population percentage favoring each candidate is *not* the same from town to town.

The null hypothesis could also be expressed like this:

H_0: The candidate preference of voters is *independent* of their town of residence.

Thus, this type of χ^2 hypothesis test is frequently referred to as a **test of independence** because the H_0 tested is essentially that sample data are being classified in independent ways. And the alternative hypothesis is then:

H_1: The candidate preference of voters is dependent on (or related to) their place of residence.

Select the Level of Significance   This step doesn't differ from similar steps in other tests. As noted earlier, though, a detailed table of χ^2 distributions could quickly become very large. Thus, there are only a limited number of values in Appendix 7 (not yet, not yet . . .). We'll assume in our example problem that Bound's campaign manager has specified that the test be made at the .05 level of significance.

Determine the Observed (Sample) Frequencies (f_o)   The actual data from the random samples of voters taken in the three towns are shown in Figure 11-3.

Compute the Expected Frequencies (f_e)   If our null hypothesis is true, and if the population percentage of voters favoring Bound is the same in each of the three towns, we should be able to compute the number of sample responses favoring Bound that would be expected in each of the locations. A total of 310 voters were polled, and 125 of these voters expressed a preference for Bound. Since the 125 "votes" received by Bound is 40.32 percent of the total cast in the three towns— $(125/310) \times 100 = 40.32$ percent—Bound should be favored by 40.32 percent of those interviewed in each of the three towns if the H_0 is true. Thus, of the 100 people interviewed in White Lightning, we would expect 40.32 percent of them to favor Bound. Similarly, of the 130 persons polled in Casino City, we would anticipate that 40.32 percent (or 52.42) of them would prefer Bound. And we would expect 32.26 people in Smugglersville (40.32 percent of the 80 persons interviewed) to be in Bound's corner.

The same analysis can also be used to compute the f_e values for the other candidates in each of the towns. (Since 100/310 or 32.26 percent of all those sampled showed a preference for Lux, for example, we would expect that Lux would be favored by about 32.26 of the 100 persons interviewed in the White Lightning sample.) But *the computations of the expected frequencies are somewhat easier to follow* if we refer to the contingency table in Figure 11-3 and compute the hypothetical or expected frequencies for each *cell* in the table. A **cell** is formed by the intersection of a column and a row. Since there are 3 rows and 3 columns in Figure 11-3, there are 3 $\times$ 3 or 9 cells. (Tables with r rows and c columns are often referred to as $r \times c$ **tables;** in our example, we have a 3 $\times$ 3 table.)

The *expected frequencies may be computed for each cell of a contingency table by the following formula:*

$$f_e = \frac{\text{(row total)(column total)}}{\text{grand total}} \qquad \text{(formula 11-2)}$$

The use of this formula may be illustrated by computing the number of persons who would be *expected* to favor Bound in the town of White Lightning if the H_0 is true. From Figure 11-3, you can see that the total for the row in which this cell is located is 125, the total for the column for this cell is 100, and the grand total is 310. Thus, the f_e value for the cell is computed as follows:

$$f_e = \frac{(125)(100)}{310} = 40.32$$

The same procedure can be used to get the f_e values for all the other cells in the table. Figure 11-4 duplicates all the observed frequencies from Figure 11-3 and shows the expected frequencies for each cell in parentheses.

FIGURE 11-4   Observed and Expected Frequencies

	TOWNS				
	WHITE LIGHTNING	CASINO CITY	SMUGGLERSVILLE		
Bound	50 (40.32)	40 (52.42)	35 (32.26)	125	
Lux	30 (32.26)	45 (41.94)	25 (25.81)	100	row totals
Nocruk	20 (27.42)	45 (35.65)	20 (21.94)	85	
	100	130	80	310	
		column totals		grand total	

Compute the χ^2 Value Using f_o and f_e The computed χ^2 value, you'll recall, is found by formula 11-1:

$$\chi^2 = \sum \frac{(f_o - f_e)^2}{f_e}$$

where f_o is the observed frequency of a cell and f_e is the hypothetical expected frequency of the cell

Figure 11-5 illustrates the use of this formula. The columns numbered 1 and 2 in Figure 11-5 reproduce the data from Figure 11-4 in a more convenient format. As a check of your arithmetic, note that $\Sigma f_o = \Sigma f_e$. The *steps to compute* χ^2 (shown in Figure 11-5) are:

1   Subtract the f_e value from the f_o value—that is, $f_o - f_e$—for each cell in the contingency table and record the difference (as shown in column 3 of Figure

FIGURE 11-5   Computation of χ^2

ROW/COLUMN (CELL)	f_o (1)	f_e (2)	$f_o - f_e$ (3)	$(f_o - f_e)^2$ (4)	$\dfrac{(f_o - f_e)^2}{f_e}$ (5)
1-1	50	40.32	9.68	93.702	2.323
1-2	40	52.42	−12.42	154.256	2.942
1-3	35	32.26	2.74	7.508	.233
2-1	30	32.26	−2.26	5.108	.158
2-2	45	41.94	3.06	9.364	.224
2-3	25	25.81	−.81	.656	.025
3-1	20	27.42	−7.42	55.056	2.008
3-2	45	35.65	9.35	87.423	2.455
3-3	20	21.94	−1.94	3.764	.171
	310	310.0	0		10.539

11-5). As another check of your math, note also that $\Sigma(f_o - f_e)$ must equal zero.

2   Square the $f_o - f_e$ differences to get $(f_o - f_e)^2$ (column 4, Figure 11-5).

3   Divide each of these squared differences—$(f_o - f_e)^2$—by the f_e value for each cell to get $(f_o - f_e)^2/f_e$ (column 5, Figure 11-5).

4   Add the $(f_o - f_e)^2/f_e$ values to get the computed χ^2 value (column 5, Figure 11-5). As you can see, the computed χ^2 value for our example problem is 10.539.

Compare the Computed χ^2 Value with the Table χ^2 Value   It was pointed out earlier that the computed χ^2 value should not differ significantly from zero if the H_0 is to be accepted. Our computed χ^2 value of 10.539 is obviously not zero, but is it close enough to zero to fall into the area of acceptance shown in Figure 11-2, page 418? Or does our value of 10.539 exceed the critical value point on the χ^2 scale in Figure 11-2 and fall into the area of rejection? To answer these questions, we must know the χ^2 table value from Appendix 7 that separates the area of acceptance from the area of rejection. We will then be in a position to compare the computed χ^2 value with the table χ^2 value and make a decision about the H_0.

Now is the time you've been waiting for (?). To *locate the table value* in Appendix 7 and thus make the comparison, *you must know* (1) the desired level of significance and (2) the number of degrees of freedom for the particular problem. We already know that α is .05 for our example problem, but what about the number of degrees of freedom? The answer to this question is that *for an r × c contingency table the df value is found by the following formula:*

$$df = (r - 1)(c - 1) \hspace{3cm} \text{(formula 11-3)}$$

where r = number of rows in the table
c = number of columns in the table

FIGURE 11-6   The 15 df in a 6 × 4 Contingency Table (A df Is Indicated by a Check Mark.)

	(1)	(2)	(3)	(4)	ROW TOTALS
(1)	✔	✔	✔	X	100
(2)	✔	✔	✔	X	120
(3)	✔	✔	✔	X	90
(4)	✔	✔	✔	X	230
(5)	✔	✔	✔	X	180
(6)	X	X	X	X	145
Column totals	200	150	180	335	865

Thus, in a 6 × 4 table, the *df* equals $(6 - 1)(4 - 1)$, or 15. Figure 11-6 may give you an intuitive understanding of the meaning of this *df* figure of 15. For a 6 × 4 table, it's actually necessary to compute only 15 f_e values (for the cells in Figure 11-6 indicated by the check marks). The other nine f_e values for the nine remaining cells in a 6 × 4 table (the ones indicated by the X's in Figure 11-6) are then automatically determined by the row and column totals. In other words, the first three cells in *row 1* are "free" to accept many different values, but the last cell in row 1 is constrained by the row total of 100. Likewise, the first five cells in *column 1* are "free," but the last cell in the column is constrained by the column total. Thus, there are 15 *df* in a 6 × 4 table.

For our example table with 3 rows and 3 columns, we have

$$
\begin{aligned}
df &= (r - 1)(c - 1) \\
&= (3 - 1)(3 - 1) \\
&= 4
\end{aligned}
$$

From Appendix 7, then, we can find the critical χ^2 table value at the intersection of the .05 column (the α specified in our example) and the 4 *df* row. This table value is 9.488, and, of course, it's less than our computed χ^2 value of 10.539. The table value of 9.488 means that if the H_0 is true, the probability of obtaining a computed χ^2 value as large as 9.488 is only .05. Therefore, the probability of getting a computed χ^2 value as large as 10.539 *is less than .05*.

Make the Statistical Decision   The *decision rule* in a χ^2 hypothesis test is as follows:

Accept the H_0 if the computed χ^2 value $\leq$ the appropriate table value.

or

Reject the H_0 if the computed χ^2 value $>$ the appropriate table value.

Since our computed χ^2 value of 10.539 > our table value of 9.488, we reject the H_0 and conclude that the population percentage favoring each candidate for sheriff of Lawless County is not the same from town to town. Bound's campaign manager may decide, as a result of this conclusion, to conduct campaign activities in a more selective fashion.

Computers Make It Easy

In virtually every case in this book where we've encountered a *standardized* test or procedure that requires numerous calculations, we've also located a program in a statistical software package that takes over the chore of carrying out those calculations. Thus, you'll not be surprised now to learn that programs are readily available to process contingency tables and carry out *k*-sample hypothesis tests of percentages.

THE NEW STATISTICS

What type of consumer, by age, sex, and income, is likely to purchase a laundry soap that includes a fabric softener?

How might a company's stock value fare if management fails to increase the company's share in the sugar market?

What's the probability that a married, 45-year-old male union member with three children will vote for a Republican in the next presidential election?

Until recently, getting answers to questions like these required the use of an expensive mainframe computer running sophisticated statistics programs. If you were the analyst at a corporation seeking such truths, chances are you would have to wait hours, if not days, for enlightenment as the DP department dealt with more mundane chores—like the company payroll. By the time your query was processed, that 45-year-old potential Republican might be the father of four, at which point you would have to revise your assumptions and start all over again.

But that's beginning to change. Although PC-based statistics programs can't handle as much data or as many variables as their mainframe equivalents, they offer nearly all the same functions at a fraction of the cost. Data entry is usually easier, and the interactive nature of the PC—and the fact that only one user is in line—means the answers come more quickly. As a result, companies big and not so big are increasingly turning to PC statistics packages to tame traditional business problems.

Of course, statistics programs for the PC are no panacea. Many are minor variations of ungainly, RAM-hungry mainframe packages that are difficult to learn and master. Even those specially designed for the PC can be imposing; either way, the program requires formal statistical training. Despite these constraints, PC-based statistical software is helping businesses save money and bringing them advanced analytical tools.

FILLING THE COFFERS
Market research, one of the most common statistical software applications, helps businesses target their products more effectively. That's what happened when World Vision, a Christian humanitarian group in Monrovia, California, used statistical software to fine-tune its direct-mail campaign. World Vision, which solicits donations for causes such as Ethiopian famine relief, pulls down $20 million a year in direct-mail donations, but it wanted to make the most of its investment in solicitation schemes.

"Planning direct-mail campaigns was pretty much trial and error," says World Vision marketing researcher Jon Van Wyk. "We would try one thing—like improving the art in a direct-mail kit—and if that seemed to bring in more money, we'd do more of it." World Vision tracked response rates with *SuperCalc*, but like most spreadsheets, the program lacked sophisticated tools that could predict how, based on historical data, changes in direct-mail packaging would affect donations.

In January 1987, World Vision began to target its appeals by experimenting with *SPSS/PC +* , a PC version of the SPSS mainframe statistics package. The organization collected extensive background information on a control group of donors, asked them to evaluate different direct-mail pitches, and ran the data through *SPSS/PC +*'s t-tests and regressions. The generated analysis enabled World Vision to better target donors in specific age, income, and other categories and thus tailor mailings for maximum return.

World Vision originally selected *SPSS/PC +* because younger members of the data processing staff had used the mainframe version in college. They soon discovered that the PC version offered features not found on mainframes. Unlike its forebears, *SPSS/PC +* can be linked to *Microsoft Chart* to generate graphs that even nontechnical managers can understand. This made life easier for World Vision's statistics gurus—and got essential information to the right decision makers.

TAKING THE PULSE OF THE PEOPLE
Populist, a small firm in Connecticut, also uses a PC-based statistical package to conduct market research—specifically, opinion analysis. When Populist got started two-and-a-half years ago, compet-

ing with the likes of Lou Harris required major-league money for the necessary mainframe hardware and software, according to co-owner John Fiedler. Having just wrapped up a consulting job with the Republican National Committee, Fiedler decided to start his own opinion analysis firm; the first step was to become a beta tester of *SPSS/PC +* .

"Without statistics on the PC, we probably couldn't have gone into business at all," says Fiedler.

Although its $1 million in annual bookings places it in the minors, Populist has survived—and managed to land some major accounts, including 7-Up and the Republican National Committee. Fiedler credits his success in large measure to the interactive nature of PC-based statistics. "Frankly, when you do a lot of statistical analysis you make a lot of mistakes," he says. "You may set the test up wrong or use the wrong test or function. In the mainframe environment, it can take an entire day to discover one mistake, and it can take three weeks to make twenty, because you're competing for time on the system. With a PC, I can make mistakes quicker, and that makes me more productive."

According to Fiedler, the big guns in the opinion research business aren't trading in their mainframes for PCs. "These companies have invested a lot in equipment, and they're reluctant to change." But that day may yet come—if only to keep the smaller, nimbler firms from whittling away at their market.

MORE FOR THE MONEY

Although statistical analysis is commonly used to track and profile a product or population, the cost-effectiveness of PC-based statistical packages is just being discovered—particularly by government agencies in search of savings in an era of tax revolts and budget cuts.

That's what spurred the financially strapped public health department in California's San Bernardino County to break old habits. The agency, which has used mainframe-based statistics programs for years, recently decided to save time and money by extending its statistical analysis activity to PCs. "We didn't have the funds for analyzing mental health statistics on the mainframe, yet our department director wanted detailed information

very quickly," says Betty Kettering, a biostatistician for the county.

The county settled on *StatPac Gold,* a menu-driven statistics program with an integrated tutorial feature. The product was comparatively simple to set up and use, and at $595, was easier on the pocketbook than packages like *SAS PC*, whose annual license fee starts at $2500.

Better still, Kettering has discovered that *StatPac Gold*'s subset of mainframe statistical functions is more than adequate. Other county employees, tired of waiting for mainframe time, have come around to the PC approach and have begun borrowing Kettering's *StatPac Gold*. The movement has taken root; several other county departments have written PC statistical packages into their budgets for next year to expand data analysis and, in the long run, to save money.

MAKING A CASE BY THE NUMBERS

In addition to helping cut costs, PC-based statistical software can be used creatively to improve a firm's products or services. Skadden, Arps, Slate, Meagher, and Flom, a Manhattan-based law firm specializing in antitrust cases, bought *Stata*, a PC-based package, to analyze product pricing and other market factors in defending corporate clients charged with running afoul of monopoly laws. But Duncan Cameron, the economist hired for the job, soon discovered an unexpected legal application—determining damages in a class-action liability lawsuit against a client whose products were allegedly responsible for widespread salmonella poisoning in a large midwestern city.

In an effort to decrease the damages that might be awarded to the 20,000 people involved in the suit, Cameron established a statistical basis for settling each claim. He looked at comparable settlements based on at least eight different variables, such as time spent by a victim in the hospital. Then, with the aid of a sorting function, Cameron identified and segregated families that had been affected by the alleged poisoning, since settlements are usually larger when more than one family member is involved.

At press time, the case was still pending, but Cameron claims this is the first time statistical analysis has been used to reckon damages in a product liability case. "People used to rely on a seat-of-the-pants approach to establish damages," says Cameron. "They would come up with a total settlement figure, then decide how to divide it up. It was backwards, really." By quantifying the process of awarding settlements, Skadden, Arps, and company may be able to make a stronger case for lower compensation—and save its client money.

A SELECT AUDIENCE

Regardless of how inexpensive and effective statistical programs for the PC become, they'll never be for everyone. "There's a scare barrier when it comes to statistics, even though some packages are accessible and well supported," says biostatistician Kettering.

But for those with the training or the willingness to learn, PC-based statistical software is an attractive alternative to mainframe-based packages. The PC's interactive nature means data can be analyzed more quickly, providing managers with much-needed results. Moreover, the low cost of PC-based packages allows firms that wouldn't normally consider performing statistical analysis in-house to squeeze the most out of raw data.

—Jeff Moad, "The New Statistics," *PC World*, October 1987, pp. 253–255. Reprinted with permission.

Figure 11-7 shows how the *Minitab* package processes the example problem we've just considered. The voters in White Lightning preferring Bound, Lux, and Nocruk are stored in that order in column 1 (C1) of a *Minitab* worksheet, just as they were shown in Figure 11-3. The data for the other two towns are similarly stored in columns 2 and 3, as you can see in the top lines of Figure 11-7. Once these facts are entered and available to the program, the user merely enters the CHISQUARE C1-C3 command, and the software produces (1) the observed and expected frequencies found in Figure 11-4, (2) the results in column 5 of Figure 11-5, and the computed χ^2 value which is the total of those column 5 results, and (3) the degrees of freedom needed (4 in this case) to use the χ^2 table in Appendix 7. The user then finds the appropriate χ^2 table value at the intersection of the selected α column and the 4 *df* row. Finally, the last step (which we've already done) is to compare the computed χ^2 value with the table value and make the statistical decision.

A summary of the steps in a *k*-sample hypothesis test of percentages is outlined in Figure 11-8.

Self-Testing Review 11-1

1 a If a χ^2 distribution has 8 degrees of freedom, the mean of this distribution is 8, and the mode is 6. True or false?

  b If a test is being conducted at the .01 level, the critical χ^2 table value for the distribution described in a is 18.475. True or false?

2 The H_0 in a *k*-sample hypothesis test of percentages is basically that there's no significant difference between the percentages that are of interest in the several populations under study. True or false?

3 The computed χ^2 value is zero if the H_0 is true. True or false?

4 If the computed χ^2 value < the table χ^2 value, it falls into the area of rejection. True or false?

5 The student government at Radical University conducted a sample survey to determine

```
MTB > PRINT C1

WHTLTNG
    50     30     20

MTB > PRINT C2

CASCITY
    40     45     45

MTB > PRINT C3

SMGVILLE
    35     25     20

MTB > CHISQUARE C1-C3

Expected counts are printed below observed counts

          WHTLTNG  CASCITY SMGVILLE     Total
    1          50       40       35       125
            40.32    52.42    32.26

    2          30       45       25       100
            32.26    41.94    25.81

    3          20       45       20        85
            27.42    35.65    21.94

 Total        100      130       80       310

 ChiSq =   2.323 +  2.942 +  0.233 +
           0.158 +  0.224 +  0.025 +
           2.008 +  2.455 +  0.171 = 10.539
 df = 4
```

FIGURE 11-7
The Minitab
solution to the k-
sample hypothesis
test example
problem.

student opinion on a proposed new constitution. The results of the survey are summarized below. Is there a significant difference, at the .05 level, between the percentages of each class approving the proposal?

	FRESHMEN	SOPHOMORES	JUNIORS	SENIORS
Approve	40	30	30	20
Disapprove	30	30	20	30

ANOTHER CHI-SQUARE TEST: GOODNESS OF FIT

A second important analysis procedure is Karl Pearson's test for goodness of fit that was mentioned in the opening vignette. This test is used, for example, to see if a population under study "fits" or follows one with a known distribution of

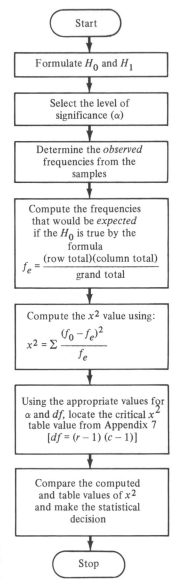

FIGURE 11-8
Procedure for a k-sample hypothesis test of percentages.

values. That is, a **goodness of fit test** is one that may be used to evaluate how well a population under study follows a normal, binomial, Poisson, or uniform distribution. We've looked at the first three distributions in earlier chapters, and a **uniform distribution** is one in which all outcomes considered have an equal or uniform probability.

Since the purpose of a goodness of fit test is to decide if the sample results are consistent with the results that would have been obtained if a random sample had been selected from a population with a known distribution, the *null hypoth-*

esis is essentially that the population being analyzed fits some distribution pattern. For example, the H_0 might be that the population distribution is uniform. The *alternative hypothesis* then is simply that the population doesn't fit the specified distribution. Thus, the H_1 might be that the population distribution is *not* uniform.

The following problem shows how χ^2 concepts can be used to test the goodness of fit between sample data and a uniform distribution. The test for goodness of fit between sample data and a uniform distribution is the simplest and perhaps the most useful. The procedure in testing other known distributions differs from our example primarily in the ways the f_e and *df* values are determined. We'll not consider tests of other distributions. As you'll see in our example, the steps in the testing procedure follow a familiar pattern.

The Bitter Bottling Company has developed "Featherweight," the cola with fewer calories and less taste. To evaluate this new product, the marketing manager gives a taste test to 300 people. Each person in the sample tastes Featherweight and four other diet cola brands. To avoid prejudice, the actual brand labels are replaced by the letters A, B, C, D, and E. The results of the sample are shown in Figure 11-9.

Steps in the Goodness of Fit Testing Procedure

Formulate the Null and Alternative Hypotheses The *null hypothesis* is:

H_0: The population distribution is uniform—that is, the cola brands are preferred by an equal percentage of the population.

And the *alternative hypothesis* is:

H_1: The population distribution is *not* uniform—that is, the taste preference frequencies are not equal.

Select the Level of Significance Let's assume that Bitter's marketing manager wants to conduct the test at the .05 level of significance.

FIGURE 11-9   Results of Bitter Bottling's Taste Test

BRAND	NUMBER PREFERRING BRAND
A	50
B	65
C	45
D	70
E	70
	300

Determine the Observed (Sample) Frequencies (f_o)   The f_o values are given in Figure 11-9.

Compute the Expected Frequencies (f_e)   If the null hypothesis is true, we would expect an equal or uniform number of people to prefer each of the five brands. That is, there's no significant difference in taste preference, and one-fifth or 20 percent of the tasters should prefer brand A, 20 percent should prefer brand B, and so on. Thus, the f_e value for each brand should be 20 percent of 300, or 60.

Compute the x^2 Value Using f_o and f_e Once again, the appropriate formula is:

$$x^2 = \sum \frac{(f_o - f_e)^2}{f_e}$$

Figure 11-10 shows the computation of x^2 for our goodness of fit example problem. The computed value is 9.168. The *Minitab* software package doesn't have a special command to prepare the computed x^2 value for a goodness of fit test. But the simple program in Figure 11-11 shows how the user can supply observed and expected frequencies and a chi-square formula line to arrive at the computed x^2 value.

Compare the Computed x^2 Value with the Table x^2 Value   The number of degrees of freedom for our example is 4. This is because the sum of the sample observations is 300, and this total must also be the sum of the expected frequencies. Therefore, one of the five brand preference categories is constrained by the total. In fitting a uniform distribution, then, the *df* value is one less than the number of classes or categories.

Once the *df* value is known, the x^2 table value can be determined. At the .05 level, and with 4 degrees of freedom, the table value found in Appendix 7 is 9.488—a figure that is slightly larger than our computed x^2 statistic of 9.168.

Make the Statistical Decision   The decision rule for the goodness of fit test is the same as the rule given earlier for the *k*-sample test of percentages. Thus, we will *accept* our H_0 because the computed x^2 value < the table value. At the .05 level, we cannot reject the hypothesis that all the cola brands are preferred by an

FIGURE 11-10   Computation of x^2 for Goodness of Fit Test

BRAND	NUMBER PREFERRING (f_o)	f_e	$f_o - f_e$	$(f_o - f_e)^2$	$\dfrac{(f_o - f_e)^2}{f_e}$
A	50	60	−10	100	1.667
B	65	60	5	25	.417
C	45	60	−15	225	3.750
D	70	60	10	100	1.667
E	70	60	10	100	1.667
	300	300	0		9.168

```
MTB > PRINT C1

C1
    50    65    45    70    70

MTB > PRINT C2

C2
    60    60    60    60    60

MTB > LET K1 = SUM ((C1 - C2) **2/C2)
MTB > PRINT K1
K1        9.16667
```

FIGURE 11-11
Minitab help with
the goodness of fit
test example.

equal percentage of the population. We must conclude that at the .05 level Featherweight is not significantly better tasting than the other brands.

Self-Testing Review 11-2

1   The chi-square goodness of fit test follows the same steps as the k-sample test, although there are some differences within the steps. True or false?

2   In a test to determine if a population distribution is uniform, the df value is one less than the number of classes being considered. True or false?

3   If the null hypothesis is that the sales produced in six sales districts are equal or uniform, and if the test is to be made at the .10 level, the χ^2 table value in this test is 11.070. True or false?

4   In the preceding question, should the H_0 be accepted if the computed χ^2 value had been 10.625?

5   An examination of an automobile agency's sales records shows that on 70 business days only one new truck was sold. However, two trucks per day were sold on each of 60 business days; three trucks per day were sold on 40 days; and there were daily sales of four trucks on 30 days. That is,

TRUCK SALES PER DAY	NUMBER OF SALES DAYS
1	70
2	60
3	40
4	30

At the .01 level, test the hypothesis that the demand for trucks is uniformly distributed.

LOOKING BACK

1   The purpose of a k-sample hypothesis test of percentages is to test the hypothesis that three or more independent samples have all come from populations that have

the same percentage of a given characteristic. When such a test is made, the chi-square distribution is used. This test is thus based on a continuous probability distribution with a shape that depends on the number of degrees of freedom in a problem.

2   The purpose of the goodness of fit test is to see if a population under study fits or follows one with a known distribution of values. Our focus in this chapter was to test the goodness of fit between sample data and a uniform distribution, but the technique is also applicable to other probability distributions.

3   Regardless of the application, though, the steps followed in one χ^2 testing situation are similar to those followed in another. Of course, the null and alternative hypotheses are phrased differently depending on the type of application. And the expected frequencies and number of degrees of freedom are computed in different ways depending on the logic of the situation. But the χ^2 test statistic is generally computed in the same way. And this computed value is always compared to a critical χ^2 table value. The statistical decision is then to accept the H_0 if the computed value $\leq$ the table value.

KEY TERMS AND CONCEPTS

Chi-square distribution   *416*

$$\chi^2 = \sum \frac{(f_o - f_e)^2}{f_e}$$

(formula 11-1)   *417*

k-sample hypothesis test of percentages   *418*

Contingency table   *418*

Test of independence   *419*

Cell   *420*

$r \times c$ tables   *420*

$$f_e = \frac{\text{(row total)(column total)}}{\text{grand total}}$$

(formula 11-2)   *420*

$df = (r - 1)(c - 1)$ (formula 11-3)   *422*

Goodness of fit test   *428*

Uniform distribution   *428*

PROBLEMS

1   The following table shows the number of good and defective parts produced on each work shift at a manufacturing plant. Using the .05 level of significance, test the hypothesis that there's no significant difference between the percentages of defective parts produced on the three shifts.

	FIRST SHIFT	SECOND SHIFT	THIRD SHIFT
Good	90	70	60
Defective	10	20	20

2   A survey of rural and urban television viewing populations is made to determine television programming preferences, and the following results are obtained:

| | TYPES OF PROGRAMS PREFERRED | | | |
	WESTERN	COMEDY	MYSTERY	VARIETY
Urban	80	100	100	60
Rural	70	70	50	40

At the .05 level, test the hypothesis that there's no difference in program preference between urban and rural residents.

3   A sample of 300 consumers was asked to choose among six brands of coffee. The results of the survey are as follows:

BRAND	NUMBER PREFERRING
A	40
B	55
C	60
D	40
E	60
F	45
	300

Using a .01 level of significance, test the hypothesis that a uniform distribution describes the coffee-drinking population.

4   Charlie "Crank" Schaff, the automotive writer for the local newspaper, believes that front fenders of cars are more likely to be damaged in accidents than rear fenders. In a sampling of body shops, Charlie observed the following damage:

FENDER DAMAGED	NUMBER OF DAMAGED FENDERS
Right front	150
Left front	120
Right rear	125
Left rear	105
	500

At the .10 level, would you agree with Charlie?

5   George Gullible has a bet with a fraternity brother that if the 1, 2, or 3 face of a die turns up on a single roll, he will pay the brother 10 cents. If, on the other hand, the 4, 5, or 6 face is rolled, George will win 10 cents. The brother is to supply the die. A third fraternity member has borrowed and experimented with the die, and the results were as follows:

DIE FACE	FREQUENCY OF APPEARANCE
1	115
2	100
3	125
4	95
5	85
6	80

At the .05 level, is George going to be rolling with a fair die?

6   Integrated circuits are supplied to a computer manufacturer by two vendors. Each circuit is tested by the manufacturer for four common defects before it is accepted. The following data, representing test results over a 2-week period, have been supplied to the purchasing department by the quality control manager:

	DEFECTIVE CIRCUITS BY TYPE OF DEFECT			
VENDOR	1	2	3	4
A	60	80	40	30
B	30	32	25	20

At the .01 level, would you accept the hypothesis that the percentages of common defects are the same for both vendors?

7   Seeking public office, Polly Tician is planning her campaign strategy on a particular issue and wants to know if there is a significant difference in the percentage of voters who favor the issue between urban, suburban, and rural voters. The following data have been gathered:

	URBAN	SUBURBAN	RURAL
In favor	78	65	60
Opposed	42	45	50

Should the null hypothesis that the percentage favoring the issue is the same for the urban, suburban, and rural voters be accepted at the .10 level?

8   A botanist is interbreeding two types of plants, and she expects four classes of hybrid results to appear in the Mendelian ratio of 9:3:3:1. The results of her experiment show that there are 860 plants of one class, 350 of another, 300 of a third, and 90 of a fourth class. At the .05 level, are these results in accordance with the Mendelian ratio?

9   To determine the thinking of union workers about a proposed change in the union constitution, the ruling council sent questionnaires to a random sample of 100 members in three local unions. The survey results are as follows:

	UNION LOCAL		
OPINION	X	Y	Z
Favor change	17	23	10
Oppose change	9	13	8
No response	4	4	12
	30	40	30

At the .05 level, is there a significant difference in the reactions of the workers in the three locals to the proposed change?

10   The sales manager of the Tackey Toy Company has hired a researcher to gather and evaluate opinion data about Tackey products from people in different parts of the country. The researcher randomly sampled toy buyers in each of four regions. Each

respondent was first asked if he or she had ever heard of Tackey toys. Those who were familiar with Tackey products were then asked if Tackey toys were competitively priced or overpriced. The following sample results were obtained:

GEOGRAPHIC REGION	UNFAMILIAR WITH TACKEY TOYS	TOYS PRICED COMPETITIVELY	TOYS OVERPRICED	TOTAL
Atlantic	64	28	106	198
Pacific	84	42	76	202
Gulf Coast	56	14	130	200
Central	60	20	120	200
Totals	264	104	432	800

At the .01 level, would you accept the hypothesis that there are no regional differences in the way toy buyers responded to the Tackey survey?

TOPICS FOR REVIEW AND DISCUSSION

1   In what ways are the t, F, and χ^2 probability distributions similar?
2   "In any hypothesis-testing situation, we are always concerned with the computation of the value of some critical ratio or statistic for which we know the appropriate probability distribution." Discuss this statement in the context of χ^2 procedures.
3   a   Discuss the general steps to be followed in conducting a k-sample hypothesis test of percentages.
    b   In conducting a goodness of fit test.
4   "From the standpoint of verifying the null hypothesis, a χ^2 computed value of zero would be ideal." Explain why this statement is true.
5   Given the statement in question 4, explain why it's very unlikely that you would get a computed value of zero even if the H_0 is true.
6   A critical χ^2 table value is found to be 9.236 in a hypothesis test conducted at the .10 level.
    a   Explain the meaning of 9.236.
    b   Assume that the computed χ^2 value in the test is 12.312. Make the appropriate statistical decision, and explain the reason for your decision.
7   Discuss the procedure for computing the expected frequencies in the 6 × 4 contingency table shown in Figure 11-6.
8   a   Discuss the purpose of a k-sample hypothesis test of percentages.
    b   Discuss the purpose of a goodness of fit test.

PROJECTS/ISSUES TO CONSIDER

1   Refer back to the project involving campus issues that was presented at the end of Chapter 9 (see page 373). You'll recall that class teams identified a campus issue of interest, and then picked random samples from two populations to see if the two identified groups had different views about the issue. Now it's possible to separate

campus people into more than two groups and then undertake a similar project. The following steps should be carried out:

a   One or more teams (or perhaps the entire class) should identify a campus issue of interest to team participants.

b   Team members should determine the null hypothesis and alternative hypothesis to be tested.

c   Members should next prepare the question(s) that will be used to solicit an opinion from the respondents. For example, refer back to question 5, Self-Testing Review 11-1. It's obvious that the pollsters at Radical University asked students if they approved of a proposed new constitution. In addition to providing space to record "approve" and "disapprove" comments, team members might also want to allow room for "no opinion" remarks.

d   Randomly sample at least 40 students from each population to see how each group views the issue that has been identified. Refer to step d in the Chapter 9 project instructions for sampling suggestions.

e   After the data have been gathered, draw up a contingency table similar to the one shown in the Radical University example in question 5, Self-Testing Review 11-1. If a "no opinion" answer is possible, a third row in the table will be needed.

f   Conduct a k-sample hypothesis test of percentages that leads to the acceptance or rejection of the null hypothesis. Team members will pick the level of significance to use in this test.

g   Prepare a class presentation of the study methodology and test results obtained.

2   In the projects described in Chapters 8 and 10, you were asked to team up with other class members and then locate journal articles that contained published values for statistical tests. Review the steps taken to complete those projects and follow the same procedures again. That is, locate a few journal articles of interest to team members that contain chi-square test results. If possible, recompute the test figures presented in the articles, note any discrepancies you might find, and prepare a brief team presentation for the class.

ANSWERS TO SELF-TESTING REVIEW QUESTIONS

11-1

1   a   This statement is true.

b   This is incorrect. The table value is 20.090 when $\alpha = .01$ and $df = 8$.

2   This is a true statement.

3   This is false. Sampling variation is likely to lead to discrepancies between f_o and f_e even when the H_0 is true.

4   Another false statement. The computed value would be in the area of acceptance, and the H_0 would be accepted.

5   a   The hypotheses are:

H_0: The population percentage of students approving the proposed constitution is the same from class to class—that is, student approval is independent of class standing.

H_1: The population percentage of students approving the proposed constitution is not the same from class to class—that is, student approval is dependent on (or related to) class standing.

b   A .05 value of α is specified.
c   The f_e values for each cell are shown below in parentheses:

	FRESHMEN	SOPHOMORES	JUNIORS	SENIORS
Approve	40	30	30	20
	(36.52)	(31.30)	(26.09)	(26.09)
Disapprove	30	30	20	30
	(33.48)	(28.70)	(23.91)	(23.91)

d   The computed χ^2 value is 5.003.
e   The df value is computed as follows:
$$df = (r - 1)(c - 1)$$
$$= (2 - 1)(4 - 1)$$
$$= 3$$
Therefore, the χ^2 table value at $df = 3$ and $\alpha = .05$ is 7.815.
f   The statistical decision is:

Accept H_0, since the computed χ^2 value of 5.003 < the table value of 7.815.

You can verify these results by examining the following computer output for this problem:

```
MTB > PRINT C1

FRESHMEN
   40     30

MTB > PRINT C2

SOPHMORE
   30     30

MTB > PRINT C3

JUNIORS
   30     20

MTB > PRINT C4

SENIORS
   20     30

MTB > CHISQUARE C1-C4
```

```
     Expected counts are printed below observed counts

          FRESHMEN SOPHMORE  JUNIORS  SENIORS    Total
      1       40       30       30       20       120
            36.52    31.30    26.09    26.09

      2       30       30       20       30       110
            33.48    28.70    23.91    23.91

   Total      70       60       50       50       230

   ChiSq =  0.331 +  0.054 +  0.587 +  1.420 +
            0.361 +  0.059 +  0.640 +  1.549 = 5.003
   df = 3
```

11-2

1 This is true.

2 This is also true.

3 Finally, a false statement. The table value of 11.070 would be correct for the .05 level and 5 df. At the .10 level, however, the figure should be 9.236.

4 No. With a computed value of 10.625, and with a correct table value of 9.236, the H_0 should be rejected.

5 a The hypotheses are:

H_0: The population distribution is uniform.
H_1: The population distribution is not uniform.

b The level of significance is specified at .01.

c If the H_0 is true, there would be 50 days in which one truck was sold, 50 days in which two trucks were sold, etc.

d The computed χ^2 value is 20.00.

e The df value = 3. Therefore, the χ^2 table value at $df = 3$ and $\alpha = .01$ is 11.345.

f The statistical decision is:

Reject H_0 since the χ^2 computed value > the table value.

A CLOSER LOOK

MORE ABOUT KARL PEARSON AND OTHER PROMINENT STATISTICIANS

We've seen that Karl Pearson was a towering figure in the statistical world in 1900. Ian Hacking, the author of the opening vignette, continues his discussion of Pearson and other prominent statisticians with these words:

Pearson was one of those prodigiously energetic Victorians tackling all sorts of projects simultaneously. Mildly socialist in inclination and a strong supporter of rights for women, he gets bad press for his enthusiasm for eugenics. He thought that selectively breeding middle-class European stock would help prevent what he saw as the decay of civilization. In this and many other respects he was the spiritual heir of Francis Galton, another energetic Victorian. To Galton we attribute the modern theory of correlation, a way to quantify the degree or connection between two variables, such as incidence of lung cancer and cigarettes smoked per day.

Galton's theory of correlation originated in the 1880s as part of a study of inheritance. He wanted measures for and explanations of his thesis that "good" families pass down talent while "bad" families pass down vice. Whatever we think of his motivation, his measures are applicable to any pair of variables. Independently wealthy, he endowed a chair of eugenics at University College, London, where statistics was applied to the study of eugenics. It was there that Pearson became what could be called the first professor of mathematical statistics.

At the beginning of the 19th century, K. F. Gauss and P. S. de Laplace had put firmly in place the curve of errors. This is the familiar bell-shaped curve, to which, for example, the grades of students from A to F are still supposed to conform. By mid-century there arose the idea that almost all human and biological phenomena—heights, weights, intelligence, endurance, and so on—could be described by curves of this form. Pearson himself gave the curve its common English language designation—

we call it the *normal* curve because it was once thought to be a picture of what was normal.

When people had the idea that just one curve, with no rivals, can describe most phenomena, the question of goodness of fit did not readily arise—the curve had to fit because it was the only one in town. Pearson and many others came to realize that this was fantasy. He devised rival curves—ones that were fat on the left and thin on the right or ones that looked like the crest of a wave rather than the symmetrical bell curve—because they were better suited to a distribution of heights or to an ability to run a mile or other types of data. Then he could seriously ask how well they fit the data. Although most readily understood in cases where a population falls naturally into discrete cells such as heads or tails or number of packs per day, the chi-square test can be applied to these other curves too in a standardized way.

Other tests were soon to follow. After World War I, R. A. Fisher came to dominate the scene, at least in the English-speaking world. He created an extraordinarily rich array of statistical ideas. He also directed an experimental farm at Rothamsted, which in addition to its daily work of statistically testing fertilizers, chicken feed, and the like, was the breeding ground for an entire statistical methodology.

The growth of statistical methodology has not been a story of tranquil progress. Karl Pearson's own two best pupils were his son Egon and a young Rumanian-born student at the laboratory, Jerzy Neyman. Where Fisher seemed to supersede Karl Pearson, the ideas of Neyman and E. S. Pearson widely replaced Fisher's. But not without polemics and arm-waving. Fisher jeered at Neyman and Pearson's theory as having nothing to do with scientific inference. Neyman was not gracious in return. But the substance of the debates was perhaps less important than the institutions and practices that ensued. Neyman, for example, moved to Berkeley, which under his guidance became for a long while the premier American school of statistics.

Statistics has not yet aged into a stable discipline with complete agreement on foundations. All the statisticians mentioned here assumed that the key to probability lay in the relative frequency with which different kinds of events occur—the kind of odds the weather bureau gives when they say that there is a 90 percent chance of rain tomorrow. But what does that mean? Some say nothing, for probability is concerned with subjective degrees of belief, and that subjective approach only gives a reasonable degree of certainty. Work emanating from F. P. Ramsey in England, Bruno de Finetti in Italy, and L. J. Savage in the United States has turned such subjectivity into a serious scientific approach. Today we have vigorous, sometimes violent, disagreement on these matters. But perhaps battles about first principles are less important than the large-scale application of many competing methods.

Statistical talk has now invaded many aspects of daily life, creating sometimes spurious impressions of objectivity. Policy makers demand numbers to measure every risk and hope. Thus the 1975 *Reactor Safety Study* of the Nuclear Regulatory Commission attached probabilities to various kinds of danger. Some critics suggest that this is just a way of escaping the responsibility of admitting ignorance. Other critics point to the increasing use of complex models generated to "fit" reams of data in some statistical sense—models that can become a fantasyland without any connection to the real world.

For better or worse, statistical inference has provided an entirely new style of reasoning. The quiet statisticians have changed our world—not by discovering new facts or technical developments but by changing the ways we reason, experiment, and form our opinions about it.

—From Ian Hacking, "Trial By Number," *Science 84,* November, 1984, pp. 69–70. Copyright 1984 by the AAAS. Reprinted with permission.

PART THREE

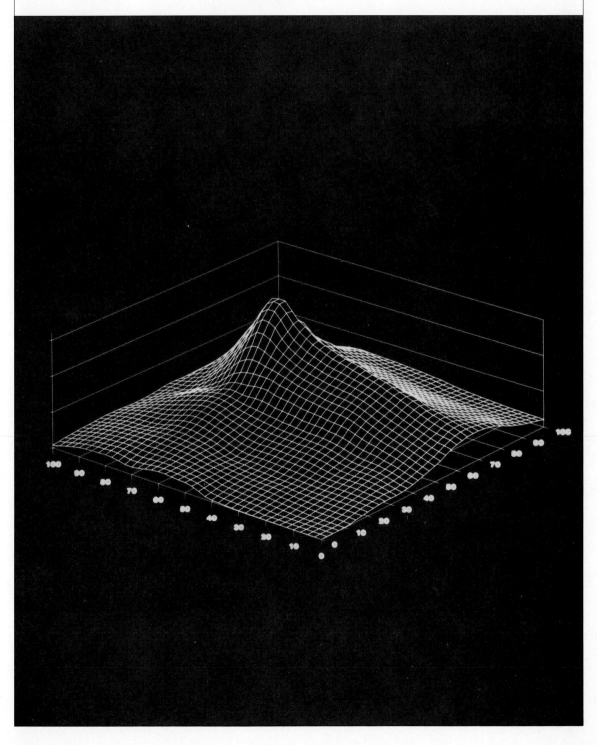

PART THREE

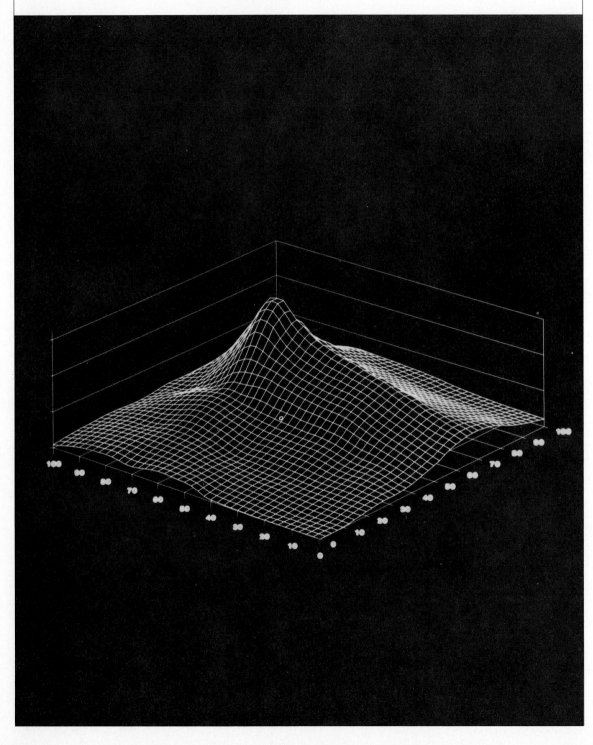

COPING WITH CHANGE

You saw in Chapter 1 that statistical theories and methods are useful in (1) describing relationships between variables, (2) helping with the decision-making process, and (3) providing a conceptual foundation to analyze and cope with change. There are sections in one or more of the chapters in this part that deal with *each* of these uses of statistical techniques. As indicated by the part title, however, we are particularly interested here in the *measurement* and *prediction of change*.

To plan is to decide in advance on a future course of action. Therefore, virtually all plans and decisions are based on expectations about future events and/or relationships. And thus virtually all planners and decision makers are required to employ some forecasting process or technique to arrive at a future expectation. *One* forecasting approach, of course, is to assume that nothing will change and that no new plans are needed. In the short run, this may frequently be the approach followed, but the assumption about a lack of change seldom remains valid for long, although it is possible for gradual change to go unnoticed for some time. In Chapter 12, therefore, we look at procedures which have been developed to *measure relative changes* in economic conditions over time.

A *second* forecasting approach consists of (1) analyzing past empirical data to detect reasonably dependable patterns and then (2) projecting these patterns into the future to arrive at future expectations. In Chapter 13 we examine some of the techniques employed in this approach.

Finally, a *third* forecasting approach is to identify and analyze reasonably predictable independent factors which are having a significant influence on the dependent variable to be predicted. A prediction of values of the dependent variable may then be made on the basis of (1) the closeness of the relationship between the variables and (2) the values of the independent variables. A brief introduction to this forecasting approach involving the use of only two variables is the subject of Chapter 14.

The chapters included in Part 3 are:

12

MEASURING CHANGE: INDEX NUMBERS

The Changes Behind the CPI's New Look

[The Consumer Price Index (CPI) has a new look.] The Bureau of Labor Statistics has revised the fixed market basket of goods and services that make up the index so that it may better reflect changing consumption behavior. While the old CPI reflected household spending patterns in 1972–73, the new one is based on consumer budgets in the 1982–84 period.

Consumers alter their spending patterns for a variety of reasons, of course, including income changes, shifts in relative prices, demographic trends, and the development of new products and services (video recorders and personal financial services have been added to the index). But the bottom line is that now some consumption categories will carry considerably more weight in measuring inflation than in the past. Here are some of the more important changes:

Reflecting the national shift toward a service economy, services will become the larger of the two major categories, services and goods. Services' weight will rise to 52% of the index from about 46%, compared with 48% for goods or commodity prices. That implies that wage trends will have a more direct impact on consumer inflation from here on in, since labor costs are the chief determinants of most service prices.

Meanwhile, housing, one of the largest subcategories in the index, will become even larger, accounting for nearly 43% of the index, compared with 38% previously. And energy costs, which run the gamut from gasoline and fuel oil to electric utility bills, will now affect 10% of the index instead of 7.7%. Thus, future changes in energy costs will have a larger impact on inflation than they have had in the previous 10 years.

Although medical costs have soared over the past decade, government programs and employer-sponsored health insurance have picked up so much of the added bill that household medical outlays will still account for only about 5% of the CPI. Unfortunately, however, the revised index doesn't reflect the current trend among employers to cut back insurance coverage and pass on more of the cost directly to their employees. Thus, it may well understate the impact of medical costs on consumer pocketbooks in coming years.

As for major categories that will have less influence, food will decline from about 19% of the consumer market basket to 16% and apparel from 7.5% to 6.5%. For those who are wondering how inflation will stack up in the new CPI against the old index in the months ahead, the Labor Dept. will publish both measures for the next six months.

—From "The Changes Behind the CPI's New Look." Reprinted from March 2, 1987 issue of *Business Week*, p. 24, by special permission, copyright © 1987 by McGraw-Hill, Inc.

LOOKING AHEAD

The subject of our opening vignette is the Consumer Price Index, and the subject of this chapter is how to construct and use price and quantity index numbers to measure economic changes. We'll also look at some important index number series (the Consumer Price Index is the one that's best known), and we'll examine some of the problems and pitfalls that accompany the construction and use of index values.

Thus, after studying this chapter you should be able to:

- Explain the uses and advantages of index numbers.
- Describe the types of index numbers in current use.
- Construct index numbers by the use of the aggregative and average of relatives methods.
- Discuss some of the possible problems and pitfalls associated with the use of index numbers.

CHAPTER OUTLINE

LOOKING AHEAD
WHAT'S AN INDEX NUMBER?
ADVANTAGES OF INDEX NUMBERS

WHAT'S AN INDEX NUMBER?

People need a convenient measure of the changes that occur in prices, quantities, or other variables over time. There's thus a need for index numbers. Let's assume, for example, that the famous overworked pedagogue, Professor Staff, has an unhealthy interest in the price of chalk he uses in teaching his statistics course. Figure 12-1 shows the prices for a box of chalk for the years 1985–1989.

In its most basic form, an index number is simply a *percentage relative* that shows the relationship between two values. Thus, we can compare the prices of chalk from Figure 12-1 by (1) dividing each year's price by the price in 1985 and (2) converting each relative figure obtained into a percentage as follows:

YEAR	PRICE IN YEAR ÷ PRICE IN 1985	PRICE RELATIVE	PERCENT RELATIVE OR INDEX NUMBER
1985	$0.60 ÷ $0.60 =	1.000 × 100 =	100.0
1986	0.60 ÷ 0.60 =	1.000 × 100 =	100.0
1987	0.65 ÷ 0.60 =	1.083 × 100 =	108.3
1988	0.70 ÷ 0.60 =	1.167 × 100 =	116.7
1989	0.75 ÷ 0.60 =	1.250 × 100 =	125.0

The rightmost column shows the relationship between yearly chalk prices in a simple index number form.

Several points can be illustrated by this example:

1 A **base period** (the year 1985) was *arbitrarily* selected, and this *base period has an index number value of 100*.
2 The computed index numbers are percentages, but the percent sign is seldom, if ever, used.
3 The index number value for 1987 (108.3) does not show the price of chalk in either 1985 or 1987; nor does it show the fact that there was a $0.05 difference in price. Rather, the index number only shows the *relative change*.
4 The *interpretation* of the 1987 index number (108.3) is that the price of chalk in 1987 was 108.3 percent of the 1985 price; alternatively, we could say that there was an 8.3 percent *increase* in chalk price between 1985 and 1987. (Between 1985 and 1989, of course, there was a 25 percent increase in price.)

A series of index numbers can be constructed from a single variable (after all, we've just used data from a single variable to produce a series of chalk price indexes). Index numbers prepared from changes in a single data item are called **simple index numbers.** But most of the index numbers commonly used today are **composite index numbers**—that is, they are constructed from the changes that occur in a number of different items. In general usage, then, an **index number** is a measure of how much a composite group changes on a relative basis with respect to a base period that's usually defined to have a value of 100.

For the remainder of this chapter, the term "index number" refers to a composite index number. These values are prepared in different ways, as we'll see later, but the points illustrated by the chalk example will usually remain valid.

FIGURE 12-1

YEAR	PRICE OF CHALK PER BOX
1985	$0.60
1986	0.60
1987	0.65
1988	0.70
1989	0.75

ADVANTAGES OF INDEX NUMBERS

Index numbers allow people to measure, summarize, and communicate the changes that occur over time. When compared with raw data, *index numbers may have the following advantages:*

1 *They may simplify data and aid in communication.* A *single* composite index number is used, for example, to give an indication of some overall change in an economic variable. The **Consumer Price Index (CPI)** discussed in the opening reading is a popular index number series that periodically measures the cost of about 400 dissimilar things that people buy. (We'll discuss two such CPI series later in the chapter.) Thus, if the CPI for a particular month is 116.5, and if the base period is 1982–1984, this single index value indicates the average relationship between prices in the two time periods. In this example, it takes $11.65 to buy the same quantities of goods and services that could have been purchased for $10.00 in the 1982–1984 period. Obviously, the single value of 116.5 is simpler to understand and deal with than the prices of 400 items which, in the case of the CPI, come from over 80 different urban areas.

2 *They may facilitate comparisons.* Changes in series of items expressed in different absolute measurements (for example, in dollars, tons, cubic feet, bales) may be compared by converting the different measurements to relative values.

3 *They may be used to show typical seasonal variations.* A series of index numbers may be prepared by a company to show the variations in retail store sales during each month for several years. When compared with an average monthly sales index of 100, the variations in seasonal sales patterns become obvious. An index of 145 for December shows, for example, that retail sales during this peak month are 45 percent greater than they are during an average month of the year. We'll consider this topic again in the next chapter.

TYPES OF INDEX NUMBERS

Index numbers are classified as price, quantity, or value indexes.

Price Indexes

The first recorded **price index**—a series that shows how selected prices change over time—was prepared in Italy in 1764 by G. R. Carli to compare Italian oil, grain, and wine prices in 1750 with the prices in 1500 for the same items. Price indexes are useful to decision makers such as (1) business managers who are keenly aware of the important role of price in consumer buying decisions, (2) government economists and policy makers who must attempt to plan budgets and avoid the perils of both price inflation and economic recession, and (3) union leaders who are sensitive to the effect of price changes on the economic well-being of union members.

Let's suppose a union leader wants to know what has been happening to the *purchasing power* of the dollar before entering into wage negotiations. A proposed raise in the money earnings of workers may be misleading, unless price changes affecting what the earnings will buy are known. What's needed by the union leader are data on **real earnings**—dollar earnings adjusted for changes in price levels.

Fortunately for the union negotiator, a purchasing power index may be easily computed by using an appropriate price index. "How?" you ask (with commendable curiosity). In this way:

1   The union leader first finds the "all items" index figure in a recent Consumer Price Index. In March 1988, this figure was 116.5, but it changes each month. The base of 100 is the period from 1982–1984. Thus, as we've seen, it takes about $11.65 in March 1988, to buy the things that would have cost $10.00 in 1982–1984.
2   Convert the CPI figure (let's use the 116.5 value) back into a price relative. Thus, the CPI figure of 116.5 becomes 1.165.
3   Calculate the reciprocal of 1.165—that is, 1/1.165 gives a reciprocal value of 8584.
4   Prepare a **purchasing power index number** by multiplying the reciprocal value of .8584 by 100. Thus, the purchasing power index number is 85.84 when the CPI value is 116.5.

What does this number of 85.84 mean? It means that there's no difference in purchasing power between $100 in March 1988, and $85.84 in the 1982–1984 period. The union leader would, of course, want to convert old wage rates and proposed new wage rates into *comparable* real earnings terms before reaching any agreements.

Self-Testing Review 12-1

Before January, 1988, Consumer Price Index figures were based on the year 1967. That is, 1967 = 100, rather than 1982–1984 = 100. Thus, during the 1970s and most of the 1980s, the base year was 1967. Let's see how prices have behaved during this period. The monthly money wages of a worker for 1970, for 1974, and for March 1988, along with the CPI values (1967 = 100) for these periods are:

TIME PERIOD	MONEY WAGES (MONTHLY)	CONSUMER PRICE INDEX (1967 = 100)
1970	$600	116.3
1974	750	147.7
March 1988	2,000	349.0

1 Prepare a purchasing power index for each time period.
2 What has happened to the worker's purchasing power between 1970 and 1974? Between 1970 and March 1988?

Quantity Indexes

Price indexes, as we've just seen, have important uses. There are numerous variables subject to change, however, that cannot and/or should not be expressed in terms of price changes. Rather, these variables should perhaps be measured in terms of relative change in the physical volume of such activities as production and construction. A **quantity index,** therefore, is one that shows relative change in physical units. For example, production of some good—say electric motors— is a measure of industrial output which may be best expressed as a physical count of motors produced if that is the information needed by a decision maker. The number of books sold by a publisher and the number of housing starts within some geographic area are additional examples for which physical activity values may be more useful than price data. The methods used to prepare price indexes are easily adapted to quantity index construction.

Value Indexes

Since a measure of *value* is obtained by multiplying price by quantity, a **value index** deals with a variable which has elements of change in both price and quantity. Dollar sales volume, for example, is the product of price and quantity sold. When dollar volume changes, the change could be the result of a change in price, a change in quantity sold, or a change in both of these components. Similarly, changes in income received could result from changes in the rate of pay, from changes in the level of output, or from changes in both factors. An example of a value index is the one showing the value of construction contracts awarded, which is published monthly by the F. W. Dodge Corporation, a McGraw-Hill division.

Regardless of whether they deal primarily with price, quantity, or value, most index numbers are constructed by either an *aggregative* or an *average of relatives* method. Let's now examine each of these techniques used to prepare index number series.

AGGREGATIVE METHOD OF INDEX NUMBER CONSTRUCTION

The term "aggregate" means the sum of a series of values. An **aggregative index method,** then, is one that bases the computation in each time period on the sum of the units being considered.

Implicitly Weighted Aggregative Indexes

Price Index Let's assume that the mild-mannered statistics instructor we met earlier, Professor Ogive Staff, is interested in measuring how the price of certain

FIGURE 12-2   Professor Staff's Essential Commodities for Teaching Statistics

COMMODITY	UNIT OF PURCHASE	UNIT PRICE				
		1985	1986	1987	1988	1989
Chalk	12-piece box	$ 0.60	$ 0.60	$ 0.65	$ 0.70	$ 0.75
Red pencils	1/2 dozen box	0.72	0.75	0.80	0.85	0.92
Statistics book	1 workbook	9.50	10.00	10.00	10.50	11.00
Aspirin	1 bottle	0.59	0.59	0.65	0.69	0.75
		$11.41	$11.94	$12.10	$12.74	$13.42

commodities has changed during his past 5 years of teaching statistics. Figure 12-2 gives the items regarded as essential by Professor Staff and the prices of these items from 1985 to 1989.

An implicitly weighted aggregative price index may be found by using the following formula:

$$PI_n = \frac{\Sigma\, p_n}{\Sigma\, p_0} \cdot 100 \qquad\qquad \text{(formula 12-1)}$$

where PI_n = price index for a given time period n
p_n = prices in period n of the components in the series
p_0 = prices in the base period of the components in the series

Thus, if we select the prices of Professor Staff's commodities in 1985 to represent the base period we can compute the price index for selected time periods covered by the series as shown below:

$$PI_{1985} = \frac{\Sigma p_{1985}}{\Sigma p_0} \cdot 100 = \frac{\$11.41}{\$11.41} \cdot 100 = 100.0$$

$$PI_{1989} = \frac{\Sigma p_{1989}}{\Sigma p_0} \cdot 100 = \frac{\$13.42}{\$11.41} \cdot 100 = 117.6$$

The complete price index series for our example is shown in Figure 12-3.

Quantity Index   An implicitly weighted aggregative quantity index *could* be computed using the formula $QI = \Sigma q_n/\Sigma q_0 \cdot 100$, where q_n and q_0 obviously refer to *quantities* in different periods rather than prices. However, this formula is seldom used to measure changes in quantities because (to take one possible case) it would be meaningless to add together the quantities in a time period that are expressed in different units such as ounces, pounds, tons, and bales.

Self-Testing Review 12-2

1   Compute an implicitly weighted aggregative price index series for the following data, using the prices in 1987 as the base.

COMMODITY	1987 UNIT PRICE	1988 UNIT PRICE	1989 UNIT PRICE
A	$21	$23	$24
B	40	44	48
C	10	9	10
D	25	25	28

2 Interpret the meaning of your price index series.

The Need for Explicit Weights

Perhaps you've wondered what was meant by the words "implicitly weighted" in the preceding paragraphs. Let's see now if we can clarify this matter. In our example of the prices of Professor Staff's commodities, we did not attach any *relative importance* to the various commodities. But, of course, Professor Staff probably does *not* consider each item to be equally important. Rather, he may attach a greater priority or *weight* to aspirin than to red pencils. But as you can see from Figure 12-2, the price of a unit of pencils is greater than the price of a unit of aspirin. Thus, by using formula 12-1 we have automatically or *implicitly* given *greater* weight to red pencils than to aspirin.

When weights are *not* assigned to the various items in a series, the index produced is often called an unweighted index. It's more appropriate to call it an implicitly weighted index, however, because (in the case of a price index) greater weight is implicitly given to higher-priced items than to lower-priced items. To summarize, if explicit weights are *not* assigned to the individual components in a composite consumer price series in a logical way—on the basis of the relative importance of the components—an item such as lawn mower blades is given more influence than bread or milk. Obviously, however, people buy much greater *quantities* of bread than lawn mower blades, and thus they consider an increase in the price of bread more significant than an increase in the price of blades. To give bread its proper place in a composite price index series, a *system of explicit weights* is needed that takes into account the *typical quantities* of each item

FIGURE 12-3   Implicitly Weighted Aggregative Price Index for Professor Staff's Commodities

YEAR	PRICE INDEX (1985 = 100)
1985	100.0
1986	104.6
1987	106.0
1988	111.7
1989	117.6

consumed. In the next section we'll examine methods of constructing weighted aggregative indexes.

Weighted Aggregative Indexes

Price Index   To assign proper priority to each of the items included in an index number series, a logical weighting system must be used. The weights employed depend on the nature and purpose of the computed index. In the case of price indexes, as we saw above, the usual weighting scheme is to take into account the *typical quantities used*[1] of each of the items in the series. By multiplying the price of an item by the quantities of the item consumed in a time period, we get a dollar value that's indicative of the overall importance placed on the item.

Let's see how a weighted aggregative price index may be prepared by referring once again to Professor Staff's commodities. The data in Figure 12-4 are the same as those in Figure 12-2, with one important addition: We've now included a column indicating the typical annual consumption of each of the items. (You'll note the much greater consumption of aspirin than of red pencils.)

The formula for computing a weighted aggregative price index may be expressed as follows:

$$PI_n = \frac{\Sigma(p_n q_t)}{\Sigma(p_0 q_t)} \cdot 100 \qquad\qquad \text{(formula 12-2)}$$

where PI_n = price index in a given time period n
          p_n = price in a given period n
          p_0 = price in the base period
          q_t = typical number of units produced or consumed during the time periods considered

The application of this formula is not difficult. To get the *numerator*, we merely:

1   Multiply each commodity in a *given period* by the corresponding typical quantities consumed.
2   Total the products of prices and quantities for the period.

And to get the *denominator*, we:

1   Multiply each commodity price in the *base period* by the corresponding typical quantities consumed.
2   Total the products of base period prices and typical quantities.

---

[1] In some situations, the base period quantities (q_o) are used to represent typical quantities; in other situations, the quantities in the given time period (q_n) are used as weights. The technical considerations involved in selecting the appropriate weights are generally beyond the scope of this book, but we'll briefly consider weighting problems in a later section.

FIGURE 12-6

DISLIKE UNCLE SAM REVISING DATA? NOW THERE'S AN INDEX JUST FOR YOU

There's revisionist history. There's revisionist Shakespeare. And now meet Edward J. Hyman. He's a kind of revisionist economist. "You might say that I'm attempting to make sweet use out of adversity," he says.

Economists hate statistical revisions. After each month, the government begins gathering data and releases dozens of statistics on things such as the money supply and industrial production; then, as more data come in from the earlier month, it revises the figures. At the end of last December, for instance, durable-goods orders were reported up 0.2%. But when the revised figure was released a month later, it was 1.7%.

REVISIONS INDEX
To economists, who base economic forecasts on the statistics before they're revised, such revisions can mean the difference between a correct prediction and snickering behind their back.

Mr. Hyman, chief economist at Cyrus J. Lawrence Inc., a New York securities firm, has developed a way to use those irritating revisions to his advantage. Over the years, Mr. Hyman says, he noticed that when numbers were repeatedly revised upward, the economy tended to gain strength. When the revisions were repeatedly downward, so was the economic trend. The upshot of this correlation is Mr. Hyman's new indicator of where the economy may head next: the revisions index.

To construct it, he simply takes a given month's preliminary statistical report on, say, industrial production. A month later, he notes how much the statistic is revised. Using a half-dozen economic indicators, he totals the pluses and minuses and comes up with a trend figure that supposedly indicates strength or weakness in the economy.

A 'DUBIOUS PASTIME'
Mr. Hyman has been using his system since 1981, and a close look indicates that it has been pretty successful in predicting future trends. So far this year, the readings have progressively eased from +3.96 to +1.75. To Mr. Hyman, that suggests a slowing in economic growth, but no recession.

At the Commerce Department, a chronic reviser of data, chief economist Robert Ortner says the revisions index is "probably a dubious pastime and practice. I think watching the revisions and trying to create an index out of them is probably trying to be cute."

But Arthur Laffer, the economist whose curve made him famous, disagrees. "My guess is that it's a nice contribution," he says. "It's neither earth-shattering nor giggleable."

—From Ashok Chandrasekhar, "Dislike Uncle Sam Revising Data? Now There's an Index Just for You," *The Wall Street Journal*, June 6, 1984, p. 35. Reprinted by permission of *The Wall Street Journal*, © Dow Jones & Company, Inc. 1984. All Rights Reserved Worldwide.

AVERAGE OF RELAT

Figure 12-5 shows the results of these computations. Thus, if we designate 1985 *as the base year*, we can put the totals for selected time periods from Figure 12-5 into formula 12-2 as shown below:

$$PI_{1985} = \frac{\Sigma(p_{1985}q_t)}{\Sigma(p_0q_t)} \cdot 100 = \frac{\$15.80}{\$15.80} \cdot 100 = 100.0$$

$$PI_{1989} = \frac{\Sigma(p_{1989}q_t)}{\Sigma(p_0q_t)} \cdot 100 = \frac{\$18.96}{\$15.80} \cdot 100 = 120.0$$

FIGURE 12-4

amount typically spent on the item during the time periods considered. And since a dollar amount spent is the product of price and quantity, the weight used in an average of relatives index is customarily expressed as the *value* of an item that is consumed, bought, or sold.[2]

The general formula for a weighted average of relatives price index is:

$$PI_n = \frac{\sum\left[\left(\frac{p_n}{p_0} \cdot 100\right)(p_t q_t)\right]}{\sum(p_t q_t)}$$ (formula 12-6)

where $p_n/p_0 \cdot 100$ = price relative for period n
p_t = typical price during the time periods considered
q_t = typical number of units produced or consumed during the time periods studied
$p_t q_t$ = typical dollar expenditures (value weights) on the items during the time periods studied

Although you were hoping it wouldn't happen, we are once again going to use Professor Staff's essential commodities to illustrate the use of formula 12-6. Only the price data for the years 1985—the base period—and 1989 are used. Figure 12-8 shows the computational procedures. You'll notice in column 2 that we are assuming that a typical price is the price in the base period—that is, $p_t = p_0$. The *numerator* for the price index for the year 1985 is the total in column 8; the *numerator* for the index for 1989 is the total in column 9; and the *denominator* for the index for both years is the total in column 7. By now you should be able to interpret the meaning of the index of 120.0 for 1989.

If you compare the 1989 index number just computed with the price index for 1989 computed by the weighted aggregative method (see Figure 12-6), you'll notice that they are equal. This isn't surprising because formulas 12-2 and 12-6 are algebraically identical *if* the typical prices (p_t) are assumed to be the prices in the base year (p_0). This can be easily shown below if we change the p_t symbols in formula 12-6 to p_0's and cancel the p_0's in the numerator:

FIGURE 12-5

$$PI_n = \frac{\sum\left[\left(\frac{p_n}{\cancel{p_0}} \cdot 100\right)(\cancel{p_0} q_t)\right]}{\sum(p_0 q_t)} = \frac{\sum(p_n q_t)}{\sum(p_0 q_t)} \cdot 100$$

Formula 12-6 = formula 12-2

Why, then, if these two index construction methods yield the same results, do we bother with the weighted average of relatives approach? (The formula is, after all, a real bear!) The answer, of course, is that the two methods are *not* identical

---

[2] We can't simply use quantities alone as weights as we did earlier in the chapter because if we *multiply relatives* (which are likely to be in percentage form and which have no units) *by the quantities* in a series expressed in such units as pounds, tons, or bales, the resulting products of these multiplications would also be in the different units and could thus not be added logically.

FIGURE 12-8 Computation of Weighted Average of Relatives Price Index for Professor Staff's Essential Commodities (Base = 1985 = 100)

COMMODITY (1)	PRICES		PRICE RELATIVES		TYPICAL ANNUAL CONSUMPTION (q_t) (6)	VALUE WEIGHTS $(p_t q_t)$ (col. 2 × col. 6) (7)	WEIGHTED PRICE RELATIVES	
	1985 $(p_0$ and $p_t)$ (2)	1989 (p_n) (3)	$\left(\dfrac{p_{1985}}{p_0} \cdot 100\right)$ (4)	$\left(\dfrac{p_{1989}}{p_0} \cdot 100\right)$ (5)			1985 (col. 4 × col. 7) (8)	1989 (col. 5 × col. 7) (9)
Chalk	$0.60	$ 0.75	100.0	125.0	4	$ 2.40	240	300
Red pencils	0.72	0.92	100.0	127.8	0.5	0.36	36	46
Statistics book	9.50	11.00	100.0	115.8	1	9.50	950	1,100.1
Aspirin	0.59	0.75	100.0	127.1	6	3.54	354	449.9

$$\Sigma\left[\left(\frac{p_n}{p_0} \cdot 100\right)(p_t q_t)\right] \quad \longrightarrow \quad 1,580 \longrightarrow 1,896$$

$$\Sigma(p_t q_t) \quad \longrightarrow \$15.80$$

Therefore, $PI_{1985} = \dfrac{1,580}{\$15.80} = 100.0$ and $PI_{1989} = \dfrac{1,896}{\$15.80} = 120.0$

if $p_t \neq p_0$, and it's often desirable to use prices other than those in the base period to represent what is typical during the time periods studied.

Quantity Index   You'll recall that exchanging the p's and q's in the weighted aggregative price index formula produced the weighted aggregative quantity index formula. We'll get a similar result if we exchange the p's and q's in formula 12-6. In other words, the general formula for a weighted average of relatives quantity index is:

$$QI_n = \frac{\Sigma\left[\left(\dfrac{q_n}{q_0} \cdot 100\right)(q_t p_t)\right]}{\Sigma(q_t p_t)} \qquad \text{(formula 12-7)}$$

With the exception that quantity relatives are computed in place of price relatives in the numerator of formula 12-7, the arithmetic *procedures* are identical to those used in preparing a price index. Therefore, it's not necessary here to work through an example quantity index problem. However, you may test your understanding of the procedure by preparing the quantity index values in Self-Testing Review 12-5.

Computers and Index Numbers

Statistical software needs are met in much the same way that people satisfy their clothing needs. General-purpose descriptive measures, common graphic analyses, and standardized test results are processed with prewritten off-the-shelf packages, just as general clothing needs may be met with off-the-rack clothes. All statistical software and clothing needs of many people are satisfied in this way. But not all statistical software or clothing needs are general; some are unique and cannot be met by existing products. Just as a custom-tailored garment is designed to fit a specific person, a **custom-made statistical program**—one designed to carry out unique tasks—is often needed to process a specific job. Index number calculations usually fall into this "specific job" category; weights are selected in different ways for different index series, and processing techniques also vary. The general approach used by these custom programs follows the processing concepts introduced in this chapter, but the details usually call for special programs. Thus, we'll not duplicate any of our examples in this chapter with computer printouts.

A custom-made statistical program to prepare an index-number series can be written by using the built-in programming capability included in general-purpose statistical packages such as *Systat, Minitab*, and many others. Or a separate **programming language** that includes all the symbols, characters, and usage rules that permit people to communicate with computers may be used. The federal government uses custom programs and large mainframe computers to process data and prepare the Consumer Price Index and other series discussed in the next section.

Self-Testing Review 12-5

1　Using the price and quantity data in Figure 12-4, compute the weighted average of relatives price index for Professor Staff's commodities for the years 1987 and 1988, using 1987 as the base period and the prices in 1987 as typical prices.

2　Compute a weighted average of relatives quantity index for the data given below. (Assume that the base year is 1985 and the typical quantities produced are those in 1985.)

| | | | QUANTITY PRODUCED | |
COMMODITY	UNIT OF PRODUCTION	TYPICAL PRICE	1985	1986
X	Dozen	$1.00	48	52
Y	Pound	2.00	860	900
Z	Carton	1.50	550	500

IMPORTANT INDEX NUMBERS IN CURRENT USE

We frequently learn about the movements of certain index number series by reading newspapers or watching television. News reporters pass along index number information because they correctly believe that the public has a right to know the nature of general economic changes and the ways in which such changes may be affecting individuals. (Voters frequently use such information about economic changes as an important decision-making factor on Election Day.) And, of course, economists and administrators with a need for more detailed and specialized index number information for planning, decision making, and control purposes may choose from a wealth of published index number series. Sources of such series are the *Survey of Current Business, Federal Reserve Bulletin, Monthly Labor Review*, and *Business Conditions Digest*, to name just a few. Let's now very briefly describe a few of the more commonly used index number series.

Price Indexes

The *Consumer Price Index* (CPI) published by the United States Bureau of Labor Statistics measures the average change in prices of many types of consumer goods and services. Actually, two CPIs are available. One index for *all* urban consumers (the CPI-U) represents about 80 percent of the population. This is the index that the news media follow, and it's the best measure of inflation. The other index (the CPI-W) includes only urban wage earners and clerical workers. Professional, managerial, and technical workers, retirees, and self-employed and unemployed persons are included in the CPI-U but not in the CPI-W. A sample of the prices of about 400 items (a fixed market basket of goods and services) is taken each period (usually monthly) from about 85 urban areas. (These areas are given weights according to the size of the working population in each area.) Price data from the urban areas are then combined to form a composite national index. The items in the current series were selected because of their frequency of purchase and their importance relative to total expenditures. However, the composition of the market basket *does* gradually change, as you saw in the reading that opened this chapter. Old items are dropped when they're no longer sold in volume, and new items are added when they become sufficiently important.

The weighted average of relatives method is used to construct the CPI, and constant weights based on studies made in 1982–1984 are employed. The current base period is also 1982–1984. The weighting system and base years have been revised several times in the past to improve the usability of the index.

The CPI is the most widely known and used of all index numbers; it may even be the most important statistic published regularly by the federal government. Social Security benefits are pegged to the CPI. And the CPI-W series also serves as a basis for adjusting union wage rates to take into account changes in consumer prices. A clause—often called an **escalator clause** or a cost-of-living adjustment (COLA)—may be written into a union contract. The clause may specify that an automatic wage rate change of a given amount will be made to a union member's pay when the CPI-W changes by a specified amount.

The inflation rate between 1975 and 1987, as measured by the year-to-year changes in the CPI, is shown in Figure 12-9. The inflation rate and the unemployment rate are sometimes added to get an unofficial measure called a **discomfort index.** Created in the 1970s by economist Arthur Okun, this measure shows how the national economy is faring. During the Ford Administration in the mid-1970s, this index was in the teens, and that's "high discomfort" territory. Jimmy Carter substituted the term "misery index" for discomfort index and used the value against Ford in the 1976 presidential election campaign. Under Carter,

FIGURE 12-9
Percentage growth
in the Consumer
Price Index between
1975 and 1987.
(Source: *Federal
Reserve Bulletin*,
January 1988, p. 2.
Reprinted with
permission.)

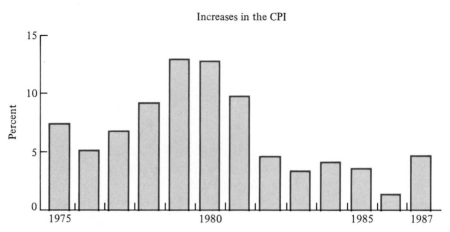

Increases in the CPI

Growth from fourth quarter to fourth quarter, except for 1987,
when data shown are through the third quarter.

though, the index climbed to even more intolerable heights, reaching a peak of 22 in the first quarter of 1980. Ronald Reagan was then able to use this development against Carter in the 1980 election.

The **Producer Price Index (PPI)** published by the U. S. Bureau of Labor Statistics measures changes in the general price level of goods at their *first commercial transaction level*—that is, in prices received at the producer level. For years the PPI was referred to as the *Wholesale Price Index*, but this name often led to the erroneous assumption that it measured prices received by wholesalers and others in the channels of distribution.

The PPI is calculated as a weighted average of relatives price index, has a current base period of 1982, and currently includes several thousand commodities. Commodity data are broken down into groups and subgroups by type of finished product, by durability of product, by state of processing, and by other classifications. The value weights for the composite indexes are derived from the shipments (sales) of commodities in a particular year. These weights are revised periodically. The PPI is used for a number of purposes, including market analysis and the preparation of escalator clauses in long-term commodity purchase and sales contracts.

Quantity Index

The **Index of Industrial Production (IIP)** is published monthly by the Board of Governors of the Federal Reserve System and measures changes in the output quantities of plants, mines, and electrical and gas utilities. It does not cover farm production, construction, or transportation activities or the various trade and service industries. Component series are combined using the weighted average of relatives quantity method to prepare the IIP. The value weights used are based on the concept of *value added*—the difference between the value of production and the cost of materials or supplies consumed. The base period is currently

1977. As an indicator of changes taking place in the economy's output, the IIP is widely used to support administrative planning decisions. An industrial producer, for example, can compare the company's performance with that of competitors and make those decisions that seem appropriate. More information on another industrial production index is presented in the Closer Look reading at the end of this chapter

PROBLEMS AND PITFALLS WITH INDEX NUMBERS

Although most index number series are *not deliberately misleading*, they are far from perfect. It's beyond the scope of this book to go into the limitations of index numbers in great detail; however, there are (1) a number of major *problems associated with the construction of indexes* and (2) *several pitfalls associated with their use* that should be mentioned. As potential consumers of statistical data, you should be aware of these problems and pitfalls.

Problems of Index Number Construction

Some of the major problems encountered by statisticians in constructing index numbers are:

1 *The problem of selecting a sample.* Because the most popular price and quantity indexes are used in such a large number of ways, it's difficult (if not impossible) to select the items to be included in an index in such a way that the results are equally meaningful to those who use (or unknowingly misuse) the index. The CPI versions, we've seen, focus attention on prices paid by *urban* persons for about 400 items (out of an identified total of about 2,000 items). Thus, the CPI is not equally applicable to small-community families and farm families. And the decision on which 400 items will be in the sample selected is based on the *judgment* of government statisticians; thus the sample is not a random sample.

2 *The problem of assigning appropriate weights.* In the weighted index formulas presented in this chapter, we've used typical prices and quantities for weighting purposes. But what's typical or appropriate for one period and for one purpose may quickly become inappropriate for another time and for another use. There's often a lag between the time that a weighting system should be changed and the time when it's actually revised. Isn't it possible, for example, that the typical quantities purchased (and used to weight a price index) could change significantly during periods of rapid price increases when consumption patterns are being altered? How typical would the quantity weights then be? To illustrate, at this writing the CPI is based on spending habits for 1982–1984, and many people may have revised their buying habits since then.

3 *The problem of choosing an appropriate base period.* The criteria for determining an appropriate base period include (1) the recency and normalcy of the period—i.e., how well it is remembered and how well it avoids both the

WHAT IS REAL INFLATION RATE?
Official CPI Sometimes Fails to Reflect Actual Cost of Living

"Dear Mr. Burns,

"You keep writing about lower inflation, and I keep wondering if we live in the same country. Have you bought any toothpaste lately? Been to the dry cleaners? Bought some pet food? Had your shoes repaired? Checked the premium on your health insurance? For the things I buy as a retired person, I don't see any let-up. I don't believe inflation is 'only' 4 percent—and I don't know why a guy like you believes those government numbers."

Those comments, which came from a retired reader in Phoenix several months ago, led me to visit Irwin Kellner on a recent trip to New York. Kellner, who is chief economist for Manufacturers Hanover, achieved some notoriety last year when he announced his alternative to the government's monthly Consumer Price Index.

Like the Phoenix reader (and several others who wrote with the same theme), Kellner was troubled by the apparent gap between his day-to-day experience of prices and the monthly change in the CPI reported by the Department of Labor. In his case, the precipitating event was a haircut.

"In October 1986, I was getting a haircut at the local unisex shop," he said. "The price was $22. The barber told me the next haircut would be $26. I did a quick calculation: It was an 18 percent increase. So I asked how they could charge that much when inflation was only 1 percent.

"He answered that little things were costing a lot now. Some people might have taken that at face value and forgotten about it, but, well, you just shouldn't say things like that to economists. I decided to see how much of that was true.

TABLE 1 The Original Nuisance Index

ITEM	JAN. '85	JAN. '87	% CHANGE
Ground coffee (1 lb)	$1.79	$3.19	78.2%
Liquid detergent (22 oz)	1.29	1.89	46.5
Toothpaste (5 oz)	0.99	1.39	40.4
Taxi (2-mile ride)	3.10	4.25	37.1
Orange juice (1/2 gal)	1.69	2.29	35.5
Pack of gum	0.45	0.60	33.3
Pizza slice	0.95	1.25	31.6
Bananas (1 lb)	0.39	0.49	25.6
Ground round beef (1 lb)	1.99	2.79	40.2
Woman's haircut	20.00	25.00	25.0
Pint of ice cream	1.09	1.45	33.0
Movie	5.00	6.00	20.0
Shoeshine	1.25	1.50	20.0
Peanut butter (18 oz)	1.79	2.39	33.5
Paperback book	3.75	4.35	16.0
Bottle of imported beer	1.30	1.50	15.4
Dry cleaning of a suit	6.00	6.75	12.5
News magazine	1.75	1.95	11.4
Fast-food burger and shake	2.63	2.88	9.5

Source: Manufacturers Hanover Corp.

"So we developed a survey, unscientific but weighted like the Consumer Price Index. And we found, to our amazement, that while the CPI was up modestly, our list was up 25 percent. We called it the Nuisance Index because it was a list of those little pesty things you hate to buy." (See Table 1.)

Kellner was quick to say that he was amazed at how people, particularly the media, responded to the index. Within a few weeks, he was appearing on TV regularly.

"We figured we were on to something. So we made it a national index—something more scientific, more representative of the country.

"We took the national CPI and removed the big, infrequent expense items—the purchase of a house, the price of new cars and appliances, etc., and that left us with the everyday items of expense.

"Now the American Association of Retired Persons is interested in our developing an index that is appropriate for retired people."

Kellner observed that no single index could represent the purchasing patterns of *all* people and that your experience of inflation would depend on your personal consumption habits.

Elderly people are seldom home buyers and have often paid off the mortgages on their houses. This means that the price of houses and the cost of financing them—a major component of the CPI—isn't very important to them. Similarly, elderly people tend to be infrequent buyers of cars and new appliances—but are very affected by the rising price of everyday items.

The difference is substantial. For the three years ending in November 1987, for instance, the Manufacturers Hanover version of the cost-of-living index rose 13.43 percent while the CPI rose only 9.58 percent.

Since Social Security benefits are pegged to the CPI, elderly people are correct when they feel that real prices are rising faster than the index—and their income.

"When people see pizza going up 10 cents on a 40 cent slice, they know inflation isn't going up at 1 percent," Kellner observed.

Does this mean that elderly Americans are taking a hosing because their income is regulated by an index that is unrelated to their actual spending patterns?

Sometimes, yes. And sometimes, no.

Kellner pointed out that the CPI could overstate inflation as well as understate it. From 1980 until late in 1982, for instance, the Consumer Price Index was rising much faster than the Manufacturers Hanover index, largely because home prices and interest rates were rising rapidly. During this period, Social Security income rose faster than the price of Kellner's "Nuisance Index."

In 1984 and 1985, the two indexes rose at virtually identical rates, so income and expenses were nearly matched.

The mismatch that brings the reader letters of protest began in 1986 with the abrupt decline in interest rates—while the cost of actual items purchased continued to rise, the CPI rose only 1.1 percent in 1986 due to declining energy prices. In the same year, however, medical costs rose 7.7 percent—the largest amount since 1982—and food prices rose 3.8 percent.

To compound matters, interest *income*, which is also very important to older people, also declined, catching them in a double bind—income was declining while expenses were rising faster than the CPI.

Is there a remedy?

Not really. It all comes down to a "sometimes you win, sometimes you lose" proposition. Right now, elderly Americans are on the losing side of the income/expense equation.

—Scott Burns, "What Is Real Inflation Rate?" *The Dallas Morning News*, February 2, 1988, pp. 1D; 19D. Reprinted with permission.

peaks and troughs that periodically occur in economic activity—and (2) the need for comparability with other popular indexes. The difficulty of defining a fairly recent normal period for a composite index made up of hundreds of items is obvious. Yet some period must be used, even if it's not particularly suited for some items.

Pitfalls in the Use of Index Numbers

Some of the pitfalls encountered by users of index numbers may be attributed to their failure to:

1 *Obtain adequate knowledge of indexes.* Few serious users of index numbers have as good an understanding of the kinds of indexes available and of the methods employed in their construction as they should have. (Careful study of the publications of the organizations that print the indexes is necessary to gain a specialized understanding.) If, by summarizing a few of the problems of index construction, we've motivated you to make a more thorough study of an index before you use it, we've performed a useful service.

2 *Consider possible bias introduced through the passage of time.* This pitfall is related to the problem of assigning appropriate weights discussed in the preceding section. As time passes, an index number may understate or overstate the extent of the change being measured because of shifts in the relative importance of (but not the weights of) the items being studied. And, of course, the use of a biased index could lead to unfortunate decisions.

3 *Consider qualitative changes.* Technological change over time may lead to quality changes in products being compared. When this happens, it's very difficult, if not impossible, to make satisfactory adjustments in a price index to reflect adequately the qualitative differences. Automobiles are not the same today as they were 10 years ago; nor, for that matter, are computers the same today as they were in 1982–1984. A clone of the computer the author paid $5,000 for in 1983 can be purchased at this writing for less than $1,500. And for $5,000 now you can put the equivalent of a 1983 minicomputer on your desktop. But the CPI doesn't reflect these qualitative changes. Thus, some analysts believe that over a period of years the increases in the CPI may overstate actual price increases in terms of a truly *fixed* basket of items.

LOOKING BACK

1 An index number is a measure of how much a composite group changes on a relative basis with respect to a base period that's usually defined to have a value of 100. Index numbers are generally used to measure changes over some period of time, and so they provide a way to summarize and communicate the nature of those changes that occur. They are also useful in simplifying data, facilitating comparisons, and showing typical seasonal variations.

2 Most index numbers are classified as (a) price indexes, (b) quantity indexes, or (c) value indexes. One use of a price index is to serve as a basis for preparing a purchasing power index number. A purchasing power index number may be found by first locating a recent "all items" figure from the CPI-U series. The next steps are to convert this figure back into a price relative, find the reciprocal of this price relative value, and then multiply this reciprocal value by 100 to get the purchasing power index number.

3   Just as a price index shows the relative change that takes place in the price of goods and services over time, a quantity index shows relative change in physical units. Regardless of whether they deal primarily with price, quantity, or value, though, most index numbers are constructed by either an aggregative or an average of relatives method. Examples of implicitly weighted aggregative and average of relatives price indexes are given in the chapter. Another term sometimes used in place of "implicitly weighted" is "unweighted," but this isn't strictly correct. A system of explicit weights is generally needed to take into account the relative importance of the various items in an index. Determining exactly how to weight a series is a complex matter that's generally beyond the scope of this book, but we've looked at some simple examples of weighted aggregative and weighted average of relatives indexes in the chapter. Published index series that employ sophisticated weighting schemes are processed with custom-made statistical programs.

4   Important *price* indexes in current use are the Consumer Price Index and the Producer Price Index, and an important *quantity* index is the Index of Industrial Production. The Consumer Price Index is actually two series—the CPI-U and the CPI-W. The CPI-U measures the prices paid by all urban consumers, while the CPI-W measures the prices paid by urban wage earners and clerical workers. What measures the prices paid by farmers? Nothing; the CPI doesn't survey the prices paid by about 20 percent of the population. But Social Security benefits are pegged to the CPI, and union wage rates are tied to changes in the CPI-W by escalator clauses in contracts. A "discomfort index" that links the inflation rate (as measured by the CPI) and the unemployment rate has been used by politicians as a measure of how the national economy is faring.

5   There are problems associated with the construction of indexes, and there are several pitfalls associated with their use. Some of the vexing issues of index number construction are the (a) problem of selecting a sample of items to be included in the index series, (b) difficulty of assigning appropriate weights, and (c) problem of choosing a representative base period. And some of the pitfalls of using index numbers may be attributed to a user's failure to (a) obtain adequate knowledge of indexes, (b) consider possible bias introduced through the passage of time, and (c) consider qualitative changes.

KEY TERMS AND CONCEPTS

PROBLEMS

1 Using 1985 as a base, compute a simple price index for 1985–1990 with the data given below.

YEAR	PRICE PER POUND OF PRODUCT
1985	$0.10
1986	0.09
1987	0.11
1988	0.13
1989	0.13
1990	0.15

2 An old survey has recently been found which shows the consumption habits of a member of the Signa Phi Nuthen fraternity at a nondescript college. The following data had been gathered, and in a fit of nostalgia you decide to apply your substantial expertise in the area of index numbers.

COMMODITY	UNIT OF PRICE QUOTATION	1970		1971		1972	
		PRICE	QUANTITY	PRICE	QUANTITY	PRICE	QUANTITY
Shirts	1 each	$6.00	10	$6.67	11	$7.55	10
Books	1 each	8.75	11	9.35	12	10.15	13
Beer	1 mug	0.30	105	0.30	127	0.35	153
Gasoline	1 gallon	0.28	380	0.32	130	0.39	364

Assuming that the base year is 1970 and the typical quantities are the quantities in 1970:

a Construct a price index series using the implicitly weighted aggregative method.

b Construct a price index series using the weighted aggregative method.

3 The Bureau of Business Research of a university must prepare a report for a statewide publication. The purpose of this report is to show how prices and quantities of three

products have changed over time. The bureau obtained the following information from its old files:

	1983		1984		1985		1986	
PRODUCT	UNIT PRICE	QUANTITY SOLD	UNIT PRICE	QUANTITY SOLD	UNIT PRICE	QUANTITY SOLD	UNIT PRICE	QUANTITY SOLD
A	$0.30	960	$0.35	975	$0.45	900	$0.47	965
B	0.75	135	0.70	100	0.80	115	0.78	140
C	1.00	290	1.05	280	1.07	285	1.05	295

Using 1983 as a base, compute the following for all years:

a   A simple price index series for product A
b   An implicitly weighted aggregative price index
c   A weighted aggregative price index (weighted with base period quantities)
d   A weighted aggregative quantity index (weighted with base period prices)
e   An implicitly weighted average of relatives price index
f   A weighted average of relatives price index (weighted with prices and quantities in 1983)
g   A weighted average of relatives quantity index (weighted with prices and quantities in 1983)

4   The following table gives the amount of oil produced by an independent oil company for the years shown. Using 1985 as the base period, compute a simple quantity relative index for oil production.

YEAR	BARRELS OF OIL PRODUCED
1975	10,113
1977	12,276
1979	15,372
1981	19,450
1983	17,320
1985	15,000
1987	13,342
1989	11,133
1991	8,732
1993	10,111

5   The data below show prices and quantities of guns and butter for a 1988–1990 period:

	1988		1989		1990	
COMMODITY	PRICE	QUANTITY	PRICE	QUANTITY	PRICE	QUANTITY
Guns	$600.00	10	$650.00	9	$750.00	12
Butter	8.00	5,000	8.50	6,000	9.00	8,000

a   Construct a price index series by the implicitly weighted aggregative method.
b   Construct a price index series using the weighted aggregative method, and assume that the typical quantities are those in 1988.

A CLOSER LOOK

U.S. INDUSTRY IS WASTING AWAY—BUT OFFICIAL FIGURES DON'T SHOW IT

For the past decade, as the trade deficit has widened and imports have invaded industry after industry, one odd statistic keeps providing eerie assurance that all is well with U.S. manufacturing. According to the Commerce Dept., industrial production has somehow maintained its historic share of total U.S. output—about 21% to 22%.

Armed with this statistic, many economists insist that nothing fundamental ails American industry except perhaps the federal budget deficit. The 1988 *Economic Report of the President* blithely declared: "Except for business cycle movements, the shares of real manufacturing output and final goods output have been remarkably stable for 25 years." The same statistic has enabled the Panglossian wing of the economics profession to debunk calls for industrial policy and to view falling manufacturing employment as a sign of health. If it takes fewer workers to produce a constant fraction of gross national product, that obviously means we are becoming more productive as a nation.

But new evidence suggests there may be something amiss after all—with the statistic itself. A pending study by the Office of Technology Assessment, conducted at the request of eight congressional committees and titled *Technology and the American Economic Transition,* recalculates the Commerce Dept. findings and concludes that the real manufacturing share in GNP is in fact declining. And a paper, *Manufacturing Numbers,* by economist Lawrence R. Mishel of the Economic Policy Institute, concludes that the government's own methodology tends to overstate the stability of manufacturing output. (Truth-in-packaging disclosure: I serve on the board of the institute.)

CRUNCHED NUMBERS
The statistical evidence for the claim that manufacturing thrives is a Commerce Dept. index called the Gross Product Originating (GPO) series. The GPO series estimates the value added in the manufacturing sector by calculating gross production, subtracting the costs of inputs, and applying "deflators" to adjust the result for inflation. But as Mishel points out, the methodology used by the Commerce Dept. fails to account adequately for the impact of imported components and hence overstates domestic production. Obviously, if an American auto uses an engine built in Brazil or a U.S.-made computer uses a disk drive fabricated in Japan, then all of that output is not domestic.

Among the several weaknesses in the GPO index, according to Mishel, is the fact that the series estimates the costs of components (intermediate goods) using an input-output model that was last updated in 1977. The past 11 years, of course, have seen significant substitutions of cheaper foreign components for more expensive domestic ones. But in its adjustments for inflation, the series employs the questionable assumption that prices of U.S.-made components change at the same rate as imported ones. When appropriate adjustments are made, Mishel calculates that the true share of manufacturing in total domestic output is more like 18% or 19% than 22%—which would represent a drop of unprecedented magnitude. "The problem with the methodology," says another critic, John R. Norsworthy, former chief of productivity research at the Bureau of Labor Statistics, "is that we're imposing 1977 technology on 1988 production."

DATA GAP
We need much better data on U.S. manufacturing. One of the unfortunate bits of fallout from the Administration's chronic antiplanning bias is a reluctance to have the government collect adequate information about industry. Far too much of our official statistical base is not collected directly but is cobbled together from trade association tabulations, samples, or "proxies" (a fancy word for edu-

cated guesses). These official indexes then become the grist for rigorous-looking academic studies whose ostensible precision is built on statistical sand.

A close look inside the business of government data collection reveals gross inconsistencies in the quality and methodology used to assemble data for different industries, and in different statistical series. All of America's major trading partners have far better industrial data than we do and use it to better competitive effect. Last year the American Economic Assn., not a bastion of statism, issued a plea for better official data collection on U.S. industry.

Moreover, the Panglossians ought not to be too smug about dismissing the risk of American deindustrialization. "Anecdotal" is a good word in journalism and a bad one in academia. But there is overwhelming evidence that one domestic industry after another is losing market share at home and not making it up abroad.

The trade deficit in manufactured goods last year was $170 billion. If Mishel's recalculation is correct, that just about equals the 3% to 4% of GNP that manufacturing has lost. (Japan's manufacturing share of GNP has risen from 28% to 35% in six years.) America does need fiscal discipline, but it desperately needs more imaginative trade and sectoral policies, too. And before we can intelligently fashion those, we need to know the full truth about what we are producing.

—Robert Kutner, "U.S. Industry Is Wasting Away—But Official Figures Don't Show It." Reprinted from May 16, 1988 issue of *Business Week* by special permission, copyright © 1988 by McGraw-Hill, Inc.

FORECASTING TOOLS: TIME-SERIES ANALYSIS

Calling the Economy's Turns May Be Fun—But Futile

To borrow Andy Warhol's aphorism, nowadays every economic statistic achieves celebrity status for about 15 minutes. Oh, maybe it's more like a few days or even a week or two—but the point is the same.

These celebrity numbers run the gamut of economic pulse-taking: the trade deficit, retail sales, housing starts, inventories, the leading indicators, and the unemployment rate. For several days leading up to their release—and a few days after—economic analysts and the media have a field day speculating whether this piece of data, finally, is the one that tells us if we're going into a recession. Nobody wants a slump, of course, but everyone is desperate for some guidance about how to plan, invest, spend, or save wisely—especially when the economy seems close to a turning point.

'VERY UNFORGIVING'
That desire is understandable, and it makes for good sport. Yet this fixation on the hot numbers makes little sense: No one or two measures of business activity, especially those covering a week or a month, can tell us how the entire economy is going to perform—or when. "The economy is very unforgiving of small mistakes in predicting the behavior of its components," says Robert E. Hall, director of research on economic fluctuations at the National Bureau of Economic Research.

Get a few of the myriad details wrong, or fail to consider possible strengths that might offset predicted weaknesses, and the model flunks.

Economists add to the confusion by analyzing each number to death in a day or two. They all want to be the first to proclaim that a 4.2% jump in real gross national product is actually lousy news, since most of the gain came from an inventory pileup. Or that the December trade deficit may have looked healthier at $12.2 billion but really wasn't so great because imports failed to drop.

Some of us thought we got a clear sign last Oct. 19—a 508-point drop in the Dow Jones industrial average. If that wasn't *the* number, what is? In fact, notes Victor Zarnowitz of the University of Chicago's business school, the stock market "is one of the most accurate leading indicators." But so far, most of the economy seems to have shrugged off the crash.

What we may be asking for is a number that doesn't yet exist: the date on which the current expansion will reach its peak and the economy will start to recede. And the one thing economic forecasters do worst—and they admit it—is to call turns in the business cycle.

Not only are they unable to tell us when the next recession will start, they won't even be able to tell us we're in it until six months or so after the fact. That's about the minimum time it takes the NBER, the arbiter of such matters, to date a business-cycle turn. Sometimes it takes longer: The Ford Administration launched its Whip Inflation Now campaign in the fall of 1974, only to learn during the winter that the nation had gone into a severe recession in November, 1973.

Calling a trend that's under way is far easier, because the small errors cancel one another out. Moreover, when you're dealing with trends, the past and present behavior of economic fundamentals provides guidance. But history is no help on turns, because each one is different. Turns often result from outside shocks, such as the first big OPEC oil price hike, and shocks are by definition unpredictable. Government responses to shocks can also be unpredictable and may make things worse.

DETAILS

Hall of the NBER compares calling turns with weather forecasting. "The closer you get to the day you're forecasting, the easier it gets to be right. Five days or a week before that day, the forecasts are almost worthless."

Many economists now see signs of a storm, but have no idea when it will hit. They agree that economic growth is slowing. And despite its poor record during the inflation-crazed 1970s, the profession has been surprisingly accurate in predicting annual growth rates. But all the economists and all their computers can't tell us when the growth rates will cross the line where the plus sign becomes a minus.

Should we really care so much about the details? Ray Fair, a forecaster at Yale University who sees no recession in 1988, says no: "The difference between predicting minus 0.2% in GNP and having it coming out plus 0.2% is statistically unimportant. The more serious error is to get the sign right, but say GNP will rise 3% and then have it climb 8%."

That would be great, but don't expect it. Barring some new shocks, the economy isn't likely to move sharply higher or lower in 1988. Crunching every new number for a sign of an economic turn may be fun, but it isn't very useful. People have enough to do just living with the trends.

—Norman Jonas, "Calling the Economy's Turns May Be Fun—But Futile." Reprinted from February 29, 1988 issue of *Business Week* by special permission, copyright © 1988 by McGraw-Hill, Inc.

LOOKING AHEAD

Our opening vignette dealt with the difficulties of calling the turns in the business cycle, and our subject in this chapter is time-series analysis. Since one of the components of a time series is the fluctuations that may be attributed to the business cycle, we'll have more to say about cyclical matters in later pages. First, though, we'll define a time series and consider the reasons for studying an economic series or time series. Next, we'll look at an overview of the time-series components including the cyclical element. After studying each component in more detail, we'll then look at the use of time-series analysis in forecasting, and point out the possible problems in the use of time-series analysis.

Thus, after studying this chapter you should be able to:

▶ Explain (a) the reasons for studying time series and (b) the possible problems that may be encountered in the use of time-series analysis.
▶ Identify the components of a time series.
▶ Compute and project a linear secular trend.
▶ Compute and use a seasonal index.
▶ Summarize how time-series analysis may be used in forecasting.

CHAPTER OUTLINE

LOOKING AHEAD
WHAT'S A TIME SERIES?
WHY STUDY TIME SERIES?
TIME-SERIES COMPONENTS: AN OVERVIEW
    Secular Trend
    Seasonal Variations
    Cyclical Fluctuations
    Irregular Movements
    An Approach to Time-Series Analysis

WHAT'S A TIME SERIES?

Administrators in all organizations make plans to cope with future changes. The planning function looks to the future; to plan is to make decisions in advance about a future course of action. Obviously, then, planning and decision making are based on **forecasts** or expectations of what the future holds. Thus, whether they employ complex quantitative methods or merely rely on intuitive hunches, administrators must look down the road and make these forecasts.

Every forecasting approach used today is based on several simple assumptions:

- Future occurrences depend, at least partially, on presently observable events.
- Future activities will follow patterns similar to those that have been traced in the past.
- Past relationships can be discovered by observation and study.

Regardless of the approach used, though, future expectations are needed before decision makers (1) select both short- and long-run goals, (2) develop policies and strategies to help accomplish future objectives or counter anticipated threats, and (3) revise earlier plans and decisions in the light of changing conditions.

Once plans are implemented, an organization is more or less committed to them for a variable time period. The term **planning horizon** refers to the period in which planners are committed to their decisions. In some cases, the planning horizon is as short as 1 day—for example, your plans on what to wear to class tomorrow, based on a weather forecast. But in business decisions involving, say, capital expenditures, the planning horizon is measured in months or years, and the accuracy of the forecast is a matter of much greater concern. In government, too, the planning horizon is often measured in months or years. Figure 13-1 shows the long-run forecasts of Gross National Product and unemployment by the Congressional Budget Office in 1986 and 1987. As you would expect, such forecasts are subject to substantial revision from year to year.

Among the statistical forecasting tools used by managers in planning and in coping with change are time-series analysis (the subject of this chapter) and regression analysis (the subject of the next chapter).

A **time series** is a set (or series) of numerical values of a particular variable listed in chronological order. In Figure 13-2, for example, there are two time

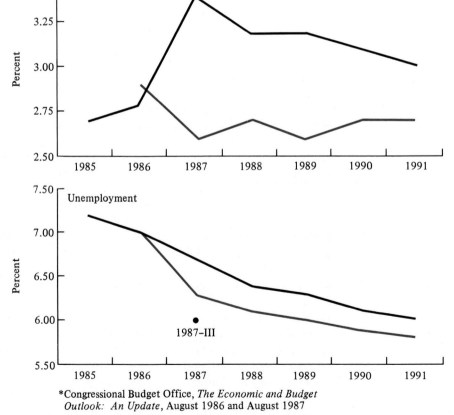

FIGURE 13-1

The Congressional Budget Office's expectations about Gross National Product and Unemployment vary from year to year, as you would expect. (Source: *Quarterly Review,* Federal Reserve Bank of New York, Autumn 1987, p. 37. Reprinted with permission.)

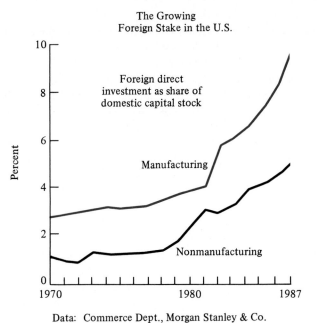

FIGURE 13-2
Two time series covering the period from 1970 to 1987. (Source: Gene Koretz, "The Buying of America: Should We Be Worried?" Reprinted from May 9, 1988 issue of *Business Week*, p. 36, by special permission, copyright © 1988 by McGraw-Hill, Inc.)

The Growing
Foreign Stake in the U.S.

Foreign direct investment as share of domestic capital stock

Manufacturing

Nonmanufacturing

Data: Commerce Dept., Morgan Stanley & Co.

series: One variable is the percentage of the investment in nonmanufacturing industries located in the United States that's owned by non-U.S. citizens; the other variable is the percent of domestic manufacturing operations that's owned by foreign nationals. As you can see, the time represented in both series is the period from 1970 to 1987, and the percentage of foreign investment in both of the listed categories has grown over this period. **Time-series analysis** involves classifying and studying the patterns of movement of the values of the variable over regular intervals of time (and the time intervals may be expressed in months or quarters as well as years).

WHY STUDY TIME SERIES?

In the chapters in Part 2, we dealt with the subject of sampling and statistical inference—a subject employing a mature theoretical base to support a number of popular and proven application techniques. In time-series analysis, however, we are dealing with a subject that (1) *will not* give us future estimate figures to which we can confidently assign a high probability and (2) uses a popular model that is *rather crude* and may be expected to yield only approximate results at best. In light of these disturbing facts, you may indeed ask yourself the question, "Why study time series?"

Time series are worthy of our attention for the following reasons:

1   *They may enhance understanding of past and current patterns of change.* A careful study of past events and experiences may give us greater insight into

the dynamic forces affecting the patterns of change. Historians are fond of saying that "those ignorant of the past are condemned to repeat it." The decision maker who understands why past changes have occurred may be less likely to repeat past mistakes.

2   *They may provide clues about future patterns to aid in forecasting.* Historians are also fond of saying that "all that is past is prologue." *If,* after studying a time series, a decision maker has reason to believe that an identifiable past pattern *will persist,* he or she can then build a forecast by projecting the past pattern into the future. Of course, subjective judgment is bound to be involved in any future estimate, but since decision makers can't choose to engage in forecasting or to abstain from it, it's considered a better approach to use thoughtfully projected time-series data than to rely solely on speculative hunches.

To summarize, it's only natural that time series are seriously studied. There's a need to understand the dynamic forces at work in the economy, and time-series analysis can enhance this understanding. And there's no choice but to make forecasts, and time-series analysis may reveal persistent patterns. Yet in our study of time series we should keep a proper perspective. Perhaps the following comment best sums up this perspective:

> *In our desire to find a logical integration of different stories told by the time series of economics, we must not be unmindful of a conversation between Alice and the White King. Alice had been asked by the King to look down the road and tell him if she could see his two messengers. "I see nobody on the road," said Alice. "I only wish I had such eyes," the King remarked in a fretful tone. "To be able to see Nobody! and at that distance too."*[1]

TIME-SERIES COMPONENTS: AN OVERVIEW

It's possible to concentrate on the broad picture presented by the total values of a particular time series. More often in time-series analysis, however, the analyst singles out one or more of the elements that make up the series. These components, superimposed and acting in unison, tend to represent the effects on the series of changes in such diverse factors as technology, population, consumer buying habits, weather conditions, capital investment, and productivity. These economic, natural, and cultural factors, of course, may account for changes in the series over time.

The four components which may be identified in a time series are (1) *secular trend,* (2) *seasonal variation,* (3) *cyclical fluctuations,* and (4) *irregular movements.* It's convenient in time-series analysis to use the symbols, T, S, C, and I to represent these trend, seasonal, cyclical, and irregular elements. Although in

---

[1] Harold T. Davis, *The Analysis of Economic Time Series,* The Cowles Commission for Research in Economics, Monograph 6, The Principia Press, Bloomington, Ind., 1951, p. 580.

later pages we'll consider each of these components in more detail, it's probably appropriate here to pause briefly to examine the nature of these elements.

Secular Trend

The **secular trend** is the smooth or regular long-term (that is, secular) growth or decline of a series. What does "long term" mean? You certainly can ask embarrassing questions. The length may be impossible to determine exactly, and it varies depending on the series. In the case of an economic time series, though, it's often argued that the period should be long enough to include two or more business cycles so that a persistent pattern may have time to emerge.

In Figure 13-3a, a time series (showing chemical production) is identified by the letter Y and is plotted over a 15-year period. A trend line is superimposed on the original data. In this case, there's a *growth trend* that's represented by a *straight line*. Other growth trends drawn with straight lines are shown in Figure 13-4 superimposed on U.S. employment data covering the years from 1965 to

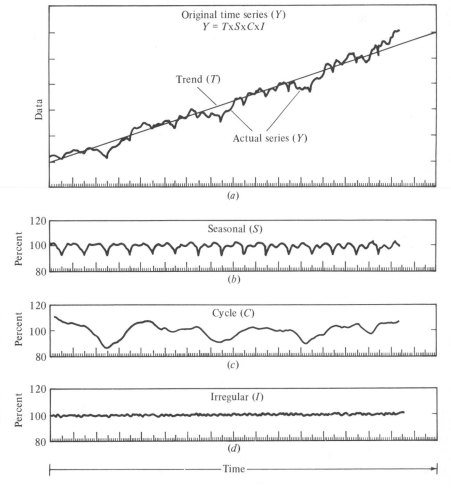

FIGURE 13-3
The components of a 15-year time series. (Adapted from Mabel A. Smith, "Seasonal Adjustment of Economic Time Series," *Survey of Current Business,* September 1962, pp. 24–32.)

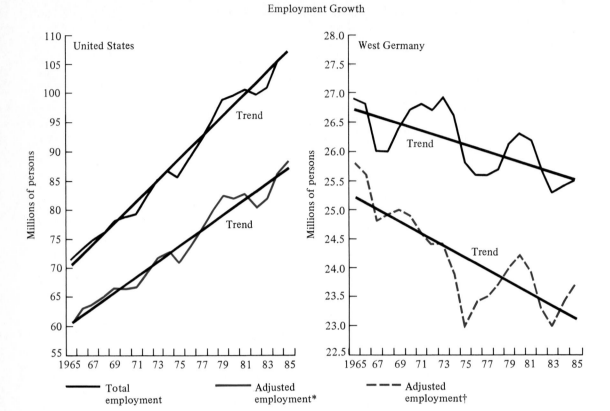

Employment Growth

*Data represent total employment (as measured by the *Current Population Survey*) minus part-time employment.

†Data represent total employment minus employment of foreign and short-term workers.

Data: U.S. Department of Labor, *Current Population Survey*; Deutsche Bundesbank, Statistical Supplement to the Monthly Report, and staff estimates.

FIGURE 13-4
Employment trends represented by straight lines. (Source: *Quarterly Review,* Federal Reserve Bank of New York, Autumn 1986, p. 13. Reprinted with permission.)

1985. In other situations, though, as you can also see in the employment chart for West Germany in Figure 13-4, the data may show a *declining* pattern. An appropriate trend line may also be *curvilinear* rather than straight. (We'll limit our study in this chapter to straight-line trends.)

Included among the underlying factors responsible for an average long-term growth or decline are (1) population changes (which have an obvious impact on those who produce food, clothing, and shelter) and (2) technological innovations (buggy makers have never recovered from the development of the automobile).

Seasonal Variations

Variations in a time series that are periodic in nature and that recur regularly within a period of 1 year (or less) are called **seasonal variations** (see Figure 13-3*b*).

As the name implies, *climatic conditions* are one important factor responsible for seasonal variations. Construction activities, sales of heating fuel and suntan lotion, the production of agricultural products—all these variables are obviously related to the weather. In addition, however, seasonal variations may be attributed to *social customs* and *religious holidays*. Any department store manager knows that there will be recurring patterns of sales activity in the days (and weeks) prior to Christmas, Easter, Labor Day, and the first day of a new school year. Figure 13-5 shows these recurring patterns in monthly retail sales for a 10-year period. The December peaks and the January–February troughs are obvious and recurring.

Cyclical Fluctuations

A **cyclical fluctuation** is periodic in nature, involves an up-and-down movement (see Figure 13-3*c*), extends over a period of several years, and *cannot* be counted on to repeat itself with relatively predictable regularity. Different factors tend to account for each new cycle. Much effort has gone into the study of the periods of prosperity or stability, recession or depression, and recovery that characterize a complete cycle, and there have probably been at least as many theories to explain the cycle as there have been cycles. Thus far, though, as you saw in the

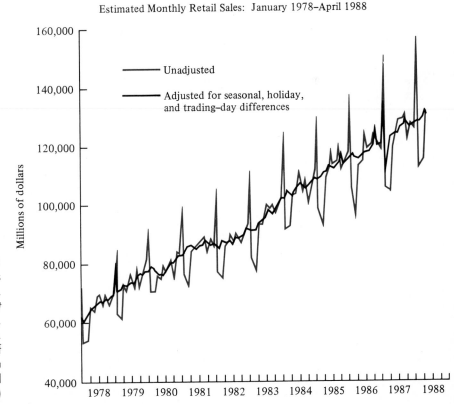

Estimated Monthly Retail Sales: January 1978–April 1988

FIGURE 13-5
Seasonal variations
in retail sales.
(Source: *Current
Business Reports,
Monthly Retail Trade,*
U.S. Department of
Commerce, Bureau
of the Census, April
1988, p. 3.)

opening reading, no satisfactory model has been developed to explain or forecast cyclical fluctuations adequately.

Irregular Movements

Irregular movements in a series occur over varying but usually brief periods of time, follow no regular pattern, and are unpredictable. These movements (see Figure 13-3*d*) may be caused by random factors or special events such as labor strikes; floods, droughts, hurricanes, and other natural disasters; armed conflicts; changes in governments; and a thousand other possibilities. As a practical matter, if a time-series variation can't be attributed to trend, seasonal, or cyclical elements, it's lumped into the irregular or residual category.

An Approach to Time-Series Analysis

From the preceding sections it's possible to arbitrarily conclude that if the original data (Y) representing a time series consist of the four components (T, S, C, and I) mentioned above, we can describe the relationship between the components and the original data as follows:

$$Y = T \times S \times C \times I \qquad \text{(formula 13-1)}$$

Using this time-series model, an analyst may (1) compute measures of, say, the trend and seasonal elements to analyze these components and then *eliminate* them from the original data so that (2) the cyclical and irregular components may be separately identified and studied. That is, the following decomposition of the time series may be made:

$$\frac{\cancel{T} \times \cancel{S} \times C \times I}{\cancel{T} \times \cancel{S}} = C \times I$$

In the following pages we'll follow this time-series analysis approach. That is, we'll consider (1) the *computation and projection of secular trend*, (2) the *measurement and use of seasonal patterns*, and (3) the *identification of cyclical and irregular fluctuations*.

TREND ANALYSIS

As noted above, an important part of time-series analysis is the study of secular trend. In this section we'll consider (1) the *reasons for measuring trend*, (2) a *method of computing the trend component*, and (3) the *use of trend in prediction or forecasting*.

Reasons for Measuring Trend

We measure trend for several of the same general reasons that we study time series. That is, the *trend component is studied for the following purposes:*

1 *To describe historical patterns.* The sales trend of one firm, for example, may be measured and compared with the trend patterns exhibited by competitors.
2 *To project persistent patterns into the future. If* (and this is a very important "if") an analyst believes that there's reason to expect a past trend to continue into the future, the measurement of that trend may be used as the basis for a future estimate or forecast.
3 *To eliminate the trend component.* An analyst may be primarily concerned with the cyclical element in a time series. To isolate the cyclical component, however, the effects of trend must be measured and eliminated from the original data.

Linear Trend Computation

There are several ways to display a linear trend. For example, you could try the freehand or eyeball method of using a ruler to draw a straight line through a graph of the data. But we'll use a more objective approach that mathematically fits the line to the data. Before looking at trend computations, though, it's necessary here to discuss briefly (1) the equation for a straight line and (2) the properties of the linear trend line. As you read these discussions, you'll gain a better understanding of why the rod and rubber band gadget described in a nearby box works as it does. And the straight-line equation and linear properties discussed below are equally applicable to the linear regression material we'll consider in the next chapter.

The Straight-Line Equation   Figure 13-6*a* shows a straight line (identified by the symbol Y_t) of the type that we'll soon be computing. To define this line, we must know *two things* about it. First, we must know the value of the *Y intercept*— that is, the value (read on the *Y* axis) of *a* in Figure 13-6*a* when *X* is at the origin or equal to zero. And second, we need to know the *slope of the line (b).* This slope is found as shown in Figure 13-6*a* by (1) taking a segment of the line, (2) measuring the change in one unit of time on the *X* axis, (3) measuring the corresponding change in Y_t on the *Y* axis, and (4) dividing the change in Y_t by the change in time. In Figure 13-6*a*, the slope of the line has a *positive* value; in Figure 13-6*b*, the slope has a *negative* value. In both cases, however, the formula for Y_t (and the formula that we will be using to compute the straight-line trend) is:

$$Y_t = a + bx \qquad \text{(formula 13-2)}$$

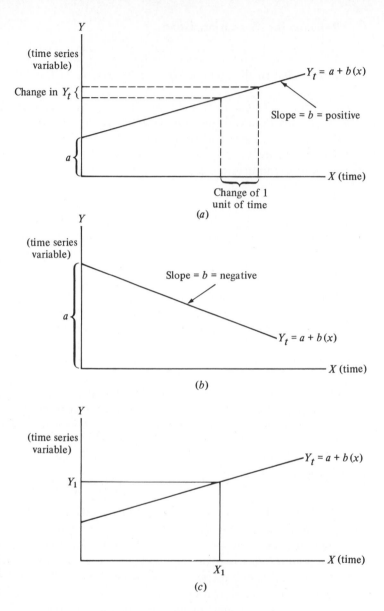

FIGURE 13-6
The straight-line
equation.

where Y_t = trend value for a given time period
     a = value of Y_t when X is at the origin
     b = slope of the line, or the increase or decrease in Y_t for each change of one
           unit of time
     x = any time period selected

Thus, if we selected a time period X_1 as shown in Figure 13-6c, drew a vertical
line up to the trend line (Y_t), and then drew a horizontal line to the Y axis, the
value of Y_1 would be the value of the trend component for the X_1 time period.

AN ANALOG GADGET

Exactly one year ago a collection of analog gadgets in these pages set off an avalanche of similar devices from inspired readers. I am still extricating myself from a vast heap of wood boards, rubber bands, strings, balls of polystyrene, fish tanks, lead weights, canisters, tubing and stopcocks. . . . The latest collection includes several ingenious new gadgets for solving problems in statistics, network theory, algebra and arithmetic. . . .

The first of the new gadgets solves a certain problem in statistics by means of a wood board, rubber bands, nails and a smooth, rigid rod. A set of data points plotted on a sheet of graph paper may present a linear trend to the eye. If a linear relation really governs the points, what straight line best displays the relation? The gadget suggested by Marc Hawley of Mount Vernon, Ind., supplies one possible answer:

Plot the data points on a wood surface and drive a nail partway into the wood at each point. Next, slip a number of uniform rubber bands onto the rod, one band for each nail. Fit the rod approximately into place and pull each band over one of the nails. When the rod is released, it wiggles and shivers quickly into an equilibrium position [*see illustration*].

The equilibrium position minimizes the total energy of the system; therefore the sum of the distances from the nails to the rod has also been minimized. In terms of such distances, the rod's final position indicates the straight line that best fits the data. It is not such distances but their squares that appear in the formulas for linear regression used by statisticians. Hawley's gadget computes something at least as complicated.

—From A. K. Dewdney, "Computer Recreations," *Scientific American*, June 1985, p. 18. Copyright © 1985 by *Scientific American*, Inc. All rights reserved.

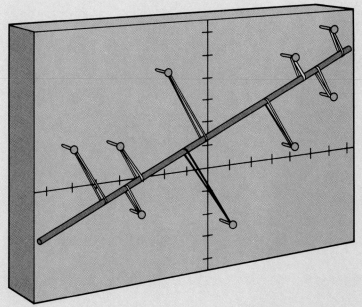

A gadget for finding the line that best fits a series of data points.

Properties of the Linear Trend Line   There are *two properties* of the straight-line trend that we'll compute. The *first* of these properties is demonstrated in Figure 13-7, which shows the trend component (Y_t) fitted to the time-series data (Y) that are plotted on a chart. At time X_1, the value of the actual data (Y) is greater than the value of Y_t. (The values of both Y and Y_t are read on the Y axis.) Thus, if we subtract Y_t from Y, we will get a *positive* value. At time X_2, however, the reverse is true—that is, the Y_t is larger than the Y value, and so the difference $Y - Y_t$ is *negative*. The fact is that the trend line is fitted to the data in such a way that the sum of the deviations of the Y values about the trend line is zero. Therefore, the *first property* of the trend line is:

$$\Sigma(Y - Y_t) = 0$$

And the sum of the *squares* of the deviations is less than would be the case if any other straight line were substituted for the Y_t line and the process of computing and squaring deviations were carried out. In other words, the *second property* is:

$$\Sigma(Y - Y_t)^2 = \text{a minimum or least value}$$

And so the name **method of least squares** is used to describe the approach we'll follow to fit a straight trend line to a time series.

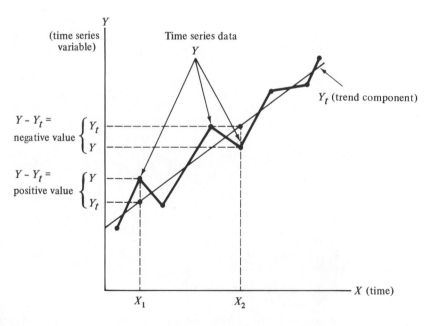

FIGURE 13-7
Properties of the linear trend line.

① $\Sigma(Y - Y_t) = 0$

② $\Sigma(Y - Y_t)^2 = \text{minimum or "least" value}$

FIGURE 13-8   Annual Sales of Buckaroo Lingerie Company (millions of dollars)

1982	10
1983	8
1984	10
1985	12
1986	16
1987	12
1988	16

Trend Computation—Odd Number of Years   We'll use the hypothetical data in Figure 13-8 to illustrate a linear trend computation. As you can see, annual sales of the Buckaroo Lingerie Company have generally been increasing, although there were slumps in 1983 and 1987. (In 1987, a new line of girdles and bras was developed, and the *former* sales manager, Hoagy "Tex" Chauvinist, approved an extensive advertising campaign to market these products built around the tactless slogan "We round up the strays." This campaign, combined with general economic conditions, led to predictable results: sales declined as shown in Fig. 13-8; Tex was removed as sales manager and given the sales territory for a radius of 20 miles around Cut and Shoot, Texas,[2] and a new sales manager—Forrestal D. Zaster—was appointed.)

Figure 13-9 shows the work sheet used to compute the trend component for the Buckaroo Company for the years 1982–1988. (You'll notice that the first two columns are unchanged from Figure 13-8.) To simplify the computations, we've *coded* the 7-year time period by restating the calendar year numbers (X) in terms of the numbers of years from the *middle of the time period*. In other words, in our example we've shifted the *origin* from the beginning of the time period to the *middle of 1985*. Then, *in column 3 of Figure 13-9, we've merely determined the deviations from the origin to the middle of each year* in yearly units. Thus, the middle of 1982 is −3 years from the middle of 1985 (the origin), and the middle of 1987 is +2 years from the origin. Having performed this little trick to code time, we can now compute the linear trend equation. The formulas for *a* and *b* in the trend equation are:

$$a = \frac{\Sigma Y}{n} \qquad \text{(formula 13-3)}$$

where Y = time-series variable and n is the number of years.

---

[2] Yes, there really is a Cut and Shoot, Texas! According to M. J. Quimby in *Scratch Ankle, USA*, "Cut and Shoot, Texas, got its name from a community dispute said to have arisen over the pattern for a new church steeple."

FIGURE 13-9   Annual Sales of Buckaroo Lingerie Company

YEARS (X) (1)	COMPANY SALES (MILLIONS OF $) (Y) (2)	CODED TIME (x) (3)	(xY) (COLS. 2 × 3) (4)	(x²) (5)	Y_t (FOR YEARS INDICATED, IN MILLIONS OF $) (6)
1982	10	−3	−30	9	8.571
1983	8	−2	−16	4	9.714
1984	10	−1	−10	1	10.857
1985	12	0	0	0	12.000
1986	16	1	16	1	13.143
1987	12	2	24	4	14.286
1988	16	3	48	9	15.429
	84	0	32	28	84.000

$$a = \frac{\Sigma Y}{n} = \frac{84}{7} = 12.00 \text{ (millions of \$)}$$

$$b = \frac{\Sigma(xY)}{\Sigma(x^2)} = \frac{32}{28} = 1.143 \text{ (millions of \$)}$$

$Y_t = 12.00 + 1.143(x)$

Origin (where $x = 0$): the middle of 1985

Coded time (x) unit: 1 year

Y data: company sales in millions of dollars

$$b = \frac{\Sigma(xY)}{\Sigma(x^2)} \qquad \text{(formula 13-4)}$$

where $x = $ *coded time values* rather than the actual years (X).

The value for a in our example is 84/7 or 12.00 (million dollars). The value for b is found in Figure 13-9 by dividing the sum of column 4 by the sum of column 5. Thus, $b = 1.143$ (million dollars). What are the meanings of these computed values of a and b? The $12 million of a means that the computed value of Y_t is $12 million in sales in the middle of 1985. Why? Because, as we saw earlier, a is the value of the Y_t line when x is zero or at the origin, and x is zero in the middle of 1985. The $1.143 million value for b means that for *each change of 1 year* in x, the sales trend component will change by $1.143 million.

Column 6 in Figure 13-9 indicates the trend values in millions of dollars for each of the 7 years. To compute Y_t for 1982, you use the following approach:

$Y_{t1982} = a + bx$

$= 12.00 + 1.143\,(-3)$

$= 12.00 + (-3.429)$

$= 8.571$, the trend for Buckaroo sales in 1982 (millions of dollars)

Other Y_t values are found in the same way by substituting the appropriate value of x in the trend equation. Alternatively, once you have the first year's trend, you can simply add to it the value of b to get the next year's trend since, by definition here, b is the change in Y_t for each change of one unit of time. You'll note that the sums of columns 2 and 6 are equal. This is always the case because of the first property of the trend line; that is, if $\Sigma(Y - Y_t) = 0$, then $\Sigma Y - \Sigma Y_t$ must also be zero.

Figure 13-10 shows the original Buckaroo sales data and the trend component found in column 6 of Figure 13-9. If you study this figure, you'll verify comments made in previous paragraphs.

Trend Computation—Even Number of Years   In the preceding paragraphs we computed the Buckaroo sales trend for an odd number (7) of years. But, of course, the available data could just as easily cover an even number of years. Let's assume, for example, that we now add the 1989 sales data for Buckaroo to the data for the previous 7 years (see Figure 13-11). Figure 13-11 shows *two alternative approaches* to computing trend for an even number of years.

The approach in Figure 13-11a is no different from the one we have just considered, and so we need not dwell at length on it here. You'll notice that since the middle of the time period is now *between* the years 1985 and 1986, the origin is shifted to January 1, 1986. Then, in the coded time column of Figure 13-11a, the deviations from the origin to the *middle of each year* in coded *yearly* units is indicated. For example, the middle of 1987 is 18 months from January 1, 1986, or 1½ yearly units. The remaining computations in Figure 13-11a are

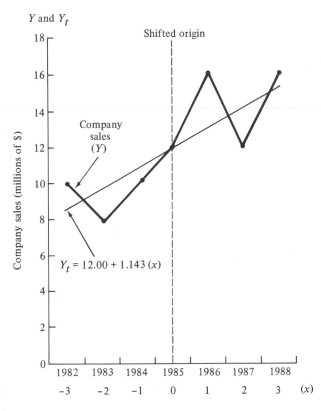

FIGURE 13-10

FIGURE 13-11   Annual Sales of Buckaroo Lingerie Company

YEARS	COMPANY SALES (MILLIONS OF $) (Y)	CODED TIME (x)	xY	x²	Y_t (FOR YEARS INDICATED, MILLIONS OF $)
1982	10.00	$-3\frac{1}{2}$	-35	12.25	8.86
1983	8.00	$-2\frac{1}{2}$	-20	6.25	9.86
1984	10.00	$-1\frac{1}{2}$	-15	2.25	10.86
1985	12.00	$-\frac{1}{2}$	-6	.25	11.86
1986	16.00	$\frac{1}{2}$	8	.25	12.86
1987	12.00	$1\frac{1}{2}$	18	2.25	13.86
1988	16.00	$2\frac{1}{2}$	40	6.25	14.86
1989	14.86	$3\frac{1}{2}$	52	12.25	15.86
	98.86	0	42	42.00	98.88

$a = 98.86/8 = 12.36$
$b = 42/42 = 1.00$
$Y_t = 12.36 + 1.00(x)$
$Y_{t1982} = 12.36 + 1.00(-3\frac{1}{2})$
$= 8.86$

(a) Origin: January 1, 1986
  x unit: 1 year
  Y unit: millions of dollars of sales

(a)

YEARS	COMPANY SALES (MILLIONS OF $) (Y)	CODED TIME (x)	xY	x²	Y_t (FOR YEARS INDICATED, MILLIONS OF $)
1982	10.00	-7	-70	49	8.86
1983	8.00	-5	-40	25	9.86
1984	10.00	-3	-30	9	10.86
1985	12.00	-1	-12	1	11.86
1986	16.00	1	16	1	12.86
1987	12.00	3	36	9	13.86
1988	16.00	5	80	25	14.86
1989	14.86	7	104	49	15.86
	98.86	0	84	168	98.88

$a = 98.86/8 = 12.36$
$b = 84/168 = .50$
$Y_t = 12.36 + .50(x)$
$Y_{t1982} = 12.36 + .50(-7)$
$= 8.86$

(a) Origin: January 1, 1986
  x unit: ½ year
  Y unit: millions of dollars of sales

(b)

performed exactly as they were in Figure 13-9. You will note that the values of *a* and *b* in the trend equation have changed as a result of the additional sales data.

An alternative to the approach shown in Figure 13-11*a* is presented in Figure 13-11*b*. Up to now we've been coding time in *yearly* units, but this is an arbitrary decision. We can just as easily use 2-year intervals or ½-year units. As a matter of fact, we *are* coding in *½-year intervals* in Figure 13-11*b*, and this is the *only difference* between the procedures in Figure 13-11*a* and *b*. By redefining the *x* unit in Figure 13-11*b* to represent ½ year, we are able to eliminate the use of fractions in the coded time column, and for many this tends to simplify the remaining computations. You'll note, however, that the computed Y_t values are exactly the same in Figure 13-11*a* and *b*. The trend equation in both approaches obviously has the same value of *a*; however, the values of *b appear* to be different. In Figure 13-11*a*, the change in Y_t for each change in time (measured in 1-year units) is 1.00; and in Figure 13-11*b*, the change in Y_t for each change in time (now measured in ½-year units) is 0.50. Thus, there is really no difference in *b*. This fact is demonstrated in both parts of Figure 13-11 by the computation of Y_t for 1982, where there's obviously no difference between multiplying 1.00 by -3.5 and multiplying 0.50 by -7. The choice of which approach to use is a matter of personal preference.

Trend Projection

Let's now assume that Forrestal D. Zaster, the Buckaroo sales manager, wishes to use the past sales pattern as an aid in developing future sales forecasts. If Mr. Zaster can assume that the past pattern is likely to persist into the future, he can use the linear trend equation to project the trend line. Suppose that Mr. Zaster wants a trend projection for 1992, given the data in Figure 13-11*b*. The coded

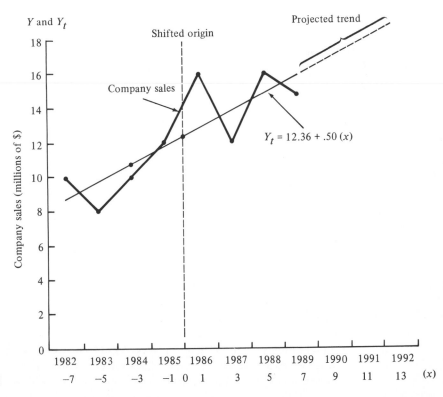

FIGURE 13-12

time value for 1989 in Figure 13-11b is 7. The value of x for 1990 is 9; for 1991 it's 11; and for 1992, x is 13. That is, the middle of 1992 is thirteen ½-year periods from the origin of January 1, 1986. Therefore,

$$Y_{t1992} = 12.36 + 0.50 \, (13)$$
$$= 12.36 + 6.5$$
$$= 18.86 \text{ million dollars of projected sales}$$

The same procedure can be used to forecast Y_t for any other future period. Of course, Mr. Zaster should realize that the very precise-looking forecast figure computed above is merely a point of departure or a beginning value that quite possibly should be modified by many subjective considerations before a final forecast figure is obtained. Figure 13-12 (page 497) shows the Buckaroo sales data through 1989 and the computed and projected trend component. A study of this figure should give you a better grasp of the comments made above.

Self-Testing Review 13-1

1 The total annual production of saddles by the Callis X. Tremity Saddle Company for the years 1979–1989 is shown below.

YEAR	ANNUAL SADDLE PRODUCTION (HUNDREDS)
1979	8
1980	9
1981	12
1982	11
1983	14
1984	17
1985	18
1986	22
1987	24
1988	25
1989	28

   a What are the trend equation values for the above data?
   b What would be the projected trend value for 1991?
   c Explain the meaning of the value computed in problem 1b.
2 Given the trend equation and the other information below, forecast the value of Y_t for 1991.

$$Y_t = 10 + 3x \qquad \text{Origin: the middle of 1987}$$
$$x \text{ unit: 1 year}$$

SEASONAL VARIATION

In addition to computing and projecting secular trend, another important part of time-series analysis is the study of seasonal variation. Since by definition, seasonal variations occur within a period of 1 year or less, we must have data reported periodically—weekly, monthly, quarterly—throughout the year. (If a time series consists of only annual data, it includes trend, cyclical, and irregular elements, but seasonal variations aren't present because they occur within a year.)

An **index of seasonal variation** measures how much a time series changes on a relative basis with respect to an average for the period (1 year or less). If a time series is reported on a quarterly (3-month) basis, four seasonal index numbers are prepared each year, and these numbers are expressed as a percentage relative to a quarterly average value. If monthly data are used, there are 12 seasonal index numbers each year, and these numbers are expressed as a percentage relative to an average monthly value. For example, a crude seasonal index could be prepared by Charley Harse, the owner of a struggling health spa, if he were to take the trouble to analyze his monthly receipts for the past year (given in Figure 13-13).

As you can see in Figure 13-13, the crude index of seasonal variation is based on average monthly receipts of $7,500. The actual receipts for each month have merely been expressed as a *percentage of the average month*. (Having read Chapter 12 on index numbers with enormous enthusiasm and astounding comprehension, you understand that the percentage sign is omitted here in good index number tradition.) Thus, the December receipts of $8,800 are 17.3 percent *greater than* (or 117.3 percent *of*) the receipts that could be expected in an average month. With an index of 82.7, however, the month of June had receipts that were only 82.7 percent of the average month. You'll notice that the total of the index numbers equals 1,200, as it must in order to give us an average index number value of 100. (The total of the quarterly index numbers for a year must be 400.)

There are numerous ways to prepare an index of seasonal variation, and our simple example in Figure 13-13 is flawed. But the method that we'll use later,

---

FIGURE 13-13   A Crude Seasonal Index

MONTH (1)	MONTH'S RECEIPTS (2)	SEASONAL INDEX [(2) ÷ AVERAGE MONTH × 100] (3)
January	$8,400	112.0
February	8,000	106.7
March	7,400	98.7
April	6,500	86.7
May	6,400	85.3
June	6,200	82.7
July	7,000	93.3
August	7,500	100.0
September	7,700	102.7
October	7,900	105.3
November	8,200	109.3
December	8,800	117.3
	$90,000	1,200.0

Average (mean) month's receipts = $90,000/12 = $7,500
Seasonal index for January = $8,400/$7,500 × 100 = 112.0

although more sophisticated, still produces seasonal index values by dividing the actual monthly data by an average monthly figure.

Reasons for Measuring Seasonal Variation

The three main reasons for measuring seasonal variation are:

1   *To understand seasonal patterns.* The seasonal patterns shown in Figure 13-5 are of obvious interest to retailers. By having a measure or index of seasonal variation, they can determine, for example, if the decline in January–February sales is more or less than the typical seasonal amount.

2   *To project existing patterns into the future.* We've seen that secular trends may be projected as an aid in the *long-term* planning and controlling of important variables. Seasonal variation patterns are also used for forecasting, but the emphasis is on *short-term* planning and control. For example, persistent seasonal patterns may be projected in order to plan for (and control) the appropriate utilization of personnel and to maintain the best levels of inventories and cash balances. A short-term forecast based on Figure 13-5 has obvious planning and control implications for scheduling sales clerks, and for maintaining and financing store inventories.

3   *To eliminate the seasonal component.* The cyclical component in a time series, like the trend and seasonal components, is also used in forecasting. Anticipated cyclical movements are frequently used in the preparation of *intermediate-term* forecasts. However, the cyclical element is often obscured by the swings in seasonal activity, and so the seasonal element must be measured and eliminated to reveal the cyclical pattern. Data published by the government in charts and tables are often on a seasonally adjusted basis for just this reason (see Figure 13-5). We'll look at a technique to "deseasonalize" data in a later section.

Seasonal Index Computation

The **ratio-to-moving average method** of computing a seasonal index that we'll use in this section is a popular approach for measuring the seasonal element in a time series. To demonstrate this method, let's use the monthly gasoline production figures (expressed in millions of gallons) of the Otto S. Guzzling Refining Company given in Figure 13-14. A *Minitab* time-series plot of these 72 monthly production figures is shown in Figure 13-15. Each January is represented by a 1 in the chart, each February by a 2, and so on. October, the tenth month, is represented by a 0, and November and December data are plotted with the letters A and B. If you draw lines to connect the plotted numbers and letters in the correct sequence, you'll see that production generally has a summer peak and a winter trough.

There are no formidable conceptual difficulties in using the ratio-to-moving-average method to compute a seasonal index for our gasoline production data. What is formidable, though, is the amount of clerical effort needed to compute the index if the work is done by hand. Fortunately, computers are ideally suited

USING STATISTICAL FORECASTING PACKAGES

When Richard Wiser goes to work, he doesn't carry a crystal ball with him. But he peeks into the future with some assurance thanks to *ESP*, a statistical analysis package. As director of business and marketing analysis at Mary Kay Cosmetics in Dallas, Wiser uses *ESP* to fill a variety of forecasting needs. When the company plans a sales promotion, he tries to predict how much additional product the promotional effort will inspire dealers to order. He regularly forecasts the size and frequency of orders, and he downloads economic information that helps him forecast recruiting figures.

John Fahey, chief economist for the Saudi Arabian embassy in London, uses another statistical program, *SORITEC*, to track the commodities market, trends in international trade, financial analysis, national welfare projections, and oil price predictions. Jeffrey Martin, a Palo Alto, California, CPA who specializes in litigation support for attorneys, enters historical sales and earnings of specified companies into the Lotus spreadsheet template *1,2,3 Forecast!* and plots a trend line. Radical deviations from this trend line can alert him to the existence of unknown factors that are affecting the company's performance. And at a Selma, Alabama, farm-im-

plement manufacturing company with the picturesque name of Bush Hog, Ernest Divelbiss uses *Forecast Plus* to combine company statistics with data from outside sources to predict farm income and other economic trends. From these figures in turn, he forecasts sales of each Bush Hog product.

These businesspeople and many others use forecasting programs to collect and analyze information to make reliable predictions about the future. Based on their findings, they'll recommend strategies to their companies.

The earliest forecasting programs available for the PC, based on mainframe versions, were difficult for anyone but FORTRAN programmers to use. Now, however, a rich assortment exists, ranging from simple, menu-driven packages for people with little or no statistical training to comprehensive programs suitable for professional economists and statisticians. Even some of the most powerful programs are surprisingly easy to use.

—Marvin Bryan, "Business Forecasting: 16 Ways to Predict the Future." Reprinted from: *PC Magazine*, August 1986, pp. 211–212. Copyright © 1986 Ziff Communications Company.

to the task of massaging seasonal data and producing the hundreds of intermediate and final calculations often required. The chances are good that a general-purpose statistical package will *not* provide built-in programs to do this work. But there are dozens of more specialized **statistical forecasting packages** and spreadsheet templates that do include preprogrammed routines to carry out the steps needed to compute a seasonal index and perform other time-series analysis operations.

Let's look now at the six steps needed to produce a seasonal index by the ratio-to-moving-average method. These steps are demonstrated in Figures 13-16 and 13-17.

1 *Compute a 12-month moving total* of the original data, and place this total opposite the *seventh*[3] of the 12 months (see column 4 of Figure 13-16). You'll

[3] The exact center of the 12-month period is, of course, midway between the sixth and seventh months, and many texts place the moving total in that position. Then, in order to center moving total figures opposite the data for a particular month, it's necessary to compute an additional 24-month moving total by adding two 12-month moving totals. For the purposes of this text, the extra work required to achieve a minor timing refinement is not necessary.

FIGURE 13-14   Production Data, Otto S. Guzzling Refining Company (millions of gallons)

MONTH	1984	1985	1986	1987	1988	1989
January	8	10	11	9	8	12
February	9	12	8	9	10	14
March	9	10	8	13	12	13
April	12	14	16	15	19	18
May	16	15	17	14	22	19
June	13	17	16	13	20	16
July	17	20	23	25	21	23
August	15	14	12	10	18	19
September	10	9	14	14	16	18
October	8	10	13	15	15	19
November	10	10	14	16	12	18
December	14	15	17	15	17	17

notice that the total for the 12 months in 1984 (a value of 141) is placed opposite the seventh month of July. The next 12-month moving total of 143 is the sum of the monthly data from February 1984 through January 1985—that is, the total value of 143 is found by dropping the value of January 1984 and adding the value of January 1985 to the previous total of 141. This procedure is continued until the last month of the last year has been included in the moving total.

2   *Compute a 12-month moving average* (see column 5 in Figure 13-16) by dividing each of the 12-month totals by 12. Thus, the first 12-month moving average of 11.8 in Figure 13-16 is found by dividing the moving total of 141 by 12, and the second moving average of 11.9 is 143 divided by 12. Again, this procedure is continued until the last average is found using the last moving total.

3   *Compute the specific seasonals.* For the month of July 1984 we have the actual production data in column 3 (17 million gallons), and we now have a yearly moving average of 11.8 (million gallons). If you recall, back in Figure 13-13 we divided the actual data by the average month's value to get a crude seasonal index. Well, this is exactly what we do in column 6 of Figure 13-16—we divide the actual data in July 1984 (17) by the yearly moving average of 11.8 (and multiply the result by 100) to get a seasonal index value—called a **specific seasonal**—of 144.1. What does the 144.1 mean? It means that gasoline production in July 1984 was 44.1 percent greater than during the average month. The second specific seasonal is found by dividing the actual value of 15 for August 1984 by the corresponding moving average of 11.9. This same procedure is continued until the data are exhausted (or until you are), as shown in Figure 13-16.

4   *Arrange the specific seasonals.* Our problem now is *not* that we lack seasonal index values; rather, our problem is that we have *five* specific seasonals for every month of the year. Obviously, we must distill from all these index values a *single* representative or typical index number for each month. We approach this task by preparing Figure 13-17. (The data in Figure 13-17 are simply the

```
MTB > PRINT 'GALLONS'

GALLONS
     8     9     9    12    16    13    17    15    10     8    10    14    10
    12    10    14    15    17    20    14     9    10    10    15    11     8
     8    16    17    16    23    12    14    13    14    17     9     9    13
    15    14    13    25    10    14    15    16    15     8    10    12    19
    22    20    21    18    16    15    12    17    12    14    13    18    19
    16    23    19    18    19    18    17

MTB > TSPLOT WITH PERIOD = 12 ON 'GALLONS'

        -
        -                                           7
    24.0+
        -                               7
GALLONS -                                                     5
        -                   7                                  67
        -                                                    4
    18.0+                                                        8
        -       7          6          5      B                       B
        -     5                      4 6                A         9
        -        8   B   45  8   B        9 A    45   90 B         0
        -      6                           0    3  6
    12.0+    4         2                 8                  3       A
        -                         1
        -         9 A 1 3      0A                    8     2
        - 123      0          9    23         12          1
        -
         +-----------+-----------+-----------+-----------+-----------+
         0          12          24          36          48          60


        -
        -
    24.0+
        -       7
GALLONS -
        -
        -     5  8 0
    18.0+    4    9 A
        -            B
        -      6
        -  2
        -   3
    12.0+ 1
        -
        -
        -
        -
         +-----------+
        60          72
```

FIGURE 13-15
A plot of the data in Figure 13-14.

FIGURE 13-16 Computation of Seasonal Index by Ratio-to-moving-average Method

YEAR (1)	MONTH (2)	PRODUCTION DATA (MILLIONS OF GALLONS) (3)	12-MONTH MOVING TOTAL (4)	12-MONTH MOVING AVERAGE [(4) ÷ 12] (5)	SPECIFIC SEASONALS [(3) ÷ (5) × 100] (6)
1984	January	8			
	February	9			
	March	9			
	April	12			
	May	16			
	June	13			
	July	17	141	11.8	144.1
	August	15	143	11.9	126.1
	September	10	146	12.2	82.0
	October	8	147	12.2	65.6
	November	10	149	12.4	80.6
	December	14	148	12.3	113.8
1985	January	10	152	12.6	79.4
	February	12	155	12.9	93.0
	March	10	154	12.8	78.1
	April	14	153	12.8	109.4
	May	15	155	12.9	116.3
	June	17	155	12.9	131.8
	July	20	156	13.0	153.8
	August	14	157	13.1	106.9
	September	9	153	12.8	70.3
	October	10	151	12.6	79.4
	November	10	153	12.8	78.1
	December	15	155	12.9	116.3
1986	January	11	154	12.8	85.9
	February	8	157	13.1	61.1
	March	8	155	12.9	62.0
	April	16	160	13.3	120.3
	May	17	163	13.6	125.0
	June	16	167	13.9	115.1
	July	23	169	14.1	163.1
	August	12	167	13.9	86.3
	September	14	168	14.0	100.0
	October	13	173	14.4	90.3
	November	14	172	14.3	97.9
	December	17	169	14.1	120.6
1987	January	9	166	13.8	65.2
	February	9	168	14.0	64.3
	March	13	166	13.8	94.2
	April	15	166	13.8	108.7
	May	14	168	14.0	100.0
	June	13	170	14.2	91.6
	July	25	168	14.0	178.6
	August	10	167	13.9	71.9
	September	14	168	14.0	100.0
	October	15	167	13.9	107.9
	November	16	171	14.2	112.7
	December	15	179	14.9	100.7

FIGURE 13-16   (continued)

YEAR (1)	MONTH (2)	PRODUCTION DATA (MILLIONS OF GALLONS) (3)	12-MONTH MOVING TOTAL (4)	12-MONTH MOVING AVERAGE [(4) ÷ 12] (5)	SPECIFIC SEASONALS [(3) ÷ (5) × 100] (6)
1988	January	8	186	15.5	51.6
	February	10	182	15.2	65.8
	March	12	190	15.8	75.9
	April	19	192	16.0	118.8
	May	22	192	16.0	137.5
	June	20	188	15.6	128.2
	July	21	190	15.8	132.9
	August	18	194	16.2	111.1
	September	16	198	16.5	97.0
	October	15	199	16.6	90.4
	November	12	198	16.5	72.7
	December	17	195	16.2	104.9
1989	January	12	191	15.9	75.5
	February	14	193	16.1	87.0
	March	13	194	16.2	80.2
	April	18	196	16.3	110.4
	May	19	200	16.6	114.5
	June	16	206	17.2	93.0
	July	23			
	August	19			
	September	18			
	October	19			
	November	18			
	December	17			

Source: Figure 13-14.

specific seasonal from column 6 of Figure 13-16 arranged in a more convenient way.)

5   *Compute the typical seasonal index values.* To find a typical seasonal index number for each month, we'll apply a measure of central tendency to the five specific seasonals for the month. We've chosen to use a trimmed or modified arithmetic mean—that is, a measure found by dropping the high and low values as indicated in Figure 13-17 and computing the mean of the three middle values. This is a standard approach, but other texts also use the arithmetic mean or the median as the measure of central tendency. The choice often appears to be rather arbitrary. Dividing the total of the three middle values for each month by 3 produces the single modified mean index value for each month shown near the bottom of Figure 13-17.

6   *Complete the seasonal index.* The modified mean or typical values in Figure 13-17 are essentially the seasonal index number series that we've been looking for. You'll notice, however, that these index values total 1,199.4; you'll also remember from Figure 13-13 that a seasonal index should, in theory, total

FIGURE 13-17   Arranging Specific Seasonals and Computing Seasonal Index

YEAR	JAN.	FEB.	MAR.	APR.	MAY	JUNE	JULY	AUG.	SEPT.	OCT.	NOV.	DEC.	
1984							144.1	~~126.1~~	82.0	~~65.6~~	80.6	113.8	
1985	79.4	~~93.0~~	78.1	109.4	116.3	~~131.8~~	153.8	106.9	~~70.3~~	79.4	78.1	116.3	
1986	~~85.9~~	~~61.1~~	~~62.0~~	~~120.3~~	125.0	115.1	163.1	86.3	~~100.0~~	90.3	97.9	~~120.6~~	
1987	65.2	64.3	~~94.2~~	~~108.7~~	~~100.0~~	~~91.6~~	~~178.6~~	~~71.9~~	100.0	~~107.9~~	~~112.7~~	~~100.7~~	
1988	~~51.6~~	65.8	75.9	118.8	~~137.5~~	128.2	~~132.9~~	111.1	97.0	90.4	~~72.7~~	104.9	
1989	75.5	87.0	80.2	110.4	114.5	93.0							
Total of middle three	220.1	217.1	234.2	338.6	355.8	336.3	461.0	304.3	279.0	260.1	256.6	335.0	Total
Modified mean	73.4	72.4	78.1	112.9	118.6	112.1	153.6	101.4	93.0	86.7	85.6	111.6	1,199.4
Seasonal index*	73.4	72.4	78.1	113.0	118.7	112.2	153.7	101.5	93.0	86.7	85.6	111.7	1,200.0

* Correction factor = 1200.00/1199.4 = 1.0005002.

1,200. To adjust a total to 1,200 we merely (*a*) divide the computed total *into* 1,200 to get a correction factor and (*b*) multiply the correction factor by each of the monthly index numbers. The *final seasonal index* in Figure 13-17 has been adjusted in this way.

The purpose of the above steps is to isolate the effects of seasonal variation from the other time-series elements present in the original data. How is this done? Very briefly, the original data in column 3 of Figure 13-16 include trend, cyclical, seasonal, and irregular movements. The rationale of the ratio-to-moving-average method is (1) that the 12-month moving average in column 5 of Figure 13-16 is an approximate estimate of the combined influences of the trend and cyclical components (because the process of yearly averaging tends to smooth out the seasonal element) and (2) that by dividing the actual data in column 3 by the moving average, the net result in column 6 of Figure 13-16 is that the trend and cyclical factors have been removed and the remaining specific seasonals in percentage form are the fluctuations occurring because of seasonal and irregular factors. In other words,

Column 3, Figure 13-16

$$\frac{\overbrace{T \times S \times C \times I}}{\underbrace{T \times C}} = S \times I\} \text{ Column 6, Figure 13-16}$$

Column 5, Figure 13-16

Finally, the irregular fluctuations are removed in Figure 13-17 through the process of eliminating high and low specific seasonal values.

Uses for the Seasonal Index

As we saw earlier, the seasonal index is used for forecasting and deseasonalizing purposes.

FIGURE 13-18 Deseasonalized Production Data, 1989, Otto S. Guzzling Refining Company
(millions of gallons)

MONTH (1)	UNADJUSTED PRODUCTION DATA, 1989 (2)	SEASONAL INDEX (3)	SEASONALLY ADJUSTED PRODUCTION [(2) ÷ (3) × 100] (4)
January	12	73.4	16.3
February	14	72.4	19.3
March	13	78.1	16.6
April	18	113.0	15.9
May	19	118.7	16.0
June	16	112.2	14.3
July	23	153.7	15.0
August	19	101.5	18.7
September	18	93.0	19.4
October	19	86.7	21.9
November	18	85.6	21.0
December	17	111.7	15.2

A Forecasting Application Sales volume, production schedules, personnel and inventory requirements, financial needs—all these variables are typically subject to short-term forecasts for planning and control purposes. A seasonal index may be used in preparing such forecasts. For example, suppose that the expected trend value for sales of a firm next February is $17,000—that is, $17,000 is the expected sales based on the past trend pattern.[4] Now let's also assume that the firm's February seasonal index value is 90.2. Thus, an analyst will need to adjust the projected trend value by multiplying it by the seasonal index for February. The result is a **seasonally adjusted forecast** for February of $15,334 ($17,000 × .902).

With an estimate of the cyclical element, the analyst can further adjust the February forecast. For example, if the cyclical factor is expected to be +4 percent in February, the forecast is $15,947 ($17,000 × .902 × 1.04). Of course, we're not so naive as to expect February sales to be exactly $15,947. After all, there are irregular movements possible, and our trend and seasonal figures are only approximations. But a forecast of $15,947 may be better than no explicit forecast at all!

A Deseasonalizing Application Original time-series data may also need to be seasonally adjusted or **deseasonalized** so that the effects of the seasonal component can be eliminated to examine the cyclical pattern. Deseasonalizing is an easy operation: You simply divide the original data by the appropriate seasonal index values. Figure 13-18 shows the seasonally adjusted monthly production figures for the Otto S. Guzzling Refining Company for 1989. You'll notice that the data in column 2 of Figure 13-18 come from Figure 13-14 and that the seasonal index

---

[4] Although we need not go into the details here, it's not difficult to modify a *yearly* trend equation such as we computed earlier in the chapter into one which can be used to project *monthly* trend values.

in column 3 is the one computed in Figure 13-17. The deseasonalized production values in column 4 of Figure 13-18 contain trend, cyclical, and irregular components. In other words,

Original data
(column 2, Figure 13-18)

$$\dfrac{\overbrace{T \times S \times C \times I}}{\underbrace{S}} = T \times C \times I\} \text{ Column 4, Figure 13-18}$$

Seasonal index
(column 3, Figure 13-18)

Self-Testing Review 13-2

The following data represent monthly receipts (in thousands of dollars) of the Natalie A. Tyred fashion model school:

YEAR	MONTH	RECEIPTS	YEAR	MONTH	RECEIPTS
1987	January	$10	1989	January	$10
	February	12		February	12
	March	14		March	15
	April	15		April	16
	May	15		May	18
	June	8		June	7
	July	9		July	9
	August	10		August	10
	September	12		September	15
	October	13		October	14
	November	15		November	16
	December	18			
1988	January	9			
	February	13			
	March	15			
	April	18			
	May	17			
	June	6			
	July	11			
	August	10			
	September	14			
	October	13			
	November	16			
	December	17			

1   Compute a seasonal index by the ratio-to-moving-average method. (Since the data are limited, you must use the arithmetic mean in place of the modified mean to compute the typical seasonal index values.)

2   Use the seasonal index computed in problem 1 above to deseasonalize the following monthly receipts (in thousands of dollars) for 1990 of the Natalie A. Tyred organization:

MONTH	UNADJUSTED RECEIPTS
January	$11
February	13
March	14
April	15
May	19
June	8
July	12
August	14
September	14
October	16
November	17
December	21

IDENTIFYING CYCLICAL AND IRREGULAR COMPONENTS

In earlier sections of this chapter we referred to a time-series model that describes the relationship between the time-series components and the original data (Y) as follows:

$$Y = T \times S \times C \times I$$

This model is frequently referred to as the **classical time-series model,** and it may be used to explain the procedure for identifying the cyclical and irregular components.

Once trend values and a seasonal index have been prepared from the original data using the methods that have now been presented in this chapter, it's possible to cancel out and thus eliminate these elements to obtain a theoretical series in which only the cyclical and irregular components are at work.[5] That is, $(T \times S \times C \times I) \div T = S \times C \times I$, and $(S \times C \times I) \div S = C \times I$. And once the trend and seasonal elements have been eliminated, the usual procedure is to then cancel out and thus *eliminate the irregular movements* by computing a moving average. The remaining or *residual* variations are considered to be the *cyclical forces* at work. Of course, it's desirable to try to measure the cyclical element because of the importance of cyclical factors in short- and intermediate-term forecasts.

The calculations required to identify the cyclical and irregular forces are not difficult, but they are tedious and better left to more advanced texts. A typical work sheet to perform the steps just described, however, might be set up as shown in Figure 13-19.

---

[5] If the original data are expressed in yearly values (or if the data are presented on a seasonally adjusted basis), the seasonal component either does not exist or has already been eliminated by deseasonalization. In this case, the procedure to identify the C and I movements is the same, except that the adjustment to eliminate the seasonal element is obviously not needed.

FIGURE 13-19   Work Sheet Format for Identifying the Cyclical Component

MONTH (1)	ORIGINAL DATA $(T \times S \times C \times I)$ (2)	TREND VALUE (T) (3)	PERCENT OF TREND $(S \times C \times I)$ [(2) ÷ (3)] (4)	SEASONAL INDEX (S) (5)	CYCLICAL IRREGULAR PERCENTAGES $(C \times I)$ [(4) ÷ (5)] (6)	MOVING TOTAL OF (6) PERCENTAGES (7)	CYCLICAL PERCENTAGES (C) [(7) ÷ NUMBER OF MONTHS (OR QUARTERS) USED IN MOVING TOTAL] (8)

TIME-SERIES ANALYSIS IN FORECASTING: A SUMMARY

Various approaches are used to forecast the future, but we've seen that these approaches are all based on a few basic assumptions. When time-series analysis is the forecasting tool used, the assumptions are that past tendencies will *persist*, and forthcoming activities are related to measurable past fluctuations that are *regular* and *recurring* and can thus be projected into the future. Forecasts using time-series data are made for the following periods:

1 *For long-term forecasting.* In making forecasts several years into the future, the analysis, and then projection, of secular trend is an important element. We've seen how it's possible to use a trend equation to make such projections. A long-term trend forecast often does not take cyclical forces into account, and the seasonal element is of no concern in the projection of annual data.

2 *For intermediate-term forecasting.* The cyclical component may be incorporated into forecasts of shorter duration by multiplying the projected trend value by an estimate of the expected percentage change in the data as a result of cyclical forces. That is, projected trend times the cyclical percentage equals the forecast value. Of course, the assumptions that relationships or patterns which existed in the past will persist into the future with recurring regularity are very shaky when applied to the cyclical component because successive cycles vary widely in timing, pattern, and percentage changes.

3 *For short-term forecasting.* We've seen how a seasonal index may be used

with a monthly trend projection to produce a short-term seasonally adjusted forecast. And we've also seen how a cyclical estimate can be incorporated into the short-term forecast. In short-term forecasts, as in longer-term projections, an attempt to predict irregular movements is usually *not* made.

PROBLEMS IN THE USE OF TIME-SERIES ANALYSIS

Economists state their GNP growth projections to the nearest tenth of a percentage point to prove that they have a sense of humor.
                                                            —*E. R. Fiedler*

Give them a number or give them a date, but never both.
                                                            —*E. R. Fiedler*

It is surprising that one soothsayer can look at another without smiling.
                                                            —*Cicero*

I have been increasingly moved to wonder whether my job is a job or a racket, whether economists . . . should cover their faces or burst into laughter when they meet on the street.
—*Frank H. Knight, in a 1951 address to the American Economic Association*

The fact that a particular forecast made with the aid of time-series analysis techniques may prove to be incorrect does not discredit the use of the techniques discussed in this chapter. After all, the incorrect forecast might have been much closer to the actual results than a hunch.[6] If a time-series analysis leads to forecasts that reduce the *avoidable* risk, it will have been worthwhile. There are, however, potential problem areas associated with the use of time-series analysis. (See the Closer Look reading at the end of this chapter for an account of the forecasting perils faced by marketing research firms.) And these problem areas lead to questions that cannot be safely ignored. A few of the *precautionary questions that should be considered* in a particular situation are:

1 *How appropriate is the classical model?* A model is an abstract representation of reality. In the classical model, each component represents the theoretical effects of a myriad of causal factors that have been grouped together. Obviously, when we create a model and apply it to the analysis of actual data,

---

[6] An alternative forecasting approach would be for you to sit in a trance on a tripod above a chasm that emitted noxious vapors and make oracular utterances for someone to record. This was the approach used by the Greeks in ancient Delphi.

we can expect results that are proportional to the accuracy and sophistication of the model itself. Unfortunately, our time-series model is not too sophisticated. It assumes, for example, that the four components are *independent of each other*, and this hardly conforms to reality in many cases. (There are many people who will probably alter their seasonal Christmas spending habits during a cyclical recession.)

2   *How valid are our assumptions about persistence and regularity?* If there are *no persuasive independent reasons* to support these assumptions, the rather mechanical mathematical procedures that produce a trend equation and a seasonal index can lead to a very precise-looking forecast that may be completely out of touch with reality. The assumptions must always be kept in mind, and the analyst must adjust computations in light of the subjective and qualitative factors that are almost always present.

3   *How reliable are the input data?* Lack of data is often a problem, and when adequate data are available, they may not be strictly comparable. During the rather long intervals of time over which data should be gathered before usable patterns can emerge, a number of things can happen, for example, to change the quality of a measured variable. It's not inconceivable that the quality of a successful product could deteriorate to the point that the past trend could be significantly altered in a short period of time.

The above questions are merely a few of the ones that the analyst should consider in using time-series analysis techniques. Though the limitations and pitfalls are many, time-series analysis can be a very useful tool when used to provide an initial and approximate forecast.

LOOKING BACK

1   Planning and decision making involve expectations of what the future will bring, and so administrators are required to make forecasts. One forecasting tool used by managers is time-series analysis. A time series is worth studying because (1) it may enhance understanding of past and current patterns of change and (2) it may provide clues about future patterns to aid in forecasting. It's only natural that time series are seriously studied. There's a need to understand the dynamic forces at work in the economy, and time-series analysis can enhance this understanding. And there's no choice but to make forecasts, and time-series analysis may reveal persistent patterns.

2   The four components which may be identified in a time series are secular trend, seasonal variation, cyclical fluctuations, and irregular movements. Secular trend is the long-term growth or decline of a series. Seasonal fluctuations are periodic in nature and recur regularly within a period of 1 year or less. Cyclical fluctuations are also periodic, but they extend over an interval of several years and they cannot be counted on to occur on a regular basis. Irregular movements take place over varying but usually brief time periods, follow no regular pattern, and are unpre-

dictable. A classical time-series model is used to describe the relationship between these components.

3   The first component studied in this chapter, secular trend, is examined to describe historical patterns in a series and to project persistent patterns into the future. Once the trend element is identified, it may be removed from the original data to reveal the movement of other components. A linear trend equation ($Y_t = a + bx$) is discussed and computed in the chapter. By using this equation, it's possible to project a trend line into the future.

4   The second component, seasonal variation, is measured by constructing an index of seasonal variation. A ratio-to-moving-average method is employed in the chapter to prepare such an index. The reasons for measuring seasonal variation are to (1) understand seasonal patterns, (2) project existing patterns into the future, and (3) eliminate the seasonal component from the time series by a process called deseasonalization.

5   Once trend and seasonal factors are measured, it's then possible to eliminate these elements to obtain a theoretical series in which only the cyclical and irregular components are at work. The purpose of measuring trend, seasonal, and cyclical components, as noted above, is often for forecasting.

6   When forecasts depend on time-series analysis, the assumptions are that past tendencies will persist and forthcoming activities are related to measurable past fluctuations that are regular and recurring. A limitation of time-series analysis, of course, is that the assumptions about persistence and regularity may not always hold. The classical model may also be inappropriate at times, and the reliability of the input data is subject to variation over the years of a time series.

KEY TERMS AND CONCEPTS

PROBLEMS

1 The Igloo Swimming Pool Company of Fairbanks, Alaska, is planning an expansion program. The company needs a forecast of annual sales for each of the next 5 business years in order to plan this expansion. An analysis of the company's records indicated the following:

YEAR	ANNUAL SALES (THOUSANDS)
1977	110
1978	125
1979	135
1980	150
1981	170
1982	185
1983	196
1984	216
1985	230

   a   Construct a time-series graph with the data given above.

   b   Compute the secular trend equation with the least-squares method. Plot this equation on the time-series graph prepared in problem 1a.

   c   Compute the needed 5-year forecast with the equation found in problem 1b.

2 Radio station WINO of Sterling, Illinois, is concerned about an apparent secular trend decline in the number of persons listening to AM radio in its market area. FM listenership is apparently on a secular upswing. The data given below were prepared by a marketing research agency.

YEAR	AM RADIO AVERAGE AUDIENCE SIZE (HUNDREDS)	FM RADIO AVERAGE AUDIENCE SIZE (HUNDREDS)
1977	31	3
1978	32	3
1979	33	6
1980	30	7
1981	29	10
1982	30	11
1983	28	14
1984	26	14
1985	24	17

   a   Construct a time-series graph with the data given above for both the AM and FM audience size series.

   b   Compute the secular trend equation with the least-squares method for both AM and FM.

   c   Forecast the total audience size for both AM and FM in 1987.

3 The security chief of Horace Cints University is making plans to assign personnel and schedule vacations next year in the traffic division. Official records indicate the total issuance of traffic tickets over the past 3 years:

	TRAFFIC TICKETS ISSUED		
MONTH	1987	1988	1989
January	100	90	110
February	90	110	120
March	80	90	90
April	90	110	100
May	110	130	140
June	130	120	140
July	140	150	140
August	160	170	160
September	200	200	220
October	210	210	220
November	180	200	240
December	240	220	280

    a   Determine seasonal indexes by the ratio-to-moving-average method. (You must use the arithmetic mean to average the specific seasonals.)

    b   Comment on how the seasonal indexes might be used by the chief.

4 The chief assistant to the assistant chief in charge of university reports must give the trustees of Horace Cints a report on traffic tickets issued during the first 6 months of 1990. Actual issuances are recorded below:

MONTH	TICKETS ISSUED
January	120
February	120
March	100
April	90
May	130
June	150

    a   Deseasonalize the ticket issuances with the seasonal index values obtained in problem **3**.

    b   Comment on the apparent increase or decrease in tickets issued thus far in 1990.

5 The following table shows the annual production of petroleum by Fundamental Oil Company (in millions of barrels):

YEAR	PETROLEUM PRODUCTION (MILLIONS OF BARRELS)
1985	5
1986	8
1987	12
1988	15
1989	20

    a   What's the trend equation for this time series?

    b   What's the trend projection for 1991?

    c   What's the trend value on January 1, 1989?

    d   The sum of the Y_t values for the years 1985 through 1989 is _____.

6 Assuming that Fundamental Oil Company now knows that its production for 1990 is 25 million barrels, compute the following answers:

a $Y_t =$ _____ + _____ (x)

b $Y_{t1991} =$ _____

c Y_t for January 1, 1989 = _____

7 The monthly sales data for the Crispy Suntan Lotion Company (in thousands of cases) are given below for the years 1988–1990:

	SALES OF CRISPY LOTION		
MONTH	1988	1989	1990
January	84	93	101
February	92	102	109
March	103	112	119
April	122	123	130
May	125	131	138
June	124	135	141
July	117	133	139
August	108	126	132
September	99	117	122
October	92	107	113
November	91	100	106
December	89	98	103
	1,246	1,377	1,453

a Find the secular trend equation for total yearly sales using the method of least squares.

b Compute a seasonal index by the ratio-to-moving-average method. (Use the arithmetic mean to average the specific seasonals.)

c Using the secular trend equation computed in a above, forecast suntan lotion sales for 1992.

d Suppose the trend value for expected sales of suntan lotion in an October is 134 (thousand cases). Use the seasonal index from part c above to seasonally adjust this forecast value.

8 A personal computer retailer is planning an expansion program. The retailer needs a forecast of annual sales for each of the next 6 years to plan this expansion. Total annual sales (in hundreds of computers) are given in the table below for the past 6 years:

YEAR	ANNUAL SALES
1985	14
1986	19
1987	25
1988	27
1989	31
1990	37

a Compute the least-squares trend equation for this time series.

b Using this trend equation, prepare a 6-year forecast.

9 Monthly production data for the Otto S. Guzzling Refining Company (in millions of gallons) are given below for the year 1990. Using the seasonal index computed in

Figure 13-17 in the chapter, deseasonalize the monthly production data for 1990. (*Hint:* See Figure 13-18.)

MONTH	1990 PRODUCTION
January	13
February	12
March	15
April	23
May	20
June	20
July	30
August	19
September	16
October	15
November	17
December	19

10 The sales data for Krinkle Gum toothpaste (in thousands of tubes) are shown below:

YEAR	TOTAL SALES
1986	81
1987	66
1988	49
1989	34
1990	18

a   Using the least-squares method, find the linear trend equation for this time series.
b   What is the trend projection for 1991?
c   Can the trend be projected for 1994?

11 An oil pipeline is capable of carrying 500,000 barrels of oil a day. The average daily flow of oil through the pipeline (in thousands of barrels) is given below:

YEAR	AVERAGE DAILY FLOW
1981	250
1983	276
1985	300
1987	327
1989	349

a   Find the trend equation for the average daily flow of oil through the pipeline.
b   What is the trend projection for 1992?
c   In what year does the linear trend projection for average daily flow exceed the carrying capacity of the pipeline?

TOPICS FOR REVIEW AND DISCUSSION

1   "All administrators must make forecasts." Comment on this statement.
2   Why should we study time series?
3   Identify and discuss the components found in a time series.

4   What's the classical time-series model, and how may it be used?

5   Compare the reasons for measuring trend with the reasons for measuring seasonal variation.

6   Why are the assumptions of persistence and regularity important in forecasting?

7   Apply the straight-line equation to a firm whose *total cost* curve is a straight line for a particular product.[7]

8   In economics, the consumption function is often drawn as a straight line. Explain why the slope of the consumption function is therefore the marginal propensity to consume.[7]

9   Explain the assumptions made when a secular trend is projected into the future.

10   A seasonal index for January is 112.6. What does this number mean?

11   Discuss the steps in the ratio-to-moving-average method of computing a seasonal index.

12   Explain how a short-term forecast could be adjusted to take the seasonal element into account.

13   Why is the method of identifying cyclical fluctuations sometimes referred to as the residual method?

14   Summarize how time-series analysis may be used in forecasting.

15   What problems are there in the use of time-series analysis?

PROJECTS/ISSUES TO CONSIDER

1   Gather information about available statistical forecasting packages from library holdings, computer retail stores, members of a computer-user group, or other sources. Identify three forecasting packages, describe the features they offer, and note the equipment needed to run the programs you've selected. Write a brief report outlining your findings.

2   Team up with some classmates and find or collect a business, economic, or other time series that interests the group—one that the team would like to understand and predict. Apply the concepts learned in this chapter to prepare a 1-year trend projection of the team's time series. Using the projected value as a point of departure, the team may want to revise this figure to take into account any subjective considerations that it believes should be recognized. Prepare a line chart similar to the one shown in Figure 13-12 to plot the time-series data. Show the trend projection on this chart, and indicate any changes made to this projection. This visual aid can then be used to summarize the team's project in a class presentation.

ANSWERS TO SELF-TESTING REVIEW QUESTIONS

13-1

1   a $a = \dfrac{188}{11} = 17.09$

$b = \dfrac{226}{110} = 2.05$

$Y_t = 17.09 + 2.05x$

[7] This is an optional and tough question. Don't worry if you're stumped by it.

  b $Y_t = 17.09 + 2.05(7)$
       $= 17.09 + 14.35$
       $= 31.44$

  c If past production patterns persist, a tentative forecast of saddle production in 1991 would indicate that approximately 3,144 saddles would be produced.

2 Since the origin is the middle of 1987, and since x is in 1-year units, the value of x in 1991 is 4. Thus,

   $Y_t\ 1991 = 10 + 3(4)$
            $= 22$

13-2

1

SPECIFIC SEASONALS												
YEAR	JAN.	FEB.	MAR.	APR.	MAY	JUNE	JULY	AUG.	SEPT.	OCT.	NOV.	DEC.
1987							71.4	80.0	95.2	103.2	116.3	137.4
1988	69.8	99.2	114.5	136.4	128.8	45.1	83.3	75.2	106.1	98.5	122.1	128.8
1989	75.8	91.6	114.5	121.2	136.4	53.0						
Mean	72.8	95.4	114.5	128.8	132.6	49.0	77.4	77.6	100.6	100.8	119.2	133.1
Seasonal index*	72.7	95.3	114.3	128.6	132.4	48.9	77.3	77.5	100.4	100.6	119.0	132.9

*Correction factor = 1200.0/1201.8 = .9985

2

MONTH	UNADJUSTED RECEIPTS	SEASONAL INDEX	SEASONALLY ADJUSTED RECEIPTS
January	$11	72.7	$15.1
February	13	95.3	13.6
March	14	114.3	12.2
April	15	128.6	11.7
May	19	132.4	14.4
June	8	48.9	16.4
July	12	77.3	15.5
August	14	77.5	18.1
September	14	100.4	13.9
October	16	100.6	15.9
November	17	119.0	14.3
December	21	132.9	15.8

A CLOSER LOOK

SORRY, WRONG NUMBER

Submitted for your approval: A world where friendly androids patrol homes, robots run factories, and home computers have replaced the daily newspaper. This Twilight Zone scene is the world of 1988 as predicted by technology-market forecasters.

Even though these prognostications are more often wrong than right, they are widely read and quoted. High-technology industries seem to need to be soothed as they embark on inherently risky ventures. To quench this thirst for numbers, reports and studies flow from market-research organizations such as Dataquest, the Gartner Group, and International Data Corp.

Sales projections often have a tantalizing aura of authority. Executives and managers gobble them up before committing company resources to grand new projects, and entrepreneurs use them to win investments from venture capitalists. Meanwhile, reporters disseminate the figures until the numbers take on a life of their own.

Erroneous forecasts can hurt companies that rely on them. The Knight-Ridder newspaper company, for example, lost $55 million in an attempt to market videotex, a system for delivering graphics-oriented information to the home via terminals connected to television sets. In 1983, Creative Strategies International, one of many to overestimate the potential market, predicted that sales of videotex equipment and services would expand more than 90 percent annually, reaching $7 billion by 1987. But a trial system convinced Knight-Ridder that the pundits were living in a fantasy world. After two years, the company declared defeat and walked away. "We saw no way to make the service profitable," says Virginia Fields, Knight-Ridder's vice president for news and circulation research.

Why do the scenarios painted by market researchers bear so little resemblance to conditions in the real world? The answer lies partly in the means used to produce the reports: researchers rely heavily on numbers from vendors, who routinely overstate their sales projections. "Extrapolated growth based on manufacturers' data will always be high," says Richard Arons, director of the market-research firm Prognos.

"All reports are overoptimistic," admits Gene Selvin, president of Electronic Trend Publications. The reason, he says, is that researchers get their data from companies with an obvious interest in seeing their sales figures inflated, mainly to attract investors.

Prab Robot president Walter Weisel knows only too well the fallibility of technology market forecasts. His company was one of many that expected the robot business to soar to the billion-dollar level by 1990. However, after a quick start, robot sales have sputtered. Industry revenues for 1987 amounted to only about $300 million—just one of numerous wrong calls that have damaged the credibility of forecasters.

Good researchers discount overoptimistic sales projections by subtracting a "fudge factor"—the level by which a vendor consistently exaggerates its claims. Reputable research firms also balance their results by polling potential customers, gaining data that is more reliable but far more difficult to gather. The problem: As few as 3 percent of the users polled respond to surveys, says International Data spokeswoman Shelley Bakst. Generally, it is much more time-consuming and expensive to survey a technology's potential customer base than to poll the much smaller set of vendors.

Misleading information flows from laboratories as well as from sales offices. Scientists and engineers, infatuated with their technologies, sometimes delude themselves about commercial practicality. Materials scientists, for example, are now touting the wonderful qualities theoretically achievable with a new class of metal-matrix composites. However, "if you say, 'make me a piston,' they scratch their heads," says Richard Bryant, senior industry analyst at Business Communications.

Usually, the fault lies with the market researchers' naivete and with their need to unearth good news. Buyers of market research typically want the numbers to support a decision to enter a new business area; the more optimistic the predictions, they feel, the more useful the report. "We regularly get pressure to say a market is bigger than it is," says Ralph Finley, executive vice president of Dataquest.

Indeed, says Finley, it is difficult to sell reports that make conservative projections. Dataquest found only about a half-dozen customers for a 1984 report foreseeing slow sales of home computers. "Bad news doesn't sell," he says.

Market research's happy-news tendency serves an important function, say its practitioners. Without a willing suspension of disbelief, they claim, bold ventures might never happen. "If people looked realistically at new technologies at their inception, they would never make the investment," says Les Cowan, senior industry analyst at Rothchild Consultants. "Self-delusion" is necessary, argues Cowan, before embarking on any long, risky enterprise. "It's the same as in a war," he says. "You can't win unless every soldier has convinced himself he's not going to be the one killed."

Market forecasts often assume that existing conditions will persist into the future, which is like trying to drive by looking in the rear-view mirror. "It's dangerous to assume present trends will continue," says Meg Lewis, vice president of Future Computing. Such an assumption has afflicted several firms trying to project the personal-computer market, Future Computing prominent among them. From 1978 to 1983, personal-computer sales grew rapidly, as pioneering users absorbed the first wave of products. But after these early adopters had bought their fill, the industry slumped. The 1985 slowdown caused "lots of finger-pointing," recalls Lewis; manufacturers accused forecasters of misleading them about the extent of demand.

Crystal-ball gazers get blindsided with such regularity that it's a wonder anyone still pays attention to them. For example, in 1983 Dataquest predicted that the market for computer-aided design (CAD) systems would rise to $6.3 billion by 1986. The market actually hit more than $7.3 billion. Why the underestimate? "We didn't expect personal computers to become part of the CAD market," says Beth Tucker Romig, associate director of Dataquest's industrial-automation research group. The firm's 1983 prediction accounted only for expected sales of minicomputer-based CAD; personal computers were relatively new on the scene. . . .

Overlooking the competition is a common flaw of market research. Companies developing a new technology tend to become its boosters, unrealistically ignoring the improvements still possible in existing products. "People underestimate the life still left in old technologies," says Finley of Dataquest.

Such parochialism led to embarrassingly inaccurate projections for sales of optical-memory disks, which hold vast quantities of information—as much as a few billion digital bits. Expecting these disks to steal market share from magnetic disks, a number of research firms, including Rothchild Consultants and Business Communications, predicted in 1983 and 1984 that optical data-storage systems would produce annual revenues of several hundred million dollars by 1990.

Alas, the market has come nowhere near that level. One reason is that it took longer than expected to bring optical disks from lab to market. But more importantly, magnetic disks underwent continual improvement. Floppy disks, which once held only 360 kilobytes, today can store a megabyte or more, and magnetic hard disks are pushing toward 100 megabytes.

Particular success was projected for compact disc read-only memories (CD-ROMs). These cousins of audio compact-disc players were expected to become essential attachments to personal computers. In 1986, several firms estimated that as many as one million CD-ROM drives would be in use by the end of the decade. But industry insiders say only about 50,000 CD-ROM drives have so far been built.

The best forecasting looks sideways as well as ahead. "The things that traumatize business come from the outside," says John Varston, president of Technology Futures. Forecasters often don't see a new technology's most profound implications, he says, because "they're looking in the wrong places." Varston points out that one of the century's pivotal inventions—the transistor—underwhelmed early analysts, who saw it only as a replacement for vacuum tubes in radio receivers.

One of the more notable success stories of market research illustrates the value of being aware of activity in other fields. In 1981, Dataquest correctly predicted that low-cost laser printers would arrive by 1985 or 1986, an assumption based on an analyst's knowledge of Japan's photocopier industry.

The analyst saw that Canon had developed a copying system based on a cheap, compact semiconductor laser; until that time, bulky and expensive gas-tube lasers were used. Dataquest correctly noted that the same laser system could serve as a printer engine.

Such a direct hit is rare among long-term predictions. Indeed, some researchers admit that they include long-range forecasts in their reports primarily to assure press coverage. Such forecasts are especially suspect for technologies not yet on the market. Still, research firms persist in publishing predictions on superconductors, neural-network computers, and other technologies with no commercial history. It's "crazy" to make forecasts on embryonic technologies, says Selvin of Electronic Trend. "We put in forecast numbers to get people's attention," admits Portia Isaacson, founder of Future Computing and now president of Future Think. "Everybody knows we're pulling numbers out of a hat."

In general, the better reports come from companies with full-time staff researchers: Dataquest, International Data Corp., and Arthur D. Little, for example. Spottier quality comes from organizations that rely heavily on free-lance contributions, such as Frost & Sullivan and Business Communications. Some free-lance researchers have a deep understanding of the industry they cover, but others do not, and their reports lean on information drawn from back issues of trade magazines. Worse, some free-lancers recycle the sales forecasts of competing number-mongers. . . .

For the most part, you get what you pay for. Off-the-shelf reports—as opposed to studies done for a limited number of clients—typically cost $1,000 to $3,000. Money buys research time, and research time usually results in better accuracy. "A $600 report won't have much beyond published data," says Robert Butler, president of Business Communications, whereas $3,000 buys two months of telephone interviewing.

High-cost reports can nevertheless offer foolish predictions, as was the case with estimates in the early 1980s of industrial-robot market growth. The reports ignored the obvious: that U.S. manufacturing—i.e., the robot market—was moving overseas in search of cheap labor. In another case, some gushing CAD forecasts predicted straight-line 50 percent annual growth, until CAD systems would outnumber all architects and engineers, the users of such systems. Most research firms "are led by wishful thinking," says Texas Instruments vice president Eugene W. Helms, who manages corporate strategic planning.

Corporate strategy makers say they don't rely greatly on published market forecasts. General Electric, for example, uses third-party research only to supplement its own strategic analysis, says spokesman George Jameson. At Texas Instruments, outside research is "just one of many pieces of information," says Helms. "It's certainly not the decisive element."

Even as they promote their far-sightedness, market researchers warn against relying on that vision—especially if it proves wrong. A company or investor would be "crazy to use a market-research report to decide where to commit its resources," says Business Communications' Butler. Kenneth Bosomworth, president of International Resource Development, says the numbers his firm or others put out should be used only as rough guides. Anybody who started, say, a robotics company based on published market-research reports "is a fool," says Bosomworth.

These firms may overestimate their customers' sophistication. Retail brokerage houses, which sell stocks to small investors, often accept official-looking market forecasts uncritically. Venture capitalists have been burned repeatedly in recent years, swallowing predictions for robotics, machine vision, and personal computers. Of all these, technology producers have arguably been the biggest victims. Overly optimistic reports of coming growth in the home-computer market were one reason Texas Instruments cut the price of its home computer to below production costs in the early 1980s. The company hoped a price tag of less than $100 would fuel enough demand to bring production costs below that cut-rate level. The demand never materialized, and Texas Instruments took a bath.

The company was similarly burned by rosy predictions about electronic-speech products. In 1980, International Resource Development figured that annual sales of speech-synthesis systems would mushroom from $25 million to $1 billion by mid-decade. "We saw such huge potential," says president Bosomworth. The artificial-speech hardware market today reaps revenues of only about $50 million.

Market-research companies get defensive when challenged on their past accuracy. Electronic Trend's Selvin calls such queries a "waste of time," because any forecast is grounded in the knowledge available at the time. "We're not in the prophecy business," says analyst Barry Bartlett of Market Intelligence Research. Shrugs Bosomworth of International Resource Development, "Sometimes we hit the nail on the head, other times we bang the hammer on our thumb. It's absurd to say that we can predict the future."

Few in the Twilight Zone of market forecasting would disagree.

—From Herb Brody, "Sorry, Wrong Number." Reprinted with permission, HIGH TECHNOLOGY BUSINESS magazine (September 1988, pp. 24–28). Copyright © 1988 by Infotechnology Publishing Corporation, 270 Lafayette Street, Suite 705, New York, NY 10012. Editorial assistant Jennifer Christensen contributed to this article.

FORECASTING TOOLS: SIMPLE LINEAR REGRESSION AND CORRELATION

The Regression Effect

A few years ago a school in New Jersey tested all its fourth graders to select students for a gifted program. Two years later the students were retested, and the school was shocked to find that the scores of the gifted group had dropped in comparison with the rest of the students.

A mathematics supervisor got a grant to improve mathematics education in her city. She tested all the students and placed those with the lowest achievement scores in a special program. After a year, she retested them and was gratified to see that the students in the special program improved in comparison with the rest of the students.

Has the gifted program been detrimental? Has the remedial program worked? The answer to both questions is "not necessarily." In both situations, the regression effect (sometimes called regression toward the mean) may be causing the results. The regression effect says this:

Suppose a test score or measurement is determined for each individual. A second score on another test is also determined. Then, as long as these scores reflect some element of chance, the people who had the lowest scores on the first test will tend to do better on the second test—and the highest scorers will fall back. . . .

REGRESSION REVEALED IN ROCK'S REDDISH RECESS

Owing to his interest in heredity, Sir Francis Galton (1822–1911) was the first to recognize the regression effect. Early in his studies, he felt that the mean height of the offspring of a couple should tend to equal the mean height of the parents. However, his observations did not verify this assumption. He wrote the following in his memoirs (Galton 1908):

A temporary shower drove me to seek refuge in a reddish recess in the rock by the side of the pathway. There the idea flashed across me, and I forgot everything else for a moment in my great delight.

The following question had been much in my mind. How is it possible for a population to remain alike in its features, as a whole, during many successive generations, if the average produce of each couple resemble their parents? Their children are not alike, but vary: therefore some would be taller, some shorter than their average height; so among the issue of a gigantic couple there would be usually some children more gigantic still. Conversely as to very small couples. But from what I could thus far find, parents had issue less exceptional than themselves. I was very desirous of ascertaining the facts of the case.

At the suggestion of his half-cousin, Charles Darwin, he began an experiment with sweet peas. He divided sweet-pea seeds into seven sizes. Ten seeds of each weight were sent to nine friends (including Darwin) in different parts of England. The seeds were planted and harvested according to precise instructions. Two crops were failures. Galton's results (see data below) were presented in 1877. The largest parent seeds tended to produce smaller daughter seeds, and the smallest parent seeds produced larger daughter seeds. Galton called the phenomenon "reversion."

Later, Galton was able to reproduce these results with 205 human couples and their offspring. He found that children deviate from the mean height by about two-thirds as much as their parents. For example, if the parents are three inches taller than average, their children will tend to be only two inches taller than average.

—From Ann E. Watkins, "The Regression Effect; or, I Always Thought that the Rich Get Richer . . . ," *Mathematics Teacher*, November 1986, pp. 644–647. Reprinted with permission.

Sweet Pea Experiment

	DIAMETER (0.01 INCH) OF SEED						
Parent	15	16	17	18	19	20	21
Daughter	15.4	15.7	16.0	16.3	16.6	17.0	17.3

LOOKING AHEAD

The equations that describe the relationship that exists between two or more variables are usually developed with the help of computers or electronic calculators. But the name given to the study methodology that uses these modern tools is *regression analysis*—an anachronistic term dating back to Sir Francis Galton's study (discussed in the opening vignette) of the relationship between parents' heights and children's heights. Since Galton's findings were that the stature of the offspring tended to "regress" toward the mean population height, he was legitimately engaged in a regression analysis. Subsequent studies, of course, had nothing to do with Galton's regression analysis, but the name has stuck. Thus, after some introductory concepts have been presented, you'll learn about regression analysis and the standard error of estimate. Then, you'll study correlation analysis—a related subject. Finally, you'll read about the common errors and limitations involved in using and interpreting regression and correlation measures.

Thus, after studying this chapter you should be able to:

 ◗ Explain the purposes of regression and correlation analysis.
 ◗ Compute the regression equation and the standard error of estimate and then use these measures to prepare interval estimates of the dependent variable for forecasting purposes.
 ◗ Compute (and explain the meaning of) the coefficients of determination and correlation.
 ◗ Point out several errors that may be made in the use of regression and correlation techniques.

CHAPTER OUTLINE

INTRODUCTORY CONCEPTS

You know that it's often necessary to prepare a forecast before a decision can be made. For example, it may be necessary to predict revenues before a budget can be prepared. And a university must predict enrollment before making up class schedules. These and other decisions can be made easier if a relationship can be established between the variable to be predicted and some other variable that is either known or is significantly easier to anticipate.

For our purposes, the term "relationship" means that changes in two (or more) variables are *associated with each other*. For example, we might find a high degree of relationship between the consumption of fuel oil and the number of cold days during the winter, or between gasoline sales and the number of registered motor vehicles. That is, we might logically expect a change in the number of cold days to be accompanied by a change in the consumption of fuel oil, or a change in the number of registered automobiles to be accompanied by a change in the demand for gasoline.

Regression and Correlation Analysis: A Preview

The tools of *regression analysis* and *correlation analysis* have been developed to study and measure the statistical relationship that exists between two or more variables. We'll only consider the relationship between *two* variables in the body of the chapter. The term **simple regression and correlation** refers to those studies dealing with just two variables. When three or more variables are considered, the study deals with **multiple regression and correlation.** We'll look at a multiple regression and correlation situation in the Closer Look reading at the end of the chapter.

In **regression analysis,** an estimating equation is developed to describe the pattern or functional nature of the relationship that exists between the variables. As the name implies, an analyst prepares an **estimating (or regression) equation** to make estimates of values of one variable from given values of the other. The variable to be estimated is called the **dependent variable** and is customarily plotted on the vertical (or Y) axis of a chart. The dependent variable is therefore identified by the symbol Y. The variable that presumably exerts an influence on or explains variations in the dependent variable is customarily plotted on the horizontal (or X) axis, is termed the **independent variable,** and is identified by the symbol X.

Let's assume that Hiram N. Hess, personnel manager of the Tackey Toy Manufacturing Company (widely known for its motto "The little monsters deserve Tackey quality"), finds that there's a close and logical relationship between the productive output of employees in a certain department of the company and their earlier performance on an aptitude test. Thus, *if* Hiram computes an estimating equation (as we'll do in a few pages), and *if* he has the aptitude test score of a job applicant, *then* he may use his estimating equation to predict the future output (the dependent variable) of the applicant based on test results (the independent variable). Included in the techniques used to make estimates of the dependent variable are regression analysis procedures that measure the dependability of these measures.

In **correlation analysis,** the purpose is to measure the strength or closeness of the relationship between the variables. In other words, regression analysis asks "What is the pattern of the existing relationship?" and correlation analysis asks "How strong is the relationship described in the regression equation?" Although it's possible to be concerned with only regression analysis, or with only an analysis of correlation, the two are typically considered together.

To summarize, in the following pages we'll concentrate on three main topics (the first two of which are a part of regression analysis):

1　The computation of the regression or estimating equation and its use in providing an estimate of the value of the dependent variable (Y) from a given value of the independent variable (X).
2　The computation of measures that indicate the possible errors that may be involved in using the estimating equation as a basis for forecasting.
3　The preparation of measures that will indicate the closeness of the association or correlation that exists between the variables.

A Logical Relationship: The First Step

Your sales last year just paralleled the sales of rum cokes in
Rio de Janeiro, as modified by the sum of the last digits
of all new telephone numbers in Toronto. So, why bother with
surveys of your own market? Just send away for the data
from Canada and Brazil.

　　　　　　　　　　　　　　　　—Lydia Strong

Two series may vary together for several reasons. And analysts may correctly assume a causal relationship in their interpretation of correlation measures. But just the fact that two variables are associated in a statistical sense does not guarantee the existence of a causal relationship. In other words, the existence of a causal relationship usually does imply correlation, but, as the above quote adequately shows, *statistical correlation alone does not in any way prove cau-*

sality. (If you believe that it does, you should indeed send away for the data from Canada and Brazil!)

Causal relationships may fit into *cause-and-effect* or *common cause* categories. A **cause-and-effect relationship** exists if a change in one variable causes a change in the other variable. For example, a change in the temperature of a chemical process may cause a change in the yield of the process, and a change in the level of output may cause a change in total production cost. Alternatively, two series may vary together because of a **common cause factor** that impacts both series in the same way. One could probably find a close relationship between jewelry sales and compact disc sales, but one of these series is not the cause and the other the effect. Rather, changes in both series are probably a result of changes in consumer income.

Of course, as we saw in the quote at the beginning of this section, some relationships are purely accidental. So if a relationship were found between furniture sales in the United States and the average temperature in Tanzania, for example, it would be a meaningless exercise to analyze the data. Relationships such as this one are known as *spurious correlations.*

Probably the first step in the study of regression and correlation, therefore, is to determine if a logical relationship may exist between the variables to support further analysis. Unfortunately, presenting a summary of many types of variables, and the forces at work on those variables, is beyond the scope of this book. Fortunately, though, such a summary is to be found in the business, economics, and other social science courses that you've taken or will take. It's only through the use of reason and judgment (along with the application of knowledge about the variables and the forces at work) that an analyst may assume causality. Yet without this assumption, there's not much point in proceeding with regression and correlation analyses.

Other students (not you) sometimes go to one of two extremes at about this point in their introduction to regression and correlation: (1) They fail to realize the importance of determining if a logical relationship exists, mechanically apply the statistical procedures of the following pages to the data, and arrive at possibly spurious correlations to which they erroneously assign interpretations of cause and effect. Or (2) they think that since one cannot prove causality from correlation it's necessary to conclude that there's no connection at all between correlation and causality. You, of course, will avoid either of these extremes.

The Scatter Diagram

After it has been determined that a logical relationship may exist between variables to support further analysis, perhaps the next step is to use a chart to plot the available data. This chart—called a **scatter diagram**—shows plotted points. Each point represents an item for which we have a value for both the dependent variable and the independent variable. The *scatter diagram serves two purposes:* (1) It helps determine if there's a useful relationship between the two variables, and (2) it helps to determine the type of equation to use to describe the relationship.

We can illustrate the purposes of a scatter diagram by using the data in Figure

LIES, DAMNED LIES . . .

Spreadsheets are wonderful tools. So many businesses now depend on them so completely that it's hard to remember the bad old days when we did things by hand. But the spreadsheet that is your trusted friend when you quantify the past may be your worst enemy when you predict the future.

Spreadsheets were designed to help people do repetitive arithmetic. Recordkeeping and accounting are a lot easier when a computer does the math. But before long, people started using spreadsheets to play what-if games and to build forecasting models. The day of the cheap computer projection had arrived. Now anyone with a PC and a little imagination can do sophisticated forecasting that used to be possible only with mainframes.

Forecasting fever has gripped corporate America. Not even the fast-trackers can now walk into the boss's office with nothing but an idea; they have to have a fistful of numbers. In some offices, you can hardly order a new coffee machine without running a 3-year costs/benefits analysis. All business decisions, large or small, must be put to the numbers test. Make a few assumptions, build them into a model, and let'er rip!

In large companies, projections are run by specialists who may know something about forecasting. But in small companies, they're done by amateurs. All of us have some notion of whether an assumption is plausible, but how many *1-2-3* or *SuperCalc* users know anything about probability? Do they know whether forecast results will be more accurate or less accurate as they increase the number of variables? Have they any idea how to gauge the accuracy of their results? Probably not.

But the beauty of spreadsheet forecasting is that technical competence doesn't matter. Spreadsheets have done away with the "objective" forecast. Since anyone can now do projections, who's going to wait for an independent expert to do one? It's the person who cares most about a project—its most ardent supporter—who now does the projection.

Is he going to do a coolly unbiased forecast? Of course not. He probably thought up the project; he loves it like an only child. His projection is not going to be an aid to objective decision-making; it's going to be a lawyer's brief. A lot of spreadsheet "forecasts" are thus not forecasts at all. They're justifications for something somebody has already decided to do.

The trick is to build a forecasting model backwards. The forecaster knows what the results of the projection have to be: fat profits. So he tweaks the assumptions and tunes the variables until the "forecast" looks plausible and gives the right results. With creative fiddling, any project can be made to look profitable. Accountants have always been able to cook the books after the fact; now you can cook them in advance.

Some people are fooled by this. They don't understand that letting an advocate run a projection is like letting a criminal write a law. They don't realize that speculative assumptions can't produce results correct to the fourth decimal place. Even if they do, they may think they have to go along with the ritual of numerical analysis. It's with columns of figures that we offer up our prayers to the gods of capitalism.

LOOKING AT THE DOWNSIDE

At the same time—and more usefully—not even the most tendentious forecaster can completely ignore costs, losses, competition, or bad debts. He may finesse the unpleasantness with an artful choice of assumptions, but at least he has to think about it. And that may be the sole value of most projections: forecasters must identify costs in order to be taken seriously. They may then go on to eliminate them by sleight of hand, but they've had to worry about them.

There's another nasty issue that spreadsheet forecasters don't like to talk about: mistakes. A persuasive forecast has lots of variables—the more the better. A large number of variables makes it look as though the forecaster thought of everything, at the same time that it gives him many more ways to fudge the results. But even apart from fudging, handling multiple variables can be fiendishly complicated. The more numbers a forecaster has to juggle, the more he's likely to drop. The computer won't ever make a mistake, but he will.

Many mistakes are never found. Once a model has more than a dozen or so variables, it's too complicated to troubleshoot just by working backwards from results. Somebody has to go over the whole thing, formula by formula, to make sure everything's right. That's such an awful bore that hardly anyone does it. Spreadsheet auditing programs make this chore much, much easier, but since their sales are only a fraction of *1-2-3*'s, it's a good guess that few people are using them. These programs also cannot spot faulty logic. Only smart, patient people can do that.

There are businesses, like banking or accounting, that live on tedious, error-prone arithmetic. They have come up with formal procedures to catch mistakes. Amateur forecasting is even more prone to error, but it hasn't developed any standards of accuracy and probably never will.

A spreadsheet program and a microcomputer are powerful tools that can be used for great good. But if you're not careful, these same tools can be used against you. So the next time someone hands you a projection, be sure to ask a few questions.

Find out who built the model. Chances are it was someone with a vested interest. If the model is at all complicated, find out how carefully it's been checked. Has anybody but the original forecaster looked it over? If no one has, it's probably got mistakes. Finally, you should pay more attention to costs than to revenues. Revenues are mostly fantasy, but costs can be estimated. Make sure they're realistic.

Whatever you do, don't assume that just because somebody used a computer, the results are somehow unassailable. As Benjamin Disraeli pointed out long ago, there are lies, damned lies—and computer projections.

—Jared Taylor, "Lies, Damned Lies. . . ." Reprinted from: *PC Magazine*, October 27, 1987, pp. 142–143. Copyright © 1987 Ziff Communications Company.

14-1. This figure gives us the output for a time period in dozens of units (the dependent variable) and the aptitude test results (the independent variable) for eight employees in a department of the Tackey Toy Manufacturing Company. If the aptitude test does what it's supposed to do, it's reasonable to assume that employees with higher aptitude scores will be among the higher producers. Our group of eight employees represents a small sample of Tackey workers. We've chosen to keep the employee count small and the data simple, though, to minimize the computational effort necessary in later sections.

The data for each employee represent *one* point on the scatter diagram shown in Figure 14-2. The points representing employees C and F are labeled to show you how the pairs of observations for the employees are used to prepare the points on the chart. As you'll notice in Figure 14-2, the eight points form a path that can be described by a *straight line*, and a high degree of relationship is

FIGURE 14-1 Output and Aptitude Test Results of Eight Employees of Tackey Toy Manufacturing Company

EMPLOYEE	OUTPUT (Y) (DOZENS OF UNITS)	APTITUDE TEST RESULTS (X)
A	30	6
B	49	9
C	18	3
D	42	8
E	39	7
F	25	5
G	41	8
H	52	10

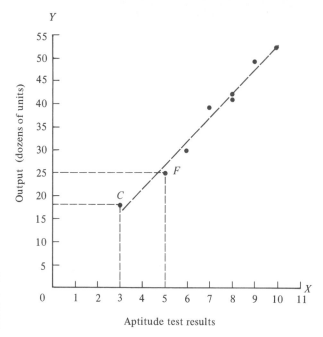

FIGURE 14-2
Scatter diagram.
(Source: Figure
14-1.)

indicated by the fact that the points are all close to this straight-line path. You'll also notice that there's a **positive** (or **direct**) **relationship** between the variables— that is, as aptitude test results *increase*, output also *increases*. Of course, it's quite possible for variables to have a **negative** (or **inverse**) **relationship** (as the X value increases, the Y value decreases).

Figure 14-3*a* shows how one statistical software package prepares a scatter diagram with the same data set used to produce Figure 14-2. The eight points must still follow a positive straight-line pattern, but the program has elected to use different scales for the two variables. Another scatter diagram that plots the relationship between per capita gross domestic product of selected nations (the independent variable) and the per capita amount spent on health care (the dependent variable) by those nations is shown in Figure 14-3*b*. This diagram also shows a positive linear relationship.

Figure 14-4 summarizes some possible scatter diagram forms. Figure 14-4*a*, *b*, and *c* shows the positive and negative linear patterns we've just been considering. But relationships need not be linear, as shown in Figure 14-4*d* to *f*. In Figure 14-4*f*, the variables might be family income and age of the head of the household. (Income tends to rise for a period and then fall off when retirement age is reached.) Finally, it's possible that a scatter diagram such as the one in Figure 14-4*g* might show *no relationship* at all between the variables.

REGRESSION ANALYSIS

The straight line in the scatter diagram in Figure 14-2 (and the straight lines in Figures 14-3 and 14-4) that describe the relationship between the variables is

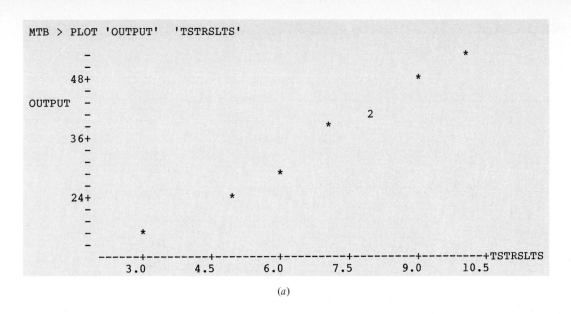

(a)

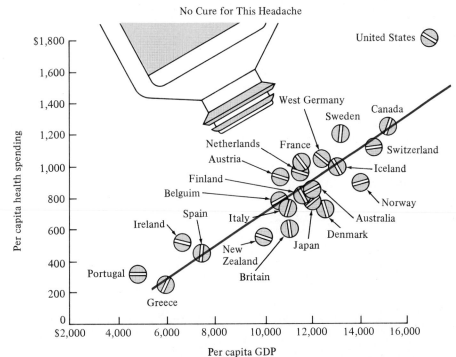

No Cure for This Headache

The line represents health care spending as a function of gross domestic product, as predicted by an empirically based economic formula. Note that U.S. health care spending exceeds prediction by nearly $400.

Data: Organization for Economic Cooperation and Development, Social Data Bank.

(b)

FIGURE 14-3   (a) A scatter diagram prepared by the *Minitab* statistical package. (b) An example of a positive linear relationship. (Source: *Insight,* August 8, 1988, p. 10. Figure *b* reprinted with permission of *Insight* magazine/M. Vey Martini.)

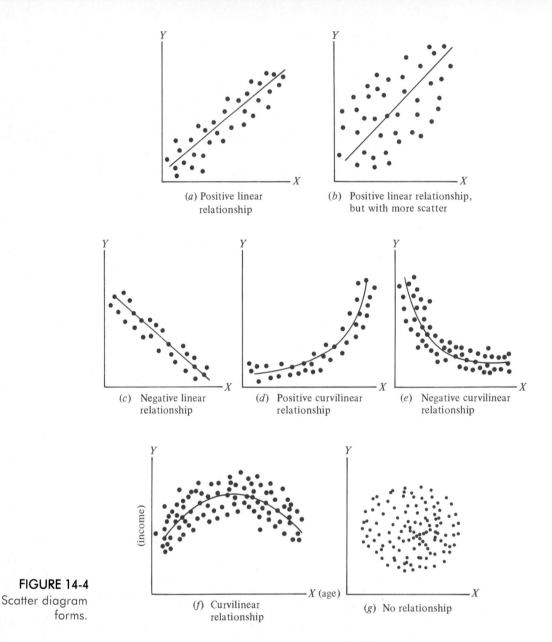

FIGURE 14-4

Scatter diagram forms.

(a) Positive linear relationship

(b) Positive linear relationship, but with more scatter

(c) Negative linear relationship

(d) Positive curvilinear relationship

(e) Negative curvilinear relationship

(f) Curvilinear relationship

(g) No relationship

called a **regression** (or **estimating**) **line.** We've seen in the opening vignette that Galton's study showed that the height of the children of tall parents tended to regress (or move back) toward the average height of the population. Galton called the line describing this relationship a "line of regression." The word "regression" has stuck with us, but other words such as "estimating" or "predictive" are probably more appropriate.

The Linear Regression Equation

We'll compute an estimating or regression equation in this section that describes the relationship between the variables. Our interest here is limited to an analysis of **simple linear regression**—that is, to the case in which the relationship between two variables can be adequately described by a straight line.

If you remember the discussion in the last chapter about linear trend computation, you should have little difficulty here with the regression equation. Why? Because the general straight-line equation presented there is also used in fitting the regression line to the data. That is, the *method of least squares* will once again be used to fit a line to the observed data. If you haven't read Chapter 13, a review of the section titled "The Straight-Line Equation" (page 489) will be helpful. As you read that section, just substitute (1) Y_c for each reference to Y_t, (2) the words "regression line" for each reference to "trend line," and (3) the dependent (X) variable for each reference to time.

The formula for the regression equation is:

$$Y_c = a + bX \qquad \text{(formula 14-1)}$$

where a = Y intercept (the value of Y_c when $X = 0$)
      b = slope of the regression line (the increase or decrease in Y_c for each change of one unit of X)
      X = a given value of the independent variable
      Y_c = a computed value of the dependent variable

The similarities between the regression line and the trend line do not end with the straight-line equation. The regression line (like the trend line and the arithmetic mean) has the following two mathematical properties:

$$\Sigma(Y - Y_c) = 0$$
and $\Sigma(Y - Y_c)^2$ = a minimum or least value

In other words, the regression line is fitted to the data in the scatter diagram in such a way that the *positive* deviations of the scatter points *above* the line in the diagram cancel out the *negative* deviations of the scatter points *below* the line, and the resulting sum is zero (see Figure 14-5).

The values of a and b in the regression equation are computed with the following formulas:

$$b = \frac{n(\Sigma XY) - (\Sigma X)(\Sigma Y)}{n(\Sigma X^2) - (\Sigma X)^2} \qquad \text{(formula 14-2)}$$

where n = number of paired observations.

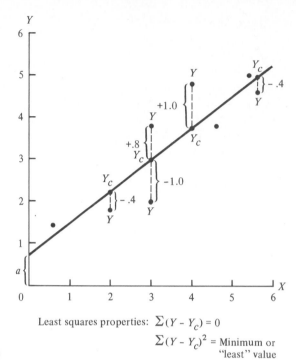

Least squares properties: $\Sigma(Y - Y_c) = 0$

$\Sigma(Y - Y_c)^2$ = Minimum or "least" value

FIGURE 14-5

$$a = \overline{Y} - b\overline{X} \qquad \text{(formula 14-3)}$$

where $\overline{Y}$ = the mean of the Y variable
$\overline{X}$ = the mean of the X variable

We're now ready to find the regression equation for the data presented in Figure 14-1. A work sheet for computing the values required to solve for b and a (using formulas 14-2 and 14-3) is given in Figure 14-6. A Y^2 column is also included in Figure 14-6. We don't need that column to calculate the regression equation, but we'll use it in just a few pages. The values of b and a are computed as follows:

$$b = \frac{n(\Sigma XY) - (\Sigma X)(\Sigma Y)}{n(\Sigma X^2) - (\Sigma X)^2} = \frac{8(2,257) - (56)(296)}{8(428) - (56)^2}$$

$$= \frac{1,480}{288} = 5.1389$$

$$a = \overline{Y} - \overline{b}X = 37 - (5.1389)(7) = 1.0277$$

Therefore, the regression equation that describes the relationship between the output of our sample of Tackey Toy employees and their aptitude test results is:

$$Y_c = 1.0277 + 5.1389(X)$$

FIGURE 14-6

EMPLOYEE	OUTPUT IN DOZENS OF UNITS (Y)	TEST RESULTS (X)	XY	X^2	Y^2
A	30	6	180	36	900
B	49	9	441	81	2,401
C	18	3	54	9	324
D	42	8	336	64	1,764
E	39	7	273	49	1,521
F	25	5	125	25	625
G	41	8	328	64	1,681
H	52	10	520	100	2,704
	296	56	2,257	428	11,920

$$\bar{Y} = \frac{\Sigma Y}{N} = \frac{296}{8} = 37; \bar{X} = \frac{\Sigma X}{N} = \frac{56}{8} = 7$$

Making Predictions with the Regression Equation

There was once a young manager named Hess
Whose forecasts were always a mess.
So his boss did appear,
And in voice loud and clear,
Said, "Hess, son, try regression, or consider another career!"

The primary use of the regression equation is to estimate values of the dependent variable given values of the independent variable. Suppose, for example, that the unfortunate Hiram Hess, personnel manager for Tackey Toys, is considering hiring an applicant who scored a 4 on the aptitude test. The supervisor of the department wants someone hired who can produce an average of 30 dozen units. Of course, it's not possible to tell exactly what the applicant's future production might be, but Hiram can use the equation computed in the preceding section to arrive at an estimate or forecast of the average amount of output produced by those who score a 4 on the aptitude test. How? By simply substituting 4 for X in the regression equation. The estimate is computed as follows:

$Y_c = 1.0277 + 5.1389(4) = 1.0277 + 20.556$
     $= 21.58$ dozen units of output

This prediction is shown graphically in Figure 14-7.

Self-Testing Review 14-1

The questions in this review are based on the following data:

Y	X
255	5
100	2
307	6
150	3

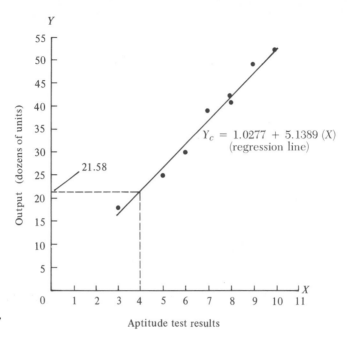

FIGURE 14-7

1 What are the values of
   a The slope of the regression line (b)?
   b The Y intercept (a)?
2 What is the estimate of Y when X is 4?
3 What is the estimate of Y when X is 5?
4 What can be said about the difference between the estimated values you've computed in parts 2 and 3 above?

The Standard Error of Estimate

When a prediction is made from a regression equation, this question naturally arises: "How dependable is the estimate?" Obviously, an important factor in determining dependability is the closeness of the relationship between the variables. Let's look at Figure 14-8 and assume that both scatter diagrams have the same scales for the variables and the same regression line. When the points in a scatter diagram are closely spaced around the regression line, as they are in Figure 14-8a, it's logical to assume that an estimate based on that relationship will probably be more dependable than an estimate based on a regression line such as that shown in Figure 14-8b, where the scatter is much greater. Therefore, if we had a measure of the extent of the spread or scatter of the points around the regression line, we would be in a better position to judge the dependability of estimates made using the line. (You just know we are leading up to something here, don't you?)

You'll not be surprised to learn that we *do have a measure* that indicates the

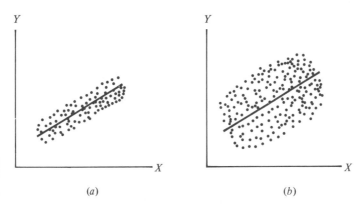

FIGURE 14-8
Varying degrees of spread or scatter.

(a)   (b)

extent of the spread, scatter, or dispersion of the points about the regression line. From an estimating standpoint, *the smaller this measure is, the more dependable the prediction is likely to be*. (The numerical value of this measure for Figure 14-8a is smaller than the value for Figure 14-8b because the dispersion is smaller in Figure 14-8a.) The name of this measure is the **standard error of estimate** (the symbol is $S_{y \cdot x}$), and, as the name implies, it's used to qualify the estimate made with the regression equation by indicating the extent of the possible variation (or error) that may be present.

Computation of $S_{y \cdot x}$

In more precise terms, the standard error of estimate is a standard deviation that measures the scatter of the observed values around the regression line. Thus, one formula that may be used to compute the standard error of estimate naturally bears a striking resemblance to the formula used to compute the standard deviation for ungrouped data. The primary difference between the two formulas lies in the fact that the standard deviation is measured from the mean, while the standard error of estimate is measured from the regression line. Both the mean and the regression line, of course, indicate central tendency. This formula for the standard error of estimate is:

$$S_{y \cdot x} = \sqrt{\frac{\Sigma(Y - Y_c)^2}{n - 2}}$$
(formula 14-4)

The $n - 2$ in the denominator is used in this case because it's assumed that we're dealing with sample data. We'll omit a detailed explanation of degrees of freedom, and why $n - 2$ is appropriate.

In using formula 14-4, we must compute a value for Y_c for each value of X by plugging each X value into the regression equation. We must then compute the difference between these Y_c values and the corresponding observed values of Y. A work sheet for calculating the standard error of estimate for our Tackey Toy

FIGURE 14-9

EMPLOYEE	OUTPUT (Y)	TEST RESULTS (X)	Y_c	$(Y - Y_c)$	$(Y - Y_c)^2$
A	30	6	31.86	−1.86	3.4596
B	49	9	47.28	1.72	2.9584
C	18	3	16.44	1.56	2.4336
D	42	8	42.14	−0.14	0.0196
E	39	7	37.00	2.00	4.0000
F	25	5	26.72	−1.72	2.9584
G	41	8	42.14	−1.14	1.2996
H	52	10	52.42	− .42	.1764
	296*		296.00*	0.0	17.3056

* Note that the sum of Y and Y_c are equal. This must always be true if $\Sigma(Y - Y_c) = 0$.

example is presented in Figure 14-9. The standard error of estimate is calculated as follows:

$$S_{y \cdot x} = \sqrt{\frac{\Sigma(Y - Y_c)^2}{n - 2}}$$

$$= \sqrt{\frac{17.3056}{6}}$$

$$= \sqrt{2.884}$$

$= 1.698$ dozens of units of output (the value of $S_{y \cdot x}$ will

always be expressed in the units of the Y variable)

Although it's helpful to use formula 14-4 to explain the nature of the standard error of estimate, a much easier formula to apply is:

$$S_{y \cdot x} = \sqrt{\frac{\Sigma(Y^2) - a(\Sigma Y) - b(\Sigma XY)}{n - 2}} \qquad \text{(formula 14-5)}$$

As you'll notice, all the values needed for this formula are available from Figure 14-6, which was used to prepare the regression equation. Using the values from Figure 14-6:

$$S_{y \cdot x} = \sqrt{\frac{(11,920) - 1.0277(296) - 5.1389(2,257)}{6}}$$

$$= \sqrt{\frac{17.304}{6}}$$

$$= \sqrt{2.884}$$

$= 1.698$ dozens of units of output

RELATIONSHIP TESTS AND PREDICTION INTERVALS

Hiram Hess now has a regression equation that gives him a central line through his sample data set. He also has a standard error of estimate that measures the scatter of the observed values of Y about the regression line. It appears that there's a close relationship between aptitude test results and employee output. But appearances can sometimes be deceiving.

Does a True Relationship Exist?

What if each point in the scatter diagram in Figure 14-10a shows how a Tackey production employee scored on the aptitude test and then performed later on the job? What if Figure 14-10a represents the entire *population* of Tackey production employees? If that were true, there's obviously no relationship between X and Y, and the slope of the population regression line (which we'll represent with the symbol B) has a value of zero. Perhaps Hiram was unfortunate enough to draw the misleading sample shown in Figure 14-10b. The positive slope of Hiram's sample regression equation indicates a relationship, but a true relationship doesn't exist if B is zero. What can Hiram do about this dilemma? He can conduct a hypothesis test using many of the same concepts we examined in Chapters 8 or 10.

A *t* Test for Slope

When statistical inferences are made using simple regression analysis techniques, there are several underlying assumptions that must apply. These assumptions are:

1   We have a population with a linear relationship between X and Y and fixed (but unknown) values of the *population* Y intercept and slope (A and B). The values of a and b calculated from sample observations selected from the population are estimates of these population values A and B.
2   For each possible value of X there's a distribution of Y values in the population scatter diagram that is normally scattered in a thin vertical slice about the regression line (see Figure 14-11).
3   Each of these distributions of Y values has the same standard deviation. Statisticians call this condition **homoscedasticity,** where *homo* means "same" and *scedastic* means "scatter." That's a terrible name, but there it is.
4   Each Y value in these distributions is independent of the others.

Let's look now at a procedure to test the value of B.

You'll recall from Chapter 8 that the first step in a hypothesis test is to formulate the null and alternative hypotheses. Hiram's interest now is to see if a true relationship exists between the X and Y variables. If no relationship exists be-

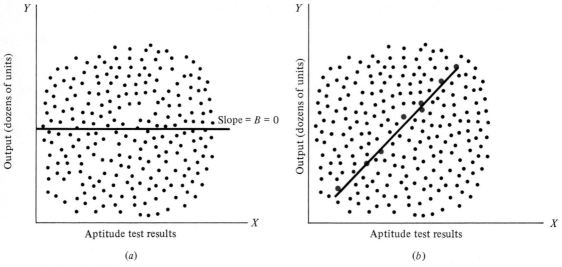

Slope = $B = 0$

Output (dozens of units)

Aptitude test results

(a)

Output (dozens of units)

Aptitude test results

(b)

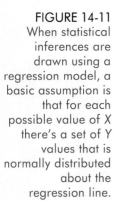

FIGURE 14-10
(a) What if this is the scatter diagram for all Tackey production workers? (b) Hiram's sample could suggest a relationship that doesn't exist in the population.

tween test results and output, the value of B (the slope of the population regression line) is zero. Thus, the null hypothesis to be tested is:

H_0: $B = 0$

And the alternative hypothesis is:

H_1: $B \neq 0$

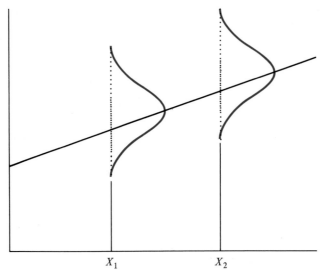

FIGURE 14-11
When statistical inferences are drawn using a regression model, a basic assumption is that for each possible value of X there's a set of Y values that is normally distributed about the regression line.

X_1 X_2

Hiram elects to conduct his test at the .05 level of significance. With the assumptions we've outlined, and with a small sample, it's appropriate to use the t table in Appendix 4. The boundaries of the rejection regions are determined by the t value corresponding to $t_{\alpha/2}$ and $n - 2$ degrees of freedom (df). In this case, the t value in the .025 column and the $8 - 2$ or 6 df row is 2.447.

The *decision rule format* is:

Accept H_0 if the standardized difference between the sample slope (b) and the hypothetical population slope (B_{H0}) falls into the acceptance region—that is, accept H_0 if the CR falls between ± 2.447.

or

Reject H_0 if the standardized difference between b and B_{H0} falls into the rejection region (beyond a t value of ± 2.447).

The next step is to make the necessary computations. The critical ratio for a t test for slope is found with this formula:

$$\text{CR} = \frac{b - B_{H0}}{s_b} \qquad \text{(formula 14-6)}$$

where b = slope of the sample regression equation
$\quad B_{H0}$ = slope of the population regression line *if* no relationship exists between the X and Y variables
$\quad s_b$ = an estimated standard error value—that is, an estimate of the standard deviation of the sampling distribution of all b values that could be calculated if repeated samples of size n were taken under controlled conditions

The following formula can be used to calculate s_b:

$$s_b = \frac{S_{y \cdot x}}{\sqrt{\Sigma(X^2) - (\Sigma X)^2/n}} \qquad \text{(formula 14-7)}$$

where n = the number of paired observations in the sample.

To compute the critical ratio for our example, we first calculate s_b:

$$s_b = \frac{S_{y \cdot x}}{\sqrt{\Sigma(X^2) - (\Sigma X)^2/n}} = \frac{1.698}{\sqrt{428 - (56)^2/8}}$$
$$= \frac{1.698}{6}$$
$$= .283$$

Then the critical ratio is:

$$\text{CR} = \frac{b - B_{H0}}{s_b} = \frac{5.1389 - 0}{.283} = 18.159$$

Conclusion: Since the CR of 18.159 is much greater than the t value of 2.447, we reject the null hypothesis that $B = 0$. This means that we accept the alternative hypothesis that there is a slope to the population regression line, and that a meaningful regression relationship *does exist* between Hiram's aptitude test scores and employee output. To Hiram's relief, the possibility described in Figure 14-10*a* apparently doesn't apply in this case. We can support this conclusion in another way by calculating a 95 percent confidence interval for the value of B as follows:

$$b - t(s_b) < B < b + t(s_b)$$
$$5.1389 - 2.447(.283) < B < 5.1389 + 2.447(.283)$$
$$5.1389 - .6925 < B < 5.1389 + .6925$$
$$4.4464 < B < 5.8314$$

Thus, we're 95 percent confident that the true population slope isn't zero, but rather has a value between 4.4464 and 5.8314.

Deliverance from Detail

You can now compute a regression equation and standard error of estimate, and you can use values of b and $S_{y \cdot x}$ to conduct a t test to see if a true relationship is likely to exist between the X and Y variables. You've dealt with many details in the last few pages, so you'll be pleased to know that many of these details can easily be turned over to a statistical software package. We've supplied two software packages with the data in our example problem—that is, with the output units produced and the aptitude test result figures for the 8 Tackey Toy employees. The *Mystat* program used that input to produce the output shown in Figure 14-12*a*, and the *Minitab* package generated the output shown in Figure 14-12*b*.

The preprogrammed output formats of the two packages vary. (We haven't yet considered some of the measures they've produced, although we will later.) But we've highlighted the following familiar values that we've computed in previous pages (minor variations are due to rounding):

$a = 1.0277$ or   1.028
$b = 5.1389$ or   5.139
$S_{y \cdot x} = 1.698$
$s_b = .283$
CR for t test (T value or t-ratio) $= 18.159$ or   18.16

The P (2 TAIL) figure of 0.000 beside the T value of 18.155 in Figure 14-12*a* allows us to reach a t-test conclusion. If the P value is *less than* the level of significance (.05 in our t test), we reject the H_0 that $B = 0$. The same P value of 0.000 beside the t-ratio value of 18.16 in Figure 14-12*b* leads to the same

```
DEP VAR:  OUTPUT      N:   8    MULTIPLE R:  .991    SQUARED MULTIPLE R:   .982
ADJUSTED SQUARED MULTIPLE R:  .979      STANDARD ERROR OF ESTIMATE:          1.698

    VARIABLE    COEFFICIENT    STD ERROR    STD COEF  TOLERANCE    T      P(2 TAIL)
    CONSTANT        1.028        2.070        0.000    .            0.496   0.637
    TSTRSLTS        5.139        0.283        0.991    .100E+01    18.155   0.000


                        ANALYSIS OF VARIANCE

    SOURCE    SUM-OF-SQUARES    DF    MEAN-SQUARE    F-RATIO      P

    REGRESSION      950.694     1       950.694      329.615    0.000
    RESIDUAL         17.306     6         2.884
```

(a)

```
MTB > REGRESS 'OUTPUT' ON 1 PREDICTOR 'TSTRSLTS';
SUBC> PREDICT 4.

The regression equation is
OUTPUT = 1.03 + 5.14 TSTRSLTS

Predictor       Coef        Stdev     t-ratio        p
Constant        1.028       2.070        0.50      0.637
TSTRSLTS        5.1389      0.2831      18.16      0.000

s = 1.698       R-sq = 98.2%      R-sq(adj) = 97.9%

Analysis of Variance

SOURCE          DF          SS          MS          F          p
Regression       1       950.69      950.69     329.61      0.000
Error            6        17.31        2.88
Total            7       968.00



    Fit  Stdev.Fit         95% C.I.              95% P.I.
 21.583      1.040   ( 19.038, 24.129)    ( 16.709, 26.458)
```

(b)

FIGURE 14-12

(a) *Mystat* output when supplied with Tackey Toy data.
(b) *Minitab* output when supplied with Tackey Toy data.

decision. The probability is 0.000 that Hiram's sample comes from a population with a slope of zero.

An Analysis of Variance Test

You'll notice that Figures 14-12*a* and *b* also have sections labeled "Analysis of Variance." Another way to test for the presence of slope in the population regression line is to use the analysis of variance (ANOVA) concepts presented in Chapter 10. An ANOVA test gives the *same results* as the *t* test we've just examined when the issue is to determine if a true relationship exists between the *X* and *Y* variables in a simple linear regression situation. If that's the case, then you may question why we bother with another testing procedure. The answer is that the ANOVA procedure, unlike the t test, can also be used in multiple regression situations where there's more than one independent variable to consider.

The ANOVA table format in Figures 14-12a and b follows the one described in Figure 10-7, page 397. We'll see how the SUM-OF-SQUARES (SS) and MEAN-SQUARE (MS) values are found in a few pages. For now, let's just look at the essence of the ANOVA test. This test also begins with a statement of the null and alternative hypotheses. Without going into unnecessary details, the ANOVA hypotheses are equivalent to those of the t test:

H_0: There's no relationship between X and Y.
H_1: There *is* such a relationship.

You'll recall that Hiram has specified a .05 level of significance. After the computer has calculated the SS and MS values shown in Figures 14-12a and b, it uses these figures to compute a critical F-ratio (CR_F)—the value of 329.61 shown in our computer printouts. The *decision rule* in an ANOVA test is to reject the H_0 if this computed F-ratio figure is *greater* than an appropriate table F value found in Appendix 6 at the back of the book. To find this table value of F, you must know the degrees of freedom for both the numerator and denominator of the computed F-ratio. These df figures in our example problem are:

$$df_{\text{num}} = m = 1$$
$$df_{\text{den}} = (n - m - 1) = 6$$

where m = the number of independent variables; always 1 in simple linear regression, but 2 or more in multiple regression situations
n = number of paired observations

Thus, the table F value ($F_{1,6,.05}$) is 5.99.

Since the CR_F value of 329.61 is far greater than the table F value of 5.99, the H_0 is rejected. The probability that Hiram's sample could come from a population that had a slope of zero is 0.000, just as it was in the t test.

Interval Estimates for Predictions—Large Samples

Just as there's a similarity between the computation of the standard deviation and the computation of the standard error of estimate, so too is there a similarity in their interpretation. We know that the standard deviation is a measure of spread or dispersion about the mean, and the standard error of estimate is a measure of scatter or dispersion about the regression line. And we've seen in many places (perhaps in too many places!) that in a normal distribution (1) the middle 68.3 percent of the distribution values lie within a range of 1 standard deviation above and below the mean, (2) the middle 95.4 percent of the values lie within ±2.00 standard deviations, and (3) the middle 99.7 percent of the values lie within ±3.00 standard deviations.

Let's assume now that we have a *large* sample (the number of paired observations is greater than 30), and the observed Y values are normally distributed about the regression line. Then approximately 68 percent of the points in the scatter diagram will fall within a range of 1 standard error of estimate above and below the regression line (see Figure 14-13). This range of $\pm 1 S_{y \cdot x}$ is represented by the *nearest* dashed lines to either side of the regression line in Figure 14-13; the range of $\pm 3 S_{y \cdot x}$ in Figure 14-13 accounts for virtually all the points.

The normal curves in Figure 14-13 have been added to help you better understand the relationship of the standard error of estimate to the regression line. If, in Figure 14-13, a value of X_1 is plugged into the regression equation, an estimate of Y_1 is obtained $(Y_1 = a + bX_1)$. It's important to note, however, that *the value of Y_1 is a point estimate*, and considering the scatter about the regression line in Figure 14-13, it's unlikely that this estimate (Y_1) will be exact. The dependability of this point estimate depends on the size of the standard error of estimate—as we've already seen, the smaller the standard error, the closer the point estimate is likely to be to the ultimate value of the dependent variable. With a knowledge of the standard error of estimate, however, we are not limited to the use of a point estimate. Rather, if the Y values are approximately normally distributed about the regression line, an interval estimate may be computed, and a probability may be assigned to this interval estimate. Thus, with large samples, an interval estimate can be computed using this form:

$$Y_c \pm Z(S_{y \cdot x}) \qquad \text{(formula 14-8)}$$

Or, as shown in Figure 14-13, the interval estimate with a predictive probability of 95.4 percent is $Y_1 \pm 2(S_{y \cdot x})$.

Interval Estimates for Predictions—Small Samples

Let's see now how the above concepts apply to our Tackey Toy example problem. A few pages earlier in this chapter, in the section entitled "Making Predictions

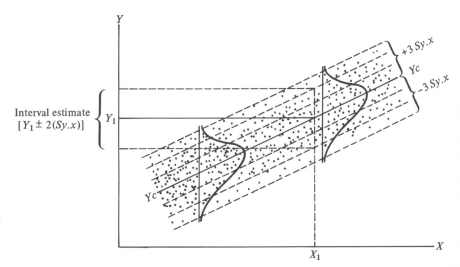

FIGURE 14-13
Interpretation and application of the standard error of the estimate with large samples.

with the Regression Equation," the question was raised of the future output of those job applicants who scored a 4 on the aptitude test. A point estimate of the average amount of output produced by those who score a 4 on the test was prepared then. Now, though, it's time to look at a small-sample *interval estimate* to which we can assign probabilities. There are actually two types of intervals that we can consider.

The *first* type of interval—one that would likely be of most interest to a personnel manager such as Hiram—is a range of the *average output* produced by *all* applicants who score a particular value (say 4) on the aptitude test. A *second* type of interval—of greater interest, perhaps, to a production supervisor— is a range of the output produced by a *specific person* who scores a 4 on the aptitude test. Since predicting the interval for a specific person is subject to more sampling variation than predicting the range for the average value of Y, the second type of interval is larger for a given level of confidence than the first.

Predicting the Range for the Average Value of Y Given X The interval estimate for small samples in this situation is found with this formula:

$$
Y_c \pm t_{\alpha/2} \left[S_{y \cdot x} \sqrt{\frac{1}{n} + \frac{(X_g - \bar{X})^2}{\Sigma(X^2) - (\Sigma X)^2/n}} \right] \qquad \text{(formula 14-9)}
$$

where Y_c = the value computed with the regression equation when a specific value of X is given

$t_{\alpha/2}$ = the appropriate t-table value for a given confidence level

X_g = the given value of X (note from the formula that the closer this value is to the $\bar{X}$, the smaller the value under the square root sign will be, and the smaller the confidence interval will be)

n = the number of paired observations in the sample

A 95 percent confidence interval for the average employee output if the aptitude test score is 4 may now be computed. We first find the Y_c value when X is 4. This was done earlier, and is:

$$Y_c = 1.0277 + 5.1389(4) = 21.583 \text{ dozen units of output}$$

Next, we determine the appropriate value from Appendix 4 for t with $n - 2$ or 6 degrees of freedom. This value is 2.447. The appropriate measure of dispersion for small samples adjusts the value of $S_{y \cdot x}$ and is called the **standard error of Y_c**. This measure is found below (the values of $\bar{X}$, $\Sigma(X^2)$, and ΣX come from Figure 14-6):

$$
s_{y \cdot x} \sqrt{\frac{1}{n} + \frac{(X_g - \bar{X})^2}{\Sigma(X^2) - (\Sigma X)^2/n}} = 1.698 \sqrt{\frac{1}{8} + \frac{(4 - 7)^2}{428 - (56)^2/8}}
$$

$$
= 1.698 \sqrt{\frac{1}{8} + \frac{9}{36}} = 1.698(.6124) = 1.04
$$

The 95 percent confidence interval is then:

$21.583 \pm 2.447 \, (1.04) = 21.583 \pm 2.545$
$19.038 < Y_c < 24.128$

Thus, we can be 95 percent confident that the average output of workers who score a 4 on the aptitude test is between 19.038 and 24.128 dozen units. Refer back now to Figure 14-12b. The PREDICT 4 subcommand in the second line of the figure specifies that a test score of 4 be used to predict confidence intervals for output. The last line in Figure 14-12b gives the 95 percent confidence interval we've just calculated. That line also gives the 95 percent prediction interval that we'll look at next.

Predicting the Range for a Specific Value of Y Given X   Suppose Hiram wanted to predict the output range of a specific applicant who scored a 4 on the aptitude test. In that case, the confidence interval is found with this formula:

$$(a + bX) \pm t_{\alpha/2} \left[S_{y \cdot x} \sqrt{1 + \frac{1}{n} + \frac{(X_g - \bar{X})^2}{\Sigma(X^2) - (\Sigma X)^2/n}} \right] \qquad \text{(formula 14-10)}$$

As you can see, the only difference is the addition of the single value, 1, under the square root sign. The 95 percent confidence interval for the output of a specific person who scored a 4 on the aptitude test—that is, that person's Y value—is:

$$21.583 \pm 2.447 \left(1.698 \sqrt{1 + \frac{1}{8} + \frac{(4 - 7)^2}{428 - (56)^2/8}} \right)$$

$$21.583 \pm 2.447 \left(1.698 \sqrt{1 + \frac{1}{8} + \frac{9}{36}} \right)$$

$21.583 \pm 2.447 \, (1.698)(1.1726)$
21.583 ± 4.872
$16.71 < Y < 26.45$

As already noted, this interval is larger than the one for the average value of Y given X because of the greater sampling variation possible. You can see in the last line of the printout in Figure 14-12b that the 95 percent P.I. (prediction interval) is the one we've just completed.

Self-Testing Review 14-2

Use the data in Self-Testing Review 14-1 to answer the following questions.

1 What's the value of $S_{y \cdot x}$ for that set of paired observations?
2 Does a true relationship exist between the X and Y variables? Conduct a t test for slope at the .05 level of significance to answer this question.

3   What's the 95 percent confidence interval for the *average value* of Y given a value of 5 for X?

4   What's the 95 percent confidence interval for a *specific value* of Y given a value of 5 for X?

5   a   A statistical software package has produced the following output when supplied with the data used in this Self-Testing Review section. Compare the results you've computed in parts 1 and 2 above with the results shown on the printout. Do they agree?

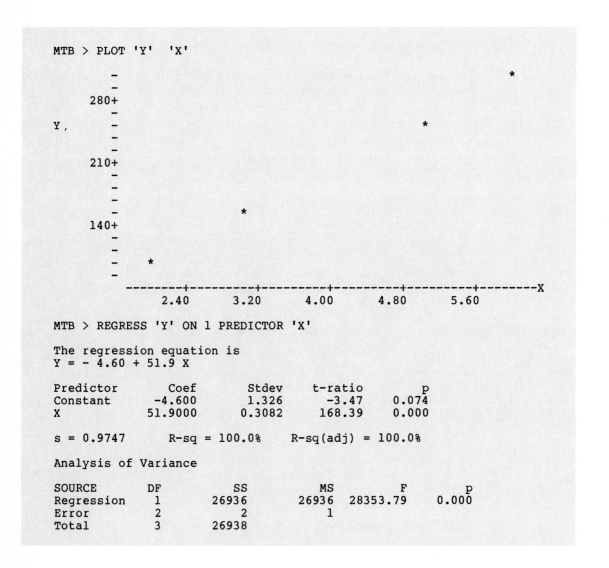

```
MTB > PLOT 'Y'  'X'

            -                                                                    *
            -
      280+
            -
 Y,         -                                             *
            -
            -
      210+
            -
            -
            -
            -                           *
      140+
            -
            -
            -         *
            -
          --------+---------+---------+---------+---------+--------X
             2.40      3.20      4.00      4.80      5.60

MTB > REGRESS 'Y' ON 1 PREDICTOR 'X'

The regression equation is
Y = - 4.60 + 51.9 X

Predictor        Coef       Stdev      t-ratio          p
Constant       -4.600       1.326       -3.47      0.074
X             51.9000      0.3082      168.39      0.000

s = 0.9747       R-sq = 100.0%     R-sq(adj) = 100.0%

Analysis of Variance

SOURCE        DF          SS          MS          F          p
Regression     1       26936       26936   28353.79      0.000
Error          2           2           1
Total          3       26938
```

b   In an ANOVA test to determine if there's a true relationship between the X and Y variables, the critical F-ratio value of 28353.79 shown in the printout is compared to an F-table value. At the .05 level, what is the appropriate F-table value to use in this situation?

CORRELATION ANALYSIS

In addition to regression and dispersion equations that allow us to measure and test relationships between variables, we also need measures that will indicate the closeness of the association or correlation that exists between the variables. In this section we'll briefly examine two of these correlation measures: the coefficient of determination and the coefficient of correlation.

The Coefficient of Determination

Before we define the coefficient of determination (its symbol is r^2), it's desirable to consider the several terms and concepts illustrated in Figure 14-14. If the mean of the Y variable $(\overline{Y})$ alone had been used to estimate the dependent variable, we would, of course, expect to find quite a bit of possible deviation between our estimate and the value of Y. A single point in Figure 14-14—let's label it Y^*—has been used to indicate the considerable *total deviation* that exists in this case between Y^* and its mean of $\overline{Y}$. But when the regression line is used as the basis for estimating the dependent variable, we can expect to have a closer estimate of Y. As indicated in Figure 14-14, our regression line is indeed closer to most of the points in the scatter diagram. Thus, in the case of the single point (Y^*) in Figure 14-14, the regression line explains or accounts for part of the deviation between Y^* and $\overline{Y}$—that is, the explained deviation is $Y_c - \overline{Y}$. Unfortunately, the regression line doesn't account for all the deviation, since the distance between Y^* and Y_c is still unexplained.

In other words, we have the following situation for point Y^* in Figure 14-14:

$Y - \overline{Y}$	$=$	$(Y_c - \overline{Y})$	$+$	$(Y - Y_c)$
Total deviation		explained deviation		unexplained deviation

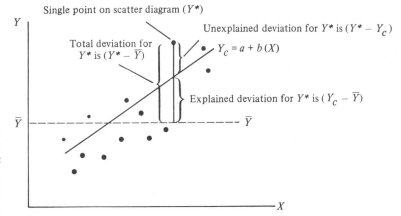

FIGURE 14-14
Conceptual representation of total, explained, and unexplained deviations.

And if we consider the *total variation* in an entire scatter diagram (and not just the deviation from a single point), we have the following situation:

$$\Sigma(Y - \overline{Y})^2 \qquad = \qquad \Sigma(Y_c - \overline{Y})^2 \qquad + \qquad \Sigma(Y - Y_c)^2$$

Total variation   explained variation   unexplained variation

It's customary in correlation analysis to use some convenient abbreviations. Since the **total variation** is the sum of the squared deviations of the Y values about their mean, the reference that's used is sum of squares (total), or **SST** in abbreviated form. Likewise, the **explained variation** is the sum of the squared deviations of the regression line values about $\overline{Y}$, and the reference that's used is sum of squares (explained by) regression, or **SSR**. Finally, the **unexplained variation** is the sum of the squared deviations of the Y values about the regression line, the reference that's used is sum of squares due to unexplained error, and the abbreviation is **SSE.**

Now that we've dazzled you with all these thoughts, we can define r^2. The **coefficient of determination** is a measure of the portion of the total variance in the Y variable that's explained or accounted for by the introduction of the X variable (and thus the regression line). That is:

$$r^2 = \frac{\text{explained variation}}{\text{total variation}} = \frac{\text{SSR}}{\text{SST}} = \frac{\Sigma(Y_c - \overline{Y})^2}{\Sigma(Y - \overline{Y})^2} \qquad \text{(formula 14-11)}$$

The calculations needed to produce r^2 with formula 14-11 are shown in Figure 14-15. As you can see, the value of SSR is 950.683, and SST is 968. If SST = SSR + SSE, then the difference between SST and SSR gives the value of SSE. We've seen in Figure 14-9 that SSE—the $\Sigma(Y - Y_c)^2$ value—is 17.3056. And we see now that SST − SSR is 17.317 (the slight discrepancy is due to rounding differences).

Calculating r^2 with formula 14-11 is tedious work. But there's a more convenient formula for r^2 that uses values we've already computed in earlier pages:

$$r^2 = \frac{a(\Sigma Y) + b(\Sigma XY) - n(\overline{Y})^2}{\Sigma(Y^2) - n(\overline{Y})^2} \qquad \text{(formula 14-12)}$$

To compute the coefficient of determination for our Tackey Toy problem with formula 14-12, we need only refer to the regression equation and Figure 14-6 to get the necessary data. Thus,

$$r^2 = \frac{a(\Sigma Y) + b(\Sigma XY) - n(\overline{Y})^2}{\Sigma(Y^2) - n(\overline{Y})^2}$$

$$= \frac{1.0277(296) + 5.1389(2,257) - 8(37)^2}{11,920 - 8(37)^2} = \frac{304.1992 + 11,598.497 - 10,952}{11,920 - 10,952}$$

$$= \frac{950.696}{968}$$

$$= .982$$

FIGURE 14-15

Y	Y_c	$(Y_c - \bar{Y})$	$(Y_c - \bar{Y})^2$	$(Y - \bar{Y})$	$(Y - \bar{Y})^2$
30	31.861	−5.139	26.409	−7	49
49	47.278	10.278	105.635	12	144
18	16.444	−20.555	422.533	−19	361
42	42.139	5.139	26.409	5	25
39	37.000	0	0	2	4
25	27.722	−10.278	105.635	−12	144
41	42.139	5.139	26.409	4	16
52	52.416	15.416	237.653	15	225
296	295.999	0	950.683	0	968

$$\bar{Y} = \frac{296}{8} = 37 \qquad r^2 = \frac{SSR}{SST} = \frac{\Sigma(Y_c - \bar{Y})^2}{\Sigma(Y - \bar{Y})^2} = \frac{950.683}{968} = .982$$

What does the r^2 value of .982 mean? Congratulations on yet another incisive question. It means that 98.2 percent of the variation in the Y variable is explained or accounted for by variation in the X variable. Or, in our example, we can conclude that 98.2 percent of the variation in output is explained by variation in test score results. Since it's obvious that the value of r^2 cannot exceed 1.00—after all, you can't explain more than 100 percent of the variation in Y!—the value of .982 is quite high. Such a high value is, of course, desirable for forecasting purposes because the higher the value of r^2, the smaller the value of $S_{y\cdot x}$. (Can you figure out why this is true?)

The Computer Printouts Revisited

Figures 14-16a and b repeat the *Mystat* and *Minitab* printouts for our Tackey Toy example that were shown in Figures 14-12a and b. The highlighted values in Figures 14-16a and b should now be clearer to you. The .982 (or 98.2 percent) value of r^2 is shown on both printouts, and the intermediate variation values are shown in the SS (sum of squares) columns. As you can see, the explained (SS Regression), unexplained (SS Error), and total (SS Total) variation figures correspond to the values we've just calculated.

You saw a few pages earlier that the ANOVA test gave the same results as the t test for slope—that is, it showed if there was likely to be a true relationship between the variables. We've seen that the null hypothesis in the t test is that no relationship exists—that is, $B = 0$. The null hypothesis in an ANOVA test is equivalent to this, but if you've read Chapter 10 you know that an ANOVA test of population means depends on two estimates of the population variance. A computed F ratio based on these two estimates is then used to test the null hypothesis.

When used with relationship tests in regression analysis, the computed F ratio depends on the explained (SSR) and unexplained (SSE) values that we've just examined. The computed critical F ratio (CR_F) formula is:

$$CR_F = \frac{SSR/m}{SSE/(n - m - 1)} \qquad \text{(formula 14-13)}$$

where m = the number of independent variables, always 1 in the simple linear case
n = number of paired observations

Thus, in Figures 14-16a and b, the CR_F value is:

$$CR_F = \frac{SSR/m}{SSE/(n - m - 1)} = \frac{950.69/1}{17.31/(8 - 1 - 1)} = \frac{950.69}{2.88} = 329.61$$

And we've seen earlier that this CR_F value is then compared to a table F value to arrive at a test decision.

```
DEP VAR:  OUTPUT       N:    8    MULTIPLE R:  .991    SQUARED MULTIPLE R:  .982
ADJUSTED SQUARED MULTIPLE R:  .979      STANDARD ERROR OF ESTIMATE:       1.698

    VARIABLE    COEFFICIENT   STD ERROR   STD COEF  TOLERANCE     T    P(2 TAIL)
CONSTANT           1.028       2.070       0.000     .         0.496    0.637
TSTRSLTS           5.139       0.283       0.991    .100E+01  18.155    0.000


                         ANALYSIS OF VARIANCE

    SOURCE    SUM-OF-SQUARES   DF   MEAN-SQUARE     F-RATIO       P

REGRESSION         950.694      1      950.694      329.615     0.000
  RESIDUAL          17.306      6        2.884
```

(a)

```
MTB > REGRESS 'OUTPUT' ON 1 PREDICTOR 'TSTRSLTS'

The regression equation is
OUTPUT = 1.03 + 5.14 TSTRSLTS

Predictor        Coef       Stdev     t-ratio        p
Constant        1.028       2.070        0.50      0.637
TSTRSLTS        5.1389      0.2831      18.16      0.000

s = 1.698       R-sq = 98.2%      R-sq(adj) = 97.9%

Analysis of Variance

SOURCE        DF          SS          MS          F          p
Regression     1      950.69      950.69      329.61      0.000
Error          6       17.31        2.88
Total          7      968.00
```

(b)

FIGURE 14-16   (a) The *Mystat* printout again. (b) The *Minitab* printout again.

The Coefficient of Correlation

Figure 14-16a also has a measure identified by the letter R. The **coefficient of correlation** (r) is simply the square root of the coefficient of determination (r^2). Thus, for our Tackey Toy problem, the coefficient of correlation is .991:

$$r = \sqrt{r^2} \sqrt{.982} = .991$$

The *Mystat* software package will also calculate r separately, as shown in Figure 14-17.

The coefficient of correlation isn't as useful as the coefficient of determination, since it's an abstract decimal and isn't subject to precise interpretation. (As the square root of a percentage, it cannot itself be interpreted in percentage terms.) But r does provide a scale against which the closeness of the relationship between X and Y can be measured. In other words, the value of r is on a scale between zero and ± 1.00. When r is zero, there is *no* correlation, and when $r = +1.00$ or -1.00, there is *perfect* correlation. (The algebraic sign of r is always the same as that of b in the regression equation.) Thus, the closer r is to its limit of ± 1.00, the better the correlation, and the smaller the value of r, the poorer the relationship between the variables.

A Graphical Summary

Let's now graphically summarize, by means of scatter diagrams, some of the relationships discussed in this chapter:

1   In Figure 14-18a, we have an example of *perfect positive correlation,* with all points in the diagram falling on the regression line. Therefore, $r = +1.00$, $r^2 = 1.00$, and $S_{y \cdot x} = 0$ because there is an absence of spread or scatter about the line.
2   In Figures 14-18b, we have an example of *perfect negative correlation.* The values of various measures are as indicated.
3   In Figures 14-18c and d, there is positive correlation, but the values of r and r^2 are less in d than in c. Assuming the same scales for X and Y and the same regression line, the value of $S_{y \cdot x}$ is greater in d than in c.

```
           PEARSON CORRELATION MATRIX

                           OUTPUT      TSTRSLTS

              OUTPUT         1.000
              TSTRSLTS       0.991          1.000

           NUMBER OF OBSERVATIONS:    8
```

FIGURE 14-17
Mystat calculation of r.

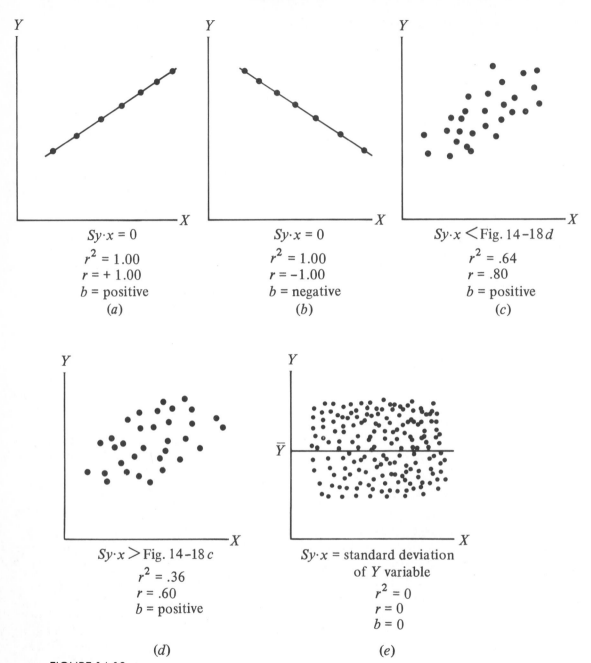

$Sy \cdot x = 0$
$r^2 = 1.00$
$r = +1.00$
b = positive
(a)

$Sy \cdot x = 0$
$r^2 = 1.00$
$r = -1.00$
b = negative
(b)

$Sy \cdot x <$ Fig. $14-18d$
$r^2 = .64$
$r = .80$
b = positive
(c)

$Sy \cdot x >$ Fig. $14-18c$
$r^2 = .36$
$r = .60$
b = positive

(d)

$Sy \cdot x$ = standard deviation
of Y variable
$r^2 = 0$
$r = 0$
$b = 0$

(e)

FIGURE 14-18
A graphical
summary.

4   In Figure 14-18e, there's no correlation. The regression line is simply a horizontal line drawn at $\bar{Y}$ with no slope—that is, $b = 0$. Both r and r^2 are zero, and $S_{y \cdot x}$ equals the standard deviation of the Y variable (this is the upper limit for $S_{y \cdot x}$).

Self-Testing Review 14-3

Using the answers obtained in Self-Testing Reviews 14-1 and 14-2, perform the following:

1 Compute the explained variation.
2 Compute the total variation.
3 Compute the coefficient of determination.
4 Interpret the coefficient of determination.
5 Compute the coefficient of correlation.

COMMON ERRORS AND LIMITATIONS

As is true with all statistical methods, regression and correlation analysis is subject to misuses and misinterpretations. *Some of the more common mistakes are briefly summarized below.*

1 *Correlation analysis is sometimes used to prove the existence of a cause-and-effect relationship.* The coefficient of determination tells us nothing about the type of relationship between the two variables. Rather, it indicates the proportion of variation that is explained *if* there is a causal relationship.

2 *The coefficient of correlation is sometimes interpreted as a percentage.* This can be a serious mistake. For example, if a coefficient of correlation of .7 is interpreted as meaning that 70 percent of the variation in Y is explained, this is significantly above the 49 percent that actually is explained.

3 *The coefficient of determination is also subject to misinterpretation.* It is sometimes interpreted as the percentage of the variation in the dependent variable *caused* by the independent variable. This is simply nonsense. It should always be remembered that it's the variation in the dependent variable that is being explained or accounted for (but not necessarily caused) by the X variable.

4 *When making estimates from the regression equation, it's incorrect to make estimates beyond the range of the original observations.* There's no way of knowing what the nature of the regression equation would be if we encountered values of the dependent variable larger or smaller than those we've observed. For example, in the Tackey Toy illustration used in this chapter it would be ridiculous to assume that an employee's output would increase indefinitely as his or her test results increased. There would obviously be some upper limit to aptitude test scores, and regardless of aptitude, the speed of productive equipment and physical endurance would also set a limit to productive output. Since the largest value of X we observed was 10, we could place no reliance on estimates we might make of the output of employees who might have test scores of 11 or 12.

5 *When time-series data are correlated, the coefficient of determination cannot be interpreted as precisely as we've indicated above.* This is because the pairs of observations are not independent, a fact which tends to inflate the value

of r^2. There are methods for dealing with this problem, but they're beyond the scope of this book.

LOOKING BACK

1   In regression analysis, an estimating equation is developed to describe the pattern or the relationship that exists between the variables. As the name implies, an analyst prepares an estimating or regression equation to make estimates of values of one variable from given values of one or more other variables. The variable to be estimated is called the dependent (Y) variable. The variable that presumably exerts an influence on or explains variations in the dependent variable is termed the independent or predictive variable, and is identified by the symbol X. In correlation analysis, the purpose is to measure the strength or closeness of the relationship between the variables. In short, the purpose of regression analysis is to determine the pattern of the existing relationship, and the purpose of correlation analysis is to determine the strength of the relationship described in the regression equation.

2   The fact that two variables are associated in a statistical sense does not guarantee the existence of a causal relationship. An early step in the study of regression and correlation is to see if a logical relationship may exist between the variables to support further analysis. After it has been determined that a logical relationship may exist, a scatter diagram is often prepared. Each point on this chart represents an item for which we have a value for both the Y and X variables. Such a chart can help determine if there's a useful relationship between the variables, and it can suggest the type of equation to use to describe the relationship. Our focus in this chapter has been on relationships that can be expressed with a straight line, but other possibilities exist.

3   The method of least squares is used to fit a simple linear regression line to the observed data. Once the values for a (the Y intercept) and b (the slope of the line) are computed, the regression equation may be used to make a point estimate of the value of the Y variable given a value for the X variable. The standard error of estimate is a standard deviation that's used to qualify the estimate made with the regression equation by indicating the extent of the possible variation that may be present. You've learned how to compute this measure in the chapter.

4   To determine if a true statistical relationship exists between the X and Y variables, a t test for slope may be made. The null hypothesis in this test is that the slope of the population regression line (B) is zero. If this hypothesis is true, then there's no meaningful regression relationship between the variables. The steps needed to conduct this test aren't particularly difficult, and they've been outlined in the chapter. But you've seen in the chapter that it's easy to obtain the regression equation, the standard error of estimate, the values needed to make a t test and an ANOVA test, and other useful information with the help of a statistical software package. The ANOVA test gives the same results as the t test for slope when only two variables are used, but the ANOVA procedure can also be used in multiple regression situations, as you'll see in the Closer Look reading at the end of this chapter.

5   A probability value can be assigned to an interval estimate of the Y variable that can be produced using the regression equation, a given value of the X variable, and other measures. In the case of large samples, the Z distribution and the standard error of estimate may be used to produce this estimate. For small samples, though, the t distribution is needed, and the standard error of estimate is adjusted to produce the appropriate measure of dispersion. Two types of interval estimates are described in the chapter—one that predicts the range for an average value of Y given X, and another that predicts the range for a specific value of Y given X. Both types of intervals are easily produced by statistical programs.

6   The coefficient of determination (r^2) measures the percentage of the variation in the Y variable that's explained or accounted for by variation in the X variable. The numerator of r^2 is the sum of squares explained by regression (SSR), and the denominator is the sum of the square deviations of the Y values about their mean—the SST amount. These values, along with the unexplained variation in the regression situation—abbreviated SSE—are produced by statistical programs as a part of the ANOVA test. Of course, the value of r^2 is also shown by these programs. The coefficient of correlation (r) is simply the square root of the coefficient of determination. The value of r is on a scale between zero and ± 1.00. When r is zero, there's no correlation, and when r is 1.00, there's perfect correlation.

7   Regression and correlation techniques have proved their worth in countless studies, but, like all statistical methods, they are subject to misuses and misinterpretations. Some of the common mistakes to guard against are: (a) don't use correlation analysis to "prove" the existence of a cause-and-effect relationship, (b) don't interpret the coefficient of correlation as a percentage, and (c) don't use the regression equation to make estimates beyond the range of the original observations.

KEY TERMS AND CONCEPTS

$Y_c \pm Z(S_{y \cdot x})$ (formula 14-8)   547

$$Y_c \pm t_{\alpha/2} \left[S_{y \cdot x} \sqrt{\frac{1}{n} + \frac{(X_g - \bar{X})^2}{\Sigma(X^2) - (\Sigma X)^2/n}} \right]$$

(formula 14-9)   548

Standard error of Y_c 548

$$(a + bX) \pm t_{\alpha/2} \left[S_{y \cdot x} \sqrt{1 + \frac{1}{n} + \frac{(X_g - \bar{X})^2}{\Sigma(X^2) - (\Sigma X)^2/n}} \right]$$

(formula 14-10)   549

Total variation (SST)   552

Explained variation (SSR)   552

Unexplained variation (SSE)   552

Coefficient of determination   552

$$r^2 = \frac{\text{explained variation}}{\text{total variation}} = \frac{\text{SSR}}{\text{SST}} = \frac{\Sigma(Y_c - \bar{Y})^2}{\Sigma(Y - \bar{Y})^2}$$

(formula 14-11)   552

$$r^2 = \frac{a(\Sigma Y) + b(\Sigma XY) - n(\bar{Y})^2}{\Sigma(Y^2) - n(\bar{Y})^2}$$ (formula 14-12)   552

$$CR_F = \frac{\text{SSR}/m}{\text{SSE}/(n - m - 1)}$$ (formula 14-13)   554

Coefficient of correlation (r) 555

PROBLEMS

1   A United States military organization wanted to know if there was a relationship between the number of recruiting offices located in particular cities and the total number of persons enlisting in those cities. Total enlistment data were obtained for 1 month from 10 cities and are given below:

CITY	TOTAL ENLISTMENT IN EACH CITY FOR 1 MONTH	NUMBER OF RECRUITING OFFICES IN EACH CITY
Austin	20	1
Pittsburgh	40	2
Chicago	60	4
Los Angeles	60	3
Denver	80	5
Atlanta	100	4
Cleveland	80	5
Louisville	50	2
New Orleans	110	5
Kansas City	30	1

a   Construct a scatter diagram.
b   Compute a least-squares regression equation, and fit it to the scatter diagram.
c   Compute the standard error of estimate.
d   Prepare a point estimate of Y_c for a city with 3 recruiting offices.
e   Conduct a t test for slope. Does a true relationship exist between the variables at the .05 level?
f   What's the 95 percent confidence interval for the value of the population slope (B) in this example?
g   At the 95 percent level of confidence, what's the *average* monthly enlistment in cities with 3 recruiting offices?
h   Compute the coefficients of determination and correlation, and explain their meaning.
i   What other factors might influence enlistment other than the number of recruiting offices located in a particular city?

2 Gus Gas operates Slochoke Service Stations on traffic arteries in Texas. Gus wants to know if there's any relationship between the total gallons of gasoline pumped for a period at each station and the average number of cars that traveled each day on the streets where his service stations are located. Gus obtained the following data from the Texas Highway Department and Slochoke Company records:

SERVICE STATION LOCATION	TOTAL GALLONS OF GASOLINE PUMPED (THOUSANDS)	AVERAGE DAILY TRAFFIC COUNT AT LOCATION (THOUSANDS)
E. Regular Street	100	3
S. Lead Street	112	4
N. Main Street	150	5
Highway 606	210	7
Baker Boulevard	60	2
E. High Street	85	3
Country Club Road	77	2

a Construct a scatter diagram.
b Compute a least-squares regression equation, and fit it to the scatter diagram.
c Compute the standard error of estimate.
d Prepare a point estimate of Y_c for a location with a daily traffic count of 6 (thousand) vehicles.
e Conduct a t test for slope. Does a true relationship exist between the variables at the .05 level?
f What's the *average* annual gasoline sales in locations with daily traffic counts of 6 (thousand) vehicles? Use the 95 percent confidence level.
g Compute the coefficient of determination, and explain its meaning.

3 The head nurse of Mortality Memorial Hospital is interested in determining if a relationship exists between the number of days a patient has been in the hospital and the number of times a patient requests a nurse during a 24-hour period. The head nurse recorded the following data:

PATIENT	NUMBER OF DAYS IN MORTALITY MEMORIAL	TOTAL NUMBER OF REQUESTS FOR NURSES
Mr. Chill Blains	2	2
Mr. Leow B. Count	4	3
Ms. Hedda Spell	5	3
Ms. Cara Reck	6	4
Mr. A. Penn Dix	3	2
Ms. Anne S. Teesha	8	6
Mr. I. M. Hurting	15	10
Mr. Hy P. Dermik	7	5
Ms. Ivy Pellagra	15	11
Ms. Ruth O'Brien	2	1

a Compute a least-squares regression equation.
b Compute the standard error of estimate.
c Conduct a t test for slope at the .05 level. Does a true relationship exist between X and Y?
d What's the 95 percent confidence interval for the value of B?

e   What's the range of the number of requests for a nurse during a 24-hour period by a *specific* patient who has been in the hospital 7 days? Use the 95 percent level of confidence.

f   Compute the coefficient of determination, and explain its meaning.

4   The Rip-Off Vending Machine Company operates coffee vending machines in office buildings. The company wants to study the relationship, if any, which exists between the number of cups of coffee sold per day and the number of persons working in each building. Data for this study were collected by the company and are presented below:

LOCATION OF COFFEE MACHINE	NUMBER OF CUPS OF COFFEE SOLD AT THIS LOCATION	NUMBER OF PERSONS WORKING AT THIS LOCATION
1	10	5
2	20	6
3	30	14
4	40	19
5	30	15
6	20	11
7	40	18
8	40	22
9	50	26
10	10	4

a   Compute a least-squares regression equation.

b   Compute the standard error of estimate.

c   Conduct a t test for slope at the .05 level. Is there a true relationship between coffee sales and building population?

d   Construct a 95 percent confidence interval for the *average* cups sold in buildings with 16 occupants.

e   Compute the coefficients of determination and correlation, and explain their meaning.

5   The Mary Jane Cosmetic Company found that the following regression equation applied to a random sample of 49 salespeople:

$$Y_c = 4 + 16(X)$$

where X = number of sales calls made by a salesperson
Y = dollar sales made by salesperson

a   What's the point estimate of the sales amount realized from two calls?

b   If the standard error of estimate is $4, and if the Y values are normally distributed about the regression line, what's the 99 percent confidence interval estimate of the sales amount realized in two calls?

6   Rancher Brown has conducted experiments on numerous tracts of his land. The following table gives the amount of fertilizer (in hundreds of pounds) applied to nine of these tracts, and the hay output (in tons) harvested in the first cutting following the fertilizer application:

TRACT	HAY OUTPUT (TONS)	FERTILIZER (HUNDREDS OF POUNDS)
1	2	1
2	3	2
3	4	4
4	7	8
5	6	6
6	5	5
7	6	6
8	7	9
9	6	7

a   What is the least-squares regression equation for this data set?
b   What is the standard error of estimate?
c   Conduct a t test for slope. Does a true relationship exist between the variables at the .05 level?
d   What's the range of hay output for a *specific* tract if 3 hundred pounds of fertilizer have been applied to that tract? Use the 95 percent level of confidence.
e   Compute the coefficient of determination, and explain its meaning.

7   A regional development council, composed of members from several local chambers of commerce, gathered the following data from selected firms in the region that produced similar products:

FIRM	PRODUCTION COST (THOUSANDS OF DOLLARS)	PRODUCTION OUTPUT (THOUSANDS OF UNITS)
A	150	40
B	140	38
C	160	48
D	170	56
E	150	62
F	162	75
G	180	70
H	165	90
I	190	110
J	185	120

a   Compute the least-squares regression equation.
b   What is the standard error of estimate?
c   Conduct a t test for slope. Does a true relationship exist between the variables at the .05 level?
d   At the 95 percent level of confidence, what's the range of the *average* production cost of a firm that produces 60 thousand units?
e   Compute the coefficient of determination, and explain its meaning.

8   The Pull & Wool Advertising Agency has noticed the following relationship between advertising and promotion costs (in thousands of dollars) and sales performance (in hundreds of thousands of dollars):

ACCOUNT	SALES PERFORMANCE (HUNDREDS OF THOUSANDS OF DOLLARS)	ADVERTISING AND PROMOTION COSTS (THOUSANDS OF DOLLARS)
1	6	45
2	7	80
3	9	70
4	9	85
5	7	60
6	8	55
7	6	75
8	12	90

a   Compute the least-squares regression equation for this data set.

b   What is the standard error of estimate?

c   Prepare a point estimate of Y_c for an account that spends \$65,000 for advertising and promotion.

d   At the 95 percent level of confidence, what's the range of the *specific* sales performance of an account when \$65,000 is spent for advertising and promotion?

e   Compute the coefficient of determination and explain its meaning.

f   Conduct a *t* test for slope at the .05 level. Does a true relationship exist between the variables?

9   Katie Dvorak trains employees to use a specific statistical software package. Selected trainees have turned in the following performances in recent weeks:

TRAINEE	ERRORS IN USING SOFTWARE	AMOUNT OF TRAINING (IN HOURS)
A	6	1
B	3	4
C	2	6
D	1	8
E	5	2
F	4	3
G	7	1

a   Compute a least-squares regression equation for this data set.

b   What is the standard error of estimate?

c   Prepare a point estimate of Y_c for a trainee who has received 5 hours of training.

d   Conduct a *t* test for slope. Does a true relationship exist between the variables at the .05 level?

e   At the 95 percent level of confidence, what's the range of the *average* number of software-using errors expected from trainees who have received 5 hours of instruction?

f   Compute the coefficients of determination and correlation and explain their meaning.

TOPICS FOR REVIEW AND DISCUSSION

1   What is the purpose of regression and correlation analysis?

2   What type of relationship exists between the following pairs of variables?

    a   Gasoline sales and the number of automobiles.

    b   Gasoline sales and motor oil sales.

    c   Gasoline sales and the number of books in the New York Public Library.

3   What is the purpose of the scatter diagram?

4   Discuss the use of the regression equation in forecasting.

5   How is the standard error of estimate used when making predictions from the regression equation?

6   What is meant by "simple linear regression analysis"?

7   What is the interpretation of the coefficient of determination?

8   Discuss the common errors made in interpreting the results of correlation analyses.

9   What is the relationship between the values of r^2 and $S_{y \cdot x}$?

10   "The standard error of estimate is similar in many respects to the standard deviation." Discuss this statement.

11   There are several underlying assumptions that must apply when statistical inferences are made using regression analysis techniques. What are these assumptions?

12   Discuss the t-test procedure that's used to evaluate the slope of the population regression line.

13   Discuss the ANOVA test procedure that's used to evaluate the relationship between X and Y.

14   How are large-sample interval estimates constructed to predict the value of Y given a value of X?

15   a   How does one predict the range for the average value of Y given X when data are available from a small sample?

    b   How does predicting the range for a specific value of Y given X differ from this procedure?

16   Explain the meaning of these abbreviations: SSR, SSE, and SST.

PROJECTS/ISSUES TO CONSIDER

1   It's possible here to carry out a project similar to those described in earlier chapters. Team up with other students and locate journal articles of interest to team members that use regression analysis techniques. Prepare a team presentation for the class that outlines:

    a   The nature of the problem tackled in the article.

    b   The way regression analysis techniques were used.

2   Working with the same team members, identify a Y variable that the team would like to predict, and locate another variable (X) that might logically be used to predict the Y variable. Use the measures and techniques discussed in this chapter to analyze these variables. Here's just one example: The dependent variable might be the final numerical grade average in a statistics course. The team could sample previous students in the course, determine their semester grade averages and SAT math scores, and then prepare a regression equation to try to predict statistics course grades of the team members by using their SAT math scores as the predicting variable. Whatever the topic, though, prepare a class presentation to describe the problem tackled, the steps the team took to gather and analyze the data, and the results obtained.

ANSWERS TO SELF-TESTING REVIEW QUESTIONS

14-1

Y	X	(XY)	(X²)	(Y²)
255	5	1,275	25	65,025
100	2	200	4	10,000
307	6	1,842	36	94,249
150	3	450	9	22,500
812	16	3,767	74	191,774
$\bar{Y} = 203$	$\bar{X} = 4$			

a   b $b = \dfrac{n(\Sigma XY) - (\Sigma X)(\Sigma Y)}{n(\Sigma X^2) - (\Sigma X)^2} = \dfrac{4(3,767) - (16)(812)}{4(74) - (16)^2} = \dfrac{2,076}{40} = 51.9$

   b $a = \bar{Y} - b(\bar{X}) = 203 - 51.9(4) = -4.6$

2   $Y_c = -4.6 + 51.9(4)$
     $= -4.6 + 207.6$
     $= 203$

3   $Y_c = -4.6 + 51.9(5)$
     $= -4.6 + 259.5$
     $= 254.9$

4   The Y_c value has changed by 51.9 with a change in X from 4 to 5. This change in Y_c that corresponds to a change of 1 unit in X is, of course, the definition of b, the slope of the linear regression line.

14-2

1   $S_{y \cdot x} = \sqrt{\dfrac{\Sigma(Y^2) - a(\Sigma Y) - b(\Sigma XY)}{n - 2}} = \sqrt{\dfrac{191,774 - (-4.6)(812) - 51.9(3767)}{4 - 2}}$

     $= \sqrt{\dfrac{1.9}{2}} = .9747$

2   $H_0 : B = 0$
     $H_1 : B \neq 0$
     $\alpha : .05; t_{.025} = 4.303.$
     Decision rule: Accept H_0 if CR falls between ± 4.303.

     $\text{CR} = \dfrac{b - B_{H0}}{s_b} = \dfrac{(51.9 - 0)}{.9747/\sqrt{74 - (16)^2/4}} = \dfrac{51.9}{.9747/\sqrt{10}}$

                          $= \dfrac{51.9}{.9747/3.1623} = \dfrac{51.9}{.3082} = 168.39$

     Conclusion: Reject H_0; the CR of 168.39 is much greater than 4.303. A meaningful relationship does exist between the X and Y variables.

3   The 95 percent confidence interval for the average value of Y given a value of 5 for X is:

     $Y_c \pm t_{\alpha/2}\left(S_{y \cdot x}\sqrt{\dfrac{1}{n} + \dfrac{(X_g - \bar{X})^2}{\Sigma(X^2) - (\Sigma X)^2/n)}}\right) = 254.9 \pm 4.303\left(.9747\sqrt{\dfrac{1}{4} + \dfrac{(5 - 4)^2}{74 - (16)^2/4}}\right)$

                          $= 254.9 \pm 4.303(.5766)$

                          $= 252.42 < Y < 257.38$

4 The 95 percent confidence interval for a specific value of Y given a value of 5 for X is:

$$254.9 \pm 4.303 \left(.9747 \sqrt{1 + \frac{1}{4} + \frac{(5-4)^2}{74 - (16)^2/4}} \right)$$

$$= 254.9 \pm 4.303\,(.9747\sqrt{1 + .25 + .1})$$

$$= 254.9 \pm 4.873$$

$$= 250.03 < Y < 259.77$$

5 a The computations shown in parts 1 and 2 above agree with the computer printout.

    b The F-table value requires df_{num} and df_{den}. In this example, df_{num} is 1 and df_{den} is 2. Thus, the table value of $F_{1,2,.05}$ is 18.5. Since 28353.79 is obviously greater than 18.5, the F-test conclusion is to reject H_0 and accept the alternative that a meaningful regression relationship does exist between the variables. (Not surprising when you look at the scatter diagram that the program has plotted!)

14-3

1 Explained variation = SSR = $a(\Sigma Y) + b(\Sigma XY) - n(\bar{Y})^2$

$$= -4.6(812) + 51.9(3,767) - 4(203)^2$$

$$= 26,936.1$$

2 Total variation = SST = $\Sigma(Y^2) - n(\bar{Y})^2 = 191,774 - 164,836$

$$= 26,938$$

3 $r^2 = \dfrac{26,936.1}{26,938.0} = .999$

4 We can say that 99.9 percent of the variation in Y is explained or accounted for by the variation in X.

5 $r = \sqrt{.999} = .999$

A CLOSER LOOK

MULTIPLE LINEAR REGRESSION AND CORRELATION

We saw in the chapter that Hiram was able to make predictions about employee output production because a single X variable—aptitude test scores—explained practically all of the variation in output. But prediction needs aren't always so easily met. Let's assume that Selam Hess, Hiram's sister and Sales Manager of Tackey Toys, needs to predict sales of Tackey products in selected market areas. Selam believes that advertising expenditures can be used to predict sales and has gathered sample toy sales and advertising cost data (shown in Figure CL14-1) for six market areas. A statistical software package has produced a simple linear regression equation for this data set (see Figure CL14-2). Selam notes that

FIGURE CL14-1   Tackey Toy Sales and Advertising Cost Data for a Sample of Six Market Areas

MARKET AREA	TOY SALES (THOUSANDS OF DOLLARS) (Y)	ADVERTISING EXPENDITURES (THOUSANDS OF DOLLARS) (X)
A	100	1.0
B	300	5.0
C	400	8.0
D	200	6.0
E	100	3.0
F	400	10.0

```
MTB > REGRESS 'SALES' ON 1 PREDICTOR 'ADVTSNG'

The regression equation is
SALES = 39.3 + 38.3 ADVTSNG

Predictor      Coef       Stdev     t-ratio       p
Constant       39.25      54.87       0.72      0.514
ADVTSNG        38.318      8.767      4.37      0.012

s = 64.13      R-sq = 82.7%     R-sq(adj) = 78.4%

Analysis of Variance

SOURCE         DF          SS          MS          F         p
Regression     1         78551       78551      19.10     0.012
Error          4         16449        4112
Total          5         95000
```

FIGURE CL14-2   Selam is right; there is a relationship between sales results and advertising costs.

the printout shows that there's a true relationship between sales and advertising expenditures at the .05 level of significance. And the r^2 value shows that 82.7 percent of the change in sales is explained by changes in advertising expenditures. That's not a bad regression fit, but Selam would like to find another variable that can further explain changes in sales before she submits her next sales forecast to her boss. In short, Selam wants to base her next sales forecast on a multiple regression model.

Although a study of multiple regression involves three (or more) variables, the same assumptions, concepts, and measures that we studied in the chapter still apply. There's still a single dependent (Y) variable that we're interested in predicting, but now there are *two (or more)* independent (X) variables that explain the variations that occur in the Y variable. In the chapter, we first calculated the simple linear regression equation, and then found the standard error of estimate. We followed up on those measures with relationship tests and interval estimates, and then computed the coefficient of determination (r^2) near the end of the chapter. Let's briefly look at these same topics in the context of multiple linear regression and correlation.

THE MULTIPLE LINEAR REGRESSION EQUATION

Let's assume that Selam has identified another predicting variable—the population in each market area—and now has the data shown in Figure CL14-3. The dependent variable is still identified by Y. But since we now have two independent variables, we'll call advertising costs X_1 and population X_2. The simple linear regression equation ($Y_c = a + bX$) must now be changed to the following **multiple linear regression equation:**

$$Y_c = a + b_1X_1 + b_2X_2$$

where Y_c = estimated value of the dependent variable
      a = Y intercept
     X_1 = value of the first independent variable
     X_2 = value of the second independent variable
     b_1 = slope associated with X_1 (the change in Y_c if X_2 is held constant and X_1 varies by one unit)
     b_2 = slope associated with X_2 (the change in Y_c if X_1 is held constant and X_2 varies by one unit)

In the simple linear regression case, the relationship between X and Y was shown with a straight regression line that displayed length and width. Now, with three variables, we

FIGURE CL14-3 Selam Has Added a Second Predicting Variable
—Population—to Her Sample Data

MARKET AREA	TOY SALES (THOUSANDS OF DOLLARS) (Y)	ADVERTISING COSTS (THOUSANDS OF DOLLARS) (X_1)	POPULATION (THOUSANDS) (X_2)
A	100	1.0	200
B	300	5.0	700
C	400	8.0	800
D	200	6.0	400
E	100	3.0	100
F	400	10.0	600

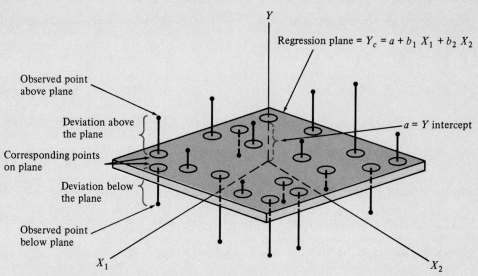

FIGURE CL14-4   A regression plane replaces a regression line in multiple regression. If you face the corner of your room, hold this book at arm's length, and tilt the book in various ways, you'll get an idea of how different planes might appear in a three-variable situation.

have a **three-dimensional plane** that shows length, width, and depth (see Figure CL14-4). Just as the points in a two-variable regression problem fall above and below the regression line shown in a two-dimensional scatter diagram, so, too, do the points in a three-variable regression problem fall above and below the regression plane shown in Figure CL14-4. And just as the simple linear regression equation is used to get the best straight-line fit to the points in a scatter diagram, so, too, does the multiple regression equation identify the plane that gives the best fit to the data.

Computing the Multiple Linear Regression Equation   The next step is to compute the values of a, b_1, and b_2 in the multiple regression equation. The table in Figure CL14-5 repeats the Y, X_1, and X_2 values given in Figure CL14-3 and then presents other intermediate figures needed to compute these regression equation values. Although our sample covers only six market areas, and the data are deliberately simplified, the work needed to

FIGURE CL14-5   Intermediate Figures Needed to Compute a, b_1, and b_2

Y	X_1	X_2	X_1Y	X_2Y	X_1X_2	X_1^2	X_2^2	Y^2
100	1	200	100	20,000	200	1	40,000	10,000
300	5	700	1,500	210,000	3,500	25	490,000	90,000
400	8	800	3,200	320,000	6,400	64	640,000	160,000
200	6	400	1,200	80,000	2,400	36	160,000	40,000
100	3	100	300	10,000	300	9	10,000	10,000
400	10	600	4,000	240,000	6,000	100	360,000	160,000
1,500	33	2,800	10,300	880,000	18,800	235	1,700,000	470,000

calculate these values by hand is daunting. For example, one way to calculate b_1 is with the following formula:

$$b_1 = \frac{[n(\Sigma X_1 Y) - (\Sigma X_1)(\Sigma Y)] \, [n\Sigma X_2^2 - (\Sigma X_2)^2] - [n\Sigma X_1 X_2 - (\Sigma X_1)(\Sigma X_2)] \, [n\Sigma X_2 Y - (\Sigma X_2)(\Sigma Y)]}{[n\Sigma X_1^2 - (\Sigma X_1)^2] \, [n\Sigma \, X_2^2 - (\Sigma X_2)^2] - [n\Sigma X_1 X_2 - (\Sigma X_1)(\Sigma X_2)]^2}$$

You *could* plug in the values from Figure CL14-5 into this formula to calculate the value of b_1 for Selam's regression equation (if you do, you'll find it's 20.49209). And you *could* use an equally formidable formula to calculate b_2, and then find a. But Selam didn't engage in such masochistic behavior. Rather, she simply keyed the sample data into a statistical software package, entered the appropriate command, and waited a few seconds for the program to do the work for her. Figures CL14-6*a* and *b* show the preprogrammed formats used by *Mystat* and *Minitab* to present the results of Selam's multiple regression analysis. These formats, of course, are the same as those used in the chapter to deal with Hiram's simple regression problem.

As you can see, Selam's multiple regression equation has the following values:

$a = 6.397$ or 6.40
$b_1 = 20.492$
$b_2 = .28049$ or $.280$

Interpreting the Values in the Equation   What do these values mean? The a figure is still the Y intercept—the value of Y_c when X_1 and X_2 are both zero. The b_1 value of 20.492 shows that if one market area spends \$1,000 more on advertising Tackey Toys (a change of one unit in X_1) than another market area, and if the markets have the same population (X_2 is held constant), then the estimated toy sales in the higher advertising cost area exceed those of the second area by about \$20,492. And the b_2 value of .28049 means that if a first market area has 1,000 more people than a second area (a change of one unit in X_2), and if the same amount has been spent on advertising in the two areas (X_1 is held constant), then the toy sales of the larger market area exceed those of the smaller market by about \$280. The b_1 and b_2 constants are called **estimated regression coefficients** in multiple regression terminology.

Making Predictions with the Multiple Regression Equation   Selam now has the following multiple regression equation:

$Y_c = 6.397 + 20.492X_1 + .28049X_2$

And let's assume that Selam needs a sales forecast for a market area. Tackey Toys has recently spent \$4,000 advertising in this market, which has a population of 500,000 people. Selam's point estimate of toy sales in the market is:

$Y_c = 6.397 + 20.492(4) + .28049(500)$
$\quad = 6.397 + 81.968 + 140.245$
$\quad = 228.610$ or \$228,610 in toy sales

The Multicollinearity Problem   **Multicollinearity** is the name statisticians give to the situation when predictive variables X_1 and X_2 are closely intercorrelated. If this is the case, the values of b_1 and b_2 tend to be unreliable and an estimate made with an equation that uses these values also tends to be unreliable. This is because, in closely correlated

```
DEP VAR:   SALES     N:   6    MULTIPLE R:  .987   SQUARED MULTIPLE R:  .974
ADJUSTED SQUARED MULTIPLE R:  .956      STANDARD ERROR OF ESTIMATE:       28.883

    VARIABLE    COEFFICIENT    STD ERROR    STD COEF TOLERANCE    T    P(2 TAIL)
    CONSTANT          6.397       25.986      0.000    .         0.246    0.821
    ADVTSNG          20.492        5.882      0.486 0.4506574     3.484    0.040
    POPULTN           0.280        0.069      0.571 0.4506574     4.089    0.026


                       ANALYSIS OF VARIANCE

     SOURCE     SUM-OF-SQUARES    DF   MEAN-SQUARE     F-RATIO       P

  REGRESSION        92497.364      2    46248.682      55.440      0.004
    RESIDUAL         2502.636      3      834.212
```

(a)

```
    MTB > REGRESS 'SALES' ON 2 PREDICTORS 'ADVTSNG'  'POPULTN';
    SUBC> PREDICT 4    500.

    The regression equation is
    SALES = 6.4 + 20.5 ADVTSNG + 0.280 POPULTN

    Predictor        Coef        Stdev     t-ratio        p
    Constant         6.40        25.99        0.25      0.821
    ADVTSNG         20.492        5.882        3.48      0.040
    POPULTN        0.28049      0.06860        4.09      0.026

    s = 28.88      R-sq = 97.4%     R-sq(adj) = 95.6%

    Analysis of Variance

    SOURCE        DF          SS           MS        F         p
    Regression     2       92497        46249     55.44     0.004
    Error          3        2503          834
    Total          5       95000

    SOURCE        DF      SEQ SS
    ADVTSNG        1       78551
    POPULTN        1       13946



        Fit   Stdev.Fit       95% C.I.          95% P.I.
       228.6       15.9   (  178.1,   279.1)  (  123.7,   333.5)
```

(b)

FIGURE CL14-6   (a) The *Mystat* printout when supplied with Selam's data. (b)
The *Minitab* printout when supplied with Selam's data.

series, variables in X_2 don't necessarily remain constant while X_1 changes. If two independent variables are closely correlated—that is, if they have an r value close to ± 1.00—a simple solution is to use just one of them in a multiple regression model.

THE STANDARD ERROR OF ESTIMATE FOR MULTIPLE REGRESSION
You saw in this chapter that the standard error of estimate for a simple linear regression model is a measure of the extent of the scatter or dispersion of the sample data points

about the regression line. In a multiple linear regression situation, we still have a standard error of estimate, but now it measures the dispersion or scatter of the sample data points about the multiple regression plane. One way to calculate this measure (we'll use the symbol $S_{y \cdot 12}$) is with a formula that's analogous to formula 14-4:

$$S_{y \cdot 12} = \sqrt{\frac{\Sigma(Y - Y_c)^2}{n - 3}}$$

Of course, an easier formula to apply—one that uses the values given in Figure CL14-5—is:

$$S_{y \cdot 12} = \sqrt{\frac{\Sigma(Y^2) - a(\Sigma Y) - b_1\Sigma(X_1Y) - b_2\Sigma(X_2Y)}{n - 3}}$$

$$= \sqrt{\frac{470{,}000 - 6.397(1{,}500) - 20.492(10{,}300) - .28049(880{,}000)}{6 - 3}}$$

$$= \sqrt{\frac{2{,}505.7}{3}} = 28.90$$

You can see that the value 28.90 agrees with the figure shown in the printouts.

RELATIONSHIP TESTS

The underlying assumptions listed in the chapter that are necessary to apply statistical inference concepts to a simple linear regression situation still apply (although, of course, we have another predicting variable).

t **Tests for Slope**   It's still possible to conduct *t* tests for the slopes of the population regression coefficients B_1 and B_2. From the printouts in Figures CL14-6a and b, you can see the results of these *t* tests. Let's assume, for example, that Selam wants to test the null hypothesis that $B_1 = 0$ against the alternative that $B_2 \neq 0$ at the .05 level of significance. The decision rule is:

Accept H_0 [and the assumption that no relationship exists between toy sales (Y) and advertising costs (X_1)] if the critical ratio falls into the acceptance region determined by the *t*-table value that corresponds to $t_{\alpha/2}$ and $n - 3$ degrees of freedom. This *t*-table value is 3.182.

The critical ratio is found like this (the estimated standard deviation of b_1 is given in Figures CL14-6a and b):

$$CR = \frac{b_1 - B_{1H_0}}{\text{estimated standard deviation of } b_1} = \frac{20.492 - 0}{5.882} = 3.48$$

Since 3.48 is greater than 3.182, we reject the H_0 that $B_1 = 0$. There *is* a meaningful relationship between toy sales and advertising. (The *p* value that's less than .05—in this case it's .04—tells you the same thing.) You should now be able to examine Figures CL14-6a and b and explain why there's also a meaningful relationship between toy sales and population in the market area.

The Analysis of Variance Test   Each t test determines if there's a meaningful relationship between the Y variable and *one* of the predicting variables. The purpose of the ANOVA test is to examine the *total* regression effect that all the predicting (X) variables have on the Y variable. The null and alternative hypotheses are:

H_0: The total regression is not significant ($B_1 = B_2 = 0$)

H_1: The total regression is significant

Let's assume that Selam wants to make an ANOVA test at the .05 level of significance. She can examine Figures CL14-6*a* and *b* and see that the programs have calculated the variations of the data points about the regression plane. The *explained* variation (SS Regression) is 92,497, the *unexplained* variation (SS Error) is 2,503, and the *total* variation (SS Total) is 95,000. These SS figures divided by the corresponding degrees of freedom values given in the printouts produce the MS (MEAN-SQUARE) values shown. The programs then use these MS values to compute a critical F or F-RATIO (CR_F)—the value of 55.44 shown in the printouts.

The decision rule now is just as it has been in this chapter: The H_0 is rejected if the CR_F value is *greater* than an appropriate F-table value found in Appendix 6. To find this table value, you'll recall, you need df_{num} and df_{den}. These df figures are:

$$df_{num} = 2$$

$$df_{den} = n - m - 1 = 3$$

where m = number of independent variables = 2
      n = number of market areas in our example

Thus, the F-table value for our example ($F_{2,3,.05}$) is 9.55.

Since the CR_F value of 55.44 is greater than the table value of 9.55, the H_0 is rejected. The total regression *is* significant at the .05 level.

INTERVAL ESTIMATES FOR PREDICTIONS

You saw in this chapter that *small-sample* interval estimates can be constructed and probabilities can be assigned to these estimates. Two types of intervals were developed then: one that predicted the range for the average value of Y given X, and another that predicted the range of a specific value of Y given X. These same concepts can be applied in a multiple regression situation. You'll notice on the second line of Figure CL14-6*b* that a subcommand—PREDICT 4 500—is specified. This tells the program to construct 95 percent confidence ranges for both types of intervals, and these ranges are printed at the bottom of Figure CL14-6*b*.

The 228.6 value in the printout is a point estimate. Selam obtained this value when she plugged an X_1 value of 4 (thousand dollars of advertising) and an X_2 value of 500 (thousand people in a market area) into her multiple regression equation a little earlier in this reading. The 95 percent confidence interval for the average range of Y given 4 for X_1 and 500 for X_2 is found in this way:

$$228.610 \pm t_{\alpha/2} \text{ (appropriate measure of dispersion)}$$

The $t_{\alpha/2}$ value needed in Selam's estimate has $n - 3$ ($6 - 3$) or 3 degrees of freedom. The t table shows that this value is 3.182. And the appropriate measure of dispersion value shown at the bottom of Figure CL14-6b is 15.9. Thus, Selam can be 95 percent confident that the *average* sales range for Tackey Toys in market areas with these values of X_1 and X_2 is:

$$228.610 \pm 3.182(15.9)$$

or from $178,100 to $279,100.

The printout at the bottom of Figure CL14-6b further shows that the predicted sales range for a specific market area with the given values of X_1 and X_2 is:

$$228.610 \pm 3.182 \sqrt{(15.9)^2 + (28.88)^2}$$

where 28.88 is the $S_{y \cdot 12}$. This prediction interval is thus $123,700 to $333,500.

THE COEFFICIENT OF MULTIPLE DETERMINATION

The r^2 value in the chapter measures the percentage of the variation in the Y variable that's explained by the variation in the X variable. Similarly, the **coefficient of multiple determination** (R^2) is a decimal fraction or percentage that shows the variation in the Y variable that's explained by its relation to X_1 and X_2.

One way to find R^2 is:

$$R^2 = \frac{\Sigma(Y_c - \overline{Y})^2}{\Sigma(Y - \overline{Y})^2} = \frac{\text{SSR}}{\text{SST}} = \frac{92,497}{95,000} = .9736 \text{ or } 97.4\%$$

where the SSR and SST values come from the ANOVA table in the printout. Another way to find R^2 uses the data in the multiple regression equation and in Figure CL14-5:

$$R^2 = \frac{n[a(\Sigma Y) + b_1(\Sigma X_1 Y) + b_2(\Sigma X_2 Y)] - (\Sigma Y)^2}{n(\Sigma Y^2) - (\Sigma Y)^2}$$

$$= \frac{6[6.397(1,500) + 20.492(10,300) + .28049(880,000)] - (1,500)^2}{6(470,000) - (1,500)^2} = \frac{554,965.8}{570,000} = .9736 \text{ or } 97.4\%$$

And a third (and obviously best) way to get R^2 is to just read the value directly from the printouts. The .974 figure, of course, means that 97.4 percent of Tackey Toy sales in the market areas is explained by advertising expenditures and population size.

PART FOUR

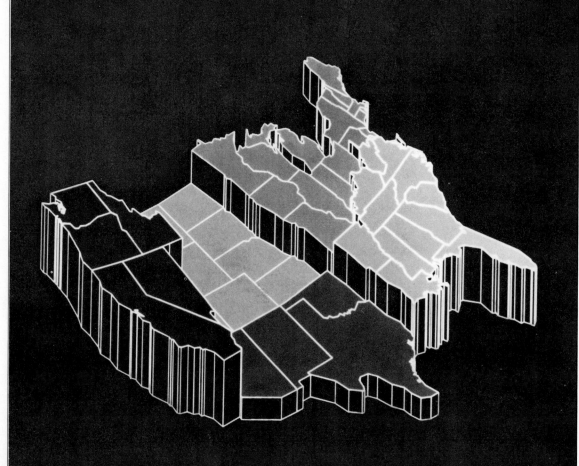

CONCLUDING TOPICS

The following comments appear on a poster in the window of a curio shop on Bourbon Street in New Orleans:

The objective of all our dedicated employees should be to analyze all situations, to sort out important problems prior to their occurrence, and to have answers to all these problems when they are called upon. . . . However, when you're up to your [anatomical reference deleted] in alligators it's difficult to remember that your initial objective was to drain the swamp.

Perhaps as you struggled to master the material found in the preceding chapters there were times when you felt bogged down by details; perhaps you wondered why you were expected to learn everything about statistics in a single course; and perhaps you felt in complete sympathy with the author of the Bourbon Street poster. But you will be chagrined (or relieved) to learn that in this book we have only outlined a few of the basic topics in the areas of statistics and quantitative methods.

Although we cannot cover in any great detail many of the important methods of quantitative analysis that have been ignored up to this point, we also cannot conclude this book without some discussion of the uses and limitations of several additional quantitative tools available to the statistician and decision maker.

In Chapters 7 through 10, inferences were made and decisions were reached through the use of estimation and hypothesis-testing techniques. But these earlier techniques are generally valid only if the assumptions that were made about the shapes of various sampling and/or population distributions are correct. However, there is an important group of statistical procedures which do not require that assumptions be made about the shape of a distribution and which are thus referred to as distribution-free procedures. These *nonparametric methods,* in other words, enable the decision maker to test hypotheses without the need to impose certain restrictive assumptions. Although it would certainly have been appropriate to include these nonparametric tests in the chapters of Part 2, we have chosen to place the material in this final part.

In addition, a few other advanced quantitative tools are briefly described in the final chapter. Thus, the chapters included in Part 4 are:

15

NONPARAMETRIC STATISTICAL METHODS

Parametric and Nonparametric Techniques

Parametric statistical techniques, such as the t test or the F test, are used when the researcher knows certain facts about the group of data from which a sample has been taken. Basically, the researcher must be working with data that has been taken independently and without bias from a group of items. The data must come from normally distributed populations which have the same variance, and the data must be measured at least at the interval level.

Nonparametric statistical techniques, such as the Mann-Whitney U test or the Kruskal-Wallis analysis of variance, are used when the researcher does not know the characteristics of the group of items from which he has taken his sample. These techniques can be applied to data measured on the ordinal and in some cases the nominal scale. Nonparametric tests are useful for working with small sizes of samples and are easier to calculate than the parametric techniques. Nonparametric techniques are also widely used to analyze data encountered in the behavioral sciences.

—From Richard T. Dué, "Predicting Results with Statistics," *Datamation*, May 1980, p. 227. Reprinted with permission of Datamation® Magazine, © copyright by Technical Publishing Company, A Dun & Bradstreet Co., 1980. All rights reserved.

LOOKING AHEAD

The use of nonparametric methods is briefly introduced in our opening excerpt, and we'll expand on this introduction and on the meaning of ordinal and nominal data early in this chapter. But there are entire books that deal with the subject of nonparametric statistics, and so we can only touch on the subject here. Thus, we'll consider the Mann-Whitney U test mentioned in the excerpt, but we'll omit the Kruskal-Wallis procedure. In order of presentation, then, the widely used nonparametric methods considered in this chapter are the (1) *sign test*, (2) *Wilcoxon signed rank test*, (3) *Mann-Whitney test*, (4) *runs test for randomness*, and (5) *Spearman rank correlation coefficient*. Although most of these techniques may be discussed in the context of both large- and small-sample situations, we'll limit our attention primarily to the small-sample cases.

Thus, after studying this chapter you should be able to:

- Identify situations that call for the use of particular nonparametric methods.
- Apply sign test procedures for both small and large samples.
- Apply the Wilcoxon signed rank test in small-sample situations.
- Use the Mann-Whitney test to determine if two small independent random samples are taken from identical populations.
- Determine if randomness exists (or if there is an underlying pattern) in a small sequence of sample data through the use of the runs test for randomness.
- Compute the Spearman rank correlation coefficient and then test this measure for significance.

CHAPTER OUTLINE

USE OF NONPARAMETRIC TECHNIQUES

A short time ago, some of your skeptical classmates (maybe even you) might have remarked that much of the material covered in Chapters 7, 8, 9, and 10 may not always be relevant because a designated probability distribution cannot always be assumed in a real-world situation. This skepticism, of course, may be justified in certain situations! The validity of the inferences made in those earlier chapters depended upon the accuracy of the assumptions[1] made about such things as (1) the shape of the various sampling distributions of sample statistics and/or the shape of the population distribution and (2) the relationship of these probability distributions to the underlying population parameters. (For those of you who are getting an anxiety attack, a quick review of the Central Limit Theorem in Chapter 6 may be in order.)

In this chapter, though, we're concerned with **nonparametric statistics**—that is, with statistics that don't require that assumptions be made about the shape of a distribution and which are thus distribution-free statistics. With nonparametric techniques, inferences may be made regardless of the shape of the population distribution; with the parametric statistics of the earlier chapters, inferences are valid only if certain restrictive assumptions are true.

It may surprise you to learn that distribution-free techniques have already been considered in this book. The chi-square methods in Chapter 11 are nonparametric in nature. The chi-square procedure is used to compare observed (sample) frequencies with expected (hypothesized) frequencies, and the expected frequencies aren't necessarily restricted to a particular type of distribution.

When should nonparametric methods be used? They should be employed in any of the following situations:

1   When the sample size is so small that the sampling distributions of the statistics don't approximate the normal distribution, and when no assumption can be made about the shape of the population distribution from which the sample is drawn.

2   When **ordinal** (or **rank**) **data** are used. (Ordinal data only tell us if one item is higher than, lower than, or equal to another item; they don't tell us the size of the difference.)

---

[1] A statistician and the statistician's spouse found themselves marooned on a remote island. When the spouse asked how they would escape from the island and get to civilization, the statistician replied, "Assuming we have a boat. . . ."

3   When **nominal data** are used. (Nominal data are simply data where names such as "male" or "female" are assigned to items and there's no suggestion in the names that one item is higher or lower than another.)

Let's look now at some widely used nonparametric techniques.

THE SIGN TEST

When you have paired ordinal measurements taken from the same subjects or matched subjects, and when you're simply interested in seeing if there are real differences regardless of the size of the differences, the sign test should be used. The **sign test** is based on the signs—negative or positive—of the differences between pairs of ordinal data, and it considers only the direction of the differences and not the magnitude of the differences.

Sign Test Procedure with Small Samples

Let's look at an example. Chicken Out, a national fast-food chain, has developed a new formula for the batter used in coating its chicken, and the marketing department wants to know if the new batter is tastier than the original batter. At the present stage of product development, the department is not interested in the degree of taste improvement.

Ten consumers are randomly selected for a taste test. Each consumer first tastes a piece of chicken with the original batter and rates the taste on a scale of 1 to 10, where 1 is very poor and 10 is very good. Then, the same consumer munches a piece of chicken fried with the new batter and rates the taste on a scale of 1 to 10. The data thus collected are presented in Figure 15-1.

What should the market research data tell us? If there is truly no difference in flavor, we would expect, in a large survey, the number of consumers who rate the new-batter taste as better than the original-batter taste to be equal to the number of consumers who think the new taste is worse than the old taste. In other words, if there's truly no difference between the taste of the original batter and the taste of the new batter, the median difference between the two taste ratings should be zero. That is, the probability of getting consumers who report better taste should be equal to the probability of selecting consumers who would report a worse taste.

Stating the Null and Alternative Hypotheses   As in the case of any hypothesis test, the first step in a sign test is to state the null and alternative hypotheses. Two-tailed or one-tailed sign tests may be conducted, and this fact, of course, determines the form of the alternative hypothesis.

The *null hypothesis* to be tested in our example is that the new ingredients have no effect on the taste of the chicken: a positive sign indicating a taste improvement is just as likely as a negative sign indicating a loss of flavor when the difference between the two taste ratings for each subject is determined. The *alternative hypothesis* in our example is that the new batter improves taste. We

FIGURE 15-1   Data for the Sign Test Taste ratings by 10 consumers of chicken coated with original batter and chicken coated with new batter (10 indicates a "very good taste," and 1 indicates a "very poor taste")

CONSUMER	TASTE RATINGS		SIGN OF DIFFERENCE BETWEEN NEW COATING AND ORIGINAL COATING $(y-x)$
	ORIGINAL BATTER (x)	NEW BATTER (y)	
R. MacDonald	3	9	+
G. Price	5	5	0
B. King	3	6	+
L. J. Silver	1	3	+
P. P. Gino	5	10	+
E. J. McGee	8	4	−
S. White	2	2	0
E. Fudd	8	5	−
Y. Sam	4	6	+
M. Muffett	6	7	+

n = number of relevant observations
   = number of plus signs + number of minus signs
   = 6 + 2
   = 8
r = the number of fewer signs
   = 2

thus have a right-tailed test, and the alternative hypothesis is that there's more than a 50 percent chance that a person will report that the new batter has a better flavor than the original batter. Thus, the statistical hypotheses are:

H_0: $p = 0.5$
H_1: $p > 0.5$

where p is the probability of getting a taste improvement.

Selecting the Level of Significance   After stating the null and alternative hypotheses, the second step is to establish a criterion for rejecting or accepting the null hypothesis. Let's assume that, for our example, the risk of erroneously rejecting the null hypothesis when it's actually true is to be limited to no more than 5 percent. Thus, the level of significance is $\alpha = .05$.

Determining the Sign of Differences between Paired Observations   After the null and alternative hypotheses are determined, and after the level of significance is selected, the next step is to systematically subtract one observation from the other observation and then record whether the difference is positive (an improvement in taste) or negative (a loss in flavor). The last column of Figure 15-1 shows the sign of the difference for each subject when the taste rating for the original batter is subtracted *from* the taste rating for the new batter. In the case of the first subject, R. MacDonald, the taste rating for the new batter is greater or *better* than the taste rating for the original batter; thus, there is a *positive* sign. In situations where there's no change in taste ratings, a zero representing a tie is recorded.

Counting the Frequency of Signs   The next step is to tally up the pluses, the minuses, and the zeros. Figure 15-1 shows 6 pluses, 2 minuses, and 2 zeros—which means that 6 consumers thought there was a taste improvement, 2 thought flavor had declined, and 2 perceived no change. After the tally, *we designate the lesser sum of the two signs as r*. In the case of Figure 15-1, we have $r = 2$ because there are only 2 negatives compared with the 6 positive signs.

Determining the Likelihood of Observed Sample Results   The only subjects or paired observations relevant for analysis are those where taste differences (positive or negative) are recorded. In our case, only 8 of the 10 pairs of data are relevant for analysis, and thus we have $n = 8$. (The responses of Price and White are not included in the analysis because they provide no indication of a difference one way or another.) From the 8 relevant subjects or paired observations, one would expect four of the differences to be positive and four of the differences to be negative *if* the null hypothesis is true. On the basis of the two negative responses in Figure 15-1 and the nature of a right-tailed test, we must ask ourselves the following question: What is the chance of having at most only 2 out of 8 subjects perceiving a negative taste change when in fact the null hypothesis is true (where 50 percent of the sample should record a positive change and where 50 percent should record a negative change)?

   Formulation of the answer begins by referring to the appropriate probability distribution. In this test, since n is small (≤ 30), that's the binomial probability distribution in Appendix 1 at the back of the book. When the sample size is relatively large—that is, when it's > 30—the normal approximation to the binomial distribution may be used. (Actually, there's little difference in results in using the normal approximation to the binomial in sign tests when the sample size is greater than 20.) Since we have 8 relevant subjects in the present test, we look for that section of the binomial distribution table in Appendix 1 where $n = 8$ and $r = 2$. After locating that section, we look in the column where $p = .50$—a value that stems from the null hypothesis. We see that the probability of getting *at most* only 2 out of 8 subjects reporting a negative change is .1445, which is a summation of the probabilities of getting 0 out of 8 (.0039), 1 out of 8 (.0312), and 2 out of 8 (.1094). In other words, if there were truly no difference in taste between the original and new batters, the chances of getting at most only 2 out of 8 subjects reporting a loss of taste is only 14.5 percent.

Drawing a Statistical Conclusion Concerning the Null Hypothesis   The question now is whether or not the sample probability result of .1445 is sufficient for us to accept the null hypothesis that there is no significant difference in consumer taste ratings. Although the probability of getting at most only 2 out of 8 consumers to report negatively on the new batter mix is not particularly high at .1445, it is higher than the .05 level of significance specified earlier in the problem. That is, the sample probability result would have had to be less than .05 for us to reject the null hypothesis.

In summary, then, the *decision rule* to follow in small-sample sign test situations in making a statistical decision is:

Accept the H_0 if $\alpha \leq$ the probability of the sample results.

or

Reject H_0 and accept H_1 if $\alpha >$ the probability of the sample results.

Since, in our example, .05 < .1445, we accept the null hypothesis; the new batter mix cannot be said to be a significant improvement over the original recipe.

If we're conducting a two-tailed test, we *double* the probabilities obtained from the binomial table before making the statistical decision. For example, if we're conducting a two-tailed test on the Chicken Out data, the sample results are 2 times .1445, or .2890. This two-tailed test result is shown in the output produced by the *Mystat* statistical package in Figure 15-2.

Sign Test Procedure with Large Samples

If the sample size is reasonably large, and if the normal approximation to the binomial distribution may be used, the same decision rules discussed in Chapter 8 apply, and the critical ratio (Z value) is computed as follows:

$$CR = \frac{2R - n}{\sqrt{n}} \qquad \text{(formula 15-1)}$$

where R = number of positive signs
n = number of relevant paired observations

FIGURE 15-2
A two-tailed sign test produced by Mystat.

```
SIGN TEST RESULTS

   COUNTS OF DIFFERENCES (ROW VARIABLE GREATER THAN COLUMN)


                        ORIGINAL            NEW

   ORIGINAL                 0                2
       NEW                  6                0


   TWO-SIDED PROBABILITIES FOR EACH PAIR OF VARIABLES


                        ORIGINAL            NEW

   ORIGINAL              1.000
       NEW                .289            1.000
```

Suppose, for example, that in our Chicken Out problem there are 35 consumers in the sample. Assume also that the following results are obtained:

$$
\left. \begin{array}{l} + \text{ differences } = 19 \\ - \text{ differences } = 13 \end{array} \right\} n = 32
$$

$$
\begin{array}{l} 0 \text{ differences } = \underline{3} \\ \text{total } = 35 \end{array}
$$

If a right-tailed test is made, the hypotheses remain unchanged. And if the .05 level of significance is used, the decision rule is stated in the following familiar format:

Accept H_0 if CR ≤ 1.64.

or

Reject H_0 and accept H_1 if CR > 1.64.

The critical ratio is computed as follows:

$$
\begin{aligned}
CR &= \frac{2R - n}{\sqrt{n}} \\
&= \frac{2(19) - 32}{\sqrt{32}} \\
&= \frac{38 - 32}{5.657} \\
&= 1.061
\end{aligned}
$$

Since $1.061 < 1.64$, the null hypothesis is accepted. In this case, the conclusion is that there's no significant difference between the taste ratings of the two batters.

Figure 15-3 summarizes the sign test procedures outlined in this section.

Self-Testing Review 15-1

1 What is a sign test?
2 "Only a one-tailed hypothesis test may be conducted in a sign test." Comment on this statement.
3 How many observations for each subject are required for a sign test?
4 What is the null hypothesis in a sign test?
5 a What probability distribution is used in testing the hypotheses of sign tests when the sample size is small?
   b When the sample size exceeds 30?
6 a If the differences between paired data used in a sign test are 5 positives, 7 negatives, and 6 ties or zeros, we have $n = 18$ and $r = 7$. True or false?
   b In a right-tailed test at the .10 level, should the null hypothesis be accepted using the data given in a?
7 If the differences between paired data used in a sign test are 16 positives, 26 negatives, and 4 zeros, what would be the statistical decision for a two-tailed test at the .05 level of significance?

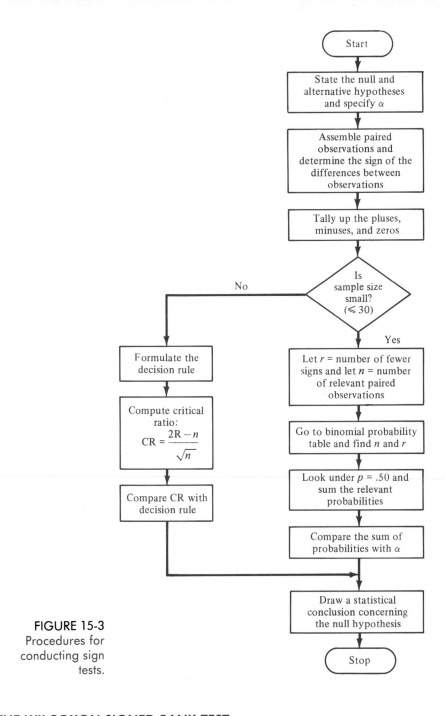

FIGURE 15-3
Procedures for conducting sign tests.

THE WILCOXON SIGNED RANK TEST

While the sign test focuses solely on the *direction* of the differences within pairs, the **Wilcoxon signed rank test** uses the *magnitude* as well as the direction of the differences to see if there are true differences between pairs of data drawn from one sample or two related samples. When we wish to incorporate the *size* of differences in addition to the direction of the differences into our decision making, the Wilcoxon signed rank test should be used.

Wilcoxon Signed Rank Test Procedure

Let's use the Chicken Out example again. Suppose that the management of the firm wants to make a decision concerning the new batter mix based not just on how many persons thought the new batter improved taste but also on the *amount* of taste improvement from the new batter. The Wilcoxon signed rank test is appropriate, and the data for analysis are drawn from Figure 15-1 and are reproduced in Figure 15-4.

Stating the Hypotheses and α As you may have anticipated, we must state the hypotheses and the desired level of significance. The null hypothesis in this case is that there's no difference between the tastes of the original and new batters. Therefore, in a large sample, the number of positive signs should equal the number of negative signs. Since this is a right-tailed test, the alternative hypothesis states that the taste of the new batter is better than the taste of the original mix. Thus, the hypotheses might be written as follows:

H_0: The two batters are equally tasty (or tasteless?).
H_1: The new batter mix is better tasting.

In addition, for this example, we'll reject the null hypothesis at the .01 level of significance.

FIGURE 15-4   Computations for Wilcoxon Signed Rank Tests

CONSUMERS	(1) ORIGINAL BATTER TASTE SCORE	(2) NEW BATTER TASTE SCORE	(3) DIFFERENCE: NEW BATTER RATING LESS ORIGINAL BATTER RATING	(4) RANK IRRESPECTIVE OF SIGN	(5) (6) SIGNED RANK	
					POSITIVE	NEGATIVE
R. MacDonald	3	9	+6	8	+8	
G. Price	5	5	0	(ignore)		
B. King	3	6	+3	4.5	+4.5	
L. J. Silver	1	3	+2	2.5	+2.5	
P. P. Gino	5	10	+5	7	+7	
E. J. McGee	8	4	−4	6		−6
S. White	2	2	0	(ignore)		
E. Fudd	8	5	−3	4.5		−4.5
Y. Sam	4	6	+2	2.5	+2.5	
M. Muffett	6	7	+1	1	+1	
					+25.5	−10.5

n = number of relevant observations
   = number of plus signs + number of minus signs
   = 6 + 2
   = 8
T = the smaller of the two rank sums
   = 10.5

Determining the Size and Sign of Differences between Paired Data   After stating the hypotheses and determining the significance level, the next step is to prepare the raw data for testing. The *size* and *sign* of the differences between the paired data are computed, and these are shown in the third column of Figure 15-4. For example, McGee initially gave the taste of the original batter an 8 score but felt that the taste of the new batter merited only a 4. Thus, the recorded difference for McGee is −4. Differences for the other consumers are recorded in similar fashion.

Ranking the Differences Irrespective of the Signs   In the next step, we temporarily *ignore* the plus and minus signs in column 3 and rank the *absolute* values of the differences. A rank of 1 is assigned to the *smallest* difference; a rank of 2 is given to the next smallest value; and so on. (Differences of zero are ignored.) Since the two taste scores for Muffett had the least difference, that *difference, irrespective of the direction*, is assigned the rank of 1. In the cases of Silver and Sam, who are tied for second and third ranks with differences of 2, we assign to each a rank of 2.5, which is the *average* of the ranks 2 and 3. This procedure is continued until all differences are ranked.

Affixing Signs to the Assigned Ranks   The next step is to *affix the sign of each difference* (as shown in column 3, Figure 15-3) *to its rank* (as shown in column 4). This step results in the figures in the last two columns of Figure 15-4. For example, the size of the difference between the data for Gino is assigned a rank of 7, and since the difference is positive, a corresponding +7 is recorded. Signed ranks are similarly produced for the other consumers.

Summing the Ranks   The last step before hypothesis testing is to add up all the positive ranks and then add up all the negative ranks. The *smaller of the two sums is designated as the computed T value*. Since the sum of the negative ranks is 10.5 and the sum of the positive ranks is 25.5, the sum of 10.5 is designated as the computed value of T. (As a check on your accuracy, the sum of the positive and negative ranks—that is, $25.5 + 10.5$—must, of course, equal the sum of the ranks in column 4 of Figure 15-4.)

Drawing a Statistical Conclusion Concerning the Null Hypothesis   We can now proceed to test the null hypothesis by comparing the computed T value with the appropriate table value of T for a given level of significance. Under the condition that the null hypothesis is true, the T table in Appendix 8 provides the values of T at corresponding α levels of .01 and .05 for both one-tailed and two-tailed tests. Since we tallied eight ranks (ties do not count), we have $n = 8$. For a one-tailed test where $n = 8$ and $\alpha = .01$, the table value of T is 1. *If the computed T value is equal to or less than the table T value, the null hypothesis should be rejected.* Since our computed T value equals 10.5, and since this statistic is greater than the table T value of 1, the null hypothesis cannot be rejected. It must be concluded that the new batter mix produces no significant taste improvement over the original batter.

For your convenience, the entire procedure for the Wilcoxon signed rank test is summarized in Figure 15-5.

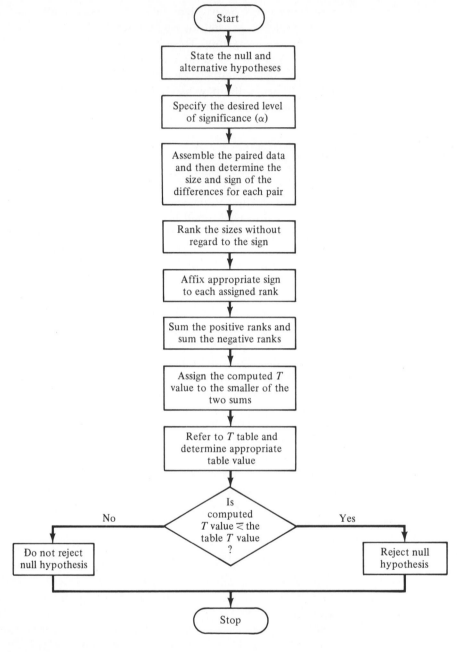

FIGURE 15-5
Procedure for the
Wilcoxon signed
rank test.

Self-Testing Review 15-2

1   How does the Wilcoxon signed rank test differ from the sign test?
2   What is the null hypothesis in the Wilcoxon signed rank test?
3   In a Wilcoxon signed rank test, when you initially rank the differences between paired data, you ignore the sign of the differences. True or false?

4   If the sum of the positive ranks and the sum of the negative ranks were 25 and 20 respectively, we would have a computed T value of 25. True or false?

5   What is the critical T-table value to use when $n = 32$, $\alpha = .05$, and a two-tailed test is being conducted?

6   If the computed T value is less than the table T value, the null hypothesis is rejected. True or false?

THE MANN-WHITNEY TEST

With the sign test and Wilcoxon signed rank test, paired data drawn from one sample or two closely related samples can be analyzed for significant differences. In contrast, the **Mann-Whitney test** is used to examine the null hypothesis that there's no true difference between two sets of data, and the data are drawn from two *independent* samples. This test is often referred to as the U test, since a statistic called U is computed for testing the null hypothesis.

The Mann-Whitney Test Procedure

Let's assume that the alumni director of a business school is compiling biographical data on alumni who graduated 10 years earlier. After receiving spotty returns from a mail survey, the director wants to know if persons who concentrated in marketing are earning more than persons who concentrated in finance. Figure 15-6 shows that the alumni director has received salary data from 8 ($n_1 = 8$) marketing majors and 12 ($n_2 = 12$) finance majors.

Stating the Hypotheses and α As with other types of hypothesis tests, the first step in conducting the Mann-Whitney test is to state the null and alternative

FIGURE 15-6   Data for the Mann-Whitney Test (salaries of 10-year graduates who majored in marketing and who majored in finance)

MARKETING MAJOR	ANNUAL INCOME, $ (IN THOUSANDS)	INCOME RANK	FINANCE MAJOR	ANNUAL INCOME, $ (IN THOUSANDS)	INCOME RANK
G. Price	22.4	15	W. Lee	21.9	14
J. Jones	17.8	3	M. Galper	16.8	1
M. Doe	26.5	16	D. Lemons	28.0	17
K. Seller	19.3	8	T. Grady	19.5	10
S. Martin	18.2	5.5	P. Davis	18.2	5.5
J. Dreher	21.1	13	D. Henry	17.9	4
B. DeVito	19.7	11	B. Ruth	35.8	19
R. Coyne	43.5	20	J. P. Getty	20.5	12
			A. Carnegie	18.7	7
			J. Carter	19.4	9
			G. Ford	17.3	2
			R. Frank	32.9	18
$n_1 = 8$		$R_1 = 91.5$	$n_2 = 12$		$R_2 = 118.5$

hypotheses and the specific level of significance. In this case, the null hypothesis is that after 10 years, there's no difference between the salaries of the marketing majors and the salaries of the finance majors—that is, H_0: The salaries of both majors are equal. Since a right-tailed test is to be made, the alternative hypothesis is that the salaries of the marketing majors are higher than the salaries of the finance majors 10 years after graduation—that is, H_1: The salaries of the marketing majors are greater than those of the finance majors. Furthermore, the alumni director desires a significance level of $\alpha = .01$.

Ranking Data Irrespective of Sample Category   After assembling the data, the next step is to *assign ranks to the entire set of income figures irrespective of major*. Since the annual salary of alumnus Galper is the lowest of the salaries of the 20 persons who responded, that salary is assigned a rank of 1. And since Coyne reported the highest income of either major, that income is assigned a rank of 20.

Summing the Ranks under Each Sample Category and Computing the U Statistic   After the ranks are assigned to all the data, the income ranks for each major are totaled. For all the marketing majors, the sum of the ranks, R_1, is 91.5, and the sum of the ranks of all the finance majors, R_2, is 118.5. We are now ready to compute the U statistic. Each of the following formulas is used to calculate U:

$$U = n_1 n_2 + \frac{n_1(n_1 + 1)}{2} - R_1 \qquad \text{(formula 15-2)}$$

and

$$U = n_1 n_2 + \frac{n_2(n_2 + 1)}{2} - R_2 \qquad \text{(formula 15-3)}$$

where R_1 = the sum of ranks assigned to the sample with size n_1
      R_2 = the sum of ranks assigned to the sample with size n_2

These two formulas will most likely result in two different values for U. *The value which is selected for U in hypothesis testing is the lesser of the two values.* In using formula 15-2, we have

$$U = 8(12) + \frac{8(8 + 1)}{2} - 91.5 = 40.5$$

And with formula 15-3, we have

$$U = 8(12) + \frac{12(12 + 1)}{2} - 118.5 = 55.5$$

Consequently, the value assigned to U for testing the null hypothesis is 40.5, which is the lesser of the two computed values. To see if our computation of the U value is correct, the following formula may be used:

Lesser value of $U = n_1 n_2 -$ larger value of U (formula 15-4)

Note that in our example

$$U = 8(12) - 55.5$$
$$= 40.5$$

Drawing a Statistical Conclusion Concerning the Null Hypothesis   After computing the U statistic, we are now ready to formally test the null hypothesis. Essentially, this test involves comparing the computed U value with the expected table U value that applies if the null hypothesis is true. The tables in Appendix 9 provide values of U for the appropriate n_1, n_2, and α under the condition that the null hypothesis is valid. *The decision rule is:*

Reject the null hypothesis if the computed U value is *equal to or less than* the appropriate value in the U table.

In our example we have $n_1 = 8$, $n_2 = 12$, and a desired level of confidence of .01 in a one-tailed test. The appropriate U value from the second table in Appendix 9 is 17. Since the computed U statistic equals 40.5 and is obviously greater than 17, the null hypothesis cannot be rejected. It may be concluded that there is no real salary difference between the alumni who majored in marketing and the alumni who majored in finance.

Figure 15-7 illustrates the procedure for conducting the Mann-Whitney test.

Self-Testing Review 15-3

1   How does the Mann-Whitney test differ from the sign test?
2   How does the Mann-Whitney test differ from the U test?
3   In the Mann-Whitney test, the sizes of the two independent samples must always be equal to each other. True or false?
4   When assigning ranks to relevant data in a Mann-Whitney test, we temporarily ignore the sample category of the specific pieces of data. True or false?
5   If the computed U value is equal to or less than the table value of U, the null hypothesis is rejected. True or false?
6   What is the critical table value of U when $n_1 = 12$, $n_2 = 13$, $\alpha = .05$, and a two-tailed test is being made?

RUNS TEST FOR RANDOMNESS

A financier wants to examine the recent increases and/or declines in the daily Dow Jones Industrial Average (DJIA) to see if these changes are random or if there's an order or pattern to the changes that might affect her portfolio. To

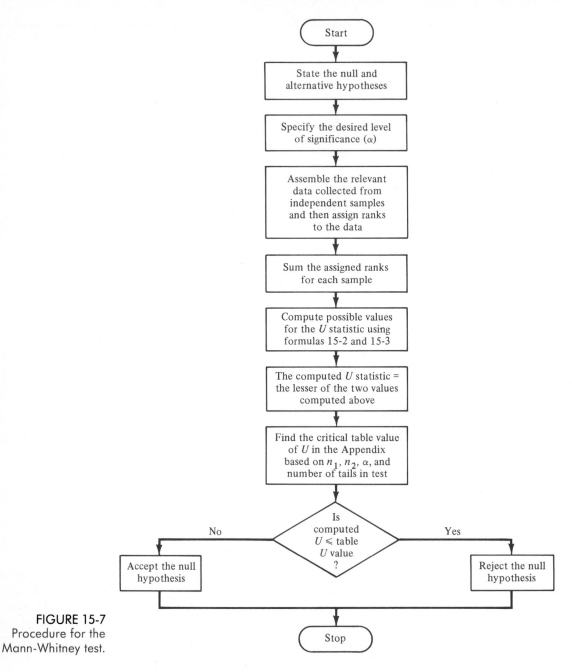

FIGURE 15-7
Procedure for the
Mann-Whitney test.

satisfy her curiosity, the financier can conduct a **runs test for randomness**—a test that can show if randomness exists or if there's an underlying pattern in a sequence of sample data. The test is based upon the number of *runs* or *series* of identical types of results in sequential data. For example, if the financier noticed that for 15 consecutive business days the DJIA had a string or run of 15 consecutive losses, she might readily conclude that there is a pattern in the behavior of the stock market. Unfortunately, decision making isn't always as clear-cut as the preceding sentence might suggest. Therefore, the runs test is another hypothesis-testing procedure that is designed to assist decision makers.

Procedure for Conducting a Runs Test

Suppose the DJIA for the most recent 15 consecutive business days reflected the following changes:

```
Day:     1  2  3  4  5  6  7  8  9  10 11 12 13 14 15
Change:  +  +  −  −  +  +  +  +  +  −  −  +  +  −  +
```

The plus signs indicate an increase over the preceding day, while the negative signs reflect a decrease over the preceding day.

Stating the Null and Alternative Hypotheses   The hypothesis for our runs test are:

H_0: There is randomness in the DJIA sequential data under analysis.
H_1: There is a pattern in the sequential DJIA data under analysis.

The runs test is designed to detect a pattern in the sequence of data, but it cannot tell us the *nature* of the pattern. Thus, in our example, the test might tell us that there is a pattern to stock market changes, but we cannot conclude from the runs test whether the pattern is in an upward or downward direction.

Counting the Number of Runs   On the basis of the sequence of signs, can the financier conclude randomness, or is there a pattern? The first step to answering this question is to count the number of runs. This is done in the following manner, using the preceding data:

```
Change: + +    − −    + + + + +    − −    + +    −    +
        └─┘    └─┘    └───────┘    └─┘    └─┘    └┘   └┘
Run:     1      2         3         4      5      6    7
```

There are seven runs in the sequence of data. The first run is a series of two pluses; the second run is a series of two minuses; the third run is a group of five pluses; and so on. Thus, we may state that r (the number of runs) = 7. Given our data, do the seven runs indicate a random movement in the stock market, or is there a possible pattern to the runs?

Counting the Frequency of Occurrence   The next step in a runs test is to first identify the number of elements of one type of data (which is labeled n_1) and then identify the number of elements of the other type of data (which is labeled n_2). For our data, we have 10 pluses (and thus $n_1 = 10$) and 5 negatives (therefore $n_2 = 5$). If there had been a case of no change in the DJIA (that is, a tie), that case would not have been counted.

Drawing a Statistical Conclusion   If n_1 and n_2 are each equal to or less than 20,[2] we begin the test of the null hypothesis by referring to the tables in Appendix 10. These tables are based on the assumption that H_0 is true, and they provide critical values of r based upon n_1, n_2, and a level of significance of .05. The following *decision rule* is used to compare the sample r value with the table r value:

The null hypothesis should be rejected if the sample r value is *equal to or less than* the appropriate r value from Appendix 10, table (*a*); the H_0 should also be rejected if the sample r value is *equal to or greater than* the table r value found in Appendix 10, table (*b*).

Since we have $n_1 = 10$ and $n_2 = 5$, the corresponding r value from Appendix 10, table (*a*), is 3, and the r value from Appendix 10, table (*b*), is 12. Appendix 10 thus tells us that in a random sequence of 15 observations where 10 pluses and 5 minuses are noted, the chances of getting 3 or less or 12 or more runs is only 5 percent. Since the sample r is 7 and thus falls between the table values, we cannot reject the null hypothesis. Seven runs are quite likely in a *random sequence* of 15 observations that are similar to our sample data. The financier may therefore conclude that there has been no detectable pattern of behavior in the stock market for the past 15 days.

Figure 15-8 summarizes the procedure for conducting a runs test for randomness.

Self-Testing Review 15-4

1   What is a runs test?
2   Does a runs test need two independent samples?
3   What is the alternative hypothesis in a runs test?
4   The runs test is only concerned with detecting a pattern; it is not concerned with the type or direction of the pattern, if any is detected. True or false?
5   If the sample r value is less than the lower table r value or greater than the upper table r value, the alternative hypothesis is accepted. True or false?
6   The tables in Appendix 10 may be used when n_1 and n_2 are each greater than 20. True or false?
7   In a runs test for randomness, there are 10 runs in the data sequence. The value of n_1 is 19, and the value of n_2 is 14. At the .05 level, should the H_0 be accepted?

SPEARMAN RANK CORRELATION COEFFICIENT

The **Spearman rank correlation coefficient** (r_s) is a measure of the closeness of association between two ordinal variables; that is, r_s is a measure of the degree of relationship between *ranked* data. The coefficient of correlation (r) found in

---

[2] Runs test procedures are available when n_1 or n_2 are > 20, but space constraints prevent us from considering these procedures in this text.

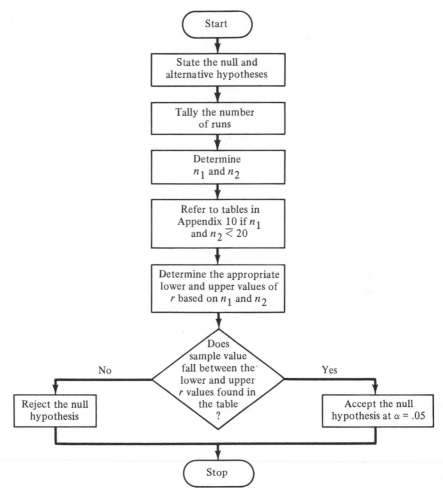

FIGURE 15-8
Procedure for the runs test.

Chapter 14 was computed using the actual values of X and Y; the Spearman measure we'll now consider uses rank values for X and Y rather than actual values.

Procedure for Computing the Spearman Rank Correlation Coefficient

The Ajax Insurance Corporation has been operating a sales refresher course designed to improve the performance of its sales representatives. A number of classes have completed the course. In an attempt to assess the value of the program, the sales training manager wants to determine if there's a relationship between performance in the program and subsequent performance in generating annual sales. Figure 15-9 shows the data collected by the sales training manager on 11 ($n = 11$) program graduates.

FIGURE 15-9   Data for the Computation of the Spearman Rank Correlation
Coefficient

SALESPERSON	COURSE PERFORMANCE RANK (1)	ANNUAL SALES RANK (2)	DIFFERENCE BETWEEN RANKS (1 − 2) D (3)	D^2 (4)
Steele	1	4	−3	9
Spier	2	6	−4	16
Devine	3	1	2	4
Hanlon	4	2	2	4
McCabe	5	7	−2	4
Braman	6	10	−4	16
Seville	7	3	4	16
McNally	8	5	3	9
Reid	9	8	1	1
Silva	10	9	1	1
Gould	11	11	0	0
			$\Sigma D = 0$	$\Sigma D^2 = 80$

$$r_s = 1 - \left(\frac{6 \Sigma D^2}{n(n^2 - 1)} \right)$$

$$= 1 - \left(\frac{6(80)}{11(121 - 1)} \right)$$

$$= 1 - .364$$

$$= .636$$

Ranking the Data   As a first step, the manager ranked each of the 11 representatives according to his or her performance in the sales course. A rank of 1 was
assigned to the person with the best performance; a rank of 2 was given to the
next best graduate; and so on. Then, each salesperson was ranked according to
sales performance in the subsequent year. A rank of 1 was assigned to the person
who had the most sales; a rank of 2 was given to the one with the next highest
sales; and so on. For example, sales representative Steele was considered the
best among the persons who attended the sales course, and Steele had the fourth
highest sales in the 12 months following completion of the program.

Computing the Differences between Ranks   The next step is the systematic computation of the differences between ranks. These differences, labeled D, are
shown in the third column of Figure 15-9. Since sales representative McCabe
achieved a rank of 5 for course performance but had a *lesser* rank of 7 for sales
performance, the difference assigned to McCabe is -2.

Computing r_s After computing D for each person, the manager is ready to
compute the Spearman measure, which is defined as follows:

$$r_s = 1 - \left[\frac{6 \Sigma D^2}{n(n^2 - 1)} \right]$$ (formula 15-5)

```
MTB > PRINT C1

COURSE
     1     2     3     4     5     6     7     8     9    10    11

MTB > PRINT C2

SALES
     4     6     1     2     7    10     3     5     8     9    11

MTB > RANK 'COURSE', PUT RANKS IN C3
MTB > RANK 'SALES', PUT RANKS IN C4
MTB > CORRELATION  C3, C4

Correlation of C3 and C4 = 0.636
```

FIGURE 15-10
A Minitab calculation of r_s.

To compute r_s, we must square the differences between ranks and then sum the squared differences—that is, perform the operations represented by ΣD^2 in the numerator of formula 15-5. The last column in Figure 15-9 shows the sum of the squared differences. The computations shown in Figure 15-9 give us a value of r_s of .636. The same result is produced by the *Minitab* statistical package in Figure 15-10 when it's supplied with our problem data.

As a basis for interpreting r_s, you should keep in mind that when r_s (like r in the last chapter) is zero, there's no correlation. And, like r in Chapter 14, when r_s is $+1.00$ or -1.00, there's perfect correlation. In our example, therefore, the manager might conclude that there's a correlation between course performance and subsequent sales activity.

Testing r_s for Significance

A more formal test may be conducted to see if there's truly a statistical relationship as suggested by r_s. A null hypothesis is established to the effect that there's no relationship between course performance and sales performance—that is, $H_0: r_s = 0$. Since the training manager prefers to believe that the course improves selling ability, a right-tailed test is appropriate, and the alternative hypothesis is that there's a positive relationship between course performance and sales performance—that is, $H_1: r_s > 0$. Let's assume that we wish to conduct the test at $\alpha = .05$. The essential question in our hypothesis test is how likely is it that we could have obtained the sample value of $r_s = .636$ if there was truly no relationship between the two variables?

If the sample size is larger than 10, we may conduct a hypothesis test by computing the critical ratio (CR) as follows:

$$CR = r_s \sqrt{\frac{n-2}{1-r_s^2}}$$

(formula 15-6)

On the basis of our data, we have

$$CR = .636 \sqrt{\frac{11 - 2}{1 - .636^2}}$$

$$= .636 \sqrt{\frac{9}{1 - .404}}$$

$$= +2.47$$

After computing the critical ratio, we're now ready to draw a conclusion based upon the following *decision rule* for a right-tailed test at the .05 level of significance:

Accept the H_0 if the CR $\leq$ the appropriate t value.

or

Reject the H_0 and accept the H_1 if the CR $>$ the appropriate t value.

What's the appropriate t value? The appropriate t value, in case you've forgotten, is found in Appendix 4. The correct df (degrees of freedom) row to select at this time is determined by using $n - 2$, since we have two variables (course performance and sales performance). The levels of significance given in the columns of Appendix 4 are for one-tailed tests. Thus, in our example the t value for $n - 2$ $(11 - 2)$ or 9 degrees of freedom at the .05 level is 1.833. Since our CR = 2.47, and since the appropriate t value = 1.833, the null hypothesis is rejected. We can conclude that there's a statistical relationship between participation in the sales course and subsequent sales performance.

Figure 15-11 summarizes the procedure for computing r_s and testing its significance.

Self-Testing Review 15-5

1   What is the Spearman rank correlation coefficient?
2   What may be concluded if $r_s = +1.36$?
3   When $n > 10$, the significance of r_s may be tested through the use of formula 15-6. True or false?
4   If $\Sigma D^2 = 566$ and $n = 16$, what is r_s?
5   a   If $r_s = .67$ and $n = 13$, then CR = 2.43. True or false?
    b   At the .01 level of significance, would the H_0 be accepted with a one-tailed test? (Use the data given in part a of this question.)

LOOKING BACK

1   A researcher is often limited by the imprecise quantitative properties of the data available for analysis purposes. For example, the data may come from only a few

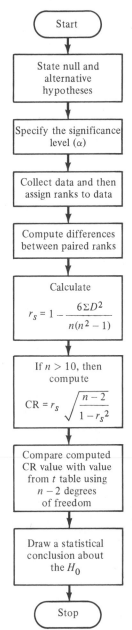

Start

State null and alternative hypotheses

Specify the significance level (α)

Collect data and then assign ranks to data

Compute differences between paired ranks

Calculate

$$r_s = 1 - \frac{6\Sigma D^2}{n(n^2 - 1)}$$

If $n > 10$, then compute

$$CR = r_s \sqrt{\frac{n-2}{1-r_s^2}}$$

Compare computed CR value with value from t table using $n - 2$ degrees of freedom

Draw a statistical conclusion about the H_0

Stop

FIGURE 15-11 Procedure for computing and testing the Spearman rank correlation coefficient.

samples, and there may be little or no knowledge of the shape of the population distribution and its effect on the sampling distribution. When such problems arise, nonparametric techniques may be used. In this chapter, we've discussed only a few of the common nonparametric techniques, and we've limited our attention primarily to small-sample cases.

2   When one wishes to find out if there are significant differences between paired rank data which are drawn from one sample or from two related samples, the sign test or the Wilcoxon signed rank test may be appropriate. If the *magnitude* or size of the differences between paired data is to be considered in decision making, the Wilcoxon signed rank test should be employed; if only the direction of differences is needed for a decision, the sign test is sufficient.

3   The Mann-Whitney test, or *U* test, should be employed when differences between paired rank data are under study *and* when the data are drawn from two independent samples.

4   In the case of a single sample with sequential data, a runs test for randomness may be conducted. This test is designed to detect the presence or absence of pattern or order in the sequential data.

5   Finally, the Spearman rank correlation coefficient was discussed. This measure is a correlation coefficient for paired rank data. The computation of r_s provides a measure of association between two variables.

KEY TERMS AND CONCEPTS

Nonparametric statistics   *581*
Ordinal (rank) data   *581*
Nominal data   *582*
Sign test   *582*
$$CR = \frac{2R - n}{\sqrt{n}} \quad \text{(formula 15-1)} \quad 585$$
Wilcoxon signed rank test   *587*
Mann-Whitney test   *591*
$$U = n_1 n_2 + \frac{n_1(n_1 + 1)}{2} - R_1 \quad \text{(formula 15-2)} \quad 592$$

$$U = n_1 n_2 + \frac{n_2(n_2 + 1)}{2} - R_2 \quad \text{(formula 15-3)} \quad 592$$
Lesser value of $U = n_1 n_2 -$ larger value of U
  (formula 15-4)   *593*
Runs test for randomness   *594*
Spearman rank correlation coefficient (r_s)   *596*
$$r_s = 1 - \left[\frac{6\Sigma D^2}{n(n^2 - 1)}\right] \quad \text{(formula 15-5)} \quad 598$$

$$CR = r_s\sqrt{\frac{n - 2}{1 - r_s^2}} \quad \text{(formula 15-6)} \quad 599$$

PROBLEMS

1   Polly Esta, owner of Natural Textiles, Ltd., is concerned about the chronically low daily output of her factory workers. Therefore, she has devised a bonus plan and wants to find out if this bonus plan will result in any improvement. (She is not overly concerned at this time with the degree of productivity improvement.) In an experi-

ment, eight workers were offered the bonus plan. Their daily output before and after the bonus plan is shown below:

WORKER	UNIT OUTPUT BEFORE BONUS PLAN	UNIT OUTPUT AFTER BONUS PLAN
Harris Tweed	80	85
Stitch N. Tyme	75	75
Les Hemm	65	71
Tom Taylor	82	79
Chuck Moore	56	68
Tex Tile	70	86
John Trim	73	71
Mat Wool	62	59

a   What is the alternative hypothesis?

b   What is n? What is r?

c   On the basis of the H_0, what's the probability of obtaining a value as high as r given the size of n?

d   If the null hypothesis is rejected, what's the level of significance?

2   The marketing director of the National Shampoo Company is wondering if producing the existing green shampoo in a darker shade will improve consumers' perception of product effectiveness. At this point, the director simply wants to determine the feasibility of developing the idea further and is just looking for some degree of improvement in perception of product effectiveness. Data have been collected from seven persons; each has rated the original green shampoo and darker-shade shampoo. A 1-to-10 scale was used where 1 stood for "very ineffective" and 10 signified "most effective." The data are shown below:

CONSUMER	EFFECTIVENESS RATING FOR ORIGINAL-COLOR SHAMPOO	EFFECTIVENESS RATING FOR DARKER-COLOR SHAMPOO
Abe Bell	4	2
Will Ling	6	6
Peg Brown	7	4
Dan D. Ruff	5	6
Sue Weese	9	8
Jack Sprat	1	3
Jim Hawkins	3	8

a   What is the alternative hypothesis?

b   What is the value of n? The value of r?

c   If the H_0 is rejected, what's the probability of error in judgment?

3   The Cal Q. Leighter Computer Company employs 500 salespeople. In an attempt to reduce the amount of time needed to close a sale, the company has produced a visual-aid package to be used in sales presentations. So far, only 10 salespeople have requested and used the visual-aid package. When each of these salespeople made the request to use the package, he or she was asked to estimate the amount of time

usually needed in a presentation to make a sale. After each one used the visual aids for 2 months, he or she was again asked to estimate how much time it took to make a sale. The data are shown below:

	AVERAGE TIME NEEDED TO MAKE A SALE	
SALESPERSON	BEFORE USE OF VISUAL AIDS	AFTER USE OF VISUAL AIDS
A	23	17
B	45	43
C	36	36
D	42	37
E	25	20
F	33	39
G	28	31
H	25	21
I	35	27
J	30	40

a What is the alternative hypothesis?

b If the .05 level is specified, would the H_0 be rejected?

4 Assume we are conducting a sign test, and in determining the differences between paired data, we have the following facts:

SUBJECT	DIFFERENCES BETWEEN PAIRED OBSERVATIONS
A	+
B	−
C	−
D	+
E	−
F	0

The alternative hypothesis is that the probability of getting a negative sign is greater than .50. If the null hypothesis is rejected, what's the level of significance?

5 Assume that you have the following facts for a sign test: $n = 15$, $r = 3$, two-tailed test, and a required $\alpha = .05$. Would the null hypothesis be rejected?

6 Conduct a sign test based upon the following data:

SUBJECT	DIFFERENCES BETWEEN PAIRED OBSERVATIONS
a	+
b	−
c	−
d	−
e	−
f	−

If we assume that the alternative hypothesis is that a negative sign is more likely than a positive sign, at what level of significance would the H_0 be rejected?

7 The Bovine Dairy Association sponsored a series of 30-second commercials promoting milk consumption. Eighteen stores were asked to record their unit sales of milk for the week prior to the campaign. After the campaign appeared on television, the same 18 stores were asked to report their sales for the week immediately following the airing of the commercials. The data were as follows:

	UNIT WEEKLY SALES	
STORE NAME	BEFORE CAMPAIGN	AFTER CAMPAIGN
Jones	124	136
Ma & Pa	107	105
Granny's	82	89
Ralph's	114	128
J & A	940	1,080
Korner	75	85
Superette	105	105
Mike's	94	95
Buy More	865	985
Value	620	820
Pete's	80	75
Foodco	750	725
Koop	330	350
Speedy	110	112
Walt's	125	120
Big Bag	400	425
Pay Now	400	450
Plus	175	215

a Conduct a sign test at the .10 level.

b Conduct a Wilcoxon signed rank test at the .05 level.

c Assume that 18 additional stores had been contacted and had recorded "before" and "after" sales data. Assume also that the following results had been obtained:

$$+ \text{ differences } = 24$$
$$- \text{ differences } = 10$$
$$\underline{0 \text{ differences } = 2}$$
$$36$$

Conduct a sign test with $\alpha = .05$.

8 The True Grit Sand Company has two operating pits in the Boston area. It has always been the suspicion of the owner of True Grit that location B is more productive than location A simply because of the geography of the areas; that is, the difference in productivity between the two areas is not attributable to differences in personnel or machines. In an effort to confirm this belief, the owner monitored the weekly output of 12 workers in location A and then transferred these workers to location B. The

output of these 12 workers in location B was also monitored for a week. The results are shown below:

	WEEKLY OUTPUT	
WORKER NAME	LOCATION A	LOCATION B
Spade	100	105
Dozer	150	145
Trukk	160	163
Graider	95	95
Levell	110	118
Bobb	87	90
Pile	135	143
Rock	125	129
Pebble	98	86
Sands	142	145
Dunes	110	85
Gravell	130	132

a   Conduct a sign test with $\alpha = .05$.

b   Conduct a Wilcoxon signed rank test with $\alpha = .01$.

9   A clinical pharmacist is wondering if a new painkiller is effective for persons with chronic pain. She believes that the new drug should significantly reduce the amount of pain. She wishes to record not only the change in pain that occurred after the dosage of the drug but also the amount of the change. Using an accepted measurement instrument, she recorded the pain level of 8 patients before and after the drug had been administered. A high score on this scale corresponds to a high degree of pain. The data are:

PATIENT	PAIN LEVEL BEFORE TAKING THE DRUG	PAIN LEVEL AFTER TAKING THE DRUG
A	14	8
B	15	9
C	10	11
D	12	10
E	11	11
F	13	9
G	12	11
H	10	10

a   What are the null and alternative hypotheses?

b   Assuming that the pharmacist specified that $\alpha = .05$, what should be concluded about the effectiveness of the new drug?

10   Assume that you are conducting a Wilcoxon signed rank test and that the differences between paired observations are as follows:

SUBJECT	DIFFERENCES BETWEEN PAIRED OBSERVATIONS
A	+3
B	0
C	−1
D	+8
E	+4
F	−2
G	+1
H	+6

a   What is the sum of the positive ranks?
b   What is the sum of the negative ranks?
c   What is the computed T value?
d   With a two-tailed test and a specified α of .05, would you reject the null hypothesis?

11   Assume that the alternative hypothesis in a test is as follows:

H_1: The probability of a decrease is greater than the probability of an increase

Conduct a Wilcoxon signed rank test at $\alpha = .01$ for the following data:

SUBJECT	DIFFERENCES BETWEEN PAIRED DATA
1	+6
2	−9
3	+2
4	−4
5	−3
6	+1
7	0
8	−1
9	−5
10	+3
11	−2
12	0
13	−7
14	−10

12   Use the data in problem 1 to conduct a two-tailed Wilcoxon signed rank test at the .05 level of significance.
13   Conduct a two-tailed Wilcoxon signed rank test on the data in problem 3. Use the .01 level of significance.

14 A high school counselor is wondering if persons who scored high on the verbal section of a college entrance exam will perform just as well in a business school as those who scored high on the math section of the same entrance exam. The grade-point average (GPA) of a sample of students was selected (a 4.0 represents an A and 1.0 represents a D), and the following data were obtained:

Students with High Verbal Scores		Students with High Math Scores	
NAME	COLLEGE GPA	NAME	COLLEGE GPA
Chipps	2.4	Boole	3.1
Hawthorne	3.2	Pythags	2.3
Walden	3.9	Chebyshev	1.9
Canterbury	1.6	Bayes	2.1
Emerson	2.2	Sine	2.7
Jones	2.5	Cosine	3.6
Smith	2.4		

Conduct a two-tailed test at the .01 level of significance.

15 The Flatt Tire Company has been testing a new emergency tire inflator which is supposed to be significantly faster than the leading competitor's inflator. Motorists were randomly selected for the product testing. Some of the subjects were assigned to use the new product, while the remainder were to use the leading competitor's product. The seconds required to inflate a tire are recorded below:

FLATT TIRE INFLATOR	LEADING COMPETITOR'S INFLATOR
17	23
16	21
21	32
19	21
15	19
14	20
16	21
16	22
23	

Assuming that you are conducting the research, what statistical decision would you make at the .05 level?

16 A job counselor believes that college graduates tend to be more satisfied in their jobs than non-college graduates. A job-satisfaction test was administered to subjects classified under each category. (A high score indicates a high degree of job satisfaction.)

The following results were obtained:

SUBJECT	NON-COLLEGE GRADUATE		SUBJECT	COLLEGE GRADUATE
a	102		aa	78
b	87		bb	93
c	93		cc	101
d	98		dd	85
e	95		ee	84
f	101		ff	77
g	92		gg	92
h	85		hh	86
i	88			
j	95			
k	97			
l	96			

Make a statistical decision at the .05 level.

17 A psychologist hypothesized that students from high school A tended to be more aggressive than students from high school B. A psychological test was administered to randomly selected students from each school. A high score on this test represented a high degree of aggression. The following results were obtained:

HIGH SCHOOL A			HIGH SCHOOL B	
STUDENT	TEST SCORE		STUDENT	TEST SCORE
Jim Jungle	43		Frank Mild	47
Mike Tuff	56		John Plain	68
Bill Bully	31		Bobby Blah	39
Sam Shove	30		Ken Kwiat	29
Tom Truant	41		Carl Calm	36
Steve Skipp	38		Dave Dull	42
			Gary Good	33
			Kurt Kind	54

Make a statistical decision at the .05 level.

18 Gary Gullible was asked by a friend to engage in flipping a coin for money. Gary's friend would win a dollar each time a head (H) appeared, while Gary would win a dollar for each tail (T) that appeared. After 20 flips, poor Gary was down by $6. Since the coin was provided by the friend, Gary began to wonder if the coin was "loaded." Here is the sequence of results:

T T T H H H T H H H H H T T T H H H H H

What can you tell Gary with $\alpha = .05$?

19 The Economics Institute has developed a new forecasting model and is anxious to learn if errors between its estimates and actual results are truly random or if there's a pattern to the errors. A series of 25 estimates was generated and compared with actual results. The errors of overestimation ($+$) and underestimation ($-$) are shown below:

$+ + - + - + - - - - - - + + - - - - + + + + + - + +$

What conclusions can you reach at the .05 level?

20   Over a 22-day period, the supervisor of a trenching crew monitored productivity to determine how the crew was performing relative to its quota. The supervisor was interested in learning if the daily work above ($+$) or below ($-$) the quota was following a random pattern. The observations were as follows:

$$- \; - \; - \; - \; - \; + \; - \; - \; - \; - \; + \; + \; - \; + \; - \; - \; - \; - \; + \; + \; -$$

Make a statistical decision at the .05 level.

21   Conduct a runs test at $\alpha = .05$ for the following data series:

$$+ \; + \; + \; + \; - \; - \; - \; - \; - \; - \; + \; + \; + \; + \; - \; - \; + \; + \; + \; + \; - \; + \; + \; + \; + \; + \; + \; +$$

22   Conduct a runs test at $\alpha = .05$ for the following data series:

H H H H H T H H H T T T T T T H H H H H

23   Abner Doubleplay, pitching coach for the Plainville Cougars, has noticed that in recent years some of the successful pitchers in the league were overweight. This has led Abner to wonder if weight affects pitching performance. Weight figures and winning percentages have been gathered for 21 league pitchers. The heaviest pitcher was assigned a weight rank of 1, and the pitcher with the highest winning percentage was assigned a winning rank of 1. The results are shown below:

PITCHER	WEIGHT RANK	WINNING RANK
a	3	6
b	7	1
c	15	21
d	10	2
e	2	9
f	13	13
g	6	8
h	21	5
i	8.5	19
j	1	12
k	12	4
l	14	14
m	17	18
n	4	20
o	18	11
p	8.5	7
q	11	16
r	20	3
s	19	15
t	16	17
u	5	10

What conclusion can you draw at the .05 level of significance?

24 Mickey Bubbles, sales manager of the Cool Cola Bottling Corporation, wants to know how strong a relationship (if any) there is between daily temperature and sales on corresponding days. Because of poor record-keeping procedures, Mickey must make do with rank data (where the hottest day has been assigned a 1 and the highest sales figure has been assigned a 1). Fifteen days were selected randomly, and the paired data are as follows:

TEMPERATURE RANK	SALES RANK
6	5
11	12
4	2
7	7
1	4
12	14
8	10
2	1
15	15
14	13
5	3
10	9
13	11
9	8
3	6

What conclusion may be drawn at the .01 level?

25 A psychologist believes that those who score high on a need-achievement test will likely have a high salary to match. To test this theory, the psychologist has administered questionnaires to 17 persons and has ranked the data in such a way that the highest value in each category has been assigned a 1. The paired data are:

NEED-ACHIEVEMENT RANK	SALARY RANK
1	3
8	7
4	2
10	12
12	9
2	1
13	11
6	6
16	17
11	13
14	15
3	5
9	10
7	8
15	14
17	16
5	4

What conclusion may be reached at the .01 level?

26   According to news reports, people in a mountain region of the country of Placebo claim that many of their neighbors live past 100 years of age. The Information Minister of Placebo claims that such longevity is related to the consumption of native cucumbers. Professor Pry doubts that there's any association (either positive or negative) between the age of a person and the annual consumption of cucumbers. The Placebo government has permitted Professor Pry to randomly select and interview 15 residents of the mountain region. Due to the lack of official records, the information provided by the subjects on age and cucumber consumption was not in a precise format. Therefore, the following data from the subjects must be converted to ordinal data for analysis:

SUBJECT	REPORTED AGE	REPORTED ANNUAL CUCUMBER CONSUMPTION
Ben Dover	102	156
Stan Strait	136	175
Al Bowe	98	134
Rip V. Winkle	110	143
Nee Kapp	106	129
I. Claude Jawn	156	164
Howard Hertz	92	124
S. Keemo	89	110
Hugo First	143	160
Rip Mend	124	109
Red Hott	94	95
Hott N. Tott	105	120
Sy N. Nara	117	133
Ive Haddit	108	119
Hal Widdit	97	101

If a rank of 1 is assigned to the lowest value in each category, and if $\alpha = .01$, what conclusion should be made?

TOPICS FOR REVIEW AND DISCUSSION

1   What are nonparametric statistics?
2   What are some examples of nominal data and ordinal data?
3   How does the sign test differ from the Wilcoxon signed rank test?
4   What do the Wilcoxon signed rank test and the U test have in common?
5   "The results of a runs test do not permit us to make conclusions about the types of patterns in sequential data if patterns are detected." Discuss this statement.
6   What is the major difference between the parametric and the nonparametric correlation coefficients?

PROJECTS/ISSUES TO CONSIDER

1   Team up with other class members and identify some experiment of interest to the team. For example, the team might consider a marketing research or consumer preference experiment similar to the Chicken Out problem discussed in the chapter. Instead of a batter mix, the team could experiment with unlabeled brands of soft drinks or other consumer products. Or the experiment could deal with a campus election or a topic suggested by one of the problems presented at the end of this chapter. Regardless of the research topic, though, each team should:
    a   Develop a rating scale or questionnaire.
    b   Select a sample and collect the sample data.
    c   Analyze the data using one or more of the procedures discussed in this chapter.
    d   Summarize the experiment, methodology, and analysis results in a class presentation.
2   Locate a journal article that makes use of a nonparametric statistical technique (journals in the behavioral sciences are likely sources). Prepare a written report that summarizes the nature of the research discussed in the article, and explain how a nonparametric tool helped the author achieve the research goal.

ANSWERS TO SELF-TESTING REVIEW QUESTIONS

15-1

1   A sign test is conducted to determine if there are real differences between paired ordinal data drawn from a single sample or two closely related samples; the test is based upon the signs of differences between pairs of data.
2   The statement is incorrect. A two-tailed test may be conducted.
3   In order to conduct a sign test, it's necessary to observe or measure each subject twice.
4   The null hypothesis in a sign test is that the probability of a positive sign occurring is equal to the probability of a negative sign occurring. In other words, the median difference between the paired data should be zero.
5   a   The binomial probability distribution should be used in a sign test when the sample size is small.
    b   The normal approximation to the binomial probability distribution may be used in this case.
6   a   False. The number of relevant data is 12 ($n = 12$). Also, r equals 5, which is the smaller sum of the two signs.
    b H_0: $p = 0.5$
        H_1: $p > 0.5$
        $\alpha = .10$

        With $n = 12$ and $r = 5$, the sum of the relevant probabilities is .3867 (.0002 + .0029 + .0161 + .0537 + .1204 + .1934). Since .10 < .3867, the H_0 is accepted.

7 H_0: $p = 0.5$
   H_1: $p \neq 0.5$
   $\alpha = .05$

Decision rule: Accept H_0 if CR falls between ± 1.96.

$$\text{CR} = \frac{2R - n}{\sqrt{n}} = \frac{2(16) - 42}{\sqrt{42}} = \frac{-10}{\sqrt{6.481}} = -1.543$$

Decision: Accept H_0, since the CR falls between ± 1.96.

15-2

1 The Wilcoxon signed rank test incorporates the magnitude as well as the direction of the differences between paired ordinal data.
2 The null hypothesis in a Wilcoxon signed rank test is that there's no real difference between the paired data.
3 True.
4 False. The statistic T is the lesser of the two sums of the ranks, and therefore we have $T = 20$.
5 The table T value is 159.
6 True.

15-3

1 The data for a Mann-Whitney test are collected from independent samples, whereas the data for a sign test are collected from one sample or two related samples.
2 The Mann-Whitney test and the U test are the same.
3 False. The samples in a Mann-Whitney test do not have to be equal.
4 True. In the initial ranking of the data, the data from the two groups are aggregated and then ranked irrespective of the sample category.
5 True.
6 The critical table value of U is 41.

15-4

1 A runs test is designed to determine the existence or nonexistence of a pattern in sequential data.
2 The data for a runs test are drawn from one sample.
3 The alternative hypothesis in a runs test is that there's a pattern in the sequence of data.
4 True.
5 True.
6 False. Use of the tables is permitted when the size of each sample is equal to or less than 20.
7 The H_0 should be rejected. The lower table r value is 11, and since the sample r of 10 falls below this table value, the H_0 cannot be accepted.

15-5

1 The r_s is a measure of association between ordinal data.

2 It must be concluded that the coefficient was computed incorrectly, since r_s may have values between -1.00 and $+1.00$ only.

3 True.

4 The answer is computed as follows:

$$r_s = 1 - \left[\frac{6\Sigma D^2}{n(n^2 - 1)}\right] = 1 - \left[\frac{6(566)}{16(256 - 1)}\right] = .1677$$

5 a False. CR $= 2.99$

$$\text{CR} = .67 \sqrt{\frac{13 - 2}{1 - .67^2}} = 2.99$$

b The H_0 would be rejected. The t value at 11 df and $\alpha = .01$ is 2.718, which is $<$ the CR of 2.99.

A CLOSER LOOK

THE ART OF LEARNING FROM EXPERIENCE

"Statistics," says Bradley Efron of Stanford University, "is quite underappreciated." Most people who think of statistics at all consider it as simply a tool—a way to tell if data are significant or to estimate confidence intervals. But statistics is a deeply philosophical subject that tries to get at how we learn from experience. It is a dynamic field, full of arguments and beginning to change its very nature as its practitioners exploit the power of large-scale computing.

Efron, 45, is one of the leaders in the new statistics. He has invented an extremely promising new statistical tool, called "the bootstrap" and which, he says, "substitutes computing for thinking."

Although statistics is often thought of as a branch of mathematics, it actually lies on the border between mathematics and philosophy. "Obviously," says Efron, "statistics has mathematical structure—that's the only way anyone has found to say things in statistics." But the subject matter of statistics does not concern itself with typical mathematical reasoning in which results are deduced from axioms. Its logic goes in the opposite direction. Statisticians start with examples of things that are and try to determine what axioms could have given rise to them. "To step backward from what you've seen to what might have given rise to it is logically, mathematically, and actually difficult," Efron remarks. "We statisticians think deduction is child's play. In a sense, statistics is the most ambitious intellectual attack."

For Efron, the decision to become a statistician came only gradually and after he realized what he believes are his limitations as a mathematician. He always wanted to be a mathematician, he says, but he had a problem. "I was a terrific 19th-century mathematician. Give me a calculus problem and I could knock it dead. But I was not a very good twentieth century mathematician. I like to compute things. Modern mathematicians don't compute. They organize their ideas to another level of abstraction beyond calculations. I was terrible at things like modern abstract algebra. I have no mind for it at all."

Efron grew up in St. Paul, Minnesota, the son of a truck driver who was also an amateur mathematician. He learned from his father how to do calculations in his head. Set on becoming a mathematician, he majored in math at the California Institute of Technology, where he graduated second in his class. Then he started graduate school at Stanford, still majoring in math. But he was suspended from Stanford when, as editor of the school humor magazine, he published an article poking fun at religion. When he returned to Stanford, he returned to the statistics department.

Efron jokes that his suspension from Stanford will haunt him to his grave. "I often say that if I cure cancer, the Stanford newspaper story will begin, 'Bradley Efron, who once was kicked out of Stanford, today discovered a cure for cancer.'" But his reentry into statistics proved providential. Here was a field after his own heart—a field where computations reign supreme and where he could invent methods that would make statisticians compute even more.

Reflecting the interplay between logic and mathematics, there are what Efron calls "two opposing currents" in statistics. One approach, initiated in the early part of this century by the British statistician Sir Ronald Fisher, is based on the idea that the way to solve statistical problems is to get at their logical basis. The problem is that of uncertain inference. For example, if you say, "John is a man and men live less long than women," does that mean that you can conclude that John will live less long than his wife?

Efron gives another example of uncertain inference. "Suppose you say that penicillin is better than sulfa drugs for treating pneumonia. Well, how do you prove something like that? What do you mean? Do you mean that every patient will do better if you give them penicillin? If so, you will be sadly surprised when you do an experiment." In the real world, most things are not always true 100 percent of the time.

The second approach to statistics, which was initiated by Jerzy Neyman, who was at the University of California at Berkeley for much of his career, is to say that the problem of solving statistics problems is basically a mathematical optimization problem. Efron explains, "Usually every possible method gives some probability of making an incorrect statement when you interpret your data. Neyman's theory says that if you can find a method of interpretation that has the smallest possible chance of making an incorrect statement, you will have done the correct thing." The idea of optimization sounds appealing but, according to Efron, it has been carried out successfully only for very simple problems.

Both of these schools of thought contribute to statistics, and, in one of the major achievements of the field, both helped demonstrate the value of taking an average. This was work that went on for nearly 50 years, concluding around 1950. The question was, What should you do if you have accumulated data that are distributed as a bell shaped curve? "An obvious thing to do is to take an average," Efron remarks. "The triumph of the logical school was its demonstration that once you have the average you can throw away the rest of the data. You might as well use the average as use all of the data. Then the optimality people finished off the problem by showing that the average really does give you the best answer."

"When I was a student, the optimality school reigned supreme [at Stanford]," Efron says. "But the biostatisticians mostly used the results of the inference group. You'd be amazed at what contention it [the debate between the two philosophies] has caused over the years."

Statistics, Efron says, is "a slow entry field." It takes time to develop a feel for it. "There has never been a great 19-year-old statistics genius," Efron remarks. "It took me a good part of 15 years to get straight in my mind what I should work on."

Efron's most recent work has been on a general question that often plagues scientists: What would be seen if there were a lot more data? He tells, as an example, of an experiment he is analyzing for a researcher in Stanford's medical school who examined 120 human-mouse hybrid cells, 40 of which made a protein of interest. She wants to know how variable her results are. Might they have occurred by chance? If she had had 10 times as many hybrid cells, would there still be about one-third of them making the protein? In this case, Efron can use well-known methods to get the standard error and answer her question. But, he says, "This is an extremely simple situation. It is common to have 1000 pieces of information and 100 unknown parameters. Then you ask how accurate is your determination of one of those parameters. You never get to see the 'real' answer. The whole point of the analysis is to know what the range of possibilities would be if you took a lot more data."

Fisher, founder of the inference school of statistics, was a leader in developing a method, called the "maximum likelihood method," that provides estimates of averages and standard deviations for general problems. But, says Efron, "The theory becomes undependable in complicated situations, particularly if there are lots of unknown parameters. The standard error it gives can be quite a bad approximation and the worst thing is that you can't always calculate it. It involves putting together a probability model assuming such things as a bell shaped curve. What I've been trying to do is to develop automated methods that don't involve such assumptions. I noticed that a number of calculations could be automated and you were barely using the model."

Efron calls his method the bootstrap because "You use the data to estimate probabilities and then you pick yourself up by your bootstraps and see how variable the data are in that framework." As an example of how the method works, Efron tells of a study he and his Stanford colleague Persi Diaconis did of correlations between grade point averages (GPA) and scores on the Law School Admission Test (LSAT). The example is a simple one and the results that Diaconis and Efron obtain with the bootstrap could easily have been obtained with classical methods. But it demonstrates the bootstrap procedures in a straightforward way.

For 15 schools in 1973, the correlation between the average LSAT score and the average GPA at each school was 0.776, which means that the LSAT scores and GPA's were highly correlated. But does that mean that LSAT's and GPA's are highly correlated for *all* law schools? The bootstrap method gives a way to find out.

The first step is to copy the data for each school in the sample an enormous number of times—a billion times, for example. This creates a universe of 15 billion data points, 1 billion for each school. The computer selects from this universe random samples of 15 data points, called bootstrap samples, and calculates the correlation coefficient for each such sample. (In practice, Efron points out, the bootstrap samples are obtained without actually creating the 15 billion point universe by using random number generators to select among the original 15 points.)

When Diaconis and Efron looked at 1000 of these computer-generated correlation coefficients, they found that 68 percent of them were between 0.654 and 0.908. The conclusion of this bootstrap analysis is that the observed value of the correlation coefficient from a random sample of 15 bootstrap samples varies from the true value by 0.127.

Since they had data from all of the law schools in this country in 1973, Efron and Diaconis were able to test the predictions of their bootstrap model. The true correlation coefficient was 0.761, with a true variability of 0.135 in samples of size 15. Since the bootstrap variability estimate was 0.127, the bootstrap was quite accurate in this case.

Efron's method seems like magic, like a sleight of hand trick. And many statisticians instinctively distrusted it, "When I presented it to people they said it wouldn't work. Some said it was too simple. Others said it was too complicated," Efron recalls. David Freedman of the University of California at Berkeley, who also has lectured on the bootstrap, says he got the same sort of reactions. "In my field, a lot of people come from Missouri and want to be shown. Some you would even call curmudgeons," Freedman says. "I never generated quite so much opposition as when I was talking about the bootstrap. People were afraid it was all done with mirrors." Frederick Mosteller of Harvard University, who says he thinks the bootstrap "is a very good idea," nonetheless sympathizes with those who tend to doubt it works. "The bootstrap is a little hard to believe," he says. "It seems incestuous. You are trying to learn about the sample error by sampling the sample." Statisticians, Mosteller remarks, "are not ordinarily involved with something as anti-intuitive as this."

But, says Freedman, the whole point about the bootstrap is that it is not done with mirrors. "There are no free lunches in statistics. To draw conclusions about data, you have to make assumptions about processes that generated them. You're really using these assumptions as well as the data when you use the bootstrap."

The bootstrap is gradually coming into use, thanks partly to the efforts of Freedman and Peter Bickel of the University of California at Berkeley who helped establish its theoretical underpinnings. "I think it's a powerful tool," says Freedman. "I see it becoming one of the standard techniques in statistics. It will have a big influence on the field."

But the bootstrap is not perfect—no statistical method is. Freedman, Bickel, Jeffry Wu of the University of Wisconsin in Madison, and others have shown that there are situations in which the bootstrap does not work. "It can give drastically wrong answers and it is hard to say in advance when it will work, although we're beginning to get some pretty good guidelines," says Freedman. A number of statisticians, including Efron, are now trying to pinpoint when the method will work and when it will not.

Richard Olshen of the University of California at San Diego gives two examples of medical statistics problems in which he used the bootstrap. In one case, the method was perfect. In the other case, it had to be slightly modified.

The first example is in a study of the evolution of gait in children. Olshen, a statistician, together with orthopedic surgeon David Sutherland, engineer Edmund Bidden, and physical therapist Marilyn Wyatt, all at Children's Hospital and Health Center in San Diego, want to develop curves to establish angles of rotation of the legs, hips, and ankles in normal children from age 1 to age 7, by which time gait is established. These will be curves much like the height and weight curves that pediatricians use to determine whether children's growth is within the normal range. But the gait curves are much more difficult to develop.

The gait curves and the percentiles showing the distributions of normal gaits will be particularly valuable, Olshen says, in assessing in a noninvasive way the physical development of children with such conditions as mild cerebral palsy or muscular dystrophy. To develop the percentiles, Olshen had data on more than 400 normal children. He recalls, "Getting the averages of the curves and the shapes of the standard deviation curves was no big deal. But getting the percentiles theoretically is a problem that I don't know how to do. It took me a long time to see the obvious—that I should use the bootstrap." Efron's method, Olshen remarks, "was tailor made for this problem."

The bootstrap's limitations were apparent in a problem Olshen worked on with Lee Goldman and Harvey Fineberg of Harvard Medical School and their associates involving diagnosis of heart attack. When patients come into an emergency room complaining of chest pain, how can you quickly determine whether they had a heart attack? The investigators had data on emergency room patients from the Yale–New Haven Hospital and they wanted to develop a set of diagnostic criteria that they then would test on patients at Peter Bent Brigham Hospital in Boston. The goal was to make a "decision tree" that would tell the physicians at each step what decisions to make in diagnosing heart attacks.

The problem, says Olshen, is that "in diagnostic techniques you can have huge biases in the estimation of error rates. The bootstrap is very good for small biases but not as good for large biases." This is not particularly surprising, Olshen notes, because the bootstrap has little variability. There is, he says, "an uncertainty principle in statistics. Many techniques with little bias have much variability and vice versa."

In the heart attack study, Olshen was able to get around the limitation of the bootstrap by doing what is called "bias adjusting." He used the subjective knowledge of the physicians to "prune" the decision tree beyond what the bootstrap suggested. The resulting tree was so good that the investigators using it did better than physicians who used their medical judgment alone to determine whether patients with chest pain had heart attacks.

The problem of determining the limitations of the bootstrap and how and when to augment it with other methods is inordinately difficult. "One of the troubles with statistics as a field is that it is very difficult to prove that things do or do not work," Efron says. "There isn't exactly a 'real world' of data sets. You want methods to work for drug companies, econometrics, all data sets. Trying to say what that means is the subject of quite bitter discussions."

In the meantime, Efron is working on a new aspect of the bootstrap. He believes that the method can give better confidence intervals than more traditional techniques and, to show that, he has been laboriously working away on the desk-size pad of paper he uses for his theoretical statistics work. "I hate working on this," he remarks. "It's difficult and it's very slow going." Efron says the pad of paper is, to him, "torture." It is much more exciting to calculate.

But, in the end, it is often the results of such torturous theoretical work that statisticians find convincing. And Efron would very much like to win the entire statistics community over to his view of large-scale computing as the wave of the future. Still, he says, "I've taken a tremendous amount of guff. Statisticians are hard to convince. They tend to be very conservative in practice. And they should be. This stuff is serious. People use it."

—Gina Kolata, "The Art of Learning from Experience," *Science*, July 13, 1984, pp. 156–158. Copyright 1984 by the AAAS.

16

WHERE DO WE GO FROM HERE?

Getting the Data on AIDS

Any planner knows that when disaster strikes, limiting its damage and preventing its spread require information. But when the disaster is new—like AIDS—getting information poses special problems.

The statistical techniques that can measure the prevalence and estimate the spread of AIDS are centuries old, noted Rockefeller University professor Joel Cohen at a National Academy of Sciences symposium last year. So are the techniques that can judge the likely effectiveness of proposed policies, like mandatory blood testing. There is even a classic method (randomized response) to evaluate answers to surveys that probe topics that people don't want to talk about. We have the analytical techniques. The problem is getting the information and getting it fast.

Because AIDS probably surfaced in other countries before it appeared in the U.S., other countries are the first sources of data. The Census Bureau's Center for International Research, which already has an international demographic database, has been funded by the Agency for International Development to set up an epidemiological database for developing countries. Data from developing countries are "qualitatively" different from our own data, according to Barbara Torrey who directs the center. In particular, it is difficult to make reliable estimates from small samples of data from developing countries. Nevertheless, the data indicate that in

Africa at least, AIDS is an urban disease, it is spreading at an increasing rate, and women get the virus at a younger age than men, possibly because women tend to be younger than their sexual partners.

Data on the prevalence of AIDS can only come from blood samples. In a great stroke of good fortune the U.S. has a head start on tracking the history of AIDS in this country, because of a set of blood samples gathered in San Francisco in 1979 for a study of hepatitis among homosexuals. Getting samples from a broader range of Americans is the next step, and the Census Bureau, on behalf of the National Center for Health Statistics, has found that most Americans are willing to be tested for AIDS.

Last fall the bureau added a special supplement of questions to the National Health Interview Survey to measure Americans' knowledge of AIDS as well as their willingness to be tested for it. More than two-thirds of all adults are willing to be tested, including three-fourths of the adults under age 55. Instead of being angry when they were asked questions about AIDS, as the researchers had feared, most Americans thought the questions should have been asked sooner. However, people did object when the questions became personal, such as "Have you been tested for AIDS?" or "Do you know someone who has AIDS?" Those questions are tame compared with the ones researchers need to ask to determine what contributes to the spread of the AIDS virus.

MODELING AIDS

Another avenue researchers are following in their effort to get fast answers on the impact of AIDS is statistical modeling. That method could be especially useful in areas where essential data are lacking, like Central Africa. The National Academy of Sciences' Institute of Medicine brought modeling specialists together in a workshop last year, but a basic problem quickly emerged. The databases and specialists tend to be from developed countries, where homosexuals' sexual behavior and intravenous drug use are the main vehicles for transmitting the virus. In less developed countries, transmission is usually by heterosexuals. Models based on a homosexual or intravenous-drug pattern for the spread of AIDS will not apply to a country where it is being spread through heterosexual sex. Getting data on people's behavior is crucial, says Burton Singer, the U.S. statistician who chaired the statistical-modeling workshop. That means that demographers and statisticians need the help not only of epidemiologists but also of behavioral specialists—like cultural anthropologists—who can figure out how to ask the questions that will give the information needed to begin to cope with the AIDS epidemic.

—Martha Farnsworth Riche, "Getting the Data on AIDS," *American Demographics*, April 1988, p. 8. Reprinted with permission. © *American Demographics*, April 1988.

LOOKING AHEAD

There comes a time when one asks even of Shakespeare, even of Beethoven, "Is this all?"

—Aldous Huxley

The opening vignette discussed statistical modeling. That's one of the concepts we'll consider as we present sketches of the following topics in this final chapter: (1) sample design, (2) decision theory, (3) linear programming, (4) inventory models, (5) waiting line theory, and (6) simulation. These are important statistical procedures, but we can only give a few examples of their use and briefly outline a few of their limitations in the remaining pages.

Thus, after studying this chapter you should be able to:

- Discuss some of the purposes and limitations of the quantitative techniques mentioned in the following pages.
- Give examples of how the techniques discussed in the chapter may be used.

CHAPTER OUTLINE

SAMPLE DESIGN

Although stratified and cluster samples were mentioned in Chapter 6, all discussion of estimation and statistical inference that followed was based on simple random samples. In many cases, though, it's either impossible or impractical to take a simple random sample.

If there's a large amount of variation in a population, a large random sample may be needed to give the desired sampling error. The cost of such a large sample could be prohibitive. In a case like this it may be possible to divide the population into groups of like elements so that the variation within each group is relatively small. Samples could then be taken from each of these groups. Since the variation in each group is less than the variation for the entire population, the sample size required for this *stratified sample* would be smaller than that required for a random sample.

Suppose, for example, that you wish to take a sample of stores in a particular city to estimate total retail sales in the city for the previous month. The amount of variation in this population is tremendous, ranging all the way from the very large sales figures for large department stores and supermarkets to the small sales figures for small neighborhood stores. However, if the stores are *stratified* into four or five groups according to size, the amount of variation in each group is relatively small. *Stratified sampling is used* by public accountants to estimate the value of an individual item. It's also used in personnel studies with employees being stratified according to job classification. In short, this type of design is useful *when a very heterogeneous population can be stratified into fairly homogeneous groups*.

To take either a random sample or a stratified sample, it's necessary to have some kind of complete listing of the population being sampled. *This is often impossible*. For example, we might wish to take a sample of employees of retail stores in a city. It's often impossible to draw up a complete listing of all these people, but we could draw up a list of all retail stores in the city. We could then take a sample of the stores, and the employees of the sampled stores could be interviewed. *This type of sample design is known as cluster sampling*, since we are dividing the population into clusters of elements and then drawing samples of the clusters. *This type of design is often used in market surveys and election polls*. Clusters of people are sampled by drawing a sample of city blocks and then interviewing the people who live on these blocks.

Multistage sampling is simply a variation of cluster sampling in which only a sample of the second-stage units is selected. With regard to the sample of store employees discussed above, this would mean that for each store taken into the sample, a *sample* of the employees of that store would be selected. This is often an advantageous design, since there's likely to be more variation among clusters than within clusters, and multistage sampling allows more clusters to be taken into the sample.

It should be understood that *choosing the best sample design does not assure the validity of the sample results*. Errors can come from many sources—the way a question is worded, the way an interviewer asks the question, or the manner in which the population is defined. Some years ago a large oil company spent a considerable sum of money on a survey to determine brand preferences and buying motives for motor oil. The data that were collected turned out to be quite useless, simply because the company forgot to ask the people who were interviewed if they owned an automobile. The expense incurred in designing the sample could not remedy this defect.

DECISION THEORY

In Chapter 8 procedures were explained for testing statistical hypotheses. As a result of these procedures, decisions were reached to either accept or reject the stated hypothesis. Modern **decision theory** also takes into consideration the *mon-*

MONTE CARLO SIMULATION

This simulation technique got its name from the gambling resort at Monte Carlo, where the fundamental principles of statistics were discovered by gamesmen more than 100 years ago.

The original research had a simple goal: beat the house. The project failed and the casino's profits rolled serenely in, but researchers discovered the "law of large numbers," which says that you can figure out the characteristics of any apparently random process if you make enough trials. That was difficult to do before computers came along.

Even with computers, simulations tend to run for a *long* time. There is a great deal of arithmetic involved in computing several thousand different distribution numbers. Depending on what kind of machine you have, it could run all night if you need to solve a complex problem. Once it's done, an advanced simulation package can process the results in sophisticated ways and plot different probability distributions.

Monte Carlo analysis gives answers when you don't have the exact numbers needed for linear programming or decision trees. Suppose you plan to make toasters and irons. The time needed to make a product is not fixed: some toasters and irons will be defective and will need rework, while others will sail right through.

Suppose that 20 percent of the toasters and 10 percent of the irons will need rework. Sometimes rework may take 10 minutes, sometimes 15 minutes. If you want an estimate of how many irons and toasters you can build in a day, you need to know the average production time.

Monte Carlo analysis lets you define the problem in various ways. You can say that the time to rework a toaster is equally likely to be either 10 or 15 minutes; or you can say it varies between 10 and 15 minutes, with 12 minutes most likely; or you can say it's 60 percent at 10 minutes, 30 percent at 13 minutes, and 10 percent at 15 minutes; or you can tell the program to use a built-in probability distribution function.

If you use distributions, you have to understand which one fits your problem. Experience shows, for example, that shaft and hole sizes tend to have *normal* distributions, customer arrivals at banks or barbershops have *Poisson* distributions, noise bursts on telephone lines have *shot-noise* distributions, and so on.

After you have entered the process model, the program generates appropriately distributed numbers to simulate irons and toasters passing through the factory, then it measures the time each takes. After enough trials (perhaps several thousand), it computes the average assembly and finishing time per product. These numbers can be plugged into decision trees or linear programming packages or can be used as weights in decision matrices.

—From Jared and William Taylor, "Searching for Solutions." Reprinted from: *PC Magazine*, September 15, 1987, pp. 311–315 ff. Copyright © 1987 Ziff Communications Company.

etary values of the actions which might be taken and formulates a procedure for making the best decision under conditions of uncertainty.

To illustrate this procedure, let's consider the following simple decision problem: The manager of a small grocery store must decide how many loaves of bread to stock each day. On the basis of past experiences, she knows that the store has never sold less than 11 nor more than 14 loaves per day. *A procedure for analyzing this decision problem is as follows:*

1   *Construct a payoff table.* Since sales always range from 11 to 14, the alternative courses of action available to the store manager are to stock 11, 12, 13, or 14 loaves. A **payoff** (or **profit**) **table** can be constructed which will show

the amount of profit that would be made for each alternative course of action at each possible level of demand.

2   *Compute the expected monetary value of each action.* On the basis of past sales records, a probability distribution for demand can be developed showing the probabilities of selling 11, 12, 13, and 14 loaves in a day. The expected monetary value of each action can be computed by applying these probabilities to the payoff table. The **expected monetary value** of an action is the amount of average profit one would expect to make if this action were followed day in and day out over a long period of time.

3   *Choose the action that has the greatest expected monetary value.* An important aspect of modern decision theory is the use of additional information to *revise* the original probability distribution. For example, a manager trying to decide whether or not to market a new product can assess the probabilities of success or failure of the product on the basis of experience with similar products in the past, but he would probably like to have some more information before making a decision. Such information could be obtained by taking a survey of department store buyers or consumers. This new information could then be incorporated into the analysis through the use of a particular probability theorem formulated by Thomas Bayes to revise the original probabilities. Because of the extensive use of Bayes' theorem, modern decision theory is commonly referred to as **Bayesian decision theory.**

It's obvious that a *decision analysis is no better than the probability distribution used for computing the expected monetary values.* The decision maker should remember that these probabilities are based on what has happened in the past and may not reflect the current situation. There are many external factors that can cause the relevant probability to change. The invention of nylon changed the probability distribution for the demand for silk. Improvements in the quality of a raw material will change the probability distribution for the percentage of defective products that will be produced. The use of out-of-date probabilities would obviously lead to a bad decision.

LINEAR PROGRAMMING

Linear programming is a technique for maximizing or minimizing a linear function subject to certain linear constraints. While this definition might sound like a lot of jargonese double-talk, it's really not as formidable as it sounds. We are generally trying to maximize profits or minimize costs, and when we speak of constraints, we simply mean that there are limits on the values the variables can take. For example, in a profit maximization problem, there's a limit to the amount of each product we can produce. *To illustrate the uses of linear programming, let's consider the following four examples:*

1   A firm produces three products which have different profits per unit. Given information on the production time required by each product in each de-

partment and the total productive capacity of each department, linear programming can be used to determine the product mix that will maximize profit.

2   A company must blend together three ingredients to make a cattle feed which must meet certain specifications as to protein and vitamin content. Given the cost of each of the ingredients and information on the protein and vitamin content of each ingredient, linear programming can be used to determine the minimum-cost mixture that meets the required nutritional specifications.

3   A company has three factories and six warehouses in different parts of the country. Linear programming can be used to determine which factories should ship to which warehouses to minimize the total transportation cost.

4   A factory has five orders to fill. The profitability of each order varies depending on which of the available machines is used to do the work. Linear programming can be used to assign the jobs to machines in a way that maximizes the total profit.

All these examples deal with business problems, but we've not come close to listing all the more common business applications. And outside the business area, linear programming is used in dealing with health problems, pollution problems, and problems of welfare economics. In short, linear programming is an *allocation model,* and as such it should be considered any time there's a problem concerning the allocation of scarce resources.

Since linear programming allows us to determine the combination of factors that will give us the maximum profit or the minimum cost, it's obviously a technique that greatly aids the decision-making process. But like all quantitative techniques, it's useful *only* if it is used in the proper circumstances. The *important limitation on the use of linear programming is* implied in the definition stated at the outset of our discussion; that is, *the functions must be linear.* If the cost (or profit) function is nonlinear and we try to use linear programming to minimize (or maximize) it, the results will be meaningless.

INVENTORY MODELS

Businesspeople must make many critical decisions, but perhaps no decisions are more critical or more difficult to make than those concerning inventories. In this regard, managers can commit two costly mistakes—they can have too much inventory on hand, or they can have too little inventory on hand. If they are *overstocked,* they incur the extra cost of storing the unneeded inventory, and they lose money on leftover items they cannot sell. If they are *understocked,* they lose the profit they could have made on sales that were lost.

Mathematical models can be used to aid in solving these and other problems relating to inventory control. These **inventory models** are simply equations, sometimes coupled with relevant probability distributions, which describe the particular inventory system being studied.

One important use of inventory models is the determination of the economic order quantity. This is the quantity that should be ordered each time inventory is replenished to minimize the total of ordering costs and inventory carrying costs. Each time an order is placed, certain administrative costs are incurred— for example, the clerical cost of preparing the purchase order, the bookkeeping costs, the cost of checking in the order and verifying the invoice, and the cost of storage and warehousing, insurance on the inventory, and interest on the invest- ment in the inventory. Obviously, the ordering cost could be minimized by ordering the entire quantity needed at one time, but this would mean carrying a large inventory at a high cost. Carrying costs can be kept to a minimum by ordering small quantities each time, but this requires a large number of orders and, therefore, a high ordering cost. A model that describes the total cost of the inventory system allows us to determine the order quantity that minimizes this total cost.

Inventory models are also used to determine the inventory level at which new orders should be placed. If this reorder point is placed too high, the average level of inventory and the resulting carrying costs are unnecessarily large. On the other hand, if the reorder point is set too low, the inventory might be depleted before the new order arrives, and the firm loses sales and customer goodwill. An inventory model that incorporates a probability distribution for inventory usage makes it possible to determine the reorder point that minimizes the total of expected shortage costs and carrying costs.

If an inventory model is to yield useful results, it must accurately describe the inventory system *under study*. Before using a model developed for another system, it is essential that the two systems be compared to make sure they are identical. Many management textbooks present "the economic lot size formula." Actually, there's no such thing as *the* economic lot size formula. What these books are featuring is *an* economic lot size equation for an inventory system that meets the following conditions: (1) the demand for the product is known, (2) the demand is constant, and (3) the cost of running short is so great that shortages can never be allowed to occur. If this formula is used in a situation where one or more of these assumptions isn't valid, the result will most certainly not be a quantity that minimizes costs.

WAITING LINE THEORY

If you've tried to fight your way out of a crowded supermarket or tried to make a bank deposit on Friday afternoon, you don't have to be told what a waiting line is. You might be surprised, though, to know that there's a body of theory available for dealing with such problems. **Waiting line theory** includes provisions that make it possible to answer such questions as (1) What's the average length of the waiting line? (2) What's the average time a person has to wait for service? (3) What's the probability that there will be more than a certain number of people waiting in line? (4) What's the probability that a person will have to wait in line more than a certain amount of time?

IMAGINE YOU MAKE MUFFINS

Linear programming is a technique for determining the mathematically optimal solution to a problem. It can be used in situations in which resources are combined or consumed in different ways to give different results.

Imagine that you are a muffin maker. You have enough apples for 200 apple muffins, enough blueberries for 300 blueberry muffins, and enough ingredients for 1,600 ounces of muffin batter. One apple muffin takes 5 ounces of batter and yields a 16-cent profit. One blueberry muffin takes 4 ounces of batter and yields a 15-cent profit. How many of each should you make in order to get the largest profit? You might be tempted simply to make as many higher-profit apple muffins as possible, and then use any leftover batter for blueberry muffins. Linear programming can show you that that won't get you the highest profit.

The first step in solving the problem is to find its constraints. The vertical axis in the three graphs represents the number of apple muffins, and the black dotted line at 200 on the vertical axis is the maximum number it's possible to make with the apples on hand. The black dotted line at 300 on the horizontal axis is the maximum number of blueberry muffins you can make with the blueberries on hand. However, you can't make 200 apple muffins *and* 300 blueberry muffins because you don't have enough muffin batter. That would take $(200 \times 5) + (300 \times 4) = 2,200$ ounces. Sixteen hundred ounces of batter is enough to make 320 apple muffins $(320 \times 5 = 1,600)$ and zero blueberry muffins or 400 blueberry muffins and zero apple muffins. It's also enough to make any combination of muffins along the color line on the chart that joins these two points.

Now all the limits to production are plotted. You can make any combination of quantities of apple and blueberry muffins that falls within the area bounded by A, B, C, D, and 0 on the chart. But what's the combination that will yield the greatest profit? You can figure that out by first determining the slope of the line that represents profits. If, for example, you made 150 apple muffins and no blueberry muffins, you would get a profit of \$24 (150 × \$.16). To get that same profit by making only blueberry muffins and no apple muffins, you'd have to make 160 (\$24/\$.15) of them. The short black line in the lower graph joins those two points and shows every combination of apple and blueberry muffins that will yield a profit of \$24.

Clearly, the more muffins you make and sell, the greater your profit, so you move the short black line out, maintaining the same slope, to the point on the plane ABCD0 that is farthest from 0. That point is C, which represents a combination of 300 blueberry muffins and 80 apple muffins, and gives a total profit of \$57.80. No other possible combination of muffins will yield as great a profit.

This simple example would be easy to work out by hand. However, if there were more than two kinds of muffin, or more ingredients to be shared between them, the problem would quickly become more complex. It would be impossible to graph the problem in only two dimensions, and the only quick, sure way to solve it would be linear programming. The formulas used in the process would determine the multidimensional shape that represented the limits of all combinations and would calculate the profit line that went through the point that gave the best results.

—From Jared and William Taylor, "Searching for Solutions." Reprinted from: *PC Magazine*, September 15, 1987, pp. 311–315 ff. Copyright © 1987 Ziff Communications Company.

A guide to linear programming: These three graphs (right) show a progression in analysis to determine what combination of muffins produced will yield the greatest profit. The first graph shows the maximum number of muffins of any combination that can be made. The second graph defines the limits of production. And the third graph plots the exact combination of muffins (C) that return the greatest profit given the amount of fruit and batter available.

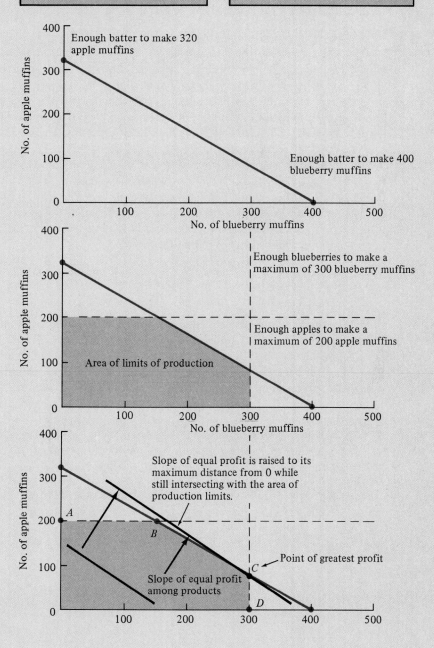

SUPPLIES

- Enough apples for 200 apple muffins
- Enough blueberries for 300 blueberry muffins
- 1,600 ounces of muffin batter

CONDITIONS

- One apple muffin takes 5 ounces of batter and yields a 16-cent profit.
- One blueberry muffin takes 4 ounces of batter and yields a 15-cent profit.

Enough batter to make 320 apple muffins

Enough batter to make 400 blueberry muffins

No. of blueberry muffins

Enough blueberries to make a maximum of 300 blueberry muffins

Enough apples to make a maximum of 200 apple muffins

Area of limits of production

No. of blueberry muffins

Slope of equal profit is raised to its maximum distance from 0 while still intersecting with the area of production limits.

Point of greatest profit

Slope of equal profit among products

Waiting line models are used to determine the number of check-out counters to keep open in a supermarket, the number of tellers' windows to keep open in a bank, the number of maintenance personnel to employ in a factory, the number of tollgates to keep open on a turnpike, or the number of nurses required at a hospital station. As in the case of all other quantitative techniques, *these methods are useful only when applied in situations in which they fit the problem under consideration.* A manager of engineering in an aerospace firm related the following story, which illustrates this point.

It seems that this manager always had a long line of engineers waiting at the blueprint room to check out materials. Since these men were in a fairly high salary bracket, the time they spent waiting in line cost the company a considerable sum of money. A research team was called in to do a waiting line analysis and determine the best number of clerks to employ in the blueprint room. This team spent several days collecting data on the number of engineers arriving at the blueprint room and the average time required for a clerk to fill a request. They then went back to their offices to feed all these numbers to their computers. Late one afternoon the personnel manager happened to walk by the blueprint room and saw the situation. The next morning he replaced the two women who had been working there with two men, and there has not been a waiting line since. So, it seems that in spite of all the numbers generated by the research team, two important figures were left out of the model.

SIMULATION

In the physical sciences, experiments may be performed in a laboratory using small models of a process or an operation. Many complex variations may be possible in these tests, and the results show the scientist what happens under certain controlled conditions. Simulation is similar to scientific experimentation. You do recall all those computer simulation experiments we carried out in Chapters 5 and 6 to demonstrate probability and Central Limit Theorem concepts, don't you?

Perhaps Figure 16-1 will clarify the meaning of simulation. At its base, Figure 16-1 rests on reality or fact. In complex situations, few people (if any) fully understand all aspects of the situation; therefore, theories are developed which

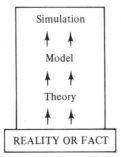

FIGURE 16-1

ECONOMIC MODELING GAINS DESPITE ACCURACY CONCERNS

"It's said that if you ask three economists a question, you'll get seven different answers. And that's why with economists, I always have one motto: trust, but verify."
—President Reagan at the City Club of Cleveland

The President's recent comment provoked an easy laugh. But that he should view economists with a level of suspicion usually reserved for the Soviets is indicative of the increasing pressure being placed on economists and their predictions. The one-day, 500-plus point plunge in the Dow Jones Industrial average on Oct. 19, 1987, evoked grim memories of 1929's Black Monday more palpably than any dire economic prediction could have.

Fears that the crash may foreshadow a sudden and cold reversal in the U.S. economy has others besides the President questioning economists' insights. Ironically, economists are finding that as they are increasingly mistrusted and criticized, greater demands are being made for their prognostications. And when their pronouncements take center stage, they turn to detailed simulations of economic behavior for verification.

ERRORS ARE SMALLER THAN IN THE RECENT PAST

Just how accurate a verification these models provide is subject to debate. By and large, the forecasts most often cited remain little more than elaborate extrapolations of present trends. Such econometric forecasts are provided by such firms as Data Resources Inc., Lexington, Mass., and the WEFA Group (formerly Wharton Econometric Forecasting Associates), Bala-Cynwyd, Pa.

What improvements there have been in the accuracy of economic forecasts are measurable only when viewed over the course of many years. According to Stephen McNees, a vice president and economist at the Federal Reserve Bank in Boston, the best that can be said of more recent work is that today's errors tend to be smaller than those of 10 or 20 years ago. McNees has studied the accuracy of econometric models and finds major im-

provements in today's projections. But more than anything else, the most striking difference in economic forecasting over the past 30 years is the development of larger models.

Sam Cole, a professor at the State University of New York at Buffalo and an econometric model builder, researcher, and critic, says the largest and newest models—the so-called global models—for the most part follow a "classical" structure based on well-established theories of capital accumulation and capital input/output ratios.

"Most, if you look at them hard, rely on one or two assumptions that determine everything that happens," says Cole. "The difference among these models is whether or not they deal with environmental or ecological limits and the linkages they show between countries," he says.

GLOBAL MODELS TAKE LONG-RANGE VIEW

While most global models are classically econometric in structure, they are a relatively recent phenomena. According to Cole, about two dozen have been developed worldwide, and their spread most often is tied to the widely publicized 1971 study, *World Dynamics* (Wright Alenn, Cambridge, Mass., 1971), which painted a pessimistic long-term economic outlook, he says. Published by MIT Sloan School of Management professor Jay W. Forrester, a computer industry pioneer and director of MIT's System Dynamics Group, the work eschews econometric methods and was the first to propose 50- to 100-year forecasts. Forrester, whose work grew out of the study of corporate decision-making on business growth, plans to release the details of the System Dynamics National Model soon, a short- and long-range model of the U.S.

Ironically, while Forrester's work proved to encourage the construction of larger global models with longer-term horizons, the System Dynamics method proposed in *World Dynamics* is considered the least computer- or technology-intensive of the major models. Its National Model can be run on larger microcomputers, although the MIT group most often uses a timesharing computer, says

Forrester. Moreover, the model's focus on the influence of policy-making on human behaviors enables its principles to be demonstrated using small groups of people, he says.

ECONOMISTS TURN TO NETWORKS, GRAPHICS

Due in large part to the growth of global models, economists have begun to explore technologies such as graphics, distributed processing, and networking to improve the underlying structure of econometric models and their analyses. Faye Duchin, director of the New York-based Institute for Economic Analysis and a New York University professor, says that the larger models, such as the institute's World Input/Output model, are too complex for the simple analyses that earlier models permitted.

"Our model cries out for an implementation using computer graphics. We need maps of the world with individual [production] sectors colored for instant comprehension," says Duchin. "I've spent months of my life trying to figure out what the numbers mean. If we were able to look at the results graphically, the information would be available at a much more intuitive level."

While such graphics presently are not part of these large modeling programs, Duchin sees movement toward developing models that address the need for more accessible simulations of economic forces. "With the next generation of models, I think we've realized the structure of the model has to bear scrutiny. It can't be a black box; that hasn't always been the case," she adds.

The need to improve and update the World I/O model has prevented it from being used for five years. Now negotiating with backers to resume work on the economic model, Duchin says that the model's data requirements and structure need refinement. So far, that work has prevented the institute from accepting even limited studies. "We've been asked to get the model running for lots of questions," she says. "We say, 'Wait until we have the appropriate data and structure to get the answer.'"

MULTICOUNTRY MODELS MIRROR INTERACTIONS

Others haven't had the same financial limitations. Lawrence Klein, a University of Pennsylvania economics professor and the originator of Project Link, says that a global model threading together econometric models from 80 countries to forecast short-range economic conditions, faster computers, and increased use of networking has benefited global forecasting. Such multicountry models lead to faster and more accurate portrayals of economic conditions, Klein says. With funding from sources as diverse as the United Nations, the Bank of Norway, and the World Bank, Project Link strings together models of countries with centrally planned, developing, and industrial economies to produce a world economic model. The present version incorporates up to 20,000 equations that Klein says work best when processed simultaneously. "We're aiming to put different country models on different processors and solve all simultaneously," he says. "We used to do a small number at a time in a serial pattern. Parallel processing speeds up the process."

Faster processing speeds enable Project Link to run "What if?" scenarios more quickly to pinpoint previously unknown interactions in world economies. Researchers then can test conditions suggested by the model simulations against existing data. Higher processing speeds also enable University of Pennsylvania researchers to study what impact random deviations—so-called stochastic shocks—in the economy of one country will have on conditions in other countries, says Klein.

What's more, Klein and his Project Link associates are using networking to analyze policy decisions almost simultaneously with world economic summits. In June, the group expects to convene an electronic assembly in conjunction with the annual summit of the seven major industrial countries. "In the days when it took a half hour to an hour to make a working projection, we could not hold a live conference. Now, we can take baseline projections, discuss, and modify them even as talk continues," Klein says.

The group's goal is to involve economists the world over in monetary or trade policy discussions as the economic summit progresses. "In the past, we've placed teams in Zurich representing the European summit countries, Tokyo representing Japan and the Far East, and one in Bedminster, N.J., representing the U.S. and Canada," explains Klein. "We make a baseline [projection] on the world economy on the eve of the summit, talk about the outcome, and ask, 'What should representatives of these countries talk about?'"

In other applications, Project Link has been used to study the impact of fluctuating world oil prices on Norway's economy, to evaluate the inflationary pressures associated with U.S.–USSR arms spending, and to consult with the government of the Philippines on its foreign debt, he says. . . .

Yet, for all the many equations large econometric models process, there is still no certain answer whether such models accurately simulate economic conditions to gauge the impact of government economic or trade policies. MIT's Forrester says that most econometric models attempt to simulate the recurring conditions found in short-range periods, or under 10-year periods that economists call the business cycle. Forrester believes that "prediction of the business cycle is, on the whole, a wasted effort. They are trying to do things that can't be done."

SIMULATIONS FORM ONE FACTOR IN ANALYSIS

Project Link's Klein cites the long list of government sponsors and requests for congressional testimony as measures of Project Link's esteem and influence. But Kenneth G. Ruffing, assistant director at the U.N.'s office of development research and policy analysis, says that simulations are but one of many factors that feed policy debate at the U.N. "We use models as part of a process of economic analysis," says Ruffing. "Nobody reads so much into the outcome of these models [that] they will make a decision solely on the forecast," he adds. "It would be a mistake to overestimate their impact. . . ."

It may be, too, that larger technology bases will serve only to raise expectations that the models today cannot satisfy. The use of supercomputers may be moving economic forecasting in the same direction as weather forecasting; thus far, however, there has not been comparable improvement in the models' predictions.

According to the U.N.'s Ruffing, "Larger models are better because they provide more detail. They probably are more accurate, but not so much more you'd bet money on the outcome." Echoing these reservations, Cole believes that although the application of increased technology, such as supercomputers, to forecasting has produced larger models, those models are not necessarily more accurate.

The reason may lie in the nature of economic analysis. MIT's Forrester says econometric models such as Project Link are inaccurate because of their reliance on statistical extrapolations. Such extrapolations have built-in limits, unlike the System Dynamics approach, he says. "Econometric models are based on trends. System Dynamics is representative of the structure of the system. There is nothing in the model that says what it will do," says Forrester.

System Dynamics endorses a long-wave theory of economic behavior that asserts overproduction leads to upheavals every 45 to 60 years. Forrester says the National Model's purpose "is to understand behavior and to understand how behavior can be changed if different policies are adapted." To view the economy as a single system independent of other factors ignores the conditions that lead to these periodic upheavals. "It's been the experience of corporate executives that if they have a problem and correct it with a policy, five years later they will have the same problem. The system is sufficiently complicated that many other policies react to change the original policy," he notes.

Others say that today, economic conditions are too complex and too dependent on human behavior to simulate accurately on a computer. "Social systems cannot be modeled with the same accuracy as physical systems," argues Ruffing. As such, those who attempt to improve accuracy solely by applying more sophisticated technology may be barking up the wrong tree. . . .

Even as debate rages over whether ever larger econometric models lead to greater accuracy, as Klein suggests, or that such models are inherently flawed, as Forrester claims, there is a changed climate among those who employ the forecasts. "There was a time when people expected models would be very accurate and able to quantify the outcomes of policy discussions very precisely," says the U.N.'s Ruffing. "Now, people are looking more realistically at what models can do."

—From Gary McWilliams, "Economic Modeling Gains Despite Accuracy Concerns," *Datamation*, April 1, 1988, pp. 43–44 ff. Excerpted from DATAMATION, © 1988 Cahners Publishing Company.

may focus attention on only part of the complex whole. In some situations, a model is built or conceived to test or represent a theory. Finally, **simulation** is the use of a model in the attempt to identify and/or reflect the behavior of a real person, process, or system. The terms **modeling** and simulation are often used interchangeably. The primary limitation of simulation, of course, is that the theory and the model may *not* represent the true reality of the situation.

In organizations, administrators may evaluate proposed projects or strategies by constructing theoretical models. They can then determine what happens to these models under certain conditions or when certain assumptions are tested. Simulation is thus a trial-and-error problem-solving approach; it's also a planning aid that may be of considerable value to organizations.

Simulation models have helped top business executives decide, for example, whether or not to expand operations by acquiring a new plant. Among the dozens of complicating variables that would have to be incorporated into such models are facts and assumptions about (1) the present and potential size of the total market, (2) the present and potential company share of this total market, (3) product selling prices, and (4) the investment required to achieve various production levels. Thus, simulation has helped top executives in their strategic planning and decision-making activities.

Simulation has also been useful in (1) establishing design parameters for aircraft and space vehicles, (2) refining medical treatment techniques, (3) teaching students, (4) planning urban improvements, and (5) planning transportation systems where variables such as expected road usage, effects of highway construction on traffic load, and use of one-way streets are considered.

LOOKING BACK

1   In this chapter we've briefly surveyed several quantitative methods of analysis. No attempt is made to explain the methodology for using these techniques, but illustrations of their use are given to show the types of problems that can be solved. Attention is also given to the limitations of these analyses. It's important to understand that a quantitative analysis will not make a decision for you. It simply provides more and better information on which to base a decision.

KEY TERMS AND CONCEPTS

Multistage sampling   623
Decision theory   623
Payoff (profit) table   624
Expected monetary value   625
Bayesian decision theory   625

Linear programming   625
Inventory models   626
Waiting line theory   627
Simulation   634
Modeling   634

TOPICS FOR REVIEW AND DISCUSSION

1 a   How may a stratified sample be used?
   b   A cluster sample?
2 a   What is Bayesian decision theory?
   b   What are the steps in a decision theory problem?
3 How may linear programming be used?
4 a   What are inventory models?
   b   How may inventory models be used?
5 a   What is simulation?
   b   How may simulation models be used?

PROJECT TO CONSIDER

1 Go to the library and locate an article that describes an application of one of the quantitative techniques discussed in this chapter. Write a brief summary of the application for your instructor.

A CLOSER LOOK

THE STARTLING DISCOVERY BELL LABS KEPT IN THE SHADOWS

It happens all too often in science. An obscure researcher announces a stunning breakthrough and achieves instant fame. But when other scientists try to repeat his results, they fail. Fame quickly turns to notoriety, and eventually the episode is all but forgotten.

That seemed to be the case with Narendra K. Karmarkar, a young scientist at AT&T Bell Laboratories. In late 1984 the 28-year-old researcher astounded not only the scientific community but also the business world. He claimed he had cracked one of the thorniest aspects of computer-aided problem-solving. If so, his feat would have meant an instant windfall for many big companies. It could also have pointed to better software for small companies that use computers to help manage their businesses.

Karmarkar said he had discovered a quick way to solve problems so hideously complicated that they often defy even the most powerful supercomputers. Such problems bedevil a broad range of business activities, from assessing risk factors in stock portfolios to drawing up production schedules in factories. Just about any company that distributes products through more than a handful of warehouses bumps into such problems when calculating the cheapest routes for getting goods to customers. Even when the problems aren't terribly complex, solving them can chew up so much computer time that the answer is useless before it's found.

HEAD START
To most mathematicians, Karmarkar's precocious feat was hard to swallow. Because such questions are so common, a special branch of mathematics called linear programming (LP) has evolved, and most scientists thought that was as far as they could go. Sure enough, when other researchers independently tried to test Karmarkar's process, their results were disappointing. At scientific conferences skeptics attacked the algorithm's validity as well as Karmarkar's veracity.

But this story may end with a different twist. Other scientists weren't able to duplicate Karmarkar's work, it turns out, because his employer wanted it that way. Vital details about how best to translate the algorithm, whose mathematical notations run on for about 20 printed pages, into digital computer code were withheld to give Bell Labs a head start at developing commercial products. Following the breakup of American Telephone & Telegraph Co. in January, 1984, Bell Labs was no longer prevented from exploiting its research for profit. While the underlying concept could not be patented or copyrighted because it is pure knowledge, any computer programs that AT&T developed to implement the procedure can be protected.

Now, AT&T may soon be selling the first product based on Karmarkar's work—to the U.S. Air Force. It includes a multiprocessor computer from Alliant Computer Systems Corp. and a software version of Karmarkar's algorithm that has been optimized for high-speed parallel processing. The system would be installed at St. Louis' Scott Air Force Base, headquarters of the Military Airlift Command (MAC). Neither party will comment on the deal's cost or where the negotiations stand, but the Air Force's interest is easy to fathom.

JUGGLING ACT
On a typical day thousands of planes ferry cargo and passengers among air fields scattered around the world. To keep those jets flying, MAC must juggle the schedules of pilots and other flight personnel. In addition, MAC must determine whether it would be more efficient to reduce cargo and top off the fuel tanks at the start of each flight or to refuel at stops along the way and pay for the costs of shipping the fuel. To get the airplanes back into the air, cargo handlers need to be standing by, and ground crews must be ready to service the craft. And even MAC's best-laid plans are constantly being disrupted by bad weather or emergency changes in priority. Getting all the pieces to play

together is a classic challenge in linear programming. If a computer could wring out just a couple of percentage points of added efficiency, it would be worth millions of dollars.

Proving that the computer program works for the Air Force would virtually guarantee immediate sales in the transportation market. American Airlines Inc., for example, is eager to put the technique to the test. Because Karmarkar claims his algorithm solves such problems 50 to 100 times faster than present techniques, the airline wants to determine whether it is able to provide a better way of rejiggering its flights during bad weather. In addition, American sees major potential in projecting crew assignments for weeks in advance.

Meanwhile, AT&T is already putting Karmarkar's mathematical feat to work. And its performance is outstripping all expectations, says Stephen Chen, head of the transmission facilities planning department at Bell Labs. The savings are difficult to quantify, adds Chen's boss, Nathan Levine, but will certainly run in the "tens of millions."

AT&T is using Karmarkar's formula to forecast the most cost-effective way to satisfy future needs of the telephone network linking 20-odd countries rimming the Pacific Ocean. To do that over a 10-year horizon, Bell Labs must estimate current and future telephone demand between every pair of switching points within the network.

NEW HORIZONS
The resulting computer model involves 42,000 variables. Considering that many factors just isn't practical with the so-called simplex method, the reigning champ for handling such problems, which was devised 40 years ago by George B. Dantzig of Stanford University. It would take four to seven hours of mainframe computer time to answer each what-if question. But with Karmarkar's approach, the answers pop out in less than four minutes.

Like many mathematical concepts, how his procedure works is difficult to describe. Perhaps the most common analogy is that an LP problem resembles a geodesic dome. Each of the dome's corners is a possible solution. The task is to find which one holds the best solution. With the simplex method, the program "lands" on one corner and inspects it. Then it scouts the adjacent corners to see if there is a better answer; if so, it heads off in that direction. The procedure has to be repeated at every corner

until the program finds itself boxed in by worse solutions.

Karmarkar's method employs a radically different tactic. It starts from a point within the structure and finds the solution by taking a shortcut that avoids the tedious surface route. From the interior vantage, it uses an arcane discipline known as projective geometry to reconfigure the structure's shape. By studying the new structure, the program determines in which direction the solution lies. Then the problem-structure is allowed to return to its original shape, and the program jumps toward the solution, pausing at intervals to repeat the exercise and home in on the answer.

Karmarkar's formula works so well, in fact, that AT&T is applying it to problems so complex that no one has ever bothered to try tackling them whole. The company's model of its domestic long-distance network consists of 800,000 variables. With traditional methods, answering even a simple question would set a mainframe to whirring for weeks. So such problems have to be carved up into more digestible pieces, then the separate answers are stitched together to yield a patchwork solution. But Karmarkar's technique weighs all the variables and still spits out answers in less than an hour. The difference between the two approaches showed AT&T how to squeeze an extra 9% to 10% of capacity out of its $15 billion system.

'TIP OF THE ICEBERG'
Optimistic results are beginning to trickle in from outside AT&T as well—from such research centers as Cornell University and the University of California–Berkeley. Karmarkar recently visited both schools to shed light on the murkier aspects of his procedure. Now more researchers agree that the power unleashed by Karmarkar provides them, for the first time, with the tools to analyze total systems that before had to be studied piecemeal. Karmarkar himself believes that the payoffs to date are "just the tip of the iceberg." Says Bell Labs' Levine: "Mental horizons have been opened."

Karmarkar brought an unusually open mind to linear programming. Born in India, he grew up hearing his father and uncle, both mathematicians, discuss mathematical principles. He came to America and studied operations research, the overall field that embraces linear programming, and earned a PhD in computer science at Berkeley. Therefore when Karmarkar took up the LP challenge, he brought a rather rare, interdisciplinary bag of tricks to bear on the research.

Fame now seems assured for Narendra Karmarkar. The algorithm that bears his name is already showing up in textbooks. And AT&T has honored his work by naming him a Bell Fellow—a position in which he can soon expect to earn $100,000 a year. Who knows? He may even win the Nobel prize.

—William G. Wild Jr. and Otis Port, "The Startling Discovery Bell Labs Kept in the Shadows." Reprinted from September 21, 1987 issue of *Business Week* by special permission, copyright © 1987 by McGraw-Hill, Inc.

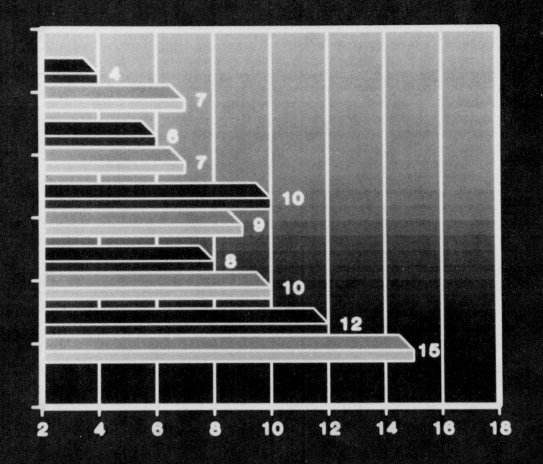

APPENDIX 1 SELECTED VALUES OF THE BINOMIAL PROBABILITY DISTRIBUTION

$$P(r) = {_nC_r}\,(p)^r(q)^{n-r}$$

Example: If $p = .30$, $n = 5$, and $r = 2$, then $P(r) = .3087$. (When p is greater than .50, the value of $P(r)$ is found by locating the table for the specified n and using $n - r$ in place of the given r and $1 - p$ in place of the specified p.)

n	r	.01	.05	.10	.15	.20	.25	p .30	.35	.40	.45	.50
1	0	.9900	.9500	.9000	.8500	.8000	.7500	.7000	.6500	.6000	.5500	.5000
	1	.0100	.0500	.1000	.1500	.2000	.2500	.3000	.3500	.4000	.4500	.5000
2	0	.9801	.9025	.8100	.7225	.6400	.5625	.4900	.4225	.3600	.3025	.2500
	1	.0198	.0950	.1800	.2550	.3200	.3750	.4200	.4550	.4800	.4950	.5000
	2	.0001	.0025	.0100	.0225	.0400	.0625	.0900	.1225	.1600	.2025	.2500
3	0	.9703	.8574	.7290	.6141	.5120	.4219	.3430	.2746	.2160	.1664	.1250
	1	.0294	.1354	.2430	.3251	.3840	.4219	.4410	.4436	.4320	.4084	.3750
	2	.0003	.0071	.0270	.0574	.0960	.1406	.1890	.2389	.2880	.3341	.3750
	3	.0000	.0001	.0010	.0034	.0080	.0156	.0270	.0429	.0640	.0911	.1250
4	0	.9606	.8145	.6561	.5220	.4096	.3164	.2401	.1785	.1296	.0915	.0625
	1	.0388	.1715	.2916	.3685	.4096	.4219	.4116	.3845	.3456	.2995	.2500
	2	.0006	.0135	.0486	.0975	.1536	.2109	.2646	.3105	.3456	.3675	.3750
	3	.0000	.0005	.0036	.0115	.0258	.0469	.0756	.1115	.1536	.2005	.2500
	4	.0000	.0000	.0001	.0005	.0016	.0039	.0081	.0150	.0256	.0410	.0625
5	0	.9510	.7738	.5905	.4437	.3277	.2373	.1681	.1160	.0778	.0503	.0312
	1	.0480	.2036	.3280	.3915	.4096	.3955	.3602	.3124	.2592	.2059	.1562
	2	.0010	.0214	.0729	.1382	.2048	.2637	.3087	.3364	.3456	.3369	.3125
	3	.0000	.0011	.0081	.0244	.0512	.0879	.1323	.1811	.2304	.2757	.3125
	4	.0000	.0000	.0004	.0022	.0064	.0146	.0284	.0488	.0768	.1128	.1562
	5	.0000	.0000	.0000	.0001	.0003	.0010	.0024	.0053	.0102	.0185	.0312
6	0	.9415	.7351	.5314	.3771	.2621	.1780	.1176	.0754	.0467	.0277	.0156
	1	.0571	.2321	.3543	.3993	.3932	.3560	.3025	.2437	.1866	.1359	.0938
	2	.0014	.0305	.0984	.1762	.2458	.2966	.3241	.3280	.3110	.2780	.2344
	3	.0000	.0021	.0146	.0415	.0819	.1318	.1852	.2355	.2765	.3032	.3125
	4	.0000	.0001	.0012	.0055	.0154	.0330	.0595	.0951	.1382	.1861	.2344
	5	.0000	.0000	.0001	.0004	.0015	.0044	.0102	.0205	.0369	.0609	.0938
	6	.0000	.0000	.0000	.0000	.0001	.0002	.0007	.0018	.0041	.0083	.0156
7	0	.9321	.6983	.4783	.3206	.2097	.1335	.0824	.0490	.0280	.0152	.0078
	1	.0659	.2573	.3720	.3960	.3670	.3115	.2471	.1848	.1306	.0872	.0547
	2	.0020	.0406	.1240	.2097	.2753	.3115	.3177	.2985	.2613	.2140	.1641
	3	.0000	.0036	.0230	.0617	.1147	.1730	.2269	.2679	.2903	.2918	.2734
	4	.0000	.0002	.0026	.0109	.0287	.0577	.0972	.1442	.1935	.2388	.2734
	5	.0000	.0000	.0002	.0012	.0043	.0115	.0250	.0466	.0774	.1172	.1641
	6	.0000	.0000	.0000	.0001	.0004	.0013	.0036	.0084	.0172	.0320	.0547
	7	.0000	.0000	.0000	.0000	.0000	.0001	.0002	.0006	.0016	.0037	.0078

Source: Adapted from Leonard J. Kazmier, *Statistical Analysis for Business and Economics,* 2d ed., copyright © 1973 by McGraw-Hill, Inc. Used by permission of McGraw-Hill Book Company.

n	r	.01	.05	.10	.15	.20	.25	p .30	.35	.40	.45	.50
8	0	.9227	.6634	.4305	.2725	.1678	.1002	.0576	.0319	.0168	.0084	.0039
	1	.0746	.2793	.3826	.3847	.3355	.2670	.1977	.1373	.0896	.0548	.0312
	2	.0026	.0515	.1488	.2376	.2936	.3115	.2965	.2587	.2090	.1569	.1094
	3	.0001	.0054	.0331	.0839	.1468	.2076	.2541	.2786	.2787	.2568	.2188
	4	.0000	.0004	.0046	.0185	.0459	.0865	.1361	.1875	.2322	.2627	.2734
	5	.0000	.0000	.0004	.0026	.0092	.0231	.0467	.0808	.1239	.1719	.2188
	6	.0000	.0000	.0000	.0002	.0011	.0038	.0100	.0217	.0413	.0403	.1094
	7	.0000	.0000	.0000	.0000	.0001	.0004	.0012	.0033	.0079	.0164	.0312
	8	.0000	.0000	.0000	.0000	.0000	.0000	.0001	.0002	.0007	.0017	.0039
9	0	.9135	.6302	.3874	.2316	.1342	.0751	.0404	.0207	.0101	.0046	.0020
	1	.0830	.2985	.3874	.3679	.3020	.2253	.1556	.1004	.0605	.0339	.0176
	2	.0034	.0629	.1722	.2597	.3020	.3003	.2668	.2162	.1612	.1110	.0703
	3	.0001	.0077	.0446	.1069	.1762	.2336	.2668	.2716	.2508	.2119	.1641
	4	.0000	.0006	.0074	.0283	.0661	.1168	.1715	.2194	.2508	.2600	.2461
	5	.0000	.0000	.0008	.0050	.0165	.0389	.0735	.1181	.1672	.2128	.2461
	6	.0000	.0000	.0001	.0006	.0028	.0087	.0210	.0424	.0743	.1160	.1641
	7	.0000	.0000	.0000	.0000	.0003	.0012	.0039	.0098	.0212	.0407	.0703
	8	.0000	.0000	.0000	.0000	.0000	.0001	.0004	.0013	.0035	.0083	.0176
	9	.0000	.0000	.0000	.0000	.0000	.0000	.0000	.0001	.0003	.0008	.0020
10	0	.9044	.5987	.3487	.1969	.1074	.0563	.0282	.0135	.0060	.0025	.0010
	1	.0914	.3151	.3874	.3474	.2684	.1877	.1211	.0725	.0403	.0207	.0098
	2	.0042	.0746	.1937	.2759	.3020	.2816	.2335	.1757	.1209	.0763	.0439
	3	.0001	.0105	.0574	.1298	.2013	.2503	.2668	.2522	.2150	.1665	.1172
	4	.0000	.0010	.0112	.0401	.0881	.1460	.2201	.2377	.2508	.2384	.2051
	5	.0000	.0001	.0015	.0085	.0264	.0584	.1029	.1536	.2007	.2340	.2461
	6	.0000	.0000	.0001	.0012	.0055	.0162	.0368	.0689	.1115	.1596	.2051
	7	.0000	.0000	.0000	.0001	.0008	.0031	.0090	.0212	.0425	.0746	.1172
	8	.0000	.0000	.0000	.0000	.0001	.0004	.0014	.0043	.0106	.0229	.0439
	9	.0000	.0000	.0000	.0000	.0000	.0000	.0001	.0005	.0016	.0042	.0098
	10	.0000	.0000	.0000	.0000	.0000	.0000	.0000	.0000	.0001	.0003	.0010
11	0	.8953	.5688	.3138	.1673	.0859	.0422	.0198	.0088	.0036	.0014	.0005
	1	.0995	.3293	.3835	.3248	.2362	.1549	.0932	.0518	.0266	.0125	.0054
	2	.0050	.0867	.2131	.2866	.2953	.2581	.1998	.1395	.0887	.0513	.0269
	3	.0002	.0137	.0710	.1517	.2215	.2581	.2568	.2254	.1774	.1259	.0806
	4	.0000	.0010	.0112	.0401	.0881	.1460	.2001	.2377	.2508	.2384	.2051
	5	.0000	.0001	.0025	.0132	.0388	.0803	.1321	.1830	.2207	.2360	.2256
	6	.0000	.0000	.0003	.0023	.0097	.0268	.0566	.0985	.1471	.1931	.2256
	7	.0000	.0000	.0000	.0003	.0017	.0064	.0173	.0379	.0701	.1128	.1611
	8	.0000	.0000	.0000	.0000	.0002	.0011	.0037	.0102	.0234	.0462	.0806
	9	.0000	.0000	.0000	.0000	.0000	.0001	.0005	.0018	.0052	.0126	.0269
	10	.0000	.0000	.0000	.0000	.0000	.0000	.0000	.0002	.0007	.0021	.0054
	11	.0000	.0000	.0000	.0000	.0000	.0000	.0000	.0000	.0000	.0002	.0005

n	r	.01	.05	.10	.15	.20	.25	p .30	.35	.40	.45	.50
12	0	.8864	.5404	.2824	.1422	.0687	.0317	.0138	.0057	.0022	.0008	.0002
	1	.1074	.3413	.3766	.3012	.2062	.1267	.0712	.0368	.0174	.0075	.0029
	2	.0060	.0988	.2301	.2924	.2835	.2323	.1678	.1088	.0639	.0339	.0161
	3	.0002	.0173	.0852	.1720	.2362	.2581	.2397	.1954	.1419	.0923	.0537
	4	0000	.0021	.0213	.0683	.1329	.1936	.2311	.2367	.2128	.1700	.1204
	5	.0000	.0002	.0038	.0193	.0532	.1032	.1585	.2039	.2270	.2225	.1934
	6	.0000	.0000	.0005	.0040	.0155	.0401	.0792	.1281	.1766	.2124	.2256
	7	.0000	.0000	.0000	.0006	.0033	.0115	.0291	.0591	.1009	.1489	.1934
	8	.0000	.0000	.0000	.0001	.0005	.0024	.0078	.0199	.0420	.0762	.1208
	9	.0000	.0000	.0000	.0000	.0001	.0004	.0015	.0048	.0125	.0277	.0537
	10	.0000	.0000	.0000	.0000	.0000	.0000	.0002	.0008	.0025	.0068	.0161
	11	.0000	.0000	.0000	.0000	.0000	.0000	.0000	.0001	.0003	.0010	.0029
	12	.0000	.0000	.0000	.0000	.0000	.0000	.0000	.0000	.0000	.0001	.0002
13	0	.8775	.5133	.2542	.1209	.0550	.0238	.0097	.0037	.0013	.0004	.0001
	1	.1152	.3512	.3672	.2774	.1787	.1029	.0540	.0259	.0113	.0045	.0016
	2	.0070	.1109	.2448	.2937	.2680	.2059	.1388	.0836	.0453	.0220	.0095
	3	.0003	.0214	.0997	.1900	.2457	.2517	.2181	.1651	.1107	.0660	.0349
	4	.0000	.0028	.0277	.0838	.1535	.2097	.2337	.2222	.1845	.1350	.0873
	5	.0000	.0003	.0055	.0266	.0691	.1258	.1803	.2154	.2214	.1989	.1571
	6	.0000	.0000	.0008	.0063	.0230	.0559	.1030	.1546	.1968	.2169	.2095
	7	.0000	.0000	.0001	.0011	.0058	.0186	.0442	.0833	.1312	.1775	.2095
	8	.0000	.0000	.0001	.0001	.0011	.0047	.0142	.0336	.0656	.1089	.1571
	9	.0000	.0000	.0000	.0000	.0001	.0009	.0034	.0101	.0243	.0495	.0873
	10	.0000	.0000	.0000	.0000	.0000	.0001	.0006	.0022	.0065	.0162	.0349
	11	.0000	.0000	.0000	.0000	.0000	.0000	.0001	.0003	.0012	.0036	.0095
	12	.0000	.0000	.0000	.0000	.0000	.0000	.0000	.0000	.0001	.0005	.0016
	13	.0000	.0000	.0000	.0000	.0000	.0000	.0000	.0000	.0000	.0000	.0001
14	0	.8687	.4877	.2288	.1028	.0440	.0178	.0068	.0024	.0008	.0002	.0001
	1	.1229	.3593	.3559	.2539	.1539	.0832	.0407	.0181	.0073	.0027	.0009
	2	.0081	.1229	.2570	.2912	.2501	.1802	.1134	.0634	.0317	.0141	.0056
	3	.0003	.0259	.1142	.2056	.2501	.2402	.1943	.1366	.0845	.0462	.0222
	4	.0000	.0037	.0349	.0998	.1720	.2202	.2290	.2022	.1549	.1040	.0611
	5	.0000	.0004	.0078	.0352	.0860	.1468	.1963	.2178	.2066	.1701	.1222
	6	.0000	.0000	.0013	.0093	.0322	.0734	.1262	.1759	.2066	.2088	.1833
	7	.0000	.0000	.0002	.0019	.0092	.0280	.0618	.1082	.1574	.1952	.2095
	8	.0000	.0000	.0000	.0003	.0020	.0082	.0232	.0510	.0918	.1398	.1833
	9	.0000	.0000	.0000	.0000	.0003	.0018	.0066	.0183	.0408	.0762	.1222
	10	.0000	.0000	.0000	.0000	.0000	.0003	.0014	.0049	.0136	.0312	.0611
	11	.0000	.0000	.0000	.0000	.0000	.0000	.0002	.0010	.0033	.0093	.0222
	12	.0000	.0000	.0000	.0000	.0000	.0000	.0000	.0001	.0005	.0019	.0056
	13	.0000	.0000	.0000	.0000	.0000	.0000	.0000	.0000	.0001	.0002	.0009
	14	.0000	.0000	.0000	.0000	.0000	.0000	.0000	.0000	.0000	.0000	.0001

n	r	.01	.05	.10	.15	.20	.25	.30	.35	.40	.45	.50
15	0	.8601	.4633	.2059	.0874	.0352	.0134	.0047	.0016	.0005	.0001	.0000
	1	.1303	.3658	.3432	.2312	.1319	.0668	.0305	.0126	.0047	.0016	.0005
	2	.0092	.1348	.2669	.2856	.2309	.1559	.0916	.0476	.0219	.0090	.0032
	3	.0004	.0307	.1285	.2184	.2501	.2252	.1700	.1110	.0634	.0318	.0139
	4	.0000	.0049	.0428	.1156	.1876	.2252	.2186	.1792	.1268	.0780	.0417
	5	.0000	.0006	.0105	.0499	.1032	.1651	.2061	.2123	.1859	.1404	.0916
	6	.0000	.0000	.0019	.0132	.0430	.0917	.1472	.1906	.2066	.1914	.1527
	7	.0000	.0000	.0003	.0030	.0138	.0393	.0811	.1319	.1771	.2013	.1964
	8	.0000	.0000	.0000	.0005	.0035	.0131	.0348	.0710	.1181	.1647	.1964
	9	.0000	.0000	.0000	.0001	.0007	.0034	.0116	.0298	.0612	.1048	.1527
	10	.0000	.0000	.0000	.0000	.0001	.0007	.0030	.0096	.0245	.0515	.0916
	11	.0000	.0000	.0000	.0000	.0000	.0001	.0006	.0024	.0074	.0191	.0417
	12	.0000	.0000	.0000	.0000	.0000	.0000	.0001	.0004	.0016	.0052	.0139
	13	.0000	.0000	.0000	.0000	.0000	.0000	.0000	.0001	.0003	.0010	.0032
	14	.0000	.0000	.0000	.0000	.0000	.0000	.0000	.0000	.0000	.0001	.0005
	15	.0000	.0000	.0000	.0000	.0000	.0000	.0000	.0000	.0000	.0000	.0000
16	0	.8515	.4401	.1853	.0743	.0281	.0100	.0033	.0010	.0003	.0001	.0000
	1	.1376	.3706	.3294	.2097	.1126	.0535	.0228	.0087	.0030	.0009	.0002
	2	.0104	.1463	.2745	.2775	.2111	.1336	.0732	.0353	.0150	.0056	.0018
	3	.0005	.0359	.1423	.2285	.2463	.2079	.1465	.0888	.0468	.0215	.0085
	4	.0000	.0061	.0514	.1311	.2001	.2252	.2040	.1553	.1014	.0572	.0278
	5	.0000	.0008	.0137	.0555	.1201	.1802	.2099	.2008	.1623	.1123	.0667
	6	.0000	.0001	.0028	.0180	.0550	.1101	.1649	.1982	.1983	.1684	.1222
	7	.0000	.0000	.0004	.0045	.0197	.0524	.1010	.1524	.1889	.1969	.1746
	8	.0000	.0000	.0001	.0009	.0055	.0197	.0487	.0923	.1417	.1812	.1964
	9	.0000	.0000	.0000	.0001	.0012	.0058	.0185	.0442	.0840	.1318	.1746
	10	.0000	.0000	.0000	.0000	.0002	.0014	.0056	.0167	.0392	.0755	.1222
	11	.0000	.0000	.0000	.0000	.0000	.0002	.0013	.0049	.0142	.0337	.0667
	12	.0000	.0000	.0000	.0000	.0000	.0000	.0002	.0011	.0040	.0115	.0278
	13	.0000	.0000	.0000	.0000	.0000	.0000	.0000	.0002	.0008	.0029	.0085
	14	.0000	.0000	.0000	.0000	.0000	.0000	.0000	.0000	.0001	.0005	.0018
	15	.0000	.0000	.0000	.0000	.0000	.0000	.0000	.0000	.0000	.0001	.0002
	16	.0000	.0000	.0000	.0000	.0000	.0000	.0000	.0000	.0000	.0000	.0000
17	0	.8429	.4181	.1668	.0631	.0225	.0075	.0023	.0007	.0002	.0000	.0000
	1	.1447	.3741	.3150	.1893	.0957	.0426	.0169	.0060	.0019	.0005	.0001
	2	.0117	.1575	.2800	.2673	.1914	.1136	.0581	.0260	.0102	.0035	.0010
	3	.0006	.0415	.1556	.2359	.2393	.1893	.1245	.0701	.0341	.0144	.0052
	4	.0000	.0076	.0605	.1457	.2093	.2209	.1868	.1320	.0796	.0411	.0182
	5	.0000	.0010	.0175	.0668	.1361	.1914	.2081	.1849	.1379	.0875	.0472
	6	.0000	.0001	.0039	.0236	.0680	.1276	.1784	.1991	.1839	.1432	.0944
	7	.0000	.0000	.0007	.0065	.0267	.0668	.1201	.1685	.1927	.1841	.1484
	8	.0000	.0000	.0001	.0014	.0084	.0279	.0644	.1134	.1606	.1883	.1855
	9	.0000	.0000	.0000	.0003	.0021	.0093	.0276	.0611	.1070	.1540	.1855

n	r	.01	.05	.10	.15	.20	.25	p .30	.35	.40	.45	.50
17	10	.0000	.0000	.0000	.0000	.0004	.0025	.0095	.0263	.0571	.1008	.1484
	11	.0000	.0000	.0000	.0000	.0001	.0005	.0026	.0090	.0242	.0525	.0944
	12	.0000	.0000	.0000	.0000	.0000	.0001	.0006	.0024	.0081	.0215	.0472
	13	.0000	.0000	.0000	.0000	.0000	.0000	.0001	.0005	.0021	.0068	.0182
	14	.0000	.0000	.0000	.0000	.0000	.0000	.0000	.0001	.0004	.0016	.0052
	15	.0000	.0000	.0000	.0000	.0000	.0000	.0000	.0000	.0001	.0003	.0010
	16	.0000	.0000	.0000	.0000	.0000	.0000	.0000	.0000	.0000	.0000	.0001
	17	.0000	.0000	.0000	.0000	.0000	.0000	.0000	.0000	.0000	.0000	.0000
18	0	.8345	.3972	.1501	.0536	.0180	.0056	.0016	.0004	.0001	.0003	.0010
	1	.1517	.3763	.3002	.1704	.0811	.0338	.0126	.0042	.0012	.0003	.0001
	2	.0130	.1683	.2835	.2556	.1723	.0958	.0458	.0190	.0069	.0022	.0006
	3	.0007	.0473	.1680	.2406	.2297	.1704	.1046	.0547	.0246	.0095	.0001
	4	.0000	.0093	.0700	.1592	.2153	.2130	.1681	.1104	.0614	.0291	.0117
	5	.0000	.0014	.0218	.0787	.1507	.1988	.2017	.1664	.1146	.0666	.0327
	6	.0000	.0002	.0052	.0301	.0816	.1436	.1873	.1941	.1655	.1181	.0708
	7	.0000	.0000	.0010	.0091	.0350	.0820	.1376	.1792	.1892	.1657	.1214
	8	.0000	.0000	.0002	.0022	.0120	.0376	.0811	.1327	.1734	.1864	.1669
	9	.0000	.0000	.0000	.0004	.0033	.0139	.0386	.0794	.1284	.1694	.1855
	10	.0000	.0000	.0000	.0001	.0008	.0042	.0149	.0385	.0771	.1248	.1669
	11	.0000	.0000	.0000	.0000	.0001	.0010	.0046	.0151	.0374	.0742	.1214
	12	.0000	.0000	.0000	.0000	.0000	.0002	.0012	.0047	.0145	.0354	.0708
	13	.0000	.0000	.0000	.0000	.0000	.0000	.0002	.0012	.0045	.0134	.0327
	14	.0000	.0000	.0000	.0000	.0000	.0000	.0000	.0002	.0011	.0039	.0117
	15	.0000	.0000	.0000	.0000	.0000	.0000	.0000	.0000	.0002	.0009	.0031
	16	.0000	.0000	.0000	.0000	.0000	.0000	.0000	.0000	.0000	.0001	.0006
	17	.0000	.0000	.0000	.0000	.0000	.0000	.0000	.0000	.0000	.0000	.0001
	18	.0000	.0000	.0000	.0000	.0000	.0000	.0000	.0000	.0000	.0000	.0000
19	0	.8262	.3774	.1351	.0456	.0144	.0042	.0011	.0003	.0001	.0000	.0000
	1	.1586	.3774	.2852	.1529	.0685	.0268	.0093	.0029	.0008	.0002	.0000
	2	.0144	.1787	.2852	.2428	.1540	.0803	.0358	.0138	.0046	.0013	.0003
	3	.0008	.0533	.1796	.2428	.2182	.1517	.0869	.0422	.0175	.0062	.0018
	4	.0000	.0112	.0798	.1714	.2182	.2023	.1491	.0909	.0467	.0203	.0074
	5	.0000	.0018	.0266	.0907	.1636	.2023	.1916	.1468	.0933	.0497	.0222
	6	.0000	.0002	.0069	.0374	.0955	.1574	.1916	.1844	.1451	.0949	.0518
	7	.0000	.0000	.0014	.0122	.0443	.0974	.1525	.1844	.1797	.1443	.0961
	8	.0000	.0000	.0002	.0032	.0166	.0487	.0981	.1489	.1797	.1771	.1442
	9	.0000	.0000	.0000	.0007	.0051	.0198	.0514	.0980	.1464	.1771	.1762
	10	.0000	.0000	.0000	.0001	.0013	.0066	.0220	.0528	.0976	.1449	.1762
	11	.0000	.0000	.0000	.0000	.0003	.0018	.0077	.0233	.0532	.0970	.1442
	12	.0000	.0000	.0000	.0000	.0000	.0004	.0022	.0083	.0237	.0529	.0961
	13	.0000	.0000	.0000	.0000	.0000	.0001	.0005	.0024	.0085	.0233	.0518
	14	.0000	.0000	.0000	.0000	.0000	.0000	.0001	.0006	.0024	.0082	.0222
	15	.0000	.0000	.0000	.0000	.0000	.0000	.0000	.0001	.0005	.0022	.0074
	16	.0000	.0000	.0000	.0000	.0000	.0000	.0000	.0000	.0001	.0005	.0018

n	r	.01	.05	.10	.15	.20	.25	.30	.35	.40	.45	.50
19	17	.0000	.0000	.0000	.0000	.0000	.0000	.0000	.0000	.0000	.0001	.0003
	18	.0000	.0000	.0000	.0000	.0000	.0000	.0000	.0000	.0000	.0000	.0000
	19	.0000	.0000	.0000	.0000	.0000	.0000	.0000	.0000	.0000	.0000	.0000
20	0	.8179	.3585	.1216	.0388	.0115	.0032	.0008	.0002	.0000	.0000	.0000
	1	.1652	.3774	.2702	.1368	.0576	.0211	.0068	.0020	.0005	.0001	.0000
	2	.0159	.1887	.2852	.2293	.1369	.0669	.0278	.0100	.0031	.0008	.0002
	3	.0010	.0596	.1901	.2428	.2054	.1339	.0718	.0323	.0123	.0040	.0011
	4	.0000	.0133	.0898	.1821	.2182	.1897	.1304	.0738	.0350	.0139	.0046
	5	.0000	.0022	.0319	.1028	.1746	.2023	.1789	.1272	.0746	.0365	.0148
	6	.0000	.0003	.0089	.0454	.1091	.1686	.1916	.1712	.1244	.0746	.0370
	7	.0000	.0000	.0020	.0160	.0545	.1124	.1643	.1844	.1659	.1221	.0739
	8	.0000	.0000	.0004	.0046	.0222	.0609	.1144	.1614	.1797	.1623	.1201
	9	.0000	.0000	.0001	.0011	.0074	.0271	.0654	.1158	.1597	.1771	.1602
	10	.0000	.0000	.0000	.0002	.0020	.0099	.0308	.0686	.1171	.1593	.1762
	11	.0000	.0000	.0000	.0000	.0005	.0030	.0120	.0336	.0710	.1185	.1602
	12	.0000	.0000	.0000	.0000	.0001	.0008	.0039	.0136	.0355	.0727	.1201
	13	.0000	.0000	.0000	.0000	.0000	.0002	.0010	.0045	.0146	.0366	.0739
	14	.0000	.0000	.0000	.0000	.0000	.0000	.0002	.0012	.0049	.0150	.0370
	15	.0000	.0000	.0000	.0000	.0000	.0000	.0000	.0003	.0013	.0049	.0148
	16	.0000	.0000	.0000	.0000	.0000	.0000	.0000	.0000	.0003	.0013	.0046
	17	.0000	.0000	.0000	.0000	.0000	.0000	.0000	.0000	.0000	.0002	.0011
	18	.0000	.0000	.0000	.0000	.0000	.0000	.0000	.0000	.0000	.0000	.0002
	19	.0000	.0000	.0000	.0000	.0000	.0000	.0000	.0000	.0000	.0000	.0000
	20	.0000	.0000	.0000	.0000	.0000	.0000	.0000	.0000	.0000	.0000	.0000
25	0	.7778	.2774	.0718	.0172	.0038	.0008	.0001	.0000	.0000	.0000	.0000
	1	.1964	.3650	.1994	.0759	.0236	.0063	.0014	.0003	.0000	.0000	.0000
	2	.0238	.2305	.2659	.1607	.0708	.0251	.0074	.0018	.0004	.0001	.0000
	3	.0018	.0930	.2265	.2174	.1358	.0641	.0243	.0076	.0019	.0004	.0001
	4	.0001	.0269	.1384	.2110	.1867	.1175	.0572	.0224	.0071	.0018	.0004
	5	.0000	.0060	.0646	.1564	.1960	.1645	.1030	.0506	.0199	.0063	.0016
	6	.0000	.0010	.0239	.0920	.1633	.1828	.1472	.0908	.0442	.0172	.0053
	7	.0000	.0001	.0072	.0441	.1108	.1654	.1712	.1327	.0800	.0381	.0143
	8	.0000	.0000	.0018	.0175	.0623	.1241	.1651	.1607	.1200	.0701	.0322
	9	.0000	.0000	.0004	.0058	.0294	.0781	.1336	.1635	.1511	.1084	.0609
	10	.0000	.0000	.0000	.0016	.0118	.0417	.0916	.1409	.1612	.1419	.0974
	11	.0000	.0000	.0000	.0004	.0040	.0189	.0536	.1034	.1465	.1583	.1328
	12	.0000	.0000	.0000	.0000	.0012	.0074	.0268	.0650	.1140	.1511	.1550
	13	.0000	.0000	.0000	.0000	.0003	.0025	.0115	.0350	.0760	.1236	.1550
	14	.0000	.0000	.0000	.0000	.0000	.0007	.0042	.0161	.0434	.0867	.1328

n	r	.01	.05	.10	.15	.20	.25	*p* .30	.35	.40	.45	.50
25	15	.0000	.0000	.0000	.0000	.0000	.0002	.0013	.0064	.0212	.0520	.0974
	16	.0000	.0000	.0000	.0000	.0000	.0000	.0004	.0021	.0088	.0266	.0609
	17	.0000	.0000	.0000	.0000	.0000	.0000	.0001	.0006	.0031	.0115	.0322
	18	.0000	.0000	.0000	.0000	.0000	.0000	.0000	.0001	.0009	.0042	.0143
	19	.0000	.0000	.0000	.0000	.0000	.0000	.0000	.0000	.0002	.0013	.0053
	20	.0000	.0000	.0000	.0000	.0000	.0000	.0000	.0000	.0000	.0001	.0016
	21	.0000	.0000	.0000	.0000	.0000	.0000	.0000	.0000	.0000	.0000	.0004
	22	.0000	.0000	.0000	.0000	.0000	.0000	.0000	.0000	.0000	.0000	.0001
30	0	.7397	.2146	.0424	.0076	.0012	.0002	.0000	.0000	.0000	.0000	.0000
	1	.2242	.3389	.1413	.0404	.0093	.0018	.0003	.0000	.0000	.0000	.0000
	2	.0328	.2586	.2277	.1034	.0337	.0086	.0018	.0003	.0000	.0000	.0000
	3	.0031	.1270	.2361	.1703	.0785	.0269	.0072	.0015	.0003	.0000	.0000
	4	.0002	.0451	.1771	.2028	.1325	.0604	.0208	.0056	.0012	.0002	.0000
	5	.0000	.0124	.1023	.1861	.1723	.1047	.0464	.0157	.0041	.0008	.0001
	6	.0000	.0027	.0474	.1368	.1795	.1455	.0829	.0353	.0115	.0029	.0006
	7	.0000	.0005	.0180	.0828	.1538	.1662	.1219	.0652	.0263	.0081	.0019
	8	.0000	.0001	.0058	.0420	.1106	.1593	.1501	.1009	.0505	.0191	.0055
	9	.0000	.0000	.0016	.0181	.0676	.1298	.1573	.1328	.0823	.0382	.0133
	10	.0000	.0000	.0004	.0067	.0355	.0909	.1416	.1502	.1152	.0656	.0280
	11	.0000	.0000	.0001	.0022	.0161	.0551	.1103	.1471	.1396	.0976	.0509
	12	.0000	.0000	.0000	.0006	.0064	.0291	.0749	.1254	.1474	.1265	.0806
	13	.0000	.0000	.0000	.0001	.0022	.0134	.0444	.0935	.1360	.1433	.1115
	14	.0000	.0000	.0000	.0000	.0007	.0054	.0231	.0611	.1101	.1424	.1354
	15	.0000	.0000	.0000	.0000	.0002	.0019	.0106	.0351	.0783	.1242	.1445
	16	.0000	.0000	.0000	.0000	.0000	.0006	.0042	.0177	.0489	.0953	.1354
	17	.0000	.0000	.0000	.0000	.0000	.0002	.0015	.0079	.0269	.0642	.1115
	18	.0000	.0000	.0000	.0000	.0000	.0000	.0005	.0031	.0129	.0379	.0806
	19	.0000	.0000	.0000	.0000	.0000	.0000	.0001	.0010	.0054	.0196	.0509
	20	.0000	.0000	.0000	.0000	.0000	.0000	.0000	.0003	.0020	.0088	.0280
	21	.0000	.0000	.0000	.0000	.0000	.0000	.0000	.0001	.0006	.0034	.0133
	22	.0000	.0000	.0000	.0000	.0000	.0000	.0000	.0000	.0002	.0012	.0055
	23	.0000	.0000	.0000	.0000	.0000	.0000	.0000	.0000	.0000	.0003	.0019
	24	.0000	.0000	.0000	.0000	.0000	.0000	.0000	.0000	.0000	.0001	.0006
	25	.0000	.0000	.0000	.0000	.0000	.0000	.0000	.0000	.0000	.0000	.0001

APPENDIX 2   AREAS UNDER THE STANDARD NORMAL PROBABILITY DISTRIBUTION

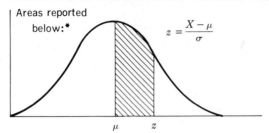

Areas reported below:*

$$z = \frac{X - \mu}{\sigma}$$

z	.00	.01	.02	.03	.04	.05	.06	.07	.08	.09
0.0	.0000	.0040	.0080	.0120	.0160	.0199	.0239	.0279	.0319	.0359
0.1	.0398	.0438	.0478	.0517	.0557	.0596	.0636	.0675	.0714	.0753
0.2	.0793	.0832	.0871	.0910	.0948	.0987	.1026	.1064	.1103	.1141
0.3	.1179	.1217	.1255	.1293	.1331	.1368	.1406	.1443	.1480	.1517
0.4	.1554	.1591	.1628	.1664	.1700	.1736	.1772	.1808	.1844	.1879
0.5	.1915	.1950	.1985	.2019	.2054	.2088	.2123	.2157	.2190	.2224
0.6	.2257	.2291	.2324	.2357	.2389	.2422	.2454	.2486	.2518	.2549
0.7	.2580	2.612	.2642	.2673	.2704	.2734	.2764	.2794	.2823	.2852
0.8	.2881	.2910	.2939	.2967	.2995	.3023	.3051	.3078	.3106	.3133
0.9	.3159	.3186	.3212	.3238	.3264	.3289	.3315	.3340	.3365	.3389
1.0	.3413	.3438	.3461	.3485	.3508	.3531	.3554	.3577	.3599	.3621
1.1	.3643	.3665	.3686	.3708	.3729	.3749	.3770	.3790	.3810	.3830
1.2	.3849	.3869	.3888	.3907	.3925	.3944	.3962	.3980	.3997	.4014
1.3	.4032	.4049	.4066	.4082	.4099	.4115	.4131	.4147	.4162	.4177
1.4	.4192	.4207	.4222	.4236	.4251	.4265	.4279	.4292	.4306	.4319
1.5	.4332	.4345	.4357	.4370	.4382	.4394	.4406	.4418	.4429	.4441
1.6	.4452	.4463	.4474	.4484	.4495	.4505	.4515	.4525	.4535	.4545
1.7	.4554	.4564	.4573	.4582	.4591	.4599	.4608	.4616	.4625	.4633
1.8	.4641	.4649	.4656	.4664	.4671	.4678	.4686	.4693	.4699	.4706
1.9	.4713	.4719	.4726	.4732	.4738	.4744	.4750	.4756	.4761	.4767
2.0	.4772	.4778	.4783	.4788	.4793	.4798	.4803	.4808	.4812	.4817
2.1	.4821	.4826	.4830	.4834	.4838	.4842	.4846	.4850	.4854	.4857
2.2	.4861	.4864	.4868	.4871	.4875	.4878	.4881	.4884	.4887	.4890
2.3	.4893	.4896	.4898	.4901	.4904	.4906	.4909	.4911	.4913	.4916
2.4	.4918	.4920	.4922	.4925	.4927	.4929	.4931	.4932	.4934	.4936
2.5	.4938	.4940	.4941	.4943	.4945	.4946	.4948	.4949	.4951	.4952
2.6	.4953	.4955	.4956	.4957	.4959	.4960	.4961	.4962	.4963	.4964
2.7	.4965	.4966	.4967	.4968	.4969	.4970	.4971	.4972	.4973	.4974
2.8	.4974	.4975	.4976	.4977	.4977	.4978	.4979	.4979	.4980	.4981
2.9	.4981	.4982	.4983	.4983	.4984	.4984	.4985	.4985	.4986	.4986
3.0	.4987	.4987	.4987	.4988	.4989	.4989	.4989	.4989	.4990	.4990
3.5	.4997									
4.0	.4999683									

* Example: For $z = 1.96$, the shaded area is 0.4750 out of the total area of 1.0000.

APPENDIX 3 A BRIEF TABLE OF RANDOM NUMBERS

10097	85017	84532	13618	23157	86952	02438	76520
37542	16719	82789	69041	05545	44109	05403	64894
08422	65842	27672	82186	14871	22115	86529	19645
99019	76875	20684	39187	38976	94324	43204	09376
12807	93640	39160	41453	97312	41548	93137	80157
66065	99478	70086	71265	11742	18226	29004	34072
31060	65119	26486	47353	43361	99436	42753	45571
85269	70322	21592	48233	93806	32584	21828	02051
63573	58133	41278	11697	49540	61777	67954	05325
73796	44655	81255	31133	36768	60452	38537	03529
98520	02295	13487	98662	07092	44673	61303	14905
11805	85035	54881	35587	43310	48897	48493	39808
83452	01197	86935	28021	61570	23350	65710	06288
88685	97907	19078	40646	31352	48625	44369	86507
99594	63268	96905	28797	57048	46359	74294	87517
65481	52841	59684	67411	09243	56092	84369	17468
80124	53722	71399	10916	07959	21225	13018	17727
74350	11434	51908	62171	93732	26958	02400	77402
69916	62375	99292	21177	72721	66995	07289	66252
09893	28337	20923	87929	61020	62841	31374	14225
91499	38631	79430	62421	97959	67422	69992	68479
80336	49172	16332	44670	35089	17691	89246	26940
44104	89232	57327	34679	62235	79655	81336	85157
12550	02844	15026	32439	58537	48274	81330	11100
63606	40387	65406	37920	08709	60623	2237	16505
61196	80240	44177	51171	08723	39323	05798	26457
15474	44910	99321	72173	56239	04595	10836	95270
94557	33663	86347	00926	44915	34823	51770	67897
42481	86430	19102	37420	41976	76559	24358	97344
23523	31379	68588	81675	15694	43438	36879	73208
04493	98086	32533	17767	14523	52494	24826	75246
00549	33185	04805	05431	94598	97654	16232	64051
35963	80951	68953	99634	81949	15307	00406	26898
59808	79752	02529	40200	73742	08391	49140	45427
46058	18633	99970	67348	49329	95236	32537	01390
32179	74029	74717	17674	90446	00597	45240	87379
69234	54178	10805	35635	45266	61406	41941	20117
19565	11664	77602	99817	28573	41430	96382	01758
45155	48324	32135	26803	16213	14938	71961	19476
94864	69074	45753	20505	78317	31994	98145	36168

Source: Leonard K. Kazmier, *Statistical Analysis for Business and Economics,* 2d ed., copyright © 1973 by McGraw-Hill, Inc. Used with permission of McGraw-Hill Book Company.

APPENDIX 4   AREAS FOR t DISTRIBUTIONS

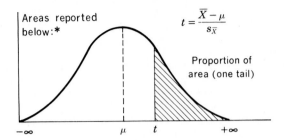

df	0.10	0.05	0.025	0.01	0.005
1	3.078	6.314	12.706	31.821	63.657
2	1.886	2.920	4.303	6.965	9.925
3	1.638	2.353	3.182	4.541	5.841
4	1.533	2.132	2.776	3.747	4.604
5	1.476	2.015	2.571	3.365	4.032
6	1.440	1.943	2.447	3.143	3.707
7	1.415	1.895	2.365	2.998	3.499
8	1.397	1.860	2.306	2.896	3.355
9	1.383	1.833	2.262	2.821	3.250
10	1.372	1.812	2.228	2.764	3.169
11	1.363	1.796	2.201	2.718	3.106
12	1.356	1.782	2.179	2.681	3.055
13	1.350	1.771	2.160	2.650	3.012
14	1.345	1.761	2.145	2.624	2.977
15	1.341	1.753	2.131	2.602	2.947
16	1.337	1.746	2.120	2.583	2.921
17	1.333	1.740	2.110	2.567	2.898
18	1.330	1.734	2.101	2.552	2.878
19	1.328	1.729	2.093	2.539	2.861
20	1.325	1.725	2.086	2.528	2.845
21	1.323	1.721	2.080	2.518	2.831
22	1.321	1.717	2.074	2.508	2.819
23	1.319	1.714	2.069	2.500	2.807
24	1.318	1.711	2.064	2.492	2.797
25	1.316	1.708	2.060	2.485	2.787
26	1.315	1.706	2.056	2.479	2.779
27	1.314	1.703	2.052	2.473	2.771
28	1.313	1.701	2.048	2.467	2.763
29	1.311	1.699	2.045	2.462	2.756
30	1.310	1.697	2.042	2.457	2.750
40	1.303	1.684	2.021	2.423	2.704
60	1.296	1.671	2.000	2.390	2.660
120	1.289	1.658	1.980	2.358	2.617
∞	1.282	1.645	1.960	2.326	2.576

*Example: For the shaded area to represent 0.05 of the total area of 1.0, the value of t with 10 degrees of freedom is 1.812.

Source: Reprinted by Hafner Press, a division of Macmillan Publishing Company, from *Statistical Methods for Research Workers*, 14th ed., abridged Table IV, by R. A. Fisher. Copyright © 1970 by University of Adelaide.

APPENDIX 5   TABLES OF SQUARES AND SQUARE ROOTS

N	N^2	$\sqrt{N}$	$\sqrt{10N}$	N	N^2	$\sqrt{N}$	$\sqrt{10N}$
				50	2 500	7.071 068	22.36068
1	1	1.000 000	3.162 278	51	2 601	7.141 428	22.58318
2	4	1.414 214	4.472 136	52	2 704	7.211 103	22.80351
3	9	1.732 051	5.477 226	53	2 809	7.280 110	23.02173
4	16	2.000 000	6.324 555	54	2 916	7.348 469	23.23790
5	25	2.236 068	7.071 068	55	3 025	7.416 198	23.45208
6	36	2.449 490	7.745 967	56	3 136	7.483 315	23.66432
7	49	2.645 751	8.366 600	57	3 249	7.549 834	23.87467
8	64	2.828 427	8.944 272	58	3 364	7.615 773	24.08319
9	81	3.000 000	9.486 833	59	3 481	7.681 146	24.28992
10	100	3.162 278	10.00000	60	3 600	7.745 967	24.49490
11	121	3.316 625	10.48809	61	3 721	7.810 250	24.69818
12	144	3.464 102	10.95445	62	3 844	7.874 008	24.89980
13	169	3.605 551	11.40175	63	3 969	7.937 254	25.09980
14	196	3.741 657	11.83216	64	4 096	8.000 000	25.29822
15	225	3.872 983	12.24745	65	4 225	8.062 258	25.49510
16	256	4.000 000	12.64911	66	4 356	8.124 038	25.69047
17	289	4.123 106	13.03840	67	4 489	8.185 353	25.88436
18	324	4.242 641	13.41641	68	4 624	8.246 211	26.07681
19	361	4.358 899	13.78405	69	4 761	8.306 824	26.26785
20	400	4.472 136	14.14214	70	4 900	8.366 600	26.45751
21	441	4.582 576	14.49138	71	5 041	8.426 150	26.64583
22	484	4.690 416	14.83240	72	5 184	8.485 281	26.83282
23	529	4.795 832	15.16575	73	5 329	8.544 004	27.01851
24	576	4.898 979	15.49193	74	5 476	8.602 325	27.20294
25	625	5.000 000	15.81139	75	5 625	8.660 254	27.38613
26	676	5.099 020	16.12452	76	5 776	8.717 798	27.56810
27	729	5.196 152	16.43168	77	5 929	8.774 964	27.74887
28	784	5.291 503	16.73320	78	6 084	8.831 761	27.92848
29	841	5.385 165	17.02939	79	6 241	8.888 194	28.10694
30	900	5.477 226	17.32051	80	6 400	8.944 272	28.28427
31	961	5.567 764	17.60682	81	6 561	9.000 000	28.46050
32	1 024	5.656 854	17.88854	82	6 724	9.055 385	28.63564
33	1 089	5.744 563	18.16590	83	6 889	9.110 434	28.80972
34	1 156	5.830 952	18.43909	84	7 056	9.165 151	28.98275
35	1 225	5.916 080	18.70829	85	7 225	9.219 544	29.15476
36	1 296	6.000 000	18.97367	86	7 396	9.273 618	29.32576
37	1 369	6.082 763	19.23538	87	7 569	9.327 379	29.49576
38	1 444	6.164 414	19.49359	88	7 744	9.380 832	29.66479
39	1 521	6.244 998	19.74842	89	7 921	9.433 981	29.83287
40	1 600	6.324 555	20.00000	90	8 100	9.486 833	30.00000
41	1 681	6.403 124	20.24846	91	8 281	9.539 392	30.16621
42	1 764	6.480 741	20.49390	92	8 464	9.591 663	30.33150
43	1 849	6.557 439	20.73644	93	8 649	9.643 651	30.49590
44	1 936	6.633 250	20.97618	94	8 836	9.695 360	30.65942
45	2 025	6.708 204	21.21320	95	9 025	9.746 794	30.82207
46	2 116	6.782 330	21.44761	96	9 216	9.797 959	30.98387
47	2 209	6.855 655	21.67948	97	9 409	9.848 858	31.14482
48	2 304	6.928 203	21.90890	98	9 604	9.899 495	31.30495
49	2 401	7.000 000	22.13594	99	9 801	9.949 874	31.46427
50	2 500	7.071 068	22.36068	100	10 000	10.00000	31.62278

N	N²	$\sqrt{N}$	$\sqrt{10N}$	N	N²	$\sqrt{N}$	$\sqrt{10N}$
100	10 000	10.00000	31.62278	150	22 500	12.24745	38.72983
101	10 201	10.04988	31.78050	151	22 801	12.28821	38.85872
102	10 404	10.09950	31.93744	152	23 104	12.32883	39.98718
103	10 609	10.14889	32.09361	153	23 409	12.36932	39.11521
104	10 816	10.19804	32.24903	154	23 716	12.40967	39.24283
105	11 025	10.24695	32.40370	155	24 025	12.44990	39.37004
106	11 236	10.29563	32.55764	156	24 336	12.40000	39.49684
107	11 449	10.34408	32.71085	157	24 649	12.52996	39.62323
108	11 664	10.39230	32.86335	158	24 964	12.56981	39.74921
109	11 881	10.44031	33.01515	159	25 281	12.60952	39.87480
110	12 100	10.48809	33.16625	160	25 600	12.64911	40.00000
111	12 321	10.53565	33.31666	161	25 921	12.68858	40.12481
112	12 544	10.58301	33.46640	162	26 244	12.72792	40.24922
113	12 769	10.63015	33.61547	163	26 569	12.76715	40.37326
114	12 996	10.67708	33.76389	164	26 896	12.80625	40.49691
115	13 225	10.72381	33.91165	165	27 225	12.84523	40.62019
116	13 456	10.77033	34.05877	166	27 556	12.88410	40.74310
117	13 689	10.81665	34.20526	167	27 889	12.92285	40.86563
118	13 924	10.86278	34.35113	168	28 224	12.96148	40.98780
119	14 161	10.90871	34.49638	169	28 561	13.00000	41.10961
120	14 400	10.95445	34.64102	170	28 900	13.03840	41.23106
121	14 641	11.00000	34.78505	171	29 241	13.07670	41.35215
122	14 884	11.04536	34.92850	172	29 584	13.11488	41.47288
123	15 129	11.09054	35.07136	173	29 929	13.15295	41.59327
124	15 376	11.13553	35.21363	174	30 276	13.19091	41.71331
125	15 625	11.18034	35.35534	175	30 625	13.22876	41.83300
126	15 876	11.22497	35.49648	176	30 976	13.26650	41.95235
127	16 129	11.26943	35.63706	177	31 329	13.30413	42.07137
128	16 384	11.31371	35.77709	178	31 684	13.34166	42.19005
129	16 641	11.35782	35.91657	179	32 041	13.37909	42.30839
130	16 900	11.40175	36.05551	180	32 400	13.41641	42.42641
131	17 161	11.44552	36.19392	181	32 761	13.45362	42.54409
132	17 424	11.48913	36.33180	182	33 124	13.49074	42.66146
133	17 689	11.53256	36.46917	183	33 489	13.52775	42.77850
134	17 956	11.57584	36.60601	184	33 856	13.56466	42.89522
135	18 225	11.61895	36.74235	185	34 225	13.60147	43.01163
136	18 496	11.66190	36.87818	186	34 596	13.63818	43.12772
137	18 769	11.70470	37.01351	187	34 969	13.67479	43.24350
138	19 044	11.74734	37.14835	188	35 344	13.71131	43.35897
139	19 321	11.78983	37.28270	189	35 721	13.74773	43.47413
140	19 600	11.83216	37.41657	190	36 100	13.78405	43.58899
141	19 881	11.87434	37.54997	191	36 481	13.82027	43.70355
142	20 164	11.91638	37.68289	192	36 864	13.85641	43.81780
143	20 449	11.95826	37.81534	193	37 249	13.89244	43.93177
144	20 736	12.00000	37.94733	194	37 636	13.92839	44.04543
145	21 025	12.04159	38.07887	195	38 025	13.96424	44.15880
146	21 316	12.08305	38.20995	196	38 416	14.00000	44.27189
147	21 609	12.12436	38.34058	197	38 809	14.03567	44.38468
148	21 904	12.16553	38.47077	198	39 204	14.07125	44.49719
149	22 201	12.20656	38.60052	199	39 601	14.10674	44.60942
150	22 500	12.24745	38.72983	200	40 000	14.14214	44.72136

N	N²	$\sqrt{N}$	$\sqrt{10N}$	N	N²	$\sqrt{N}$	$\sqrt{10N}$
200	40 000	14.14214	44.72136	250	62 500	15.81139	50.00000
201	40 401	14.17745	44.83302	251	63 001	15.84298	50.09990
202	40 804	14.21267	44.94441	252	63 504	15.87451	50.19960
203	41 209	14.24781	45.05552	253	64 009	15.90597	50.29911
204	41 616	14.28296	45.16636	254	64 516	15.93738	50.39841
205	42 025	14.31782	45.27693	255	65 025	15.96872	50.49752
206	42 436	14.35270	45.38722	256	65 536	16.00000	50.59644
207	42 849	14.38749	45.49725	257	66 049	16.03122	50.69517
208	43 264	14.42221	45.60702	258	66 564	16.06238	50.79370
209	43 681	14.45683	45.71652	259	67 081	16.09348	50.89204
210	44 100	14.49138	45.82576	260	67 600	16.12452	50.99020
211	44 521	14.52584	45.93474	261	68 121	16.15549	51.08816
212	44 944	14.56022	46.04346	262	68 644	16.18641	51.18594
213	45 369	14.59452	46.15192	263	69 169	16.21727	51.28353
214	45 796	14.62874	46.26013	264	69 696	16.24808	51.38093
215	46 225	14.66288	46.36809	265	70 225	16.27882	51.47815
216	46 656	14.69694	46.47580	266	70 756	16.30951	51.57519
217	47 089	14.73092	46.58326	267	71 289	16.34013	51.67204
218	47 524	14.76482	46.69047	268	71 824	16.37071	51.76872
219	47 961	14.79865	46.79744	269	72 361	16.40122	51.86521
220	48 400	14.83240	46.90415	270	72 900	16.43168	51.96152
221	48 841	14.86607	47.01064	271	73 441	16.46208	52.05766
222	49 284	14.89966	47.11688	272	73 984	16.49242	52.15362
223	49 729	14.93318	47.22288	273	74 529	16.52271	52.24940
224	50 176	14.96663	47.32864	274	75 076	16.55295	52.34501
225	50 625	15.00000	47.43416	275	75 625	16.58312	52.44044
226	51 076	15.03330	47.53946	276	76 176	16.61325	52.53570
227	51 529	15.06652	47.64452	277	76 729	16.64332	52.63079
228	51 984	15.09967	47.74935	278	77 284	16.67333	52.72571
229	52 441	15.13275	47.85394	279	77 841	16.70329	52.82045
230	52 900	15.16575	47.95832	280	78 400	16.73320	52.91503
231	53 361	15.19868	48.06246	281	78 961	16.76305	53.00943
232	53 824	15.23155	48.16638	282	79 524	16.79286	53.10367
233	54 289	15.26434	48.27007	283	80 089	16.82260	53.19774
234	54 756	15.29706	48.37355	284	80 656	16.85230	53.29165
235	55 225	15.32971	48.47680	285	81 225	16.88194	53.38539
236	55 696	15.36229	48.57983	286	81 796	16.91153	53.47897
237	56 169	15.39480	46.68265	287	82 369	16.94107	53.57238
238	56 644	15.42725	48.78524	288	82 944	16.97056	53.66563
239	57 121	15.45962	48.88763	289	83 521	17.00000	53.75872
240	57 600	15.49193	48.98979	290	84 100	17.02939	53.85165
241	58 081	15.52417	49.09175	291	84 681	17.05872	53.94442
242	58 564	15.55635	49.19350	292	85 264	17.08801	54.03702
243	59 049	15.58846	49.29503	293	85 849	17.11724	54.12947
244	59 536	15.52050	49.39636	294	86 436	17.14643	54.22177
245	60 025	15.65248	49.49747	295	87 025	17.17556	54.31390
246	60 516	15.68439	49.59839	296	87 616	17.20465	54.40588
247	61 009	15.71623	49.69909	297	88 209	17.23369	54.49771
248	61 504	15.74802	49.79960	298	88 804	17.26268	54.58938
249	62 001	15.77973	49.89990	299	89 401	17.29162	54.68089
250	62 500	15.81139	50.00000	300	90 000	17.32051	54.77226

N	N²	√N	√10N	N	N²	√N	√10N
300	90 000	17.32051	54.77226	350	122 500	18.70829	59.16080
301	90 601	17.34935	54.86347	351	123 201	18.73499	59.24525
302	91 204	17.37815	54.95453	352	123 904	18.76166	59.32959
303	91 809	17.40690	55.04544	353	124 609	18.78829	59.41380
304	92 416	17.43560	55.13620	354	125 316	18.81489	59.49790
305	93 025	17.46425	55.22681	355	126 025	18.84144	59.58188
306	93 636	17.49288	55.31727	356	126 736	18.86796	59.66574
307	94 249	17.52142	55.40758	357	127 449	18.89444	59.74948
308	94 864	17.54993	55.49775	358	128 164	18.92089	59.83310
309	95 481	17.57840	55.58777	359	128 881	18.94730	59.91661
310	96 100	17.60682	55.67764	360	129 600	18.97367	60.00000
311	96 721	17.63519	55.76737	361	130 321	19.00000	60.08328
312	97 344	17.66352	55.85696	362	131 044	19.02630	60.16644
313	97 969	17.69181	55.94640	363	131 769	19.05256	60.24948
314	98 596	17.72005	56.03670	364	132 496	19.07878	60.33241
315	99 225	17.74824	56.12486	365	133 225	19.10497	60.41523
316	99 856	17.77639	56.21388	366	133 956	19.13113	60.49793
317	100 489	17.80449	56.30275	367	134 689	19.15724	60.58052
318	101 124	17.83255	56.39149	368	135 424	19.18333	60.66300
319	101 761	17.86057	56.48008	369	136 161	19.20937	60.74537
320	102 400	17.88854	56.56854	370	136 900	19.23538	60.82763
321	103 041	17.91647	56.65686	371	137 641	19.26136	60.90977
322	103 684	17.94436	56.74504	372	138 384	19.28730	60.99180
323	104 329	17.97220	56.83309	373	139 129	19.31321	61.07373
324	104 976	18.00000	56.92100	374	139 876	19.33908	61.15554
325	105 625	18.02776	57.00877	375	140 625	19.36492	61.23724
326	106 276	18.05547	57.09641	376	141 376	19.39072	61.31884
327	106 929	18.08314	57.18391	377	142 129	19.41649	61.40033
328	107 584	18.11077	57.27128	378	142 884	19.44222	61.48170
329	108 241	18.13836	57.35852	379	143 641	19.46792	61.56298
330	108 900	18.16590	57.44563	380	144 000	19.49359	61.64414
331	109 561	18.19341	57.53260	381	145 161	19.51922	61.72520
332	110 224	18.22087	57.61944	382	145 924	19.54483	61.80615
333	110 889	18.24829	57.70615	383	146 689	19.57039	61.88699
334	111 556	18.27567	57.79273	384	147 456	19.59592	61.96773
335	112 225	18.30301	57.87918	385	148 225	19.62142	62.04837
336	112 896	18.33030	57.96551	386	148 996	19.64688	62.12890
337	113 569	18.35756	58.05170	387	149 769	19.67232	62.20932
338	114 224	18.38478	57.13777	388	150 544	19.69772	62.28965
339	114 921	18.41195	58.22371	389	151 321	19.72308	62.36986
340	115 600	18.43909	58.30952	390	152 100	19.74842	62.44998
341	116 281	18.46619	58.39521	391	152 881	19.77372	62.52999
342	116 694	18.49324	58.48077	392	153 664	19.79899	62.60990
343	117 649	18.52026	58.56620	393	154 449	19.82423	62.68971
344	118 336	18.54724	58.65151	394	155 236	19.84943	62.76942
345							
345	119 025	18.57418	58.73670	395	156 025	19.87461	62.84903
346	119 716	18.60108	58.82176	396	156 816	19.89975	62.92853
347	120 409	18.62794	58.90671	397	157 609	19.92486	63.00794
348	121 104	18.65476	58.99152	398	158 404	19.94994	63.08724
349	121 801	18.68154	59.07622	399	159 201	19.97498	63.16645
350	122 500	18.70829	59.16080	400	160 000	20.00000	63.24555

N	N^2	$\sqrt{N}$	$\sqrt{10N}$	N	N^2	$\sqrt{N}$	$\sqrt{10N}$
400	160 000	20.00000	63.24555	450	202 500	21.21320	67.08204
401	160 801	20.02498	63.32456	451	203 401	21.23676	67.15653
402	161 604	20.04994	63.40347	452	204 304	21.26029	67.23095
403	162 409	20.07486	63.48228	453	205 209	21.28380	67.30527
404	163 216	20.09975	63.56099	454	206 116	21.30728	67.37952
405	164 025	20.12461	63.63961	455	207 025	21.33073	67.45369
406	164 836	20.14944	63.71813	456	207 936	21.35416	67.52777
407	165 649	20.17424	63.79655	457	208 849	21.37756	67.60178
408	166 464	20.19901	63.87488	458	209 764	21.40093	67.67570
409	167 281	20.22375	63.95311	459	210 681	21.42429	67.74954
410	168 100	20.24846	64.03124	460	211 600	21.44761	67.82330
411	168 921	20.27313	64.10928	461	212 521	21.47091	67.89698
412	169 744	20.29778	64.18723	462	213 444	21.49419	67.97058
413	170 569	20.32240	64.26508	463	214 369	21.51743	68.04410
414	171 396	20.34699	64.34283	464	215 296	21.54066	68.11755
415	172 225	20.37155	64.42049	465	216 225	21.56386	68.19091
416	173 056	20.39608	64.49806	466	217 156	21.58703	68.26419
417	173 889	20.42058	64.57554	467	218 089	21.61018	68.33740
418	174 724	20.44505	64.65292	468	219 024	21.63331	68.41053
419	175 561	20.46949	64.73021	469	219 961	21.65641	68.48357
420	176 400	20.49390	64.80741	470	220 900	21.67948	68.55655
421	177 241	20.51828	64.88451	471	221 841	21.70253	68.62944
422	178 084	20.54264	64.96153	472	222 784	21.72556	68.70226
423	178 929	20.56696	65.03845	473	223 729	21.74856	68.77500
424	179 776	20.59126	65.11528	474	224 676	21.77154	68.84706
425	180 625	20.61553	65.19202	475	225 625	21.79449	68.92024
426	181 476	20.63977	65.26808	476	226 576	21.81742	68.99275
427	182 329	20.66398	65.34524	477	227 529	21.84033	69.06519
428	183 184	20.68816	65.42171	478	228 484	21.86321	69.13754
429	184 041	20.71232	65.49809	479	229 441	21.88607	69.20983
430	184 900	20.73644	65.57439	480	230 400	21.90800	69.28203
431	185 761	20.76054	65.65059	481	231 361	21.93171	69.35416
432	186 624	20.78461	65.72671	482	232 324	21.95450	69.42622
433	187 489	20.80865	65.80274	483	233 280	21.97726	69.40820
434	188 356	20.83267	65.87868	484	234 256	22.00000	69.57011
435	189 225	20.85665	65.95453	485	235 225	22.02272	69.64194
436	190 096	20.88061	66.03030	486	236 196	22.04541	69.71370
437	190 969	20.90454	66.10598	487	237 169	22.06808	69.78530
438	191 844	20.92845	66.18157	488	238 144	22.09072	69.85700
439	192 721	20.95233	66.25708	489	239 121	22.11334	69.92853
440	193 600	20.97618	66.33250	490	240 100	22.13594	70.00000
441	194 481	21.00000	66.40783	491	241 081	22.15852	70.07139
442	195 364	21.02380	66.48308	492	242 064	22.18107	70.14271
443	196 249	21.04757	66.55825	493	243 049	22.20360	70.21396
444	197 136	21.07131	66.63332	494	244 036	22.22611	70.28513
445	198 025	21.09502	66.70832	495	245 025	22.24860	70.35624
446	198 916	21.11871	66.78323	496	246 016	22.27106	70.42727
447	199 809	21.14237	66.85806	497	247 009	22.29350	70.49823
448	200 704	21.16601	66.93280	498	248 004	22.31519	70.56912
449	201 601	21.18962	67.00746	499	249 001	22.33831	70.63993
450	202 500	21.21320	67.08204	500	250 000	22.36068	70.71068

N	N^2	$\sqrt{N}$	$\sqrt{10N}$	N	N^2	$\sqrt{N}$	$\sqrt{10N}$
500	250 000	22.36068	70.71068	550	302 500	23.45208	74.16198
501	251 001	22.38303	70.78135	551	303 601	23.47339	74.22937
502	252 004	22.40536	70.85196	552	304 704	23.49468	74.29670
503	253 009	22.42766	70.92249	553	305 809	23.51595	74.36397
504	254 016	22.44994	70.99296	554	306 916	23.53720	74.43118
505	255 025	22.47221	71.06335	555	308 025	23.55844	74.49832
506	256 036	22.49444	71.13368	556	309 136	23.57965	74.56541
507	257 049	22.51666	71.20393	557	310 249	23.60085	74.63243
508	258 064	22.53886	71.27412	558	311 364	23.62202	74.69940
509	259 081	22.56103	71.34424	559	312 481	23.64318	74.76630
510	260 100	22.58318	71.41428	560	313 600	23.66432	74.83315
511	261 121	22.60531	71.48426	561	314 721	23.68544	74.89993
512	262 144	22.62742	71.55418	562	315 844	23.70654	74.96666
513	263 169	22.64950	71.62402	563	316 969	23.72762	75.03333
514	264 196	22.67157	71.69379	564	318 096	23.74868	75.09993
515	265 225	22.69361	71.76350	565	319 225	23.76973	75.16648
516	266 256	22.71563	71.83314	566	320 356	23.79075	75.23297
517	267 289	22.73763	71.90271	567	321 489	23.81176	75.29940
518	268 324	22.75961	71.97222	568	322 624	23.83275	75.36577
519	269 361	22.78157	72.04165	569	323 761	23.85372	75.43209
520	270 400	22.80351	72.11103	570	324 900	23.87467	75.49834
521	271 441	22.82542	72.18033	571	326 041	23.89561	75.56454
522	272 484	22.84732	72.24957	572	327 184	23.91652	75.63068
523	273 529	22.86919	72.31874	573	328 329	23.93742	75.69676
524	274 576	22.89105	72.38784	574	329 476	23.95830	75.76279
525	275 625	22.91288	72.45688	575	330 625	23.97916	75.82875
526	276 676	22.93469	72.52586	576	331 776	24.00000	75.89466
527	277 729	22.95648	72.59477	577	332 929	24.02082	75.96052
528	278 784	22.97825	72.66361	578	334 084	24.04163	76.02631
529	279 841	23.00000	72.73239	579	335 241	24.06242	76.09205
530	280 900	23.02173	72.80110	580	336 400	24.08319	76.15773
531	281 961	23.04344	72.86975	581	337 561	24.10394	76.22336
532	283 024	23.06513	72.93833	582	338 724	24.12468	76.28892
533	284 089	23.08679	73.00685	583	339 889	24.14539	76.35444
534	285 156	23.10844	73.07530	584	341 056	24.16609	76.41989
535	286 225	23.13007	73.14369	585	342 225	24.18677	76.48529
536	287 296	23.15167	73.21202	586	343 396	24.20744	76.55064
537	288 369	23.17326	73.28028	587	344 569	24.22808	76.61593
538	289 444	23.19483	73.34848	588	345 744	24.24871	76.68116
539	290 521	23.21637	73.41662	589	346 921	24.26932	76.74634
540	291 600	23.23790	73.48469	590	348 100	24.28992	76.81146
541	292 681	23.25941	73.55270	591	349 281	24.31049	76.87652
542	293 764	23.28089	73.62065	592	350 464	24.33105	76.94154
543	294 849	23.30236	73.68853	593	351 649	24.35159	77.00649
544	295 936	23.32381	73.75636	594	352 836	24.37212	77.07140
545	297 025	23.34524	73.82412	595	354 025	24.39262	77.13624
546	298 116	23.36664	73.89181	596	355 216	24.41311	77.20104
547	299 209	23.38803	73.95941	597	356 409	24.43358	77.26578
548	300 304	23.40940	74.02702	598	357 604	24.45404	77.33046
549	301 401	23.43075	74.09453	599	358 801	24.47448	77.39509
550	302 500	23.45208	74.16198	600	360 000	24.49490	77.45967

N	N²	√N	√10N	N	N²	√N	√10N
600	360 000	24.49490	77.45967	650	422 500	25.49510	80.62258
601	361 201	24.51530	77.52419	651	423 801	25.51470	80.68457
602	362 404	24.53569	77.58868	652	425 409	25.55386	80.80842
603	363 609	24.55606	77.65307	653	426 409	25.55386	80.80842
604	364 816	24.57641	77.71744	654	427 716	25.57342	80.87027
605	366 025	24.59675	77.78175	655	429 025	25.59297	80.93207
606	367 236	24.61707	77.84600	656	430 336	25.61250	80.99383
607	368 449	24.63737	77.91020	657	431 649	25.63201	81.05554
608	369 664	24.65766	77.97435	658	432 964	25.65151	81.11720
609	370 881	24.67793	78.03845	659	434 281	25.67100	81.17881
610	372 100	24.69818	78.10250	660	435 600	25.69047	81.24038
611	373 321	24.71841	78.16649	661	436 921	25.70992	81.30191
612	374 544	24.73863	78.23043	662	438 244	25.72936	81.36338
613	375 769	24.75884	78.29432	663	439 569	25.74879	81.42481
614	376 996	24.77902	78.35815	664	440 896	25.76820	81.48620
615	378 225	24.79919	78.42194	665	442 225	25.78759	81.54753
616	379 456	24.81935	78.48567	666	443 556	25.80698	81.60882
617	380 689	24.83948	78.54935	667	444 889	25.82634	81.67007
618	381 924	24.85961	78.61298	668	446 224	25.84570	81.73127
619	383 161	24.87971	78.67655	669	447 561	25.86503	81.79242
620	384 400	24.89980	78.74008	670	448 900	25.88436	81.85353
621	385 641	24.91987	78.80355	671	450 241	25.90367	81.91459
622	386 884	24.93993	78.86698	672	451 584	25.92296	81.97561
623	288 129	24.95997	78.93035	673	452 929	25.94224	82.03658
624	389 376	24.97999	78.99367	674	454 276	25.96151	82.09750
625	390 625	25.00000	79.05694	675	455 625	25.98076	82.15838
626	391 876	25.01999	79.12016	676	456 976	26.00000	82.21922
627	393 129	25.03997	79.18333	677	458 329	26.01922	82.28001
628	394 384	25.05993	79.24645	678	459 684	26.03843	82.34076
629	395 641	25.07987	79.30952	679	461 041	26.05763	82.40146
630	396 900	25.09980	79.37254	680	462 400	26.07681	82.46211
631	398 161	25.11971	79.43551	681	463 761	26.09598	82.42272
632	399 424	25.13961	79.49843	682	465 124	26.11513	82.58329
633	400 689	25.15949	79.56130	683	466 489	26.13427	82.64381
634	401 956	25.17936	79.62412	684	467 856	26.15339	82.70429
635	403 225	25.19921	79.68689	685	469 225	26.17250	82.76473
636	404 496	25.21904	79.74961	686	470 596	26.19160	82.82512
637	405 769	25.23886	79.81228	687	471 969	26.21068	82.88546
638	407 044	25.25866	79.87490	688	473 344	26.22975	82.94577
639	408 321	25.27845	79.93748	689	474 721	26.24881	83.00602
640	409 600	25.29822	80.00000	690	476 100	26.26785	83.06624
641	410 881	25.31798	80.06248	691	477 481	26.28688	83.12641
642	412 164	25.33772	80.12490	692	478 864	26.30589	83.18654
643	413 449	25.35744	80.18728	693	480 249	26.32489	83.24662
644	414 736	25.37716	80.24961	694	481 636	26.34388	83.30666
645	416 025	25.39685	80.31189	695	483 025	26.36285	83.36666
646	417 316	25.41653	80.37413	696	484 416	26.38181	83.42661
647	418 609	25.43619	80.43631	697	485 809	26.40076	83.48653
648	419 904	25.45584	80.49845	698	487 204	26.41969	83.54639
649	421 201	25.47548	80.56054	699	488 601	26.43861	83.60622
650	422 500	25.49510	80.62258	700	490 000	26.45751	83.66600

N	N²	√N	√10N	N	N²	√N	√10N
700	490 000	26.45751	83.66600	750	562 500	27.38613	86.60254
701	491 401	26.47640	83.72574	751	564 001	27.40438	86.66026
702	492 804	26.49528	83.78544	752	565 504	27.42262	86.71793
703	494 209	26.51415	83.84510	753	567 009	27.44085	86.77557
704	495 616	26.53300	83.90471	754	568 516	27.45906	86.83317
705	497 025	26.55184	83.96428	755	570 025	27.47726	86.89074
706	498 436	26.57066	84.02381	756	571 536	27.49545	86.94826
707	499 849	26.58947	84.08329	757	573 049	27.51363	87.00575
708	501 264	26.60827	84.14274	758	574 564	27.53180	87.06320
709	502 681	26.62705	84.20214	759	576 081	27.54995	87.12061
710	504 100	26.64583	84.26150	760	577 600	27.56810	87.17798
711	505 521	26.66458	84.32082	761	579 121	27.58623	87.23531
712	506 944	26.68333	84.38009	762	580 644	27.60435	87.29261
713	508 369	26.70206	84.43933	763	582 169	27.62245	87.34987
714	509 796	26.72078	84.49852	764	583 696	27.64055	87.40709
715	511 225	26.73948	84.55767	765	585 225	27.65863	87.46428
716	512 656	26.75818	84.61578	766	586 756	27.67671	87.52143
717	514 089	26.77686	84.67585	767	588 289	27.69476	87.57854
718	515 524	26.79552	84.73488	768	589 824	27.71281	87.63561
719	516 961	26.81418	84.79387	769	591 361	27.73085	87.69265
720	518 400	26.83282	84.85281	770	592 900	27.74887	87.74964
721	519 841	26.85144	84.91172	771	594 441	27.76689	87.80661
722	521 284	26.87006	84.97058	772	595 984	27.78489	87.86353
723	522 729	26.88866	85.02941	773	597 529	27.80288	87.92042
724	524 176	26.90725	85.08819	774	599 076	27.82086	87.97727
725	525 625	26.92582	85.14693	775	600 625	27.83882	88.03408
726	527 076	26.94439	85.20563	776	602 176	27.85678	88.09086
727	528 529	26.96294	85.26429	777	603 729	27.87472	88.14760
728	529 984	26.98148	85.32292	778	605 284	27.89265	88.20431
729	531 411	27.00000	85.38150	779	606 841	27.91057	88.26098
730	532 900	27.01851	85.44004	780	608 400	27.92848	88.31761
731	534 361	27.03701	85.49854	781	609 961	27.94638	88.37420
732	535 824	27.05550	85.55700	782	611 524	27.96426	88.43076
733	537 289	27.07397	85.61542	783	613 089	27.98214	88.48729
734	538 756	27.09243	85.67380	784	614 656	28.00000	88.54377
735	540 225	27.11088	85.73214	785	616 225	28.01785	88.60023
736	541 696	27.12932	85.79044	786	617 796	28.03569	88.65664
737	543 169	27.14774	85.84870	787	619 369	28.05352	88.71302
738	544 644	27.16616	85.90693	788	620 944	28.07134	88.76936
739	546 121	27.18455	85.96511	789	622 521	28.08914	88.82567
740	547 600	27.20294	86.02325	790	624 100	28.10694	88.88194
741	549 081	27.22132	86.08136	791	625 681	28.12472	88.93818
742	550 564	27.23968	86.13942	792	627 264	28.14249	88.99438
743	552 049	27.25803	86.10745	793	628 849	28.16026	89.05055
744	553 536	27.27636	86.25543	794	630 436	28.17801	89.10668
745	555 025	27.29469	86.31338	795	632 025	28.19574	89.16277
746	556 516	27.31300	86.37129	796	633 616	28.21347	89.21883
747	558 009	27.33130	86.42916	797	635 209	28.23119	89.27486
748	559 504	27.34959	86.48609	798	636 804	28.24889	89.33085
749	561 001	27.36786	86.54479	799	638 401	28.26659	89.38680
750	562 500	27.38613	86.60254	800	640 000	28.28427	89.44272

N	N^2	$\sqrt{N}$	$\sqrt{10N}$	N	N^2	$\sqrt{N}$	$\sqrt{10N}$
800	640 000	28.28427	89.44272	850	722 500	29.15476	92.19544
801	641 601	28.30194	89.49860	851	724 201	29.17190	92.24966
802	643 204	28.31960	89.55445	852	725 904	29.18904	92.30385
803	644 809	28.33725	89.61027	853	727 609	29.20616	92.35800
804	646 416	28.35489	89.66605	854	729 316	29.22328	92.41212
805	648 025	28.37252	89.72179	855	731 025	29.24038	92.46621
806	649 636	28.39014	89.77750	856	732 736	29.25748	92.52027
807	651 249	28.40775	89.83318	857	734 449	29.27456	92.57429
808	652 864	28.42534	89.88882	858	736 164	29.29164	92.62829
809	654 481	28.44293	89.94443	859	737 881	29.30870	92.68225
810	656 100	28.46050	90.00000	860	739 600	29.32576	92.73618
811	657 721	28.47806	90.05554	861	741 321	29.34280	92.79009
812	659 344	28.49561	90.11104	862	743 044	29.35984	92.84396
813	660 969	28.51315	90.16651	863	744 769	29.37686	92.89779
814	662 596	28.53069	90.22195	864	746 496	29.39388	92.95160
815	664 225	28.54820	90.27735	865	748 225	29.41088	93.00538
816	665 856	28.56571	90.33272	866	749 956	29.42788	93.05912
817	667 489	28.58321	90.38805	867	751 689	29.44486	93.11283
818	669 124	28.60070	90.44335	868	753 424	29.46184	93.16652
819	670 761	28.61818	90.49862	869	755 161	29.47881	93.22017
820	672 400	28.63564	90.55385	870	756 900	29.49576	93.27379
821	674 041	28.65310	90.60905	871	758 641	29.51271	93.32738
822	675 684	28.67054	90.66422	872	760 384	29.52965	93.38094
823	677 329	28.68798	90.71935	873	762 129	29.54657	93.43447
824	678 976	28.70540	90.77445	874	763 876	29.56349	93.48797
825	680 625	28.72281	90.82951	875	765 625	29.58040	93.54143
826	682 276	28.74022	90.88454	876	767 376	29.59730	93.59487
827	683 929	28.75761	90.93954	877	769 129	29.61419	93.64828
828	685 584	28.77499	90.99451	878	770 884	29.63106	93.70165
829	687 241	28.79236	91.04944	879	772 641	29.64793	93.75500
830	688 900	28.80972	91.10434	880	774 400	29.66479	93.80832
831	690 561	28.82707	91.15920	881	776 161	29.68164	93.86160
832	692 224	28.84441	91.21403	882	777 924	29.69848	93.91486
833	693 889	28.86174	91.26883	883	779 689	29.71532	93.96808
834	695 556	28.87906	91.32360	884	781 456	29.73214	94.02027
835	697 225	28.89637	91.37833	885	783 225	29.74895	94.07444
836	698 896	28.91366	91.43304	886	784 996	29.76575	94.12757
837	700 569	28.93095	91.48770	887	786 769	29.78255	94.18068
838	702 244	28.94823	91.54234	888	788 544	29.79933	94.23375
839	703 921	28.96550	91.59694	889	790 321	29.81610	94.28680
840	705 600	28.98275	91.65151	890	792 100	29.83287	94.33981
841	707 281	29.00000	91.70605	891	793 881	29.84962	94.39280
842	708 964	29.01724	91.76056	892	795 664	29.86637	94.44575
843	710 649	29.03446	91.81503	893	797 449	29.88311	94.49868
844	712 336	29.05168	91.86947	894	799 236	29.89983	94.55157
845	714 025	29.06888	91.92388	895	801 025	29.91655	94.60444
846	715 716	29.08608	91.97826	896	802 816	29.93326	94.65728
847	717 409	29.10326	92.03260	897	804 609	29.94996	94.71008
848	719 104	29.12044	92.08692	898	806 404	29.96665	94.76286
849	720 801	29.13760	92.14120	899	808 201	29.98333	94.81561
850	722 500	29.15476	92.19544	900	810 000	30.00000	94.86833

N	N²	√N	√10N	N	N²	√N	√10N
900	810 000	30.00000	94.86833	950	902 500	30.82207	97.46794
901	811 801	30.01666	94.92102	951	904 401	30.83829	97.51923
902	813 604	30.03331	94.97368	952	906 304	30.85450	97.57049
903	815 409	30.04996	95.02631	953	908 209	30.87070	97.62172
904	817 216	30.06659	95.07891	954	910.116	30.88689	97.67292
905	819 025	30.08322	95.13149	955	912 025	30.90307	97.72410
906	820 836	30.09983	95.18403	956	913 936	30.91925	97.77525
907	822 649	30.11644	95.23655	957	915 849	30.93542	97.82638
908	824 464	30.13304	95.28903	958	917 764	30.95158	97.87747
909	826 281	30.14963	95.34149	959	919 681	30.96773	97.92855
910	828 100	30.16621	95.39392	960	921 600	30.98387	97.97959
911	829 921	30.18278	95.44632	961	928 521	31.00000	98.03061
912	831 744	30.19934	95.49869	962	925 444	31.01612	98.08160
913	833 569	30.21589	95.55103	963	927 369	31.03224	98.13256
914	835 396	30.23243	95.60335	964	929 296	31.04835	98.18350
915	837 225	30.24897	95.65563	965	931 225	31.06445	98.23441
916	839 056	30.26549	95.70789	966	933 156	31.08054	98.28530
917	840 889	30.28201	95.76012	967	935 089	31.09662	98.33616
918	842 724	30.29851	95.81232	968	937 024	31.11270	98.38699
919	844 561	30.31501	95.86449	969	938 961	31.12876	98.43780
920	846 400	30.33150	95.91663	970	940 900	31.14482	98.48858
921	848 241	30.34798	95.96874	971	942 841	31.16087	98.53933
922	850 084	30.36445	96.02083	972	944 784	31.17691	98.59006
923	851 929	30.38092	96.07289	973	946 729	31.19295	98.64076
924	853 776	30.39735	96.12492	974	948 676	31.20897	98.69144
925	855 625	30.41381	96.17692	975	950 625	31.22499	98.74209
926	857 476	30.43025	96.22889	976	952 576	31.24100	98.79271
927	859 329	30.44667	96.28084	977	954 529	31.25700	98.84331
928	861 184	30.46309	96.33276	978	956 484	31.27299	98.89388
929	863 041	30.47950	96.28465	979	958 441	31.28898	98.94443
930	864 900	30.49590	96.43651	980	960 400	31.30495	98.99495
931	866 761	30.51229	96.48834	981	962 361	31.32092	99.04544
932	868 624	30.52868	96.54015	982	964 324	31.33688	99.09591
933	870 489	30.54505	96.59193	983	966 144	31.43247	99.44848
934	872 356	30.56141	96.64368	984	968 256	31.36877	99.19677
935	874 225	30.57777	96.69540	985	970 225	31.38471	99.24717
936	876 096	30.59412	96.74709	986	972 196	31.40064	99.29753
937	877 969	30.61046	96.79876	987	974 169	31.41656	99.34787
938	879 844	30.62679	96.85040	988	976 144	31.43247	99.39819
939	881 721	30.64311	96.90201	989	978 121	31.44837	99.44848
940	883 600	30.65942	96.95360	990	980 100	31.46427	99.49874
941	885 481	30.67572	97.00515	991	982 081	31.48015	99.54898
942	887 364	30.69202	97.05668	992	984 064	31.49603	99.59920
943	889 249	30.70831	97.10819	993	986 049	31.51190	99.64939
944	891 136	30.72458	97.15966	994	988 036	31.52777	99.69955
945	893 025	30.74085	97.21111	995	990 025	31.54362	99.74969
946	894 916	30.75711	97.26253	996	992 016	31.55947	99.79980
947	896 809	30.77337	97.31393	997	994 009	31.57531	99.84989
948	898 704	30.78961	97.36529	998	996 004	31.59114	99.89995
949	900 601	30.80584	97.41663	999	998 001	31.60696	99.94999
950	902 500	30.82207	97.46794	1000	1 000 000	31.62278	100.00000

APPENDIX 6 F DISTRIBUTIONS TABLES

The following tables provide critical values of F at the .05 and .01 levels of significance. The number of degrees of freedom for the *numerator* is indicated at the top of each *column,* and the number of degrees of freedom for the *denominator* determines the *row* to use.

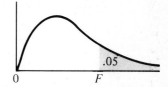

Critical Values of F_{ν_1, ν_2} for $\alpha = .05$

ν_1 = Degrees of freedom for numerator

ν_2 = Degrees of freedom for denominator

ν_2	1	2	3	4	5	6	7	8	9	10	12	15	20	24	30	40	60	120	∞
1	161	200	216	225	230	234	237	239	241	242	244	246	248	249	250	251	252	253	254
2	18.5	19.0	19.2	19.2	19.3	19.3	19.4	19.4	19.4	19.4	19.4	19.4	19.4	19.5	19.5	19.5	19.5	19.5	19.5
3	10.1	9.55	9.28	9.12	9.01	8.94	8.89	8.85	8.81	8.79	8.74	8.70	8.66	8.64	8.62	8.59	8.57	8.55	8.53
4	7.71	6.94	6.59	6.39	6.26	6.16	6.09	6.04	6.00	5.96	5.91	5.86	5.80	5.77	5.75	5.72	5.69	5.66	5.63
5	6.61	5.79	5.41	5.19	5.05	4.95	4.88	4.82	4.77	4.74	4.68	4.62	4.56	4.53	4.50	4.46	4.43	4.40	4.37
6	5.99	5.14	4.76	4.53	4.39	4.28	4.21	4.15	4.10	4.06	4.00	3.94	3.87	3.84	3.81	3.77	3.74	3.70	3.67
7	5.59	4.74	4.35	4.12	3.97	3.87	3.79	3.73	3.68	3.64	3.57	3.51	3.44	3.41	3.38	3.34	3.30	3.27	3.23
8	5.32	4.46	4.07	3.84	3.69	3.58	3.50	3.44	3.39	3.35	3.28	3.22	3.15	3.12	3.08	3.04	3.01	2.97	2.93
9	5.12	4.26	3.86	3.63	3.48	3.37	3.29	3.23	3.18	3.14	3.07	3.01	2.94	2.90	2.86	2.83	2.79	2.75	2.71
10	4.96	4.10	3.71	3.48	3.33	3.22	3.14	3.07	3.02	2.98	2.91	2.85	2.77	2.74	2.70	2.66	2.62	2.58	2.54
11	4.84	3.98	3.59	3.36	3.20	3.09	3.01	2.95	2.90	2.85	2.79	2.72	2.65	2.61	2.57	2.53	2.49	2.45	2.40
12	4.75	3.89	3.49	3.26	3.11	3.00	2.91	2.85	2.80	2.75	2.69	2.62	2.54	2.51	2.47	2.43	2.38	2.34	2.30
13	4.67	3.81	3.41	3.18	3.03	2.92	2.83	2.77	2.71	2.67	2.60	2.53	2.46	2.42	2.38	2.34	2.30	2.25	2.21
14	4.60	3.74	3.34	3.11	2.96	2.85	2.76	2.70	2.65	2.60	2.53	2.46	2.39	2.35	2.31	2.27	2.22	2.18	2.13
15	4.54	3.68	3.29	3.06	2.90	2.79	2.71	2.64	2.59	2.54	2.48	2.40	2.33	2.29	2.25	2.20	2.16	2.11	2.07
16	4.49	3.63	3.24	3.01	2.85	2.74	2.66	2.59	2.54	2.49	2.42	2.35	2.28	2.24	2.19	2.15	2.11	2.06	2.01
17	4.45	3.59	3.20	2.96	2.81	2.70	2.61	2.55	2.49	2.45	2.38	2.31	2.23	2.19	2.15	2.10	2.06	2.01	1.96
18	4.41	3.55	3.16	2.93	2.77	2.66	2.58	2.51	2.46	2.41	2.34	2.27	2.19	2.15	2.11	2.06	2.02	1.97	1.92
19	4.38	3.52	3.13	2.90	2.74	2.63	2.54	2.48	2.42	2.38	2.31	2.23	2.16	2.11	2.07	2.03	1.98	1.93	1.88
20	4.35	3.49	3.10	2.87	2.71	2.60	2.51	2.45	2.39	2.35	2.28	2.20	2.12	2.08	2.04	1.99	1.95	1.90	1.84
21	4.32	3.47	3.07	2.84	2.68	2.57	2.49	2.42	2.37	2.32	2.25	2.18	2.10	2.05	2.01	1.96	1.92	1.87	1.81
22	4.30	3.44	3.05	2.82	2.66	2.55	2.46	2.40	2.34	2.30	2.23	2.15	2.07	2.03	1.98	1.94	1.89	1.84	1.78
23	4.28	3.42	3.03	2.80	2.64	2.53	2.44	2.37	2.32	2.27	2.20	2.13	2.05	2.01	1.96	1.91	1.86	1.81	1.76
24	4.26	3.40	3.01	2.78	2.62	2.51	2.42	2.36	2.30	2.25	2.18	2.11	2.03	1.98	1.94	1.89	1.84	1.79	1.73
25	4.24	3.39	2.99	2.76	2.60	2.49	2.40	2.34	2.28	2.24	2.16	2.09	2.01	1.96	1.92	1.87	1.82	1.77	1.71
30	4.17	3.32	2.92	2.69	2.53	2.42	2.33	2.27	2.21	2.16	2.09	2.01	1.93	1.89	1.84	1.79	1.74	1.68	1.62
40	4.08	3.23	2.84	2.61	2.45	2.34	2.25	2.18	2.12	2.08	2.00	1.92	1.84	1.79	1.74	1.69	1.64	1.58	1.51
60	4.00	3.15	2.76	2.53	2.37	2.25	2.17	2.10	2.04	1.99	1.92	1.84	1.75	1.70	1.65	1.59	1.53	1.47	1.39
120	3.92	3.07	2.68	2.45	2.29	2.18	2.09	2.02	1.96	1.91	1.83	1.75	1.66	1.61	1.55	1.50	1.43	1.35	1.25
∞	3.84	3.00	2.60	2.37	2.21	2.10	2.01	1.94	1.88	1.83	1.75	1.67	1.57	1.52	1.46	1.39	1.32	1.22	1.00

Critical Values of F_{ν_1, ν_2} for $\alpha = .01$

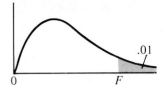

ν_2	\multicolumn{19}{c}{ν_1 = Degrees of freedom for numerator}
	1	2	3	4	5	6	7	8	9	10	12	15	20	24	30	40	60	120	∞
1	4,052	5,000	5,403	5,625	5,764	5,859	5,928	5,982	6,023	6,056	6,106	6,157	6,209	6,235	6,261	6,287	6,313	6,339	6,366
2	98.5	99.0	99.2	99.2	99.3	99.3	99.4	99.4	99.4	99.4	99.4	99.4	99.4	99.5	99.5	99.5	99.5	99.5	99.5
3	34.1	30.8	29.5	28.7	28.2	27.9	27.7	27.5	27.3	27.2	27.1	26.9	26.7	26.6	26.5	26.4	26.3	26.2	26.1
4	21.2	18.0	16.7	16.0	15.5	15.2	15.0	14.8	14.7	14.5	14.4	14.2	14.0	13.9	13.8	13.7	13.7	13.6	13.5
5	16.3	13.3	12.1	11.4	11.0	10.7	10.5	10.3	10.2	10.1	9.89	9.72	9.55	9.47	9.38	9.29	9.20	9.11	9.02
6	13.7	10.9	9.78	9.15	8.75	8.47	8.26	8.10	7.98	7.87	7.72	7.56	7.40	7.31	7.23	7.14	7.06	6.97	6.88
7	12.2	9.55	8.45	7.85	7.46	7.19	6.99	6.84	6.72	6.62	6.47	6.31	6.16	6.07	5.99	5.91	5.82	5.74	5.65
8	11.3	8.65	7.59	7.01	6.63	6.37	6.18	6.03	5.91	5.81	5.67	5.52	5.36	5.28	5.20	5.12	5.03	4.95	4.86
9	10.6	8.02	6.99	6.42	6.06	5.80	5.61	5.47	5.35	5.26	5.11	4.96	4.81	4.73	4.65	4.57	4.48	4.40	4.31
10	10.0	7.56	6.55	5.99	5.64	5.39	5.20	5.06	4.94	4.85	4.71	4.56	4.41	4.33	4.25	4.17	4.08	4.00	3.91
11	9.65	7.21	6.22	5.67	5.32	5.07	4.89	4.74	4.63	4.54	4.40	4.25	4.10	4.02	3.94	3.86	3.78	3.69	3.60
12	9.33	6.93	5.95	5.41	5.06	4.82	4.64	4.50	4.39	4.30	4.16	4.01	3.86	3.78	3.70	3.62	3.54	3.45	3.36
13	9.07	6.70	5.74	5.21	4.86	4.62	4.44	4.30	4.19	4.10	3.96	3.82	3.66	3.59	3.51	3.43	3.34	3.25	3.17
14	8.86	6.51	5.56	5.04	4.70	4.46	4.28	4.14	4.03	3.94	3.80	3.66	3.51	3.43	3.35	3.27	3.18	3.09	3.00
15	8.68	6.36	5.42	4.89	4.56	4.32	4.14	4.00	3.89	3.80	3.67	3.52	3.37	3.29	3.21	3.13	3.05	2.96	2.87
16	8.53	6.23	5.29	4.77	4.44	4.20	4.03	3.89	3.78	3.69	3.55	3.41	3.26	3.18	3.10	3.02	2.93	2.84	2.75
17	8.40	6.11	5.19	4.67	4.34	4.10	3.93	3.79	3.68	3.59	3.46	3.31	3.16	3.08	3.00	2.92	2.83	2.75	2.65
18	8.29	6.01	5.09	4.58	4.25	4.01	3.84	3.71	3.60	3.51	3.37	3.23	3.08	3.00	2.92	2.84	2.75	2.66	2.57
19	8.19	5.93	5.01	4.50	4.17	3.94	3.77	3.63	3.52	3.43	3.30	3.15	3.00	2.92	2.84	2.76	2.67	2.58	2.49
20	8.10	5.85	4.94	4.43	4.10	3.87	3.70	3.56	3.46	3.37	3.23	3.09	2.94	2.86	2.78	2.69	2.61	2.52	2.42
21	8.02	5.78	4.87	4.37	4.04	3.81	3.64	3.51	3.40	3.31	3.17	3.03	2.88	2.80	2.72	2.64	2.55	2.46	2.36
22	7.95	5.72	4.82	4.31	3.99	3.76	3.59	3.45	3.35	3.26	3.12	2.98	2.83	2.75	2.67	2.58	2.50	2.40	2.31
23	7.88	5.66	4.76	4.26	3.94	3.71	3.54	3.41	3.30	3.21	3.07	2.93	2.78	2.70	2.62	2.54	2.45	2.35	2.26
24	7.82	5.61	4.72	4.22	3.90	3.67	3.50	3.36	3.26	3.17	3.03	2.89	2.74	2.66	2.58	2.49	2.40	2.31	2.21
25	7.77	5.57	4.68	4.18	3.86	3.63	3.46	3.32	3.22	3.13	2.99	2.85	2.70	2.62	2.53	2.45	2.36	2.27	2.17
30	7.56	5.39	4.51	4.02	3.70	3.47	3.30	3.17	3.07	2.98	2.84	2.70	2.55	2.47	2.39	2.30	2.21	2.11	2.01
40	7.31	5.18	4.31	3.83	3.51	3.29	3.12	2.99	2.89	2.80	2.66	2.52	2.37	2.29	2.20	2.11	2.02	1.92	1.80
60	7.08	4.98	4.13	3.65	3.34	3.12	2.95	2.82	2.72	2.63	2.50	2.35	2.20	2.12	2.03	1.94	1.84	1.73	1.60
120	6.85	4.79	3.95	3.48	3.17	2.96	2.79	2.66	2.56	2.47	2.34	2.19	2.03	1.95	1.86	1.76	1.66	1.53	1.38
∞	6.63	4.61	3.78	3.32	3.02	2.80	2.64	2.51	2.41	2.32	2.18	2.04	1.88	1.79	1.70	1.59	1.47	1.32	1.00

ν_2 = Degrees of freedom for denominator

Source: From Maxine Merrington and Catherine M. Thompson, "Tables of the Percentage Points of the Inverted F-Distribution." *Biometrika*, vol. 33, pp. 73-88, 1943. Reprinted with the permission of the *Biometrika* Trustees.

APPENDIX 7 CHI-SQUARE DISTRIBUTION

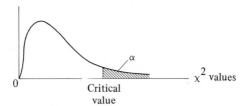

Example of how to use this table: In a chi-square distribution with 6 degrees of freedom (*df*), the area to the right of a critical value of 12.592—i.e., the α area—is .05.

Degrees of freedom (*df*)	Area in shaded right tail (α)		
	.10	.05	.01
1	2.706	3.841	6.635
2	4.605	5.991	9.210
3	6.251	7.815	11.345
4	7.779	9.488	13.277
5	9.236	11.070	15.086
6	10.645	12.592	16.812
7	12.017	14.067	18.475
8	13.362	15.507	20.090
9	14.684	16.919	21.666
10	15.987	18.307	23.209
11	17.275	19.675	24.725
12	18.549	21.026	26.217
13	19.812	22.362	27.688
14	21.064	23.685	29.141
15	22.307	24.996	30.578
16	23.542	26.296	32.000
17	24.769	27.587	33.409
18	25.989	28.869	34.805
19	27.204	30.144	36.191
20	28.412	31.410	37.566
21	29.615	32.671	38.932
22	30.813	33.924	40.289
23	32.007	35.172	41.638
24	33.196	36.415	42.980
25	34.382	37.652	44.314
26	35.563	38.885	45.642
27	36.741	40.113	46.963
28	37.916	41.337	48.278
29	39.087	42.557	49.588
30	40.256	43.773	50.892

Source: This table is abridged from Table IV of Fisher and Yates, *Statistical Tables for Biological, Agricultural and Medical Research,* published by Longman Group, Ltd., London (previously published by Oliver & Boyd, Ltd., Edinburgh). Reproduced with the permission of the authors and publishers.

APPENDIX 8 CRITICAL VALUES OF T FOR $\alpha=.05$ AND $\alpha=.01$ IN THE WILCOXON SIGNED RANK TEST

	Two-tailed test		One-tailed test	
n	.05	.01	.05	.01
4				
5			0	
6	0		2	
7	2		3	0
8	3	0	5	1
9	5	1	8	3
10	8	3	10	5
11	10	5	13	7
12	13	7	17	9
13	17	9	21	12
14	21	12	25	15
15	25	15	30	19
16	29	19	35	23
17	34	23	41	27
18	40	27	47	32
19	46	32	53	37
20	52	37	60	43
21	58	42	67	49
22	65	48	75	55
23	73	54	83	62
24	81	61	91	69
25	89	68	100	76
26	98	75	110	84
27	107	83	119	92
28	116	91	130	101
29	126	100	140	110
30	137	109	151	120
31	147	118	163	130
32	159	128	175	140
33	170	138	187	151
34	182	148	200	162
35	195	159	213	173
40	264	220	286	238
50	434	373	466	397
60	648	567	690	600
70	907	805	960	846
80	1211	1086	1276	1136
90	1560	1410	1638	1471
100	1955	1779	2045	1850

Source: Abridged from Robert L. McCormack, "Extended Tables of the Wilcoxon Matched Pair Signed Rank Statistic," *Journal of the American Statistical Association,* September 1965, pp. 866-867. Reprinted with permission.

APPENDIX 9 DISTRIBUTION OF U IN THE MANN-WHITNEY TEST

One-tailed test tables

Critical U values: $\alpha = .05$ for a one-tailed test
(and $\alpha = .10$ for a two-tailed test)

n_1 \ n_2	1	2	3	4	5	6	7	8	9	10	11	12	13	14	15	16	17	18	19	20
1																			0	0
2				0	0	0	1	1	1	1	2	2	2	3	3	3	4	4	4
3			0	0	1	2	2	3	3	4	5	5	6	7	7	8	9	9	10	11
4			0	1	2	3	4	5	6	7	8	9	10	11	12	14	15	16	17	18
5		0	1	2	4	5	6	8	9	11	12	13	15	16	18	19	20	22	23	25
6		0	2	3	5	7	8	10	12	14	16	17	19	21	23	25	26	28	30	32
7		0	2	4	6	8	11	13	15	17	19	21	24	26	28	30	33	35	37	39
8		1	3	5	8	10	13	15	18	20	23	26	28	31	33	36	39	41	44	47
9		1	3	6	9	12	15	18	21	24	27	30	33	36	39	42	45	48	51	54
10		1	4	7	11	14	17	20	24	27	31	34	37	41	44	48	51	55	58	62
11		1	5	8	12	16	19	23	27	31	34	38	42	46	50	54	57	61	65	69
12		2	5	9	13	17	21	26	30	34	38	42	47	51	55	60	64	68	72	77
13		2	6	10	15	19	24	28	33	37	42	47	51	56	61	65	70	75	80	84
14		2	7	11	16	21	26	31	36	41	46	51	56'	61	66	71	77	82	87	92
15		3	7	12	18	23	28	33	39	44	50	55	61	66	72	77	83	88	94	100
16		3	8	14	19	25	30	36	42	48	54	60	65	71	77	83	89	95	101	107
17		3	9	15	20	26	33	39	45	51	57	64	70	77	83	89	96	102	109	115
18		4	9	16	22	28	35	41	48	55	61	68	75	82	88	95	102	109	116	123
19	0	4	10	17	23	30	37	44	51	58	65	72	80	87	94	101	109	116	123	130
20	0	4	11	18	25	32	39	47	54	62	69	77	84	92	100	107	115	123	130	138

Critical U values: $\alpha = .01$ for a one-tailed test
(and $\alpha = .02$ for a two-tailed test)

n_1 \ n_2	1	2	3	4	5	6	7	8	9	10	11	12	13	14	15	16	17	18	19	20
1																				
2													0	0	0	0	0	0	1	1
3						0	0	1	1	1	2	2	2	3	3	4	4	4	5
4				0	1	1	2	3	3	4	5	5	6	7	7	8	9	9	10
5			0	1	2	3	4	5	6	7	8	9	10	11	12	13	14	15	16
6			1	2	3	4	6	7	8	9	11	12	13	15	16	18	19	20	22
7			0	1	3	4	6	7	9	11	12	14	16	17	19	21	23	24	26	28
8			0	2	4	6	7	9	11	13	15	17	20	22	24	26	28	30	32	34
9			1	3	5	7	9	11	14	16	18	21	23	26	28	31	33	36	38	40
10			1	3	6	8	11	13	16	19	22	24	27	30	33	36	38	41	44	47
11			1	4	7	9	12	15	18	22	25	28	31	34	37	41	44	47	50	53
12			2	5	8	11	14	17	21	24	28	31	35	38	42	46	49	53	56	60
13		0	2	5	9	12	16	20	23	27	31	35	39	43	47	51	55	59	63	67
14		0	2	6	10	13	17	22	26	30	34	38	43	47	51	56	60	65	69	73
15		0	3	7	11	15	19	24	28	33	37	42	47	51	56	61	66	70	75	80
16		0	3	7	12	16	21	26	31	36	41	46	51	56	61	66	71	76	82	87
17		0	4	8	13	18	23	28	33	38	44	49	55	60	66	71	77	82	88	93
18		0	4	9	14	19	24	30	36	41	47	53	59	65	70	76	82	88	94	100
19		1	4	9	15	20	26	32	38	44	50	56	63	69	75	82	88	94	101	107
20		1	5	10	16	22	28	34	40	47	53	60	67	73	80	87	93	100	107	114

Two-tailed test tables

Critical U values: $\alpha = .05$ for a two-tailed test
(and $\alpha = .025$ for a one-tailed test)

n_1 \ n_2	1	2	3	4	5	6	7	8	9	10	11	12	13	14	15	16	17	18	19	20
1																				
2								0	0	0	0	1	1	1	1	1	2	2	2	2
3					0	1	1	2	2	3	3	4	4	5	5	6	6	7	7	8
4				0	1	2	3	4	4	5	6	7	8	9	10	11	11	12	13	13
5			0	1	2	3	5	6	7	8	9	11	12	13	14	15	17	18	19	20
6			1	2	3	5	6	8	10	11	13	14	16	17	19	21	22	24	25	27
7			1	3	5	6	8	10	12	14	16	18	20	22	24	26	28	30	32	34
8		0	2	4	6	8	10	13	15	17	19	22	24	26	29	31	34	36	38	41
9		0	2	4	7	10	12	15	17	20	23	26	28	31	34	37	39	42	45	48
10		0	3	5	8	11	14	17	20	23	26	29	33	36	39	42	45	48	52	55
11		0	3	6	9	13	16	19	23	26	30	33	37	40	44	47	51	55	58	62
12		1	4	7	11	14	18	22	26	29	33	37	41	45	49	53	57	61	65	69
13		1	4	8	12	16	20	24	28	33	37	41	45	50	54	59	63	67	72	76
14		1	5	9	13	17	22	26	31	36	40	45	50	55	59	64	67	74	78	83
15		1	5	10	14	19	24	29	34	39	44	49	54	59	64	70	75	80	85	90
16		1	6	11	15	21	26	31	37	42	47	53	59	64	70	75	81	86	92	98
17		2	6	11	17	22	28	34	39	45	51	57	63	67	75	81	87	93	99	105
18		2	7	12	18	24	30	36	42	48	55	61	67	74	80	86	93	99	106	112
19		2	7	13	19	25	32	38	45	52	58	65	72	78	85	92	99	106	113	119
20		2	8	13	20	27	34	41	48	55	62	69	76	83	90	98	105	112	119	127

Critical U values: $\alpha = .01$ for a two-tailed test
(and $\alpha = .005$ for a one-tailed test)

n_1 \ n_2	1	2	3	4	5	6	7	8	9	10	11	12	13	14	15	16	17	18	19	20
1																				
2																			0	0
3									0	0	0	1	1	1	2	2	2	2	3	3
4						0	0	1	1	2	2	3	3	4	5	5	6	6	7	8
5					0	1	1	2	3	4	5	6	7	7	8	9	10	11	12	13
6				0	1	2	3	4	5	6	7	9	10	11	12	13	15	16	17	18
7				0	1	3	4	6	7	9	10	12	13	15	16	18	19	21	22	24
8				1	2	4	6	7	9	11	13	15	17	18	20	22	24	26	28	30
9			0	1	3	5	7	9	11	13	16	18	20	22	24	27	29	31	33	36
10			0	2	4	6	9	11	13	16	18	21	24	26	29	31	34	37	39	42
11			0	2	5	7	10	13	16	18	21	24	27	30	33	36	39	42	45	48
12			1	3	6	9	12	15	18	21	24	27	31	34	37	41	44	47	51	54
13			1	3	7	10	13	17	20	24	27	31	34	38	42	45	49	53	56	60
14			1	4	7	11	15	18	22	26	30	34	38	42	46	50	54	58	63	67
15			2	5	8	12	16	20	24	29	33	37	42	46	51	55	60	64	69	73
16			2	5	9	13	18	22	27	31	36	41	45	50	55	60	65	70	74	79
17			2	6	10	15	19	24	29	34	39	44	49	54	60	65	70	75	81	86
18			2	6	11	16	21	26	31	37	42	47	53	58	64	70	75	81	87	92
19		0	3	7	12	17	22	28	33	39	45	51	56	63	69	74	81	87	93	99
20		0	3	8	13	18	24	30	36	42	48	54	60	67	73	79	86	92	99	105

Source: Reprinted with permission from William H. Beyer (ed.), *Handbook of Tables for Probability and Statistics,* 2d ed., 1968. Copyright CRC Press, Inc., Boca Raton, Fla.

APPENDIX 10   CRITICAL VALUES FOR r IN THE RUNS TEST FOR RANDOMNESS

Any sample value of r which is equal to or less than that shown in table *(a)* or which is equal to or greater than that shown in table *(b)* is cause for rejection of H_0 at the .05 level of significance.

n_1 \ n_2	2	3	4	5	6	7	8	9	10	11	12	13	14	15	16	17	18	19	20
2											2	2	2	2	2	2	2	2	2
3				2	2	2	2	2	2	2	2	2	2	3	3	3	3	3	3
4			2	2	2	3	3	3	3	3	3	3	3	3	4	4	4	4	4
5			2	2	3	3	3	3	3	4	4	4	4	4	4	4	5	5	5
6		2	2	3	3	3	3	4	4	4	4	5	5	5	5	5	5	6	6
7		2	2	3	3	3	4	4	5	5	5	5	5	6	6	6	6	6	6
8		2	3	3	3	4	4	5	5	5	6	6	6	6	6	7	7	7	7
9		2	3	3	4	4	5	5	5	6	6	6	7	7	7	7	8	8	8
10		2	3	3	4	5	5	5	6	6	7	7	7	7	8	8	8	8	9
11		2	3	4	4	5	5	6	6	7	7	7	8	8	8	9	9	9	9
12	2	2	3	4	4	5	6	6	7	7	7	8	8	8	9	9	9	10	10
13	2	2	3	4	5	5	6	6	7	7	8	8	9	9	9	10	10	10	11
14	2	2	3	4	5	5	6	7	7	8	8	9	9	9	10	10	10	11	11
15	2	3	3	4	5	6	6	7	7	8	8	9	9	10	10	11	11	11	12
16	2	3	4	4	5	6	6	7	8	8	9	9	10	10	11	11	11	12	12
17	2	3	4	4	5	6	7	7	8	9	9	10	10	11	11	11	12	12	13
18	2	3	4	5	5	6	7	8	8	9	9	10	10	11	11	12	12	13	13
19	2	3	4	5	6	6	7	8	8	9	10	10	11	11	12	12	13	13	13
20	2	3	4	5	6	6	7	8	9	9	10	10	11	12	12	13	13	13	14

(a)

n_1 \ n_2	2	3	4	5	6	7	8	9	10	11	12	13	14	15	16	17	18	19	20
2											6	6	6	6	6	6	6	6	6
3				8	8	8	8	8	8	8	8	8	8	8	8	8	8	8	8
4			9	9	10	10	10	10	10	10	10	10	10	10	10	10	10	10	10
5			9	10	10	11	11	12	12	12	12	12	12	12	12	12	12	12	12
6		8	9	10	11	12	12	13	13	13	13	14	14	14	14	14	14	14	14
7		8	10	11	12	13	13	14	14	14	14	15	15	15	16	16	16	16	16
8		8	10	11	12	13	14	14	15	15	16	16	16	16	17	17	17	17	17
9		8	10	12	13	14	14	15	16	16	16	17	17	18	18	18	18	18	18
10		8	10	12	13	14	15	16	16	17	17	18	18	18	19	19	19	20	20
11		8	10	12	13	14	15	16	17	17	18	19	19	19	20	20	20	21	21
12	6	8	10	12	13	14	16	16	17	18	19	19	20	20	21	21	21	22	22
13	6	8	10	12	14	15	16	17	18	19	19	20	20	21	21	22	22	23	23
14	6	8	10	12	14	15	16	17	18	19	20	20	21	22	22	23	23	23	24
15	6	8	10	12	14	15	16	18	18	19	20	21	22	22	23	23	24	24	25
16	6	8	10	12	14	16	17	18	19	20	21	21	22	23	23	24	25	25	25
17	6	8	10	12	14	16	17	18	19	20	21	22	23	23	24	25	25	26	26
18	6	8	10	12	14	16	17	18	19	20	21	22	23	24	25	25	26	26	27
19	6	8	10	12	14	16	17	18	20	21	22	23	23	24	25	26	26	27	27
20	6	8	10	12	14	16	17	18	20	21	22	23	24	25	25	26	27	27	28

(b)

APPENDIX 11 SELECTED VALUES OF THE POISSON PROBABILITY DISTRIBUTION

					μ					
x	0.1	0.2	0.3	0.4	0.5	0.6	0.7	0.8	0.9	1.0
0	.9048	.8187	.7408	.6703	.6065	.5488	.4966	.4493	.4066	.3679
1	.0905	.1637	.2222	.2681	.3033	.3293	.3476	.3595	.3659	.3679
2	.0045	.0164	.0333	.0536	.0758	.0988	.1217	.1438	.1647	.1839
3	.0002	.0011	.0033	.0072	.0126	.0198	.0284	.0383	.0494	.0613
4	.0000	.0001	.0002	.0007	.0016	.0030	.0050	.0077	.0111	.0153
5	.0000	.0000	.0000	.0001	.0002	.0004	.0007	.0012	.0020	.0031
6	.0000	.0000	.0000	.0000	.0000	.0000	.0001	.0002	.0003	.0005
7	.0000	.0000	.0000	.0000	.0000	.0000	.0000	.0000	.0000	.0001

					μ					
x	1.1	1.2	1.3	1.4	1.5	1.6	1.7	1.8	1.9	2.0
0	.3329	.3012	.2725	.2466	.2231	.2019	.1827	.1653	.1496	.1353
1	.3662	.3614	.3543	.3452	.3347	.3230	.3106	.2975	.2842	.2707
2	.2014	.2169	.2303	.2417	.2510	.2584	.2640	.2678	.2700	.2707
3	.0738	.0867	.0998	.1128	.1255	.1378	.1496	.1607	.1710	.1804
4	.0203	.0260	.0324	.0395	.0471	.0551	.0636	.0723	.0812	.0902
5	.0045	.0062	.0084	.0111	.0141	.0176	.0216	.0260	.0309	.0361
6	.0008	.0012	.0018	.0026	.0035	.0047	.0061	.0078	.0098	.0120
7	.0001	.0002	.0003	.0005	.0008	.0011	.0015	.0020	.0027	.0034
8	.0000	.0000	.0001	.0001	.0001	.0002	.0003	.0005	.0006	.0009
9	.0000	.0000	.0000	.0000	.0000	.0000	.0001	.0001	.0001	.0002

					μ					
x	2.1	2.2	2.3	2.4	2.5	2.6	2.7	2.8	2.9	3.0
0	.1225	.1108	.1033	.0907	.0821	.0743	.0672	.0608	.0550	.0498
1	.2572	.2438	.2306	.2177	.2052	.1931	.1815	.1703	.1596	.1494
2	.2700	.2681	.2652	.2613	.2565	.2510	.2450	.2384	.2314	.2240
3	.1890	.1966	.2033	.2090	.2138	.2176	.2205	.2225	.2237	.2240
4	.0992	.1082	.1169	.1254	.1336	.1414	.1488	.1557	.1622	.1680
5	.0417	.0476	.0538	.0602	.0668	.0735	.0804	.0872	.0940	.1008
6	.0146	.0174	.0206	.0241	.0278	.0319	.0362	.0407	.0455	.0504
7	.0044	.0055	.0068	.0083	.0099	.0118	.0139	.0163	.0188	.0216
8	.0011	.0015	.0019	.0025	.0031	.0038	.0047	.0057	.0068	.0081
9	.0003	.0004	.0005	.0007	.0009	.0011	.0014	.0018	.0022	.0027
10	.0001	.0001	.0001	.0002	.0002	.0003	.0004	.0005	.0006	.0008
11	.0000	.0000	.0000	.0000	.0000	.0001	.0001	.0001	.0002	.0002
12	.0000	.0000	.0000	.0000	.0000	.0000	.0000	.0000	.0000	.0001

					μ					
x	3.1	3.2	3.3	3.4	3.5	3.6	3.7	3.8	3.9	4.0
0	.0450	.0408	.0369	.0334	.0302	.0273	.0247	.0224	.0202	.0183
1	.1397	.1304	.1217	.1135	.1057	.0984	.0915	.0850	.0789	.0733
2	.2165	.2087	.2008	.1929	.1850	.1771	.1692	.1615	.1539	.1465
3	.2237	.2226	.2209	.2186	.2158	.2125	.2087	.2046	.2001	.1954
4	.1734	.1781	.1823	.1858	.1888	.1912	.1931	.1944	.1951	.1954
5	.1075	.1140	.1203	.1264	.1322	.1377	.1429	.1477	.1522	.1563
6	.0555	.0608	.0662	.0716	.0771	.0826	.0881	.0936	.0989	.1042
7	.0246	.0278	.0312	.0348	.0385	.0425	.0466	.0508	.0551	.0595
8	.0095	.0111	.0129	.0148	.0169	.0191	.0215	.0241	.0269	.0298
9	.0033	.0040	.0047	.0056	.0066	.0076	.0089	.0102	.0116	.0132
10	.0010	.0013	.0016	.0019	.0023	.0028	.0033	.0039	.0045	.0053
11	.0003	.0004	.0005	.0006	.0007	.0009	.0011	.0013	.0016	.0019
12	.0001	.0001	.0001	.0002	.0002	.0003	.0003	.0004	.0005	.0006
13	.0000	.0000	.0000	.0000	.0001	.0001	.0001	.0001	.0002	.0002
14	.0000	.0000	.0000	.0000	.0000	.0000	.0000	.0000	.0000	.0001

Source: From *Handbook of Probability and Statistics with Tables* by Burington and May. Second Edition, Copyright © 1970 by McGraw-Hill, Inc. Used with permission of McGraw-Hill Book Company.

x	4.1	4.2	4.3	4.4	4.5	4.6	4.7	4.8	4.9	5.0
0	.0166	.0150	.0136	.0123	.0111	.0101	.0091	.0082	.0074	.0067
1	.0679	.0630	.0583	.0540	.0500	.0462	.0427	.0395	.0365	.0337
2	.1393	.1323	.1254	.1188	.1125	.1063	.1005	.0948	.0894	.0842
3	.1904	.1852	.1798	.1743	.1687	.1631	.1574	.1517	.1460	.1404
4	.1951	.1944	.1933	.1917	.1898	.1875	.1849	.1820	.1789	.1755
5	.1600	.1633	.1662	.1687	.1708	.1725	.1738	.1747	.1753	.1755
6	.1093	.1143	.1191	.1237	.1281	.1323	.1362	.1398	.1432	.1462
7	.0640	.0686	.0732	.0778	.0824	.0869	.0914	.0959	.1002	.1044
8	.0328	.0360	.0393	.0428	.0463	.0500	.0537	.0575	.0614	.0653
9	.0150	.0168	.0188	.0209	.0232	.0255	.0280	.0307	.0334	.0363
10	.0061	.0071	.0081	.0092	.0104	.0118	.0132	.0147	.0164	.0181
11	.0023	.0027	.0032	.0037	.0043	.0049	.0056	.0064	.0073	.0082
12	.0008	.0009	.0011	.0014	.0016	.0019	.0022	.0026	.0030	.0034
13	.0002	.0003	.0004	.0005	.0006	.0007	.0008	.0009	.0011	.0013
14	.0001	.0001	.0001	.0001	.0002	.0002	.0003	.0003	.0004	.0005
15	.0000	.0000	.0000	.0000	.0001	.0001	.0001	.0001	.0001	.0002

x	5.1	5.2	5.3	5.4	5.5	5.6	5 5.7	5.8	5.9	6.0
0	.0061	.0055	.0050	.0045	.0041	.0037	.0033	.0030	.0027	.0025
1	.0311	.0287	.0265	.0244	.0225	.0207	.0191	.0176	.0162	.0149
2	.0793	.0746	.0701	.0659	.0618	.0580	.0544	.0509	.0477	.0446
3	.1348	.1293	.1239	.1185	.1133	.1082	.1033	.0985	.0938	.0892
4	.1719	.1681	.1641	.1600	.1558	.1515	.1472	.1428	.1383	.1339
5	.1753	.1748	.1740	.1728	.1714	.1697	.1678	.1656	.1632	.1606
6	.1490	.1515	.1537	.1555	.1571	.1584	.1594	.1601	.1605	.1606
7	.1086	.1125	.1163	.1200	.1234	.1267	.1298	.1326	.1353	.1377
8	.0692	.0731	.0771	.0810	.0849	.0887	.0925	.0962	.0998	.1033
9	.0392	.0423	.0454	.0486	.0519	.0552	.0586	.0620	.0654	.0688
10	.0200	.0220	.0241	.0262	.0285	.0309	.0334	.0359	.0386	.0413
11	.0093	.0104	.0116	.0129	.0143	.0157	.0173	.0190	.0207	.0225
12	.0039	.0045	.0051	.0058	.0065	.0073	.0082	.0092	.0102	.0113
13	.0015	.0018	.0021	.0024	.0028	.0032	.0036	.0041	.0046	.0052
14	.0006	.0007	.0008	.0009	.0011	.0013	.0015	.0017	.0019	.0022
15	.0002	.0002	.0003	.0003	.0004	.0005	.0006	.0007	.0008	.0009
16	.0001	.0001	.0001	.0001	.0001	.0002	.0002	.0002	.0003	.0003
17	.0000	.0000	.0000	.0000	.0000	.0001	.0001	.0001	.0001	.0001

x	6.1	6.2	6.3	6.4	6.5	6.6	6.7	6.8	6.9	7.0
0	.0022	.0020	.0018	.0017	.0015	.0014	.0012	.0011	.0010	.0009
1	.0137	.0126	.0116	.0106	.0098	.0090	.0082	.0076	.0070	.0064
2	.0417	.0390	.0364	.0340	.0318	.0296	.0276	.0258	.0240	.0223
3	.0848	.0806	.0765	.0726	.0688	.0652	.0617	.0584	.0552	.0521
4	.1294	.1249	.1205	.1162	.1118	.1076	.1034	.0992	.0952	.0912
5	.1579	.1549	.1519	.1487	.1454	.1420	.1385	.1349	.1314	.1277
6	.1605	.1601	.1595	.1586	.1575	.1562	.1546	.1529	.1511	.1490
7	.1399	.1418	.1435	.1450	.1462	.1472	.1480	.1486	.1489	.1490
8	.1066	.1099	.1130	.1160	.1188	.1215	.1240	.1263	.1284	.1304
9	.0723	.0757	.0791	.0825	.0858	.0891	.0923	.0954	.0985	.1014
10	.0441	.0469	.0498	.0528	.0558	.0588	.0618	.0649	.0679	.0710
11	.0245	.0265	.0285	.0307	.0330	.0353	.0377	.0401	.0426	.0452
12	.0124	.0137	.0150	.0164	.0179	.0194	.0210	.0227	.0245	.0264
13	.0058	.0065	.0073	.0081	.0089	.0098	.0108	.0119	.0130	.0142
14	.0025	.0029	.0033	.0037	.0041	.0046	.0052	.0058	.0064	.0071
15	.0010	.0012	.0014	.0016	.0018	.0020	.0023	.0026	.0029	.0033
16	.0004	.0005	.0005	.0006	.0007	.0008	.0010	.0011	.0013	.0014
17	.0001	.0002	.0002	.0002	.0003	.0003	.0004	.0004	.0005	.0006
18	.0000	.0001	.0001	.0001	.0001	.0001	.0001	.0002	.0002	.0002
19	.0000	.0000	.0000	.0000	.0000	.0000	.0000	.0001	.0001	.0001

x	7.1	7.2	7.3	7.4	7.5	7.6	7.7	7.8	7.9	8.0
0	.0008	.0007	.0007	.0006	.0006	.0005	.0005	.0004	.0004	.0003
1	.0059	.0054	.0049	.0045	.0041	.0038	.0035	.0032	.0029	.0027
2	.0208	.0194	.0180	.0167	.0156	.0145	.0134	.0125	.0116	.0107
3	.0492	.0464	.0438	.0413	.0389	.0366	.0345	.0324	.0305	.0286
4	.0874	.0836	.0799	.0764	.0729	.0696	.0663	.0632	.0602	.0573
5	.1241	.1204	.1167	.1130	.1094	.1057	.1021	.0986	.0951	.0916
6	.1468	.1445	.1420	.1394	.1367	.1339	.1311	.1282	.1252	.1221
7	.1489	.1486	.1481	.1474	.1465	.1454	.1442	.1428	.1413	.1396
8	.1321	.1337	.1351	.1363	.1373	.1382	.1388	.1392	.1395	.1396
9	.1042	.1070	.1096	.1121	.1144	.1167	.1187	.1207	.1224	.1241
10	.0740	.0770	.0800	.0829	.0858	.0887	.0914	.0941	.0967	.0993
11	.0478	.0504	.0531	.0558	.0585	.0613	.0640	.0667	.0695	.0722
12	.0283	.0303	.0323	.0344	.0366	.0388	.0411	.0434	.0457	.0481
13	.0154	.0168	.0181	.0196	.0211	.0227	.0243	.0260	.0278	.0296
14	.0078	.0086	.0095	.0104	.0113	.0123	.0134	.0145	.0157	.0169
15	.0037	.0041	.0046	.0051	.0057	.0062	.0069	.0075	.0083	.0090
16	.0016	.0019	.0021	.0024	.0026	.0030	.0033	.0037	.0041	.0045
17	.0007	.0008	.0009	.0010	.0012	.0013	.0015	.0017	.0019	.0021
18	.0003	.0003	.0004	.0004	.0005	.0006	.0006	.0007	.0008	.0009
19	.0001	.0001	.0001	.0002	.0002	.0002	.0003	.0003	.0003	.0004
20	.0000	.0000	.0001	.0001	.0001	.0001	.0001	.0001	.0001	.0002
21	.0000	.0000	.0000	.0000	.0000	.0000	.0000	.0000	.0001	.0001

x	8.1	8.2	8.3	8.4	8.5	8.6	8.7	8.8	8.9	9.0
0	.0003	.0003	.0002	.0002	.0002	.0002	.0002	.0002	.0001	.0001
1	.0025	.0023	.0021	.0019	.0017	.0016	.0014	.0013	.0012	.0011
2	.0100	.0092	.0086	.0079	.0074	.0068	.0063	.0058	.0054	.0050
3	.0269	.0252	.0237	.0222	.0208	.0195	.0183	.0171	.0160	.0150
4	.0544	.0517	.0491	.0466	.0443	.0420	.0398	.0377	.0357	.0337
5	.0882	.0849	.0816	.0784	.0752	.0722	.0692	.0663	.0635	.0607
6	.1191	.1160	.1128	.1097	.1066	.1034	.1003	.0972	.0941	.0911
7	.1378	.1358	.1338	.1317	.1294	.1271	.1247	.1222	.1197	.1171
8	.1395	.1392	.1388	.1382	.1375	.1366	.1356	.1344	.1332	.1318
9	.1256	.1269	.1280	.1290	.1299	.1306	.1311	.1315	.1317	.1318
10	.1017	.1040	.1063	.1084	.1104	.1123	.1140	.1157	.1172	.1186
11	.0749	.0776	.0802	.0828	.0853	.0878	.0902	.0925	.0948	.0970
12	.0505	.0530	.0555	.0579	.0604	.0629	.0654	.0679	.0703	.0728
13	.0315	.0334	.0354	.0374	.0395	.0416	.0438	.0459	.0481	.0504
14	.0182	.0196	.0210	.0225	.0240	.0256	.0272	.0289	.0306	.0324
15	.0098	.0107	.0116	.0126	.0136	.0147	.0158	.0169	.0182	.0194
16	.0050	.0055	.0060	.0066	.0072	.0079	.0086	.0093	.0101	.0109
17	.0024	.0026	.0029	.0033	.0036	.0040	.0044	.0048	.0053	.0058
18	.0011	.0012	.0014	.0015	.0017	.0019	.0021	.0024	.0026	.0029
19	.0005	.0005	.0006	.0007	.0008	.0009	.0010	.0011	.0012	.0014
20	.0002	.0002	.0002	.0003	.0003	.0004	.0004	.0005	.0005	.0005
21	.0001	.0001	.0001	.0001	.0001	.0002	.0002	.0002	.0002	.0003
22	.0000	.0000	.0000	.0000	.0001	.0001	.0001	.0001	.0001	.0001

x	9.1	9.2	9.3	9.4	9.5	9.6	9.7	9.8	9.9	10
0	.0001	.0001	.0001	.0001	.0001	.0001	.0001	.0001	.0001	.0000
1	.0010	.0009	.0009	.0008	.0007	.0007	.0006	.0005	.0005	.0005
2	.0046	.0043	.0040	.0037	.0034	.0031	.0029	.0027	.0025	.0023
3	.0140	.0131	.0123	.0115	.0107	.0100	.0093	.0087	.0081	.0076
4	.0319	.0302	.0285	.0269	.0254	.0240	.0226	.0213	.0201	.0189
5	.0581	.0555	.0530	.0506	.0483	.0460	.0439	.0418	.0398	.0378
6	.0881	.0851	.0822	.0793	.0764	.0736	.0709	.0682	.0656	.0631
7	.1145	.1118	.1091	.1064	.1037	.1010	.0982	.0955	.0928	.0901
8	.1302	.1286	.1269	.1251	.1232	.1212	.1191	.1170	.1148	.1126
9	.1317	.1315	.1311	.1306	.1300	.1293	.1284	.1274	.1263	.1251

x	μ 9.1	9.2	9.3	9.4	9.5	9.6	9.7	9.8	9.9	10
10	.1198	.1210	.1219	.1228	.1235	.1241	.1245	.1249	.1250	.1251
11	.0991	.1012	.1031	.1049	.1067	.1083	.1098	.1112	.1125	.1137
12	.0752	.0776	.0799	.0822	.0844	.0866	.0888	.0908	.0928	.0948
13	.0526	.0549	.0572	.0594	.0617	.0640	.0662	.0685	.0707	.0729
14	.0342	.0361	.0380	.0399	.0419	.0439	.0459	.0479	.0500	.0521
15	.0208	.0221	.0235	.0250	.0265	.0281	.0297	.0313	.0330	.0347
16	.0118	.0127	.0137	.0147	.0157	.0168	.0180	.0192	.0204	.0217
17	.0063	.0069	.0075	.0081	.0088	.0095	.0103	.0111	.0119	.0128
18	.0032	.0035	.0039	.0042	.0046	.0051	.0055	.0060	.0065	.0071
19	.0015	.0017	.0019	.0021	.0023	.0026	.0028	.0031	.0034	.0037
20	.0007	.0008	.0009	.0010	.0011	.0012	.0014	.0015	.0017	.0019
21	.0003	.0003	.0004	.0004	.0005	.0006	.0006	.0007	.0008	.0009
22	.0001	.0001	.0002	.0002	.0002	.0002	.0003	.0003	.0004	.0004
23	.0000	.0001	.0001	.0001	.0001	.0001	.0001	.0001	.0002	.0002
24	.0000	.0000	.0000	.0000	.0000	.0000	.0000	.0001	.0001	.0001

x	μ 11	12	13	14	15	16	17	18	19	20
0	.0000	.0000	.0000	.0000	.0000	.0000	.0000	.0000	.0000	.0000
1	.0002	.0001	.0000	.0000	.0000	.0000	.0000	.0000	.0000	.0000
2	.0010	.0004	.0002	.0001	.0000	.0000	.0000	.0000	.0000	.0000
3	.0037	.0018	.0008	.0004	.0002	.0001	.0000	.0000	.0000	.0000
4	.0102	.0053	.0027	.0013	.0006	.0003	.0001	.0001	.0000	.0000
5	.0224	.0127	.0070	.0037	.0019	.0010	.0005	.0002	.0001	.0001
6	.0411	.0255	.0152	.0087	.0048	.0026	.0014	.0007	.0004	.0002
7	.0646	.0437	.0281	.0174	.0104	.0060	.0034	.0018	.0010	.0005
8	.0888	.0655	.0457	.0304	.0194	.0120	.0072	.0042	.0024	.0013
9	.1085	.0874	.0661	.0473	.0324	.0213	.0135	.0083	.0050	.0029
10	.1194	.1048	.0859	.0663	.0486	.0341	.0230	.0150	.0095	.0058
11	.1194	.1144	.1015	.0844	.0663	.0496	.0355	.0245	.0164	.0106
12	.1094	.1144	.1099	.0984	.0829	.0661	.0504	.0368	.0259	.0176
13	.0926	.1056	.1099	.1060	.0956	.0814	.0658	.0509	.0378	.0271
14	.0728	.0905	.1021	.1060	.1024	.0930	.0800	.0655	.0514	.0387
15	.0534	.0724	.0885	.0989	.1024	.0992	.0906	.0786	.0650	.0516
16	.0367	.0453	.0719	.0866	.0960	.0992	.0963	.0884	.0772	.0646
17	.0237	.0383	.0550	.0713	.0847	.0934	.0963	.0936	.0863	.0760
18	.0145	.0256	.0397	.0554	.0705	.0830	.0909	.0936	.0911	.0844
19	.0084	.0161	.0272	.0409	.0557	.0699	.0814	.0887	.0911	.0888
20	.0046	.0097	.0177	.0286	.0418	.0559	.0692	.0798	.0866	.0888
21	.0024	.0055	.0109	.0191	.0299	.0426	.0560	.0684	.0783	.0846
22	.0012	.0030	.0065	.0121	.0204	.0310	.0433	.0560	.0676	.0769
23	.0006	.0016	.0037	.0074	.0133	.0216	.0320	.0438	.0559	.0669
24	.0003	.0008	.0020	.0043	.0083	.0144	.0226	.0328	.0442	.0557
25	.0001	.0004	.0010	.0024	.0050	.0092	.0154	.0237	.0336	.0446
26	.0000	.0002	.0005	.0013	.0029	.0057	.0101	.0164	.0246	.0343
27	.0000	.0001	.0002	.0007	.0016	.0034	.0063	.0109	.0173	.0254
28	.0000	.0000	.0001	.0003	.0009	.0009	.0038	.0070	.0117	.0181
29	.0000	.0000	.0001	.0000	.0004	.0011	.0023	.0044	.0077	.0125
30	.0000	.0000	.0000	.0001	.0002	.0006	.0013	.0026	.0049	.0083
31	.0000	.0000	.0000	.0000	.0001	.0003	.0007	.0015	.0030	.0054
32	.0000	.0000	.0000	.0000	.0001	.0001	.0004	.0009	.0018	.0034
33	.0000	.0000	.0000	.0000	.0000	.0001	.0002	.0005	.0010	.0020
34	.0000	.0000	.0000	.0000	.0000	.0000	.0001	.0002	.0006	.0012
35	.0000	.0000	.0000	.0000	.0000	.0000	.0000	.0001	.0003	.0007
36	.0000	.0000	.0000	.0000	.0000	.0000	.0000	.0001	.0002	.0004
37	.0000	.0000	.0000	.0000	.0000	.0000	.0000	.0000	.0001	.0002
38	.0000	.0000	.0000	.0000	.0000	.0000	.0000	.0000	.0000	.0001
39	.0000	.0000	.0000	.0000	.0000	.0000	.0000	.0000	.0000	.0001

APPENDIX 12   ANSWERS TO SELECTED PROBLEMS

Chapter 3

3 a and b See the figures on the next page for these graphical presentations.

c

MILES TRAVELED	NO. OF TRUCKS (f)	m	fm
5,000 < 7,000	5	6,000	30,000
7,000 < 9,000	10	8,000	80,000
9,000 < 11,000	12	10,000	120,000
11,000 < 13,000	20	12,000	240,000
13,000 < 15,000	24	14,000	336,000
15,000 < 17,000	14	16,000	224,000
17,000 < 19,000	11	18,000	198,000
19,000 < 21,000	4	20,000	80,000
	100		1,308,000

Mean = 1,308,000/100 = 13,080 miles.

d Median = $13,000 + ([50 - 47]/24)2,000 = 13,250$ miles, the approximate mileage of the 50th truck in the group of 100.

e Mode = $13,000 + (4/14)2,000 = 13,571.43$ miles, an approximation of the most commonly occurring mileage.

f This is a negatively skewed distribution because the mean is the smallest of the measures of central tendency.

6 The arithmetic mean = $224/5 = 44.8$ yards/punt. The median = 45 yards (the middle value in the array).

9 a Mean = $1990/30 = 66.33$.

b Median = $60 + ([15 - 10]/10)10 = 65$.

12 a Mean(1979) = $1520/10 = 152$ pounds. Mean(1989) = $1570/10 = 167$ pounds.

b Median(1979) = 153 pounds (between 151 and 155 pounds). Median(1989) = 158.5 pounds (between 156 and 161 pounds).

c Mode(1979) = 180 pounds. Mode(1989): There is no mode in 1989.

15 a

SIZE (INCHES)	NO. CAUGHT	m	fm
10 < 12	6	11	66
12 < 14	14	13	182
14 < 16	17	15	255
16 < 18	8	17	136
18 < 20	5	19	95
	50		734

Mean = 734/50 = 14.68 inches.

b Median = $14 + ([25 - 20]/17)2 = 14.59$ inches.

c Mode = $14 + (3/12)2 = 14.50$ inches.

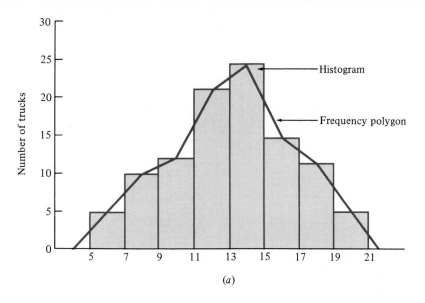

(a)

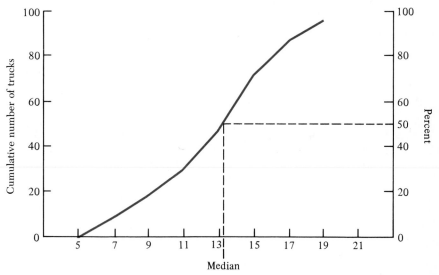

Miles traveled (in thousands)

(b)

18 a

SALES ($)	NO. TRUCKS	m	fm
300 < 350	1	325	325
350 < 400	2	375	750
400 < 450	7	525	2,975
450 < 500	12	475	5,700
500 < 550	13	525	6,825
550 < 600	9	575	5,175
600 < 650	2	625	1,250
	46		23,000

Mean = 23,000/46 = $500 daily sales.

b Mode = 550 + (1/5)50 = $560 daily sales.
c Median = 500 + ([23 − 22]/13)50 = $503.85 daily sales.

21 a The ascending array is:

12	14	15	19	22	23	25	37	41	43	51	54	55
62	65	66	68	71	73	84	85	88	92	92	96	

b The range is 96 − 12 = 84.
c Mean = 1,353/25 = 54.12. Median = 55—the $(n + 1)/2$ position.

Chapter 4

3 a Since the distribution isn't too badly skewed, we may conclude that about 20 of the 30 students had test scores in the interval of 66.33 ± 14.08 (or from about 52 to 80).

b Also, about 15 students had test scores in the interval of 65 ± 10.21 (or from about 55 to 75).

6 a The data for this problem are:

ANNUAL EARNINGS ($ IN THOUSANDS)	LANGUOR FAMILIES	FRISKYVILLE FAMILIES
5 < 8	5	2
7 < 11	40	20
11 < 14	73	32
14 < 17	52	58
17 < 20	22	35
20 < 23	8	30
23 < 26		15
26 and over		8
	200	200

Languor standard deviation:

$$\sigma = \sqrt{([38,948 − 36,720.5]/200)} = 3.33729 \text{ (in thousands of dollars)}$$
$$= \$3,337.29$$

Friskyville standard deviation: This value can't be computed because of the open-ended class.

b Languor quartile deviation:

$$Q_1 = \$11,000 + (5/73)\$3,000 = \$11,205.48$$

$$Q_3 = \$14,000 + (32/52)\$3,000 = \$15,846.15$$

$$Q_D = (\$15,846.15 - \$11,205.48)/2 = \$2,320.34$$

Friskyville quartile deviation:

$$Q_1 = \$11,000 + (28/32)\$3,000 = \$13,625.00$$

$$Q_3 = \$20,000 + (3/30)\$3,000 = \$20,300.00$$

$$QD = (\$20,300 - \$13,625)/2 = \$3,337.50$$

c Languor coefficient of variation:

$$CV = (\$3,337.29/\$13,550)(100) = 24.63 \text{ percent}$$

Friskyville coefficient of variation cannot be computed because the input data are unavailable.

d Languor coefficient of skewness:

$$Sk = 3(\$13,550 - \$13,260.27)/(\$3,337.29) = .260$$

Friskyville coefficient of skewness can't be computed because the input data are missing.

e Only the median and quartile deviation measures can be found for both communities for comparison purposes. From these data it's obvious that the "average" (median) income is higher in Friskyville. Approximately the middle 50 percent of the Friskyville families earn between $16,379 (the median earnings) ± $3,337.50, while the middle 50 percent of the Languor families only earn (approximately) between $13,260 ± $2,320.

9 a Standard deviation $= \sqrt{(1126800 - 1111320)/70}$
$$= 14.87 \text{ jobs per month}$$

b $Q_1 = 105 + ([17.5 - 4]/14)10 = 114.64$ jobs per month

$Q_3 = 135 + ([52.5 - 51]/10)10 = 136.50$ jobs per month

$QD = (136.50 - 114.64)/2 = 10.93$ jobs per month

c $CV = (14.87/126.0)(100) = 11.80\%$.

12 a $AD = |13|/6 = 2.17$ miles.

b Variance $= 41.50/6 = 6.92$.

c Standard deviation $=$ square root of variance $= 2.63$ miles.

15 Population standard deviation $= \sqrt{([100 - 1]/100)}\,(25) = 24.87$. Population variance is 618.52.

18 a   Mean = 107/12 = 8.92 cases per month.
        Standard deviation = $\sqrt{120.75 - 79.57}$ = 6.42 cases per month.
    b   Median = 9 cases per month.
    c Sk = 3(8.92 − 9.00)/6.42 = −.037.
    d CV = (6.42/8.92)(100) = 71.97%.

Chapter 5

3   a P(2 good) = (.8) × (.8) = .64.
    b P(2 defective) = (.2) × (.2) = .04.
6   a $P(4) = {}_6C_4 = (.6)^4(.4)^2$ = .3110.
    b $P(5) = {}_6C_5 = (.6)^5(.4)$ = .1866.
    c P(4 or 5) = .3110 + .1866 = .4976.
9   a ${}_{15}P_2$ = 15!/13! = 210.
    b ${}_{10}C_3$ = 10!/(7!3!) = 120.

12 a $E(X)$ = 0(.20) + 1(.30) + 2(.30)+3(.15) + 4(.05)
            = 1.55, the expected value of automobile sales

    b   200(1.55) = 310 cars may be expected to be sold in 200 days.
15 With mean = 2 and x = 4, the probability from Appendix 11 is .0902.
18 With mean = 2 and x = 0, the probability from Appendix 11 is .1353.
21 Z = ($230,000 − $200,000)/$25,000 = 1.2 = .3849 area. Then, .5000 − .3849 = .1151, the probability that a salesperson sells $230,000 or more.

24 $(.5)^4 + 4(.5)^3(.5) + 6(.5)^2(.5)^2$ = .0625 + .2500 + .3750
                = .6875

27 a P(survival) = (.95)(.95)(.95) · · · · for 20 times = .358.
    b P(survival) = (.95)(.95)(.95) · · · · for 40 times = .129.
30 a $6(1/6)(5/6)^5$ = .402.
    b $15(1/6)^2(5/6)^4$ = .201.
    c   1.00 − .402 − .201 = .397.
33 a   With mean = 2.4 and x = 0, the probability from Appendix 11 is .0907.
    b   (.0907)(.0907) = .00823.
    c   The most probable number of alarms in an hour is 2, and the probability of having exactly 2 alarms is .2613 (from Appendix 11).
    d P = 1.00 − (.0907 + .2177 + .2613 + .2090) = .2213.
    e P = .2213 − .1254 = .0959.

Chapter 6

3   a   The population mean is 77/5 or 15.4 hours. The population standard deviation is:

$$\sigma = \sqrt{\frac{70.56 + .36 + 21.16 + 11.56 + 43.45}{5}} = 5.4258 \text{ hours}$$

b The mean of the sampling distribution is the same as the population mean or 15.4 hours.

c The standard error of the mean is:

$$\sigma_x = \frac{5.4258}{\sqrt{3}} \sqrt{2/4} = 2.215$$

6 a There's a 95.4 percent chance that the sample mean will be located between the population mean ± 2(standard error). And since the standard error is equal to $15/\sqrt{45}$ or 2.24, the range is 195.52 to 204.48.

b If the sample size is 36, the standard error is 2.5, and the range is 195 to 205.
   If the sample size is 49, the standard error is 2.14, and the range is 195.71 to 204.29.
   If the sample size is 64, the standard error is 1.88, and the range is 196.24 to 203.76.

c As the sample size increases, the dispersion of the sampling distribution becomes smaller. The sample mean is likely to have a value closer to the population mean as the sample size increases.

9 a The mean of the sampling distribution is 60 percent because it is equal to the population percentage.

b Yes, the correction factor is needed. Its value is:

$$\sqrt{\frac{20 - 5}{20 - 1}} = .89$$

c Standard error = $\sqrt{(60)(40)/5}(.89) = 19.15$ percent.

12 $Z_1 = \dfrac{10,100 - 10,000}{2,000/\sqrt{400}} = 1.00 = .3413$

$Z_2 = \dfrac{10,200 - 10,000}{2,000/\sqrt{400}} = 2.00 = .4772$

.4772
$-$.3413
.1359 probability

15 $Z = \dfrac{990 - 980}{50/\sqrt{100}} = 2.00 = .4772$

.5000
$-$.4772
.0228 probability

18 $Z = \dfrac{220 - 236}{48/\sqrt{36}} = \dfrac{-16}{8} = -2.00 = .4772$

.5000
$-$.4772
.0228 probability

21   $Z = \dfrac{14.5 - 10.0}{\sqrt{(10)(90)/400}} + \dfrac{4.5}{1.5} = 3.00 = .4987$

   $\begin{aligned} &.5000 \\ -\,&.4987 \\ \hline &.0013 \end{aligned}$ probability

24   About 50 percent of all sample means will lie within $\pm.67$ standard errors of the population mean of 50 inches. The value of the standard error is:

   $\dfrac{3}{\sqrt{100}} \sqrt{\dfrac{1,000 - 100}{1,000 - 1}} = .3\,(.9492) = .285$

   Thus, the sample mean has a 50 percent chance of lying between 50 inches $\pm(.67)(.285)$ or between 49.81 inches and 50.19 inches.

27   The standard error of percentage is:

   $\sqrt{\dfrac{[53(100 - 53)]}{50}} = 7.06\%$

   Candidate X loses if he or she receives less than 50 percent of the votes—that is, Candidate X loses if the percentage of votes cast by the sample of 50 eligible voters falls into the area below 50 percent under a normal curve that has a population percentage of 53 percent. This areas has a Z value of:

   $Z = \dfrac{50 - 53}{7.06} = -.42$

   We see from Appendix 2 that a Z value of .42 corresponds to an area figure of .1628. Thus, the probability that X loses is $.5000 - .1628 = .3372$.

30   a   The number of different samples of 5 students taken 3 at a time is 10.
     b   The mean of the sampling distribution of sample means is equal to the population mean. That is:

   $\mu_x = \dfrac{92 + 83 + 70 + 53 + 77}{5} = 75$

     c   The population standard deviation is 13.16 grade points. The standard deviation of the sampling distribution of means—the standard error—is:

   $\sigma_x = \dfrac{13.16}{\sqrt{3}} \sqrt{([5 - 3]/[5 - 1])} = 5.37$

Chapter 7

3   The estimated standard error $= 6.267/\sqrt{24} = 1.28$. Since $n = 24$, the degrees of freedom are 23; therefore, the t value is 2.807, and the confidence interval is:

   $97.17 \pm (2.807)(1.28)$ or $93.58 < \mu < 100.76$

6

SAMPLE SIZE	df	CONFIDENCE LEVEL (%)	t VALUE
15	14	99	2.977
23	22	90	1.717
28	27	95	2.052
27	26	95	2.056
25	24	95	2.064
20	19	95	2.093

9   a   This is the value of s, which is 48 calls.

       b   With a 90 percent confidence level, $Z = 1.64$, and the interval is $326 \pm (1.64)(6.79)$ or 314.87 to 337.13 calls (in rounded form, 315 to 337 calls).

12   The standard error $= 17/\sqrt{36} = 2.83$ wallets. With a confidence level of 95 percent, the Z value is 1.96, and the interval estimate is $114 \pm (1.96)(2.83)$ or 108.45 to 119.55 wallets (in rounded form, 108 to 120 wallets).

15   The standard error of percentage $= \sqrt{(84)(16)/36} = 6.1\%$. With a 99 percent confidence level, the Z value is 2.58, and the interval estimate is $84 \pm (2.58)(6.1)$ or 68.26 to 99.74 percent.

18   Since $Z\sigma_p = 3\%$, and since $Z = 2.58$, the standard error of percentage $= 3\%/2.58$ or 1.16 percent. And because we have no idea what the value of the population percentage is, the necessary sample size is:

$$n = \frac{(50)(50)}{(1.16)^2} = \frac{2500}{1.35} = 1852$$

If this sample size is beyond the resources (time, money, etc.) available to the director, she will have to be content with a larger margin of error and/or a lower level of confidence.

21   The interval is 495 hours $\pm 1.96(64/\sqrt{101}$). Thus, the confidence interval (CI) is 488.63 hours to 501.37 hours.

24   The interval is 67.42 ± 1.64 ($6.28/\sqrt{50}$). The CI is thus $65.96 to $68.88.

27   The interval is 31 mg ± 4.602 ($2/\sqrt{5}$). The CI is thus 26.89 to 35.11 mg.

30   The interval is 30% ± 1.96 ($\sqrt{[(20)(80)/399]}$). The CI is then 26.08 to 33.92 percent.

33   a   Since $Z(\sigma_x) = 5$ points, the standard error $= 5/1.64$ or 3.04 in this case. Thus, $n = (10)^2/(3.04)^2 = 10.82$ or a sample of 11.

       b   The answer to this part of the question depends on the 11 scores you selected from the group. The estimated population mean score is the mean of the 11 scores you selected.

36   The standard error $= \sqrt{(4)(96)/199}$ ($\sqrt{[4,800]/[4,999]}$) $= 1.36\%$. The interval is then 4% ± 1.96 (1.36) or 1.33 to 6.67 percent.

39   The standard error $= \sqrt{(66.7)(33.3)/999} = 1.49\%$. The interval estimate is then 66.7% ± 1.64 (1.49) or 64.26 to 69.14 percent.

Chapter 8

3   a -1.64 d -2.33

       b ± 1.64 e   1.64

       c   2.33

6   H_0: $\mu = 69$

    H_1: $\mu > 69$

Since the sample size is 19, the t distribution applies, and the appropriate t value is 1.33. Therefore the decision rule (DR) is to accept the null hypothesis if the CR $\leq$ 1.33.

$$CR = \frac{75 - 69}{6}(\sqrt{19}) = 4.36$$

Conclusion: Reject the null hypothesis since 4.36 > 1.33.

9   H_0: $\pi = 33\%$

    H_1: $\pi > 33\%$

At the .01 level, the Z value is 2.33; therefore the decision rule (DR) is to accept the null hypothesis if the CR $\leq$ 2.33.

$$\text{Standard error} = \sqrt{\frac{(33)(67)}{49}} = 6.7\%$$

$$CR = \frac{35 - 33}{6.7} = .298$$

Conclusion: There's no reason to reject the null hypothesis.

12   H_0: $\pi = 90\%$

     H_1: $\pi < 90\%$

At the .05 level, the Z value is -1.64; therefore the DR is to accept the null hypothesis if the CR ≥ -1.64.

$$\text{Standard error} = \sqrt{\frac{(90)(10)}{100}} = 3\%$$

$$CR = \frac{87 - 90}{3} = -1.00$$

Conclusion: Accept the null hypothesis.

15   H_0: $\pi = 70\%$

     H_1: $\pi < 70\%$

DR: Accept H_0 if CR ≥ -1.64.

$$\text{Standard error} = \sqrt{\frac{(70)(30)}{200}} = 3.24\%$$

$$CR = \frac{65 - 70}{3.24} = -1.54$$

Conclusion: Accept the null hypothesis.

18   H_0: $\mu = 1,000$ hours
     H_1: $\mu < 1,000$ hours

     DR: Accept H_0 if CR ≥ -1.711.

     Standard error $= \dfrac{30}{\sqrt{25}} = 6.00$ hours

     CR $= \dfrac{994 - 1,000}{6.00} = -1.00$

     Conclusion: Accept the null hypothesis.

21   H_0: $\pi = 25\%$
     H_1: $\pi < 25\%$

     DR: Accept H_0 if CR ≥ -1.64.

     Standard error $= \sqrt{\dfrac{(25)(75)}{1250}} = 1.22\%$

     CR $= \dfrac{20.8 - 25.0}{1.22} = -3.44$

     Conclusion: Reject the null hypothesis and cancel the program.

24   H_0: $\mu = 75$
     H_1: $\mu > 75$

     DR: Accept H_0 if CR ≤ 2.896.

     Standard error $= \dfrac{4}{\sqrt{9}} = 1.33$

     CR $= \dfrac{80 - 75}{1.33} = 3.76$

     Conclusion: Reject the null hypothesis and Melvin's claim.

27   H_0: $\mu = 750$ hours
     H_1: $\mu < 750$ hours

     DR: Accept H_0 if the CR ≥ -1.383.

     Standard error $= \dfrac{40}{\sqrt{10}} = 12.65$ hours

     CR $= \dfrac{710 - 750}{12.65} = 3.16$

     Conclusion: Reject the null hypothesis; prolonged storage has apparently reduced the lifetime performance of the bulbs.

30 The center line (CL) value should be 50 mg—the value of the population mean when the vials are properly filled. The upper and lower control lines (UCL and LCL) are set at ± 3 standard errors from the CL. The standard error is:

$$\frac{2\,\text{mg}}{\sqrt{100}} = .2\,\text{mg}$$

Thus, the LCL and UCL are set at 49.4 mg and 50.6 mg, respectively.

33 H_0: $\mu = \$10,000$
H_1: $\mu > \$10,000$

DR: Accept H_0 if CR ≤ 1.761.

$$\text{Standard error} = \frac{\$600}{\sqrt{15}} = \$154.92$$

$$\text{CR} = \frac{\$10,575 - \$10,000}{\$154.92} = 3.71$$

Conclusion: Reject the null hypothesis. The contractor charges more than $10,000 on average.

Chapter 9

3 H_0: $\mu_1 = \mu_2$
H_1: $\mu_1 < \mu_2$

DR: Accept H_0 if CR ≥ -1.64.

$$\text{Standard error} = \sqrt{\frac{1600}{100} + \frac{900}{100}} = 5\,\text{hours}$$

$$\text{CR} = \frac{600 - 612}{5} = -2.4$$

Conclusion: Reject the null hypothesis; there has been an increase in product life.

6 H_0: $\pi_1 = \pi_2$
H_1: $\pi_1 \neq \pi_2$

DR: Accept H_0 if CR falls between ± 1.96.

$$\text{Standard error} = \sqrt{\frac{(4)(96)}{900} + \frac{(4.4)(95.6)}{1000}} = .848$$

$$\text{CR} = \frac{4.0 - 4.4}{.848} = -.472$$

Conclusion: Accept the null hypothesis. There's no significant difference in the percentage of new accounts opened in the 2 years.

9 a $H_0: \pi_1 = \pi_2$

   $H_1: \pi_1 < \pi_2$

DR: Accept H_0 if CR ≥ -1.64.

$$\text{Standard error} = \sqrt{\frac{(23.08)(76.92)}{129} + \frac{(30.47)(69.53)}{127}} = 5.52\%$$

$$\text{CR} = \frac{23.08 - 30.47}{5.52} = -1.34$$

Conclusion: Accept the null hypothesis; vitamin X doesn't significantly reduce the chances of catching cold at the .05 level.

b   If the .10 level is used to make the test, the DR is: Accept H_0 if CR ≥ -1.282. Since the CR is still -1.34, we must now reject the null hypothesis and accept the alternative. That is, we now conclude that vitamin X does help reduce the chances of catching cold at the .10 level. Of course, we're now on shakier ground since our chance of making a Type I error is now .10 rather than .05.

12 The population percentages to be tested are the launch failures of each approach. Thus,

$H_0: \pi_{TE} = \pi_{SS}$ (use Thor-Epsilon in this case since it's cheaper)

$H_1: \pi_{TE} < \pi_{SS}$ (use Space Skuttle)

DR: Accept H_0 if CR ≥ -1.64.

$$\text{Standard error} = \sqrt{\frac{(5.7)(94.3)}{34} + \frac{(10)(90)}{599}} = -4.16\%$$

$$\text{CR} = \frac{5.7 - 10}{4.16} = -1.03$$

Conclusion: The telephone company should accept H_0. Since there's no significant difference in launch failures, the less expensive Thor-Epsilon launch approach should be used.

Chapter 10

3 You saw in the chapter that a one-way ANOVA table provides a summary listing of the values needed to produce an ANOVA test. (The format of such a table is given in Figure 10-7, and an example of the use of such a table is shown in Figure 10-8.) The ANOVA table for this problem is given as follows:

```
ANALYSIS OF VARIANCE
SOURCE      DF        SS         MS        F        p
FACTOR       2       398.6      199.3     2.80    0.083
ERROR       21      1493.2       71.1
TOTAL       23      1891.8
                                   INDIVIDUAL 95 PCT CI'S FOR MEAN
                                   BASED ON POOLED STDEV
LEVEL        N       MEAN       STDEV   ----------+---------+---------+------
C1           8       98.25      11.90                (----------*----------)
C2           8       99.62       4.75                    (---------*---------)
C3           8       90.38       7.01    (-----------*----------)
                                         ----------+---------+---------+------
POOLED STDEV =       8.43                   90.0      96.0      102.0
```

As you can see, the computed F value is 2.80, and the critical table value at the .05 level is 3.47.

Conclusion: The null hypothesis cannot be rejected. There's no significant difference in sales based on shelf space.

6 The ANOVA table for this problem is given below:

```
ANALYSIS OF VARIANCE
SOURCE      DF        SS         MS        F        p
FACTOR       2      237121     118560    70.89    0.000
ERROR       27       45157       1672
TOTAL       29      282278
                                   INDIVIDUAL 95 PCT CI'S FOR MEAN
                                   BASED ON POOLED STDEV
LEVEL        N       MEAN       STDEV   ----------+---------+---------+--------
C1          10      576.90      36.13                              (--*--)
C2          10      360.00      55.80    (--*--)
C3          10      485.30      24.47                  (---*--)
                                         ----------+---------+---------+--------
POOLED STDEV =      40.90                   400       480       560
```

The computed F value is 70.89, and the critical table value at the .05 level is 3.35.

Conclusion: The null hypothesis is rejected. There is a significant difference in traffic count.

9 The computed F value is $374/93 = 4.02$. The appropriate table F value is 3.49. The null hypothesis is rejected.

12 The ANOVA table for this problem is shown as follows:

```
ANALYSIS OF VARIANCE
SOURCE       DF          SS          MS          F           p
FACTOR        3        10.73        3.58        0.65      0.591
ERROR        28       154.77        5.53
TOTAL        31       165.50
                                          INDIVIDUAL 95 PCT CI'S FOR MEAN
                                          BASED ON POOLED STDEV
LEVEL        N        MEAN        STDEV   ----+---------+---------+---------+--
C1           8       6.000        2.928       (----------*-----------)
C2           8       5.750        1.282   (----------*------------)
C3           9       7.222        2.863               (----------*-----------)
C4           7       6.429        1.718         (-----------*-----------)
                                          ----+---------+---------+---------+--
POOLED STDEV =       2.351            4.5         6.0       7.5        9.0
```

The computed F value is .65, and the critical table value at the .05 level is 2.95.
Conclusion: The null hypothesis is accepted. Consumers are insensitive to price changes.

15 The ANOVA table for this problem is shown below:

```
ANALYSIS OF VARIANCE
SOURCE       DF          SS          MS          F           p
FACTOR        2       3035.2      1517.6       15.76      0.000
ERROR        12       1155.2        96.3
TOTAL        14       4190.4
                                          INDIVIDUAL 95 PCT CI'S FOR MEAN
                                          BASED ON POOLED STDEV
LEVEL        N        MEAN        STDEV   --+---------+---------+---------+----
C1           5       142.00        6.78  (------*-----)
C2           5       152.40       14.24         (------*-----)
C3           5       176.00        6.32                         (-----*------)
                                          --+---------+---------+---------+----
POOLED STDEV =       9.81       135        150       165        180
```

The computed F value is 15.76, and the table value of F at the .05 level is 3.68.
Conclusion: We reject the null hypothesis. There is a significant difference in the life expectancy of the different brands of batteries.

Chapter 11

3 H_0: The population distribution is uniform (an equal percentage of the population favors each brand)
   H_1: The population distribution is not uniform

The chi-square computed value is:

$$(40 - 50) + (55 - 50) + (60 - 50) + (40 - 50) + (60 - 50) + (45 - 50) = 10.00$$

With $df = 5$ and $\alpha = .01$, the chi-square table value = 15.086.
Conclusion: Accept the null hypothesis.

6   H_0: The percentages of common defects are the same for both vendors
    H_1: The percentages are not the same for both vendors

The observed and expected frequencies are shown in the table below:

```
Expected counts are printed below observed counts

        DEFECT 1 DEFECT 2 DEFECT 3 DEFECT 4    Total
    1       60       80       40       30        210
          59.62    74.20    43.06    33.12

    2       30       32       25       20        107
          30.38    37.80    21.94    16.88

Total       90      112       65       50        317

ChiSq =  0.002 +  0.454 +  0.217 +  0.294 +
         0.005 +  0.891 +  0.427 +  0.578 = 2.869
df = 3
```

As you can see, the chi-square computed value is 2.869. With $df = (2 - 1)(4 - 1) = 3$ and $\alpha = .01$, the chi-square table value is 11.345.
Conclusion: Accept the null hypothesis; the computed chi-square value is less than the table value.

9   H_0: The reactions of members of the three locals are the same
    H_1: The reactions at the three locals are different

The observed and expected frequencies are shown in the table at the top of the next page. As you can see, the chi-square computed value is 11.244. With $df = 4$ and $\alpha = .05$, the table chi-square value = 9.488.
Conclusion: Reject the null hypothesis. There's a significant difference in reactions of members of the three locals to the proposed change.

Chapter 12

3   a   Simple price index for product A:

        1983 = 100.0
        1984 = 116.7—i.e., ($0.35/$.030)(100) = 116.7
        1985 = 150.0
        1986 = 156.7

```
      Expected counts are printed below observed counts

            LOCAL X   LOCAL Y   LOCAL Z    Total
      1        17        23        10        50
             15.00     20.00     15.00

      2         9        13         8        30
              9.00     12.00      9.00

      3         4         4        12        20
              6.00      8.00      6.00

Total          30        40        30       100

ChiSq =   0.267 +   0.450 +   1.667 +
          0.000 +   0.083 +   0.111 +
          0.667 +   2.000 +   6.000 = 11.244
df  =  4
```

b Implicitly weighted aggregative price index:

$$PI_{1983} = \left(\frac{\$2.05}{\$2.05}\right)(100) = 100.0$$

$$PI_{1984} = \left(\frac{\$2.10}{\$2.05}\right)(100) = 102.4$$

$$PI_{1985} = \left(\frac{\$2.32}{\$2.05}\right)(100) = 113.2$$

$$PI_{1986} = \left(\frac{\$2.30}{\$2.05}\right)(100) = 112.2$$

c Weighted aggregative price index:

PRODUCT	1983 $p_0 q_t$	1984 $p_n q_t$	1985 $p_n q_t$	1986 $p_n q_t$
A	$288.00	$336.00	$432.00	$451.20
B	101.25	94.50	108.00	105.30
C	290.00	304.50	310.30	304.50
	$679.25	$735.00	$850.30	$861.00

$$PI_{1983} = \left(\frac{\$679.25}{\$679.25}\right)(100) = 100.0$$

$$PI_{1984} = \left(\frac{\$735.00}{\$679.25}\right)(100) = 108.2$$

$$PI_{1985} = \left(\frac{\$850.30}{\$679.25}\right)(100) = 125.2$$

$$PI_{1986} = \left(\frac{\$861.00}{\$679.25}\right)(100) = 126.8$$

d Weighted aggregative quantity index:

PRODUCT	1983 $q_0 p_t$	1984 $q_n p_t$	1985 $q_n p_t$	1986 $q_n p_t$
A	$288.00	$292.50	$270.00	$289.50
B	101.25	75.00	86.25	105.00
C	290.00	280.00	285.00	295.00
	$679.25	$647.50	$641.25	$689.50

$$QI_{1983} = \left(\frac{\$679.25}{\$679.25}\right)(100) = 100.0$$

$$QI_{1984} = \left(\frac{\$647.50}{\$679.25}\right)(100) = 95.3$$

$$QI_{1985} = \left(\frac{\$641.25}{\$679.25}\right)(100) = 94.4$$

$$QI_{1986} = \left(\frac{\$689.50}{\$679.25}\right)(100) = 101.6$$

e Implicitly weighted average of price relatives index; the various price relatives $[r = (p_n/p_o)(100)]$ are as follows:

PRODUCT	r_{1983}	r_{1984}	r_{1985}	r_{1986}
A	100.0	116.7	150.0	156.7
B	100.0	93.3	106.7	104.0
C	100.0	105.0	107.0	105.0
	300.0	315.0	363.7	365.7

$$PI_{1983} = \frac{300}{3} = 100.0$$

$$PI_{1984} = \frac{315.0}{3} = 105.0$$

$$PI_{1985} = \frac{363.7}{3} = 121.2$$

$$PI_{1986} = \frac{365.7}{3} = 121.9$$

f Weighted average of price relatives index:

PRODUCT	$r_{83}p_tq_t$	$r_{84}p_tq_t$	$r_{85}p_tq_t$	$r_{86}p_tq_t$
A	28,800.0	33,609.6	43,200.0	45,129.6
B	10,125.0	9,446.6	10,803.4	10,530.0
C	29,000.0	30,450.0	31,030.0	30,450.0
	67,925.0	73,506.2	85,033.4	86,109.6

$$PI_{1983} = \frac{67,925.0}{\$679.25} = 100.0$$

$$PI_{1984} = \frac{73,506.2}{\$679.25} = 108.2$$

$$PI_{1985} = \frac{85,033.4}{\$679.25} = 125.2$$

$$PI_{1986} = \frac{86,109.6}{\$679.25} = 126.8$$

g   Weighted average of quantity relatives index:

PRODUCT	$r_{83}q_tp_t$	$r_{84}q_tp_t$	$r_{85}q_tp_t$	$r_{86}q_tp_t$
A	28,800	29,250	27,000	28,950
B	10,125	7,500	8,625	10,500
C	29,000	28,000	28,500	29,500
	67,925	64,750	64,125	68,950

$$QI_{1983} = \frac{67,925}{\$679.25} = 100.0$$

$$QI_{1984} = \frac{64,750}{\$679.25} = 95.3$$

$$QI_{1985} = \frac{64,125}{\$679.25} = 94.4$$

$$QI_{1986} = \frac{68,950}{\$679.25} = 101.6$$

6   a

COMMODITY	1988 PRICE (p_0)	1988 RELATIVE $\left(\frac{p}{p_0} \cdot 100\right)$	1989 PRICE (p)	1989 RELATIVE $\left(\frac{p}{p_0} \cdot 100\right)$	1990 PRICE (p)	1990 RELATIVE $\left(\frac{p}{p_0} \cdot 100\right)$
Guns	$600	100	$650	108.33	$750	125.00
Butter	$8.00	100	8.50	106.25	9.00	112.50
		200		214.58		237.50
Price index		100		107.29		118.75

b

COMMODITY	p_t	q_t	$(p_t)(q_t)$
Guns	$600	10	$6,000
Butter	$8	5000	$40,000
			$46,000

		1988		1989		
COMMODITY	$p_t q_t$	RELATIVE	RELATIVE × $p_t q_t$	RELATIVE	RELATIVE × $p_t q_t$	...
Guns	$6,000	100	$600,000	125.00	$750,000	
Butter	$40,000	100	$4,000,000	112.50	$4,500,000	
			$4,600,000		$4,899,980	

$$PI_{1988} = \frac{\$4,600,000}{\$46,000} = 100.00$$

$$PI_{1989} = \frac{\$4,899,980}{\$46,000} = 106.52$$

$$PI_{1990} = \frac{\$5,250,000}{\$46,000} = 114.13$$

Chapter 13

3 a

MONTH	SPECIFIC SEASONALS 1987	1988	1989	TOTALS	ARITHMETIC MEAN	SEASONAL INDEX
January		60.7	71.3	132.0	66.0	65.6
February		73.7	78.3	152.0	76.0	75.6
March		60.0	59.0	119.0	59.5	59.2
April		73.3	64.9	138.2	69.1	68.7
May		86.6	90.3	176.9	88.4	87.9
June		79.1	88.4	167.1	83.8	83.3
July	97.1	100.0		197.1	98.6	98.0
August	111.7	112.1		223.8	111.9	111.2
September	137.9	129.0		266.9	133.4	132.6
October	144.0	137.7		281.7	140.8	140.0
November	122.0	131.8		253.8	126.9	126.2
December	160.9	144.3		305.2	152.6	151.7
				Totals	1,207.0	1,200.0

Correction factor = 1200/1207.0 = .9942

b   The chief might use the data above for short-term forecasting and control purposes. He would know, for example, that his greatest personnel needs are from September to December. Perhaps he should plan as few vacations as possible during this period. Additional analysis of other divisions of the Department might enable him to shift police to traffic if the seasonal influences elsewhere permitted it and vice versa. Deseasonalization of the next year's data is an additional use of these data.

6   a $a = \dfrac{85}{6} = 14.2$ million barrels

$b = \dfrac{139}{70} = 1.99$ million barrels

$Y_t = 14.2 + 1.99(x)$

b $Y_t(1991) = 28.2$ million barrels.

c $Y_t(\text{January 1, 1989}) = 14.2 + 1.99(2) = 18.2$ million barrels.

9

MONTH	1990 UNADJUSTED PRODUCTION	SEASONAL INDEX	SEASONALLY ADJUSTED PRODUCTION (COL. 2/COL. 3)(100)
January	13	73.4	17.7
February	12	72.4	16.6
March	15	78.1	19.2
April	23	113.0	20.4
May	20	118.7	16.8
June	20	112.2	17.8
July	30	153.7	19.5
August	19	101.5	18.7
September	16	93.0	17.2
October	15	86.7	17.3
November	17	85.6	19.9
December	19	111.7	17.0

Chapter 14

3   The Y variable for this problem is the total number of requests for nurses, and the X variable is the number of days in the hospital. When supplied with the input data for this problem, the *Minitab* statistical package produced the output shown below:

```
MTB > REGRESS 'DAYS' ON 1 PREDICTOR 'REQUESTS';
SUBC> PREDICT 7.

The regression equation is
DAYS = 0.104 + 1.40 REQUESTS

Predictor      Coef        Stdev       t-ratio        p
Constant      0.1037      0.3453         0.30      0.771
REQUESTS      1.40346     0.06056       23.17      0.000

s = 0.6179      R-sq = 98.5%      R-sq(adj) = 98.3%

Analysis of Variance

SOURCE        DF          SS          MS          F         p
Regression    1        205.05      205.05      536.99     0.000
Error         8          3.05        0.38
Total         9        208.10


    Fit   Stdev.Fit         95% C.I.              95% P.I.
  9.928      0.240     ( 9.374, 10.481)     ( 8.399, 11.457)
```

a   As you can see, the regression equation is:

$$Y_c = -.004 + .702(X)$$

b   The standard error of estimate = .4371.

c   The CR for the t test for slope at the .05 level is 23.17. Since the boundaries of the rejection regions begin at a t value of ± 2.306 we must reject the null hypothesis. Stated another way, since the p value of 0.000 beside the t-ratio figure of 23.17 is less than .05, we reject the null hypothesis that $B = 0$. There is a true relationship between requests for nurses and hospital days.

d   The 95 percent confidence interval for $B = b \pm t(s_b)$. That is, $.702 \pm 2.306(.0303)$ should include the value of B 95 percent of the time. Thus, the 95 percent confidence interval for B is .6321 to .7719.

e   The answer to this part is the 95 percent prediction interval—the 95% P.I. values—shown in the last line of the *Minitab* printout. Thus, the range of number of requests for a nurse during a 24-hour period by a specific patient who has been hospitalized for 7 days is 3.853 to 5.968 (rounded to 4 to 6 requests).

f   Coefficient of determination $= \dfrac{\text{SSR}}{\text{SST}} = \dfrac{102.57}{104.10}$

$= 98.5\%$

In this problem, 98.5 percent of the variation in requests for nurses is explained by variations in the number of days patients spend in the hospital.

6   When supplied with the input data for this problem, the *Minitab* statistical package produced the following output:

```
MTB > REGRESS 'HAY' ON 1 PREDICTOR 'FERTLZER';
SUBC> PREDICT 3.

The regression equation is
HAY = 1.62 + 0.655 FERTLZER

Predictor      Coef       Stdev     t-ratio        p
Constant      1.6190     0.2791        5.80    0.000
FERTLZER      0.65476    0.04741      13.81    0.000

s = 0.3548      R-sq = 96.5%      R-sq(adj) = 96.0%

Analysis of Variance

SOURCE        DF          SS         MS         F        p
Regression     1       24.008     24.008    190.77    0.000
Error          7        0.881      0.126
Total          8       24.889


   Fit  Stdev.Fit        95% C.I.           95% P.I.
 3.583     0.162    ( 3.200, 3.966)    ( 2.661, 4.506)
```

a   As you can see, the regression equation is:

$$Y_c = 1.62 + .655(X)$$

b   The standard error of estimate = .3548 tons of hay.

c   The CR for the *t* test for slope at the .05 level is 13.81. Since the boundaries of the rejection region begin at a *t* value of ±2.365, we must reject the null hypothesis that $B = 0$. Stated another way, since the p value of 0.000 beside the t-ratio figure of 13.81 is less than .05, we reject the null hypothesis that $B = 0$. There is a true relationship between fertilizer application and hay output.

d   The answer to this part is the 95 percent prediction interval—the 95% P.I. values—shown in the last line of the *Minitab* printout. Thus, the range of hay output for a specific tract that receives a 300-pound application of fertilizer is 2.661 tons to 4.506 tons.

e   The coefficient of determination = 24.008/24.889 = 96.5%.
In this problem, 96.5 percent of the variation in hay output is explained by variations in the amount of fertilizer applied.

9   When supplied with the input data for this problem, the *Minitab* statistical package produced the following output:

```
MTB > REGRESS 'ERRORS' ON 1 PREDICTOR 'TRNGTIME';
SUBC> PREDICT 5.

The regression equation is
ERRORS = 6.83 - 0.791 TRNGTIME

Predictor        Coef        Stdev      t-ratio          p
Constant       6.8253       0.4122        16.56      0.000
TRNGTIME      -0.79110      0.09529       -8.30      0.000

s = 0.6154      R-sq = 93.2%       R-sq(adj) = 91.9%

Analysis of Variance

SOURCE        DF           SS           MS          F          p
Regression     1        26.106       26.106      68.92      0.000
Error          5         1.894        0.379
Total          6        28.000


     Fit  Stdev.Fit         95% C.I.            95% P.I.
   2.870      0.270   (  2.177,  3.563)   (  1.142,  4.597)
```

a   As you can see, the regression equation is:

$$Y_c = 6.83 - .7911(X)$$

b   The standard error of estimate = .6154 errors.

c   The point estimate is:

$Y_c = 6.83 - .7911(5) = 2.87$ errors

(See the Fit value on the last line of the *Minitab* printout.)

d   The CR for the t test for slope at the .05 level is -8.30. Since the boundaries of the rejection region begin at a t value of ± 2.571, we must reject the null hypothesis that $B = 0$. You'll note, too, that this decision is supported by the p value of 0.000 beside the t-ratio figure of -8.30. There is a true relationship between the amount of training received and the errors made in using the software.

e   The answer to this part is the 95 percent confidence interval—the 95% C.I. values—shown in the last line of the *Minitab* printout. Thus, the average range of errors expected when trainees have received 5 hours of training is 2.177 to 3.563.

f   Coefficient of determination $= \dfrac{26.106}{28.000} = 93.2\%$

Coefficient of correlation $= -.9656$

In this problem, 93.2 percent of the variation in usage errors is explained by variations in the amount of training received. The r value of $-.9656$ shows that there's a close negative relationship between the two variables.

Chapter 15

3   a   The alternative hypothesis is:

The probability of a reduction in the time needed to close a sale is greater than 50 percent.

or

A negative sign is more likely to occur than a positive sign.

b   Since $n = 9$ and $r = 3$, we have:

$P(r \le 3; n = 9) = .1641 + .0703 + .0176 + .0020 = .254$

With $\alpha = .05$, we must accept the null hypothesis.

6   Since $n = 6$ and $r = 1$, we have:

$P(r \le 1; n = 6) = .0938 + .0156 = .1094$

If the null hypothesis is rejected, there is approximately 11 percent chance of an erroneous judgment.

9 The hypotheses are:

H_0: The new painkiller has no effect on the level of pain
H_1: The probability of a decrease in pain is more likely than the probability of an increase in pain

			SIGNED RANKS	
PATIENT	DIFFERENCE	RANK	+	−
A	−6	5.5		5.5
B	−6	5.5		5.5
C	+1	1.5	1.5	
D	−2	3		3
E	0			
F	−4	4		4
G	−1	1.5		1.5
H	0		1.5	19.5

From the data, we have $n = 6$ and $T = 1.5$. At the .05 level, the table value for T is 2. The null hypothesis should be rejected.

12 The sum of the positive ranks is 22 while the sum of the negative ranks is 6. With $n = 7$ and $\alpha = .05$ in a two-tailed test, the appropriate value of T is 2. The null hypothesis cannot be rejected.

15

FLATT'S NEW INFLATOR	RANK	COMPETITOR INFLATOR	RANK
17	6	23	15.5
16	4	21	11.5
21	11.5	32	17
19	7.5	21	11.5
15	2	19	7.5
14	1	20	9
16	4	21	11.5
16	4	22	14
23	15.5		
$n_1 = 9$	$R_1 = 55.5$	$n_2 = 8$	$R_2 = 97.5$

Using formulas 15-2 and 15-3, we have:

$$U = 9(8) + \frac{9(10)}{2} - 55.5 = 61.5$$

$$U = 9(8) + \frac{8(9)}{2} - 97.5 = 10.5$$

Therefore, we have $U = 10.5$. The relevant value from the U table at the .05 level is 18. Since 10.5 is less than 18, we reject the null hypothesis. Flatt's inflator is faster.

18 TTT HHH T HHHHH TTT HHHHH

The number of runs is $r = 6$. The number of heads = 13, and the number of tails = 7. The r value from the table is 5. Thus, the null hypothesis cannot be rejected at the .05 level.

21   + + + + − − − − − − + + + + − − + + + + − + + + + + + +

The number of runs equals 7. The r value from the table is 8. The null hypothesis is rejected at the .05 level.

24

TEMPERATURE RANK	SALES RANK	DIFFERENCE BETWEEN RANKS	D^2
6	5	1	1
11	12	−1	1
4	2	2	4
7	7	0	0
1	4	−3	9
12	14	−2	4
8	10	−2	4
2	1	1	1
15	15	0	0
14	13	1	1
5	3	2	4
10	9	1	1
13	11	2	4
9	8	1	1
3	6	−3	9
			44

$$\text{Spearman rank correlation coefficient} = 1 - \frac{(6)(44)}{(15)(224)}$$
$$= .921$$

The CR value is 8.52. With 13 df and a .01 level, the t value is 2.65. The null hypothesis is rejected.

INDEX

FORMULAS

$$\mu = \frac{\Sigma X}{N}$$

(formula 3-1)

$$\mu = \frac{\Sigma fm}{N}$$

(formula 3-2)

$$Md = L_{Md} + \left(\frac{N/2 - CF}{f_{Md}}\right)(i)$$

(formula 3-3)

$$Mo = L_{Mo} + \left(\frac{d_1}{d_1 + d_2}\right)(i)$$

(formula 3-4)

$$AD = \frac{\Sigma|X - \mu|}{N}$$

(formula 4-1)

$$\sigma = \sqrt{\frac{\Sigma(X - \mu)^2}{N}}$$

(formula 4-2)

$$\sigma = \sqrt{\frac{\Sigma X^2}{N} - \left(\frac{\Sigma X}{N}\right)^2}$$

(formula 4-3)

$$\sigma = \sqrt{\frac{\Sigma f(m - \mu)^2}{N}}$$

(formula 4-4)

$$\sigma = \sqrt{\frac{\Sigma f(m)^2 - (\Sigma fm)^2/N}{N}}$$

(formula 4-5)

$$Q_1 = L_{Q_1} + \left(\frac{N/4 - CF}{f_{Q_1}}\right)(i)$$

(formula 4-6)

$$Q_3 = L_{Q_3} + \left(\frac{3N/4 - CF}{f_{Q_3}}\right)(i)$$

(formula 4-7)

$$QD = \frac{Q_3 - Q_1}{2}$$

(formula 4-8)

$$CV = \frac{\sigma}{\mu}(100)$$

(formula 4-9)

$$S_k = \frac{3(\mu - M_d)}{\sigma}$$

(formula 4-10)

$$E(X) = \Sigma[xP(x)]$$

(formula 5-1)

$$_nC_r = \frac{n!}{r!(n - r)!}$$

(formula 5-2)

$$P(r) = (_nC_r)(p)^r(q)^{n-r}$$

(formula 5-3)

$$P(x) = \frac{\mu^x e^{-\mu}}{x!} \qquad \text{(formula 5-4)}$$

$$Z = \frac{x - \mu}{\sigma} \qquad \text{(formula 5-5)}$$

$$\mu_{\bar{x}} = \frac{\Sigma(\bar{x}_1 + \bar{x}_2 + \bar{x}_3 + \cdots + \bar{x}_{NC_n})}{{}_N C_n} \qquad \text{(formula 6-1)}$$

$$\sigma_{\bar{x}} = \sqrt{\frac{\Sigma(\bar{x} - \mu_{\bar{x}})^2}{N}} \qquad \text{(formula 6-2)}$$

$$\sigma_{\bar{x}} = \frac{\sigma}{\sqrt{n}}\sqrt{\frac{N - n}{N - 1}} \qquad \text{(formula 6-3)}$$

$$\sigma_{\bar{x}} = \frac{\sigma}{\sqrt{n}} \qquad \text{(formula 6-4)}$$

$$\sigma_p = \sqrt{\frac{\pi(100 - \pi)}{n}}\sqrt{\frac{N - n}{N - 1}} \qquad \text{(formula 6-5)}$$

$$\sigma_p = \sqrt{\frac{\pi(100 - \pi)}{n}} \qquad \text{(formula 6-6)}$$

$$\underset{\substack{\text{Lower limit} \\ \text{of estimate}}}{\bar{x} - Z\sigma_{\bar{x}}} < \mu < \underset{\substack{\text{Upper limit} \\ \text{of estimate}}}{\bar{x} + Z\sigma_{\bar{x}}} \qquad \text{(formula 7-1)}$$

$$s = \sqrt{\frac{\Sigma(x - \bar{x})^2}{n - 1}} \qquad \text{(formula 7-2)}$$

$$\rightarrow \quad \hat{\sigma}_{\bar{x}} = \frac{s}{\sqrt{n}} \qquad \text{(formula 7-3)}$$

$$\hat{\sigma}_{\bar{x}} = \frac{s}{\sqrt{n}}\sqrt{\frac{N - n}{N - 1}} \qquad \text{(formula 7-4)}$$

$$\bar{x} - Z\hat{\sigma}_{\bar{x}} < \mu < \bar{x} + Z\hat{\sigma}_{\bar{x}} \qquad \text{(formula 7-5)}$$

$$\bar{x} - t_{\alpha/2}\,\hat{\sigma} < \mu < \bar{x} + t_{\alpha/2}\,\hat{\sigma}_{\bar{x}} \qquad \text{(formula 7-6)}$$

$$\underset{\substack{\text{Lower confi-} \\ \text{dence limit}}}{p - Z\hat{\sigma}_p} < \pi < \underset{\substack{\text{Upper confi-} \\ \text{dence limit}}}{p + Z\hat{\sigma}_p} \qquad \text{(formula 7-7)}$$

$$\hat{\sigma}_p = \sqrt{\frac{p(100 - p)}{n - 1}}\sqrt{\frac{N - n}{N - 1}} \qquad \text{(formula 7-8)}$$

$$\hat{\sigma}_p = \sqrt{\frac{p(100 - p)}{n - 1}} \qquad \text{(formula 7-9)}$$

$$n = \frac{\sigma^2}{\sigma_{\bar{x}}^2} \qquad \text{(formula 7-10)}$$

$$n = \frac{\pi(100 - \pi)}{\left(\dfrac{\text{desired error}}{Z}\right)^2} \qquad \text{(formula 7-11)}$$

$$CR = \frac{\bar{x} - \mu_{H_0}}{\sigma_{\bar{x}}} \qquad \text{(formula 8-1)}$$

$$CR = \frac{\bar{x} - \mu_{H_0}}{\hat{\sigma}_{\bar{x}}} \qquad \text{(formula 8-2)}$$

$$CR = \frac{p - \pi_{H_0}}{\sigma_p} \qquad \text{(formula 8-3)}$$

$$\sigma_p = \sqrt{\frac{\pi_{H_0}(100 - \pi_{H_0})}{n}} \qquad \text{(formula 8-4)}$$

$$CR = \frac{(\bar{x}_1 - \bar{x}_2) - (\mu_1 - \mu_2)}{\sigma_{\bar{x}_1 - \bar{x}_2}} \quad \text{or} \quad CR = \frac{\bar{x}_1 - \bar{x}_2}{\sigma_{\bar{x}_1 - \bar{x}_2}} \qquad \text{(formula 9-1)}$$

$$\sigma_{\bar{x}_1 - \bar{x}_2} = \sqrt{\frac{\sigma_1^2}{n_1} = \frac{\sigma_2^2}{n_2}} \qquad \text{(formula 9-2)}$$

$$\hat{\sigma}_{\bar{x}_1 - \bar{x}_2} = \sqrt{\frac{s_1^2}{n_1} + \frac{s_2^2}{n_2}} \qquad \text{(formula 9-3)}$$

$$\hat{\sigma}_{p_1 - p_2} = \sqrt{\frac{p_1(100 - p_1)}{n_1 - 1} + \frac{p_2(100 - p_2)}{n_2 - 1}} \qquad \text{(formula 9-4)}$$

$$CR = \frac{(p_1 - p_2) - (\pi_1 - \pi_2)}{\hat{\sigma}_{p_1 - p_2}} \quad \text{or} \quad CR = \frac{p_1 - p_2}{\hat{\sigma}_{p_1 - p_2}} \qquad \text{(formula 9-5)}$$

$$F = \frac{\hat{\sigma}_{\text{between}}^2}{\sigma_{\text{within}}^2} \qquad \text{(formula 10-1)}$$

$$\bar{\bar{X}} = \frac{\text{total of all sample items}}{\text{number of sample items}} \qquad \text{(formula 10-2)}$$

$$\hat{\sigma}_{\text{between}}^2 = \frac{n_1(\bar{x}_1 - \bar{\bar{X}}^2) + n_2(\bar{x}_2 - \bar{\bar{X}})^2 + \cdots + n_k(\bar{x}_k - \bar{\bar{X}})^2}{k - 1} \qquad \text{(formula 10-3)}$$

$$\hat{\sigma}_{\text{within}}^2 = \frac{\Sigma d_1^2 + \Sigma d_2^2 + \Sigma d_3^2 + \cdots + \Sigma d_k^2}{T - k} \qquad \text{(formula 10-4)}$$

$$df_{\text{num}} = k - 1 \qquad \text{(formula 10-5)}$$

$$df_{\text{den}} = T - k \qquad \text{(formula 10-6)}$$

$$\chi^2 = \sum \frac{(f_o - f_e)^2}{f_e}) \qquad \text{(formula 11-1)}$$

$$f_e = \frac{(\text{row total})(\text{column total})}{\text{grand total}} \qquad \text{(formula 11-2)}$$

$$df = (r - 1)(c - 1) \qquad \text{(formula 11-3)}$$

$$PI_n = \frac{\sum(p_n)}{\sum(p_0)} \cdot 100 \qquad \text{(formula 12-1)}$$

$$PI_n = \frac{\sum(p_n q_t)}{\sum(p_0 q_t)} \cdot 100 \qquad \text{(formula 12-2)}$$

$$QI_n = \frac{\sum(q_n p_t)}{\sum(q_0 p_t)} \cdot 100 \qquad \text{(formula 12-3)}$$

$$PI_n = \frac{\sum\left(\dfrac{p_n}{p_0} \cdot 100\right)}{n} \qquad \text{(formula 12-4)}$$

$$QI_n = \frac{\sum\left(\dfrac{q_n}{q_0} \cdot 100\right)}{n} \qquad \text{(formula 12-5)}$$

$$PI_n = \frac{\sum\left[\left(\dfrac{p_n}{p_0} \cdot 100\right)\left(p_t q_t\right)\right]}{\sum(p_t q_t)} \qquad \text{(formula 12-6)}$$

$$QI_n = \frac{\sum\left[\left(\dfrac{q_n}{q_0} \cdot 100\right)\left(q_t p_t\right)\right]}{\sum(q_t p_t)} \qquad \text{(formula 12-7)}$$

$$Y = T \times S \times C \times I \qquad \text{(formula 13-1)}$$

$$Y_t = a + bx \qquad \text{(formula 13-2)}$$

$$a = \frac{\sum Y}{n} \qquad \text{(formula 13-3)}$$

$$b = \frac{\sum(xY)}{\sum(x^2)} \qquad \text{(formula 13-4)}$$

$$Y_c = a + bX \qquad \text{(formula 14-1)}$$